J BAR M ARABIANS
BOX 1079   938-6242
OKOTOKS, ALBERTA   T0L 1T0

Breeding Management & Foal Development

ii.

Equine Research
INC.

P. O. Box 9001   Tyler, Texas 75711

WRITTEN BY:
the research staff of Equine Research, Inc.

ILLUSTRATED BY:
Patti R. Strauch, DVM
and Equine Research, Inc. staff artists

RESEARCH EDITORS:
J. Warren Evans, PhD
Richard L. Torbeck, DVM, MS

ACKNOWLEDGEMENTS

The efforts of many individuals contributed to the preparation of this book. In addition to the research editors, J. Warren Evans, Ph. D., and Richard L. Torbeck, DVM MS, Equine Research, Inc., would like to express sincere appreciation to:

Peter D. Rossdale, MA, FRCVS,

Michael J. Nolan,

Gary W. Brandt, DVM, MS,

J.M. Bowen, B. Vet. Med., FRCVS,

Linda L. Gratny, DVM,

Will A. Hadden III, DVM,

Gary S. Spence, DVM, and

Michael A. White, DVM

for their technical contributions and invaluable assistance in the preparation of this book.

TABLE OF CONTENTS

1

BREEDING FARM DESIGN

The ideal breeding farm is a carefully conceived and soundly managed business operation in which the practical elements of safety and health are skillfully integrated into an aesthetically pleasant setting. A properly planned facility should also reduce overall costs by eliminating unnecessary expenses and minimizing labor requirements. It is built literally "from the ground up," since farm site selection is one of the first concerns and can limit the success of an otherwise meticulously structured enterprise.

Farm Site Selection

The foundation of a breeding farm is the land, and well-drained, fertile soil is an important asset for a successful breeding operation. Further, if the farm is to be used for pasture or forage production, the ability of the soil to hold water and to support forage through periods of less than ideal growing conditions is an important consideration. The soil conservation service and county agricultural extension agents are good sources of information on soil analysis and management. Much of their information is gleaned from regional soil and topography maps, which are also helpful reference tools for the farm manager.

Other factors to be considered when selecting a site are an adequate water supply, good surface and subsurface drainage, availability of necessary utilities, zoning regulations and building code or legal restrictions, and adequate space for future expansion. Topography and climate are also factors. Gently rolling land with slopes in the 2 to 6 percent range (2 to 6 feet of rise or fall in 100 feet) is considered ideal for most facilities, since extreme grades require more management to prevent erosion and make facilities more difficult and expensive to construct. Contour maps showing the slope of the land are particularly useful for designing drainage systems and planning building and paddock locations.

Well-drained land with a moderate slope is ideal for the breeding farm site.

Once desirable land has been acquired it must be maintained. Plants are fed by nutrients held in the soil, and as long as the land is productive, its nutrient store is constantly being depleted. Unless proper measures are taken to replenish the soil, crop quality and yield gradually diminish. To determine the land's needs, samples may be taken and submitted for analysis to a soil laboratory. Some commercial soil analysis firms will even take soil samples and do comprehensive surveys, analyzing the major plant nutrients as well as minor soil constituents. The firm then provides detailed charts and maps of the farm along with some suggestions for proper soil management. (For specific information on pasture management refer to the Appendix.)

Farm Layout

Even before land or existing facilities are purchased, the breeder should explore potential sites with farm planning and design in mind. If a site meets initial selection requirements, the breeder's long-term goals for the operation should then be taken into consideration. Depending on the farm's size and the type of operation, several kinds of facilities may be required, including a stallion barn, outside mare barn, foaling barn, storage buildings, farm laboratory, and examination and breeding areas. In addition, a system of pastures, paddocks, and alleyways will be required. All of these structures must be taken into consideration in the farm layout.

Courtesy of Horseman's Journal

A well-planned farm layout enhances farm efficiency and safety as well as the appearance of the breeding facility.

If the site has existing facilities, it must be decided whether they are suitable to the planned operation or whether they can be renovated. Then other facilities must be planned around the existing structures. For this reason, it is sometimes more practical to purchase an undeveloped site unless the original facilities were designed for horse use. It may be helpful to study a number of prospective farm layouts and review them with an architect or building contractor before developing a final plan for construction or renovation. Additionally, visits to established, large-scale breeding farms may present new ideas.

Some specific points that should be considered in the farm layout are appearance, accessibility, safety, and drainage. Although every farm requires areas for manure disposal and equipment storage, with careful planning these locations can be screened from public view. Of course, the farm must be designed functionally, but farm structures can also be arranged for a neat and pleasing appearance. Additionally, each farm structure should be planned for accessibility. So that horses can be fed and checked easily, all-weather routes should also be provided between storage buildings, barns, sheds, and paddocks.

All farm facilities should be accessible by all-weather roads.

Paddock and pasture entries should be located near the barns they serve. This arrangement not only reduces the time spent moving horses but also reduces the chances of a horse getting loose while being led to and from turnout areas. Further, if a safety zone of at least 100 feet is allowed between some of the paddocks and the barn, the enclosures can also serve as emergency turnout areas in case horses must be evacuated during a barn fire. During emergencies and even for daily handling, alleyways are of tremendous assistance in moving horses. Alleyways between pastures and paddocks prevent fighting between horses in neighboring enclosures. They also decrease the time spent in moving large numbers of horses since an entire group can be driven through the alleyway rather than leading each horse individually.

Alleyways that separate pastures prevent direct contact between horses in adjoining enclosures.

Finally, structures should be arranged to take full advantage of the farm's topography. Good drainage should be considered in the placement of each structure by situating it on a slight elevation or on extremely well-drained soil. In addition, horse paddocks should not be located in low areas where standing water may accumulate. If the barns or paddocks create considerable runoff, the structures should also be arranged to prevent contamination of ponds or streams.

Poorly drained low spots should be avoided when constructing barns or paddocks.

Farm Facilities

Key factors in planning specific facilities are the projected goals and purposes of the operation. For instance, if the farm keeps only mares and sends them to other farms for breeding, then only paddocks, pastures and foaling stalls may be required. However, if the farm also stands a stallion, facilities for outside mares as well as additional teasing and breeding areas may be needed.

Highly specialized operations, such as stallion stations require specifically designed structures. The stallion station, for instance, usually houses several stallions while having few, if any, farm-owned mares. Although there are a number of possible set-ups, most stallion stations house all of

Courtesy of Windward Stud

Housing for outside mares should provide dry but well-ventilated shelter and adequate spaces for self-exercise. This open-sided barn design gives mares free access to shelter and to the outside paddocks.

Spacious inside alleyways are convenient for moving and feeding horses.

the farm stallions in one barn located near the breeding shed for efficiency. Because extensive facilities for outside mares are necessary, mare housing often consists of individual paddocks and shelters rather than large pastures and loafing sheds. Since unacquainted mares are likely to be aggressive toward each other, it is important that the fencing between adjoining runs be safe and strong. Also, if the stallion station has farm-owned mares, separate facilities for visiting mares should be utilized to minimize the spread of contagious disease. When designing facilities, the breeder should remember that many incoming mares may be on special light programs and that it is essential the light program not be interrupted when a mare arrives at the station. Such an interruption frequently causes irregular estrous patterns and, in some instances, the mare ceases to show signs of estrus. Additional information on specialized structures for teasing, breeding, and housing can be found in later chapters.

BARNS

Because barns represent significant long-term investments, their design should be considered carefully. Further, to create a durable, functional structure, proper construction methods should be combined with a strong foundation, high-quality building materials, proper insulation, and adequate ventilation. In some instances construction criteria are regulated by local building codes and zoning ordinances. If so, local regulations should be studied carefully when designing a horse barn. Although personal satisfaction and aesthetic value are considerations, the breeder should focus on efficient arrangement, and should place special emphasis on health and safety. Barn size should be adequate for the farm's immediate needs but also adaptable to the farm's long-term demands for stalls, traffic flow, and work areas. Well-planned areas for storage should also be incorporated into the plan. For example, by providing additional isolated storage areas for manure and hay, the breeder minimizes both barn odor problems and fire hazards. (Fire safety is discussed more thoroughly later in this chapter.)

Because most barns share many common characteristics, an existing farm building originally used for other livestock enterprises can often be converted to an equine facility with minor remodeling and the installation of box stalls. Some exceptions are specialized barns, such as those equipped with dairy stanchions or buildings with unusually low ceiling clearances. Thus, before purchasing land and existing facilities, adaptability should be carefully considered. In addition, when renovation is being considered, the remaining life span and safety of the structure should be weighed against the cost of building a new structure.

Although many breeders find that one multipurpose barn is adequate for a moderate number of animals, as mentioned earlier in this chapter, more specialized breeding farms require a greater variety of buildings to house breeding stock efficiently. For instance broodmare, stallion, yearling, hospital, and foaling barns are just a few of the many possible facilities that could be designed for a breeding operation. Because design depends

on financial and space limitations as well as goals, it would be impractical
to describe a functional barn style for one breeder and expect it to apply to
another breeder's operation. For this reason, descriptions of basic elements
in barn design are included to help the breeder plan suitable farm
structures.

Courtesy of The Thoroughbred Record

In this barn, the combination of concrete block construction and good design resulted in
a facility that is durable, safe, and well-ventilated. In addition, the barn design takes
advantage of natural sunlight for illumination.

STALLS

In general, stabling requirements are similar for stallions, mares, and
youngsters. All require roomy box stalls unless adequate shelter is available in pastures or paddocks. The minimum box stall size for any confined
horse is 12 feet by 12 feet, with a 9 to 10 foot height from floor to ceiling or
from the floor to the lowest rafters. Broodmares with foals at side require a
stall size of at least 12 feet by 14 feet, and preferably 14 feet by 14 feet. The
stallion, on the other hand, requires 14 feet by 14 feet minimum stall
dimensions to prevent boredom, and he should, in fact, have an even more
spacious stall which is not completely isolated from barn activities. Stallions with unusually vicious dispositions may require specially reinforced
walls, gates, and screens, but most stallions can be housed in normal,
sturdy box stalls.

All stall partitions should be constructed of durable, heavy-duty materials such as wood or concrete block. The lower 4 to 5 feet of the stall
partition should be solid and, from that point, bars or screens may be
installed for free air flow and observation. Screens between the stalls
should be of heavy gauge wire with a fairly close weave to prevent bickering
through partitions. In some instances, completely solid stall partitions may
be required to prevent fighting between stalls.

The inside walls of each stall should be smooth and free from projections. Concrete block stalls should be lined with 2 inch by 6 inch boards to prevent leg injuries from kicking. A concrete apron is sometimes used with clay or dirt floors to prevent holes from developing next to the stall wall. All exposed edges within the stall should be painted with creosote or covered with metal strips to discourage wood chewing. However, if paint is used in areas to which horses have access, the paint should be carefully selected to insure that nontoxic products are used on all exposed surfaces.

FLOORS

There are several important characteristics that should be considered when planning stable floors. Proper drainage is an essential prerequisite, which should be studied when the building site is selected. When the barn is constructed over well-drained soil, a clay floor over 6 inches of rock or gravel should allow sufficient drainage. Poorly drained soils, however, usually require a deeper drainage system. For example, for stalls over poorly drained soil the breeder might use a 1.5 foot layer of large rocks beneath a 1.5 foot layer of small rocks or crushed gravel with a 6 to 12 inch stall floor composed of 3 parts clay to 1 part sand. When hard, waterproof flooring materials, such as concrete or brick are used, lateral drainage must be provided. The floor should be sloped evenly at a rate of approximately 1 inch per 5 feet and should end in a gutter running to the main drain. To insure proper drainage throughout the barn, the floor level should be at least 8 to 12 inches above ground level.

In addition to sufficient drainage, stall floors should have several other characteristics including durability, resiliency, and secure footing. A floor consisting of at least 6 to 12 inches of packed clay over a gravel or rock base is the most common and probably the most suitable floor for a stall. Clay provides good footing and is durable but not damaging to the horse's feet and limbs. A mixture of 3 parts clay and 1 part sand is usually recommended since plain clay floors provide poor drainage. Wet areas of a clay floor should be removed once a year, replaced with fresh clay, releveled, and sprinkled with lime.

Because dirt floors retain moisture and odors and often become uneven, they are less desirable than packed clay for use in stalls. Concrete, asphalt and brick floors are relatively expensive but they are durable, and they can be scrubbed with cleansing agents and rinsed frequently without damaging the surface. Unfortunately, they also lack resiliency and are slippery when wet. These drawbacks can be overcome, to some extent, by the judicious use of bedding. Wood plank and wood block floors are occasionally used in stalls, but they are very difficult to keep dry and odor free. A variety of products are sometimes used in stalls and along indoor alleyways, but the packed clay stall floor and concrete walkways seem to be preferred by most horsemen.

With the exception of rubber surfaces used in foaling stalls, most stall flooring is designed to be used with bedding materials. Stall bedding should be absorbent, free from dust and mold, economical, convenient, and readily available. Wood shavings are a popular stall bedding, since they cushion the horse's feet and are absorbent and easy to handle. The reader should note that shavings (or sawdust) from some nut-bearing trees are toxic and may even cause founder, but, in general, wood shavings make excellent bedding. When it is available, dust-free wheat or barley straw is preferred for foaling stalls because it will not stick to the nostrils of a wet, newborn foal. Many other products are used for bedding, but the advantages and disadvantages of each should be examined carefully before placing them in a horse's stall.

VENTILATION

Ventilation replaces stale air with fresh air and thereby reduces high humidity and condensation, factors which contribute to respiratory problems in horses. Although ventilation is thought of primarily as a cooling method for hot weather, some air movement is also required in cold weather, particularly if the barn is completely enclosed. Proper ventilation also minimizes odor problems but should not create drafts. Ventilation can be achieved passively by employing windows, vents, screens, or dutch doors or actively by installing fans.

Although cupolas are also considered to be decorative, they provide valuable ventilation in horse barns.

Adjustable vents or windows located near the ceiling provide adequate ventilation in many barns. They should be positioned at least 6 feet from the stall floor so that air moves freely through the top of the barn without causing drafts at the horse's level. Eye-level windows can also be installed to prevent horses from becoming bored with confinement and to provide additional ventilation during hot weather. However, all glass windows that are within the horse's reach should be protected with removable welded wire, steel grates, or bars. One of the benefits of windows, unlike vents and fans, is that they also provide supplementary lighting.

Fans are frequently used for forced ventilation of the barn. To allow proper air flow for either a pressure or exhaust ventilation system, at least 1 square foot of vent or inlet space should be provided for every 750 cubic feet per minute fan capacity. Excessive vent space creates drafts, while inadequate space prevents proper ventilation.

These hinged windows provide ventilation to the barn and can be closed easily during inclement weather.

In any building where condensation is a problem, the environment becomes damp and musty and encourages fungal growth. Because this type of environment predisposes animals to respiratory infection, it is extremely important to design and construct animal housing in such a way that condensation is minimized.

Proper ventilation within any type of structure helps to minimize condensation problems since 1) the difference between the indoor and outdoor temperatures is minimized and 2) moisture does not condense as readily from moving air. Insulation within the walls and ceiling help to keep the temperature of the interior walls from dropping far below the temperature of the air inside the structure and, therefore, helps to minimize condensation on the surface of walls and ceilings. To prevent or minimize the flow of water vapor into the walls, a vapor barrier should be incorporated into, or near, the warm side of the wall (e.g. between the insulation and the inside wall). This barrier may consist of a waterproof membrane (e.g., laminated paper, plastic film, kraft-backed aluminum foil, foil-backed gypsum board) or a special paint (oil base, aluminum paint, etc.).

Because a relatively thin masonry wall allows warm air to escape readily, the interior side of the wall is very cold during severe winter weather, and condensation on the inside is often a problem. However, the use of a vapor barrier on these types of walls often aggravates condensation problems. Insulation, which helps to keep the inside of the wall warmer, reduces condensation, but if the hollow portion of a masonry wall (e.g., concrete block) is filled with insulation, it is important to prevent vapor from entering the wall and condensing on the inside. Venting the wall's cavity to the outside allows any vapor that does enter to escape.

It is also important to note that some types of floors require vapor barriers. A vapor barrier between the ground and a concrete slab, for example, prevents moisture from moving into and damaging the concrete.

LIGHTING

The value of natural light should be noted when planning a breeding farm facility. Direct sunlight helps deodorize the barn by drying out areas which would otherwise be dark and musty. It also helps to warm the barn in the winter and reduces parasite populations throughout the year. Sunlight is an essential part of a healthy environment for the horse, and skylights, windows, and dutch doors can be readily incorporated into most building designs to bring natural light indoors.

Because important activities on the breeding farm (e.g., illness, accidents, foalings) frequently occur at night, a dependable artificial light source is a necessity. A good lighting system requires sufficient wiring, and for this reason, it is very important to plan the farm's wiring capacity with future expansion in mind. Lights should illuminate all indoor facilities, outside approaches, and service areas. As a safety precaution, stall lights should either be recessed into the ceiling or covered with metal cages. Outside security lights are often a wise investment, especially since many electric companies install pole lights for a low monthly usage charge. This type of fixture can be used to light critical areas, such as the farm entrance or the barn site.

Recessed lighting prevents horses from being injured on fixtures.

HANDLING AREAS

Areas for handling horses should be planned for easy maintenance, convenience, and safety. No matter for what type of horse or what age horse the area is designed, certain guidelines always apply. Adequate space must be provided between ties to prevent horses from kicking one another. The ties should be positioned at the same height as the horse's withers and should always be immovable and unbreakable, although a quick-release snap may be used with each tie in case a tied horse pulls back or falls. The quick-release mechanism must be sturdy so that it cannot be broken by the horse but should operate easily so that it can be tripped by an attendant using one hand even with the horse's entire weight resting on the tie. If the handling area is to be used for washing or reproductive examinations, a well-drained, solid floor such as brushed concrete must be provided. Dirt or packed clay floors are suitable for ordinary handling areas, but regular maintenance will be required to repair low spots caused by wear and tear and pawing.

STORAGE AREAS

Grain should be stored in large waterproof and rodent-proof bins located in or near each barn where they can be easily filled by a truck. Each bin should be capable of holding at least a week's supply of grain for that particular area. However, due to its flammability, only small quantities of hay should be stored in livestock housing. Large amounts of hay may be stored in barns located at least 100 feet from the nearest livestock shelter. Whether the hay storage structure is an open pole barn or a completely enclosed shed, it must provide adequate shelter to prevent weathering and spoilage.

Courtesy of Windward Stud

On large farms, grain bins equipped with augers provide efficient bulk feed storage.

Each barn should have its own tack room for frequently used grooming equipment, tack, and small hand tools. Lighting and ventilation are just as important in the tack room as in the rest of the barn. All tack rooms should contain equine and human first aid kits and should be equipped with (or located near) a lavatory with a hot and cold water supply. The area should be organized and clean at all times, and if it is also used as an office or lounge, insulation, heating, and air conditioning should be considered.

Courtesy of Stubblefield Farm

An office located in or near the barn provides a convenient place for record-keeping and receiving farm guests.

If the breeding farm regularly receives visitors, then fencing materials, tools, irrigation equipment and heavy farm machinery should be housed in sheds or concealed behind walls, shrubs, or trees. If extensive machinery storage and repair work are required, a combination storage and workshop facility might be appropriate. Facilities for fuel, fertilizer, and other stored chemicals should be located in an accessible area at least 100 feet from other flammable materials such as hay and bedding.

PADDOCKS

As many stalls as possible should open from the barn into individual exercise paddocks. With free access to the outside, horses are less likely to become bored and develop vices such as cribbing, stall walking, or weaving. Stallion paddocks should be of sufficient size to allow self exercise (usually 2 to 5 acres), and the fence should be at least five to six feet high and constructed of pipe, heavy poles, mesh, or heavy board. An alleyway at least 10 feet wide should separate the stallion paddock from any adjacent paddocks to eliminate fighting or breeding over the fence. Some breeders also run electrically charged wires along the top of a stallion's fence to reduce aggressive behavior over the fence.

Sturdy fencing and wide alleyways should separate stallion paddocks.

Paddocks for other horses on the farm should also have adequate space for self exercise, but the fences need not be more than about 4.5 feet high. Some breeders also use small individual paddocks (less than an acre) for letting down mares that have arrived for breeding directly from a race training atmosphere or for turning out mares that have foaled recently. Whenever possible, adjoining paddocks should be of equal length. Occasionally, injuries occur when two horses are running in neighboring paddocks of unequal length. Since the horses are preoccupied with their play, when the horse in the longer paddock continues down the fence line, the individual in the shorter paddock fails to stop and strikes the rear fence. However, accidents of this type can be prevented easily by constructing paddocks in a uniform manner.

PASTURES

If the farm has adequate acreage and fertile soil, well-managed pastures can significantly reduce supplementary feed requirements, labor requirements, respiratory problems, and parasite infestation. Pastured horses also tend to be more content and are less likely to develop boredom-related vices than are stalled horses. A minimum of one acre of high-quality pasture per horse should be provided for broodmares and young stock. However, in many areas more than the minimum allowance will be required to support each horse due to variations in grass production, climate, and soil fertility.

The facilities required by horses at pasture are relatively simple. Sheds are often used to provide shelter, and even in severe climates, a three-sided, roofed shed opening to the south provides a dry area where the horses can escape from chilling winds of winter and the direct sun of summer. These

Well-maintained, fertile pastures are especially important to the breeding farm, because they provide supplemental nutrition and free-exercise for breeding stock and young horses.

sheds should provide adequate space for each horse, especially if feed troughs are located in the shelter. Ideally, the inside corners of the shed should be rounded to prevent individuals from being trapped by aggressive pasture mates. In addition, support posts should be padded to prevent injuries.

Pasture feeding facilities must be designed to provide sufficient feeder space per horse. Feed troughs and hay racks used in the pasture should also be designed to protect the feed from rain and snow. Despite the convenience of long troughs for bulk feeding, individual buckets have several advantages. Each horse's intake can be more easily monitored and regulated with individual feeders, and if the feeders are spaced at least 25 feet apart, they insure that timid horses will not be crowded away from the feed. In order to provide adequate feeding space, several extra feeders can be distributed in the pasture. In this arrangement a few feeders will always be unoccupied.

Loafing sheds should be provided in each pasture for shelter.

Water is one of the most important parts of the horse's diet but one of the most easily overlooked when managing pastured horses. The horse must always have a clean, fresh water supply, so water tanks must be convenient to drain and clean. There should also be enough watering space to prevent crowding and fighting, and the tank's water supply must have enough pressure to refill fast enough to supply the horse's needs. This last point may seem insignificant, but several broodmares with nursing foals at side can consume a tremendous amount of water during a hot summer day. Unfortunately, it is very easy to overlook water shortages until they are reflected in the horse's condition.

FENCES

Paddocks and pastures can be enclosed using similar materials. Both types of fencing must be sturdy, low maintenance, highly visible, and safe. Whatever type of fencing the breeder may choose, it is important that no hazardous areas be left unfenced. In addition, roadside fences should be particularly sturdy and should be checked regularly for damaged areas.

Hazardous areas such as groups of closely spaced trees should be fenced off.

Wood fencing combines visibility and attractive appearance and is a popular type of fencing. Like other wood building materials, however, wood fencing requires routine maintenance, has a limited useful life in areas where wood rot is prevalent, and is susceptible to damage by cribbers. Oak, often used in board fences, is strong and durable but has a tendency to warp more severely than other woods. It also tends to split into long spear-like splinters when broken. Fir and pine, on the other hand, are not so strong and durable, but they do not splinter like oak. To increase the life of a wood fence, lumber used for fencing should be kiln-dried and treated with a nontoxic wood preservative.

Wood fences should be protected with an application of
paint or preservative to prevent
weathering.

Although relatively expensive,
pipe fence is extremely durable.

A V-mesh fence should be supported with a pipe or board
along the top and middle of the
fence.

Although pipe is a relatively safe fencing material, it has very little
flexibility, making it a questionable choice for pasture fencing where high
speed collisions with the fence are possible. (Some new types of pipe now
have increased flexibility.) Pipe is also expensive in many areas, but
because it is extremely durable and almost maintenance free, this type of
fencing is a popular choice.

Woven V-mesh wire fences have many features which make them a popular choice for breeding farms. The small openings prevent a horse from getting a foot caught in the fence and also prevent other animals from getting into the pasture. Additionally, V-mesh wire is extremely strong and durable and is almost maintenance-free. However, mesh wire can become bowed out by horses pushing or rubbing on it and, therefore, should have a pipe or board across the top and middle for support and to increase its visibility.

Chain link fencing is sometimes used for paddocks and is a safe and exceptionally strong fence. However, the twisted ends are very sharp; hence, the bottom of the fence should be buried in concrete, and a pipe or board should be run along the top to prevent horses from pushing on the fence or hurting themselves on the fence's sharp points.

Rubber-nylon fencing (e.g., cut from strips of conveyor belt material) is sturdy, resilient, impervious to weathering, moderately priced, and reasonably attractive. However, an important drawback to this type of fencing is that horses (especially youngsters) tend to chew on it, and if they consume the rubber, they may suffer colic and impaction.

Barbed wire is the most dangerous type of fencing available. Horses that become entangled in barbed wire usually struggle, often inflicting serious and sometimes fatal wounds. Because stallions and youngsters are excitable and prone to fence-related injuries, barbed wire fencing is especially unsuitable on breeding farms. As a low cost alternative to barbed wire, twisted strands of barbless wire are sometimes used for pasture fencing. Although safer than barbed wire, smooth wire is difficult for horses to see and can also cause lacerations if the horse collides with or becomes caught in the fence.

If possible,sharp corners should be eliminated by curved fence lines.

PARKING AREAS

The parking area is another important, but frequently overlooked feature of the breeding farm operation. Every farm should have an easily accessible parking lot, and an area near the barn and accessible to the office should be set aside for this purpose. The lot must be large enough for a trailer to be turned around and to accomodate at least one trailer and five to six additional vehicles. However, a stallion station parking lot must be large enough to permit large commercial vans or trailers to maneuver in and out.

Parking lot surface is a very important consideration. Many horsemen prefer a solid gravel base due to its drainage capabilities. A concrete or blacktop surface will facilitate parking in all weather, but an additional unpaved area should be set aside for loading and unloading horses. Loading ramps and nearby holding areas are often helpful if horses are frequently delivered in large, commercial vehicles.

Fire Safety

When planning a breeding facility, it is imperative that potential fire hazards be eliminated. The forethought and expense involved in preventing fires is minimal when compared to the grim destruction and painful losses associated with stable fires. Statistics show that within three years of a fire, one out of three farm owners goes out of business. By far, the best insurance is a facility design that meets rigid fire safety standards and careful preparation in advance of a fire.

FIRE PREVENTION

When constructing or renovating a breeding facility, it is a good idea to seek the professional advice of a local fire marshal to determine whether the farm plans meet fire safety standards. For example, storage buildings and other high risk areas should be surrounded by buffer zones that are at least 100 feet wide. When feasible, the construction of several small barns instead of one large facility is also recommended to reduce the chances of losing a large number of horses in a single barn fire. In fact, some farms follow the policy of mixing extremely valuable horses with less valuable individuals in the same barn. In this manner, all of the farm's most valuable stock is unlikely to perish in a single fire.

Fire resistance must be considered when selecting building materials. Pressure-treated, fire-retardant wood, vermiculite plaster and gypsum board slow the spread of fire. Masonry and metal are even more fire retardant. All paints, sealers and other finishing materials should be fire retardant. Despite using fire retardant building materials, a fire can spread easily in most barns, because hay, bedding, wooden partitions, dust and miscellaneous supplies are capable of supporting combustion. If hay must be stored in the barn loft (separate facilities are preferable), this area must be well ventilated, and the roof must be water-tight to minimize the chances of spontaneous combustion caused by fermentation of wet hay. In addition to hay storage, special consideration must be given to the storage of bedding, feed, manure, and flammable liquids. If trash must be burned on the farm, the construction and location of the incinerator should be approved by the fire marshal.

Flammable material, such as bedding, should be stored at least 100 feet from other facilities.

It is important that all wiring and electrical fixtures be installed and inspected routinely by a licensed professional. Both immediate and projected electrical needs should be considered to insure that the system is not overloaded. When remodeling old barns, wiring should receive special scrutiny, since the original electrical system may be unable to meet increased power demands of an expanded operation.

Other fire safety features include safe installation of permanent heating systems, wire cages surrounding each light fixture, no smoking signs posted at all building entrances, and safe disposal containers conveniently located near each sign. Also lightning rods should be properly installed and maintained by a qualified company since poorly installed lightning rods are responsible for some barn fires. All wire fences should be grounded near the barn, and the system should be thoroughly inspected periodically.

Maintenance practices also play an important role in a fire prevention program. For example, thoroughly sweeping the barn floor and periodically removing settled dust and chaff from walls and rafters eliminates a very significant fire hazard. All employees should be impressed with the importance of fire safety guidelines, and cleansing solvents, gasoline, and other flammable liquids should be carefully stored in a designated area away from the stable area. Furthermore, electrical appliances must always be used with care. For example, unless appliances must be on a constant power supply, they should be unplugged after each use. Space heaters should be used only when absolutely necessary and should never be left unattended.

ALARM SYSTEMS

In the event that a fire does occur, early detection is extremely important to an effective evacuation procedure and may help to save the building and livestock from complete destruction. Although there are several different types of fire alarms, each with specific advantages and disadvantages, there are several important requirements for any system.

1. The alarm should be easy to identify as a fire alarm and should be audible from a reasonable distance. The alarm should sound in the barn, in the manager's house, and at the local fire department.
2. The alarm system should be dependable. That is, it shouldn't give false alarms and should be in working order at all times.
3. The system should be able to detect fire in any part of the building.

MINIMIZING THE SPREAD OF FIRE

When planning a breeding facility, it is important to take advantage of special features that minimize or prevent the spread of fire.

Fire walls constructed of such materials as brick, concrete, or stone and coated with fire-retardant substances, can be placed between storage and stable areas to prevent fire from spreading rapidly through a barn. They can also be placed at intervals between the stalls where horses are stabled. Ideally, fire walls should divide the building into two or more separate sections, but when this is not feasible, sliding fireproof doors can be incorporated into the wall. If the loft is used as a storage area, its floor should be constructed of three airtight layers of treated wood or other fire-retardant material. Any openings or hatchways to the loft should be equipped with automatically closing, air-tight doors. Closing off the loft in this manner gains valuable time for evacuating people and horses from the building when the roof or loft is burning.

Automatic sprinklers and chemical spraying systems can be installed in barns to minimize the spread of fire through the building. Although these systems help to protect against extensive loss of facilities, they are not usually employed in livestock structures, perhaps due to their expense or because many of the systems do not detect fires early enough to save the animals from smoke inhalation. In fact, some chemical spraying systems may actually increase the possibility of suffocation. Further, sprinkler systems can increase the chances of lung damage since large amounts of smoke form when water strikes burning materials. Ideally, chemical spraying systems should be able to suffocate the fire immediately without suffocating the horses. Regardless of whether sprinkler systems are used,

A fire hose located near areas where there is a higher than usual risk of fire may aid in early control.

all parts of the barn should be accessible to water hoses connected to hydrants that are located near barn entrances.

Portable fire extinguishers should also be placed by barn entrances and in high risk areas, such as storage buildings and shops. Extinguishers are designed to handle specified types of fires and should be chosen accordingly. Water extinguishers should be placed in hay and feed storage areas, while chemical extinguishers should be placed in workshops, machinery storage areas, and laboratories. The fire marshal may recommend that all-purpose extinguishers be placed at 50-foot intervals throughout the barn and at all entrances, since fire extinguishers must be used very soon after the fire's onset to successfully extinguish the flames. For this reason, it is extremely important that all farm employees know how to operate the extinguishers without referring to instructions or struggling with cotter pins.

Fire extinguishers should be placed near every barn entrance and employees should be instructed in proper use of the extinguisher.

EMERGENCY PROCEDURES

Although fire-related damage and injury can be minimized by the use of fire-resistant building materials and good building design, the most important aspect of fire safety (next to fire prevention) is a well-planned fire emergency procedure. Each farm owner must evaluate the entire operation carefully and formulate a plan of action designed to protect humans, livestock, and buildings in that order of importance. All employees and anyone else who is likely to be in a position of responsibility should be familiar with the plan. Fire drills may seem inconvenient and time consuming, but they are extremely helpful in educating personnel and in detecting flaws in the fire emergency procedure. The following steps are presented as guidelines for establishing individual fire emergency procedures.

1. Once a fire is detected, all persons in the immediate area should be evacuated, and fire emergency plans should be put into effect immediately.
2. If the fire is restricted to a small area, it may be extinguished or contained through prompt action. For example, a small fire may be smothered with a horse blanket or put out with a fire extinguisher. The emphasis here is on prompt action, since stable fires often spread rapidly.
3. If the fire is large or cannot be contained immediately, fire authorities should be notified (even if an alarm system notifies the fire department automatically). All farm entrances must be kept clear of people, animals, and vehicles so that incoming fire-fighting equipment is not delayed.
4. Horses should be evacuated from the building if this can be accomplished without undue risk to human life. It is at this point that forethought and planning determine whether the evacuation procedure is carried out in a state of panic and confusion or in a well-rehearsed, effective manner. Special evacuation precautions, such as outdoor exits from each stall and halters and lead ropes hung near each horse, help to speed the evacuation procedure. Stall doors should never be locked. If a horse is reluctant to leave its stall, it should be blindfolded with a shirt or rag.
5. Arrangements should be made to prevent panicky horses from re-entering the barn in a state of panic. Once horses have been removed from the barn they should be led into fenced areas so that they do not scatter or escape onto roads.
6. If possible, to help smother the flames, reduce the oxygen supply by shutting all windows and doors to the building.
7. Protect nearby buildings by hosing them down thoroughly. Move all equipment a safe distance away, plow fire lanes, and apply water around the burning structure.

These safety procedures should be reviewed periodically with all farm personnel, since preparation can save crucial minutes once a fire breaks out.

BREEDING FARM RECORDS

Detailed and systematic records contribute to the professional management, higher foaling rates, and financial success of any breeding operation. If a change in personnel occurs, a good record system guarantees minimal confusion throughout the transition by providing new personnel with a detailed history of the operation. Records can also illustrate the operation's strong and weak areas by providing statistics on breeding efficiency, rate of economic return, stallion semen quality, etc. Although a well-organized record system is extremely important to the serious breeder, it is difficult to outline a set of records for one farm and expect them to apply effectively to another farm's program. Each operation is unique with respect to services provided, responsibilities, size, cash flow, and goals. For this reason, each farm should be carefully analyzed so that **all** aspects of the operation can be represented accurately on file. As the breeder sets up or reorganizes the breeding farm records, four important guidelines for maintaining an effective system should be remembered: clarity, accuracy, accessibility, and simplicity.

A record's clarity, or its ability to be understood by all persons concerned, is its most valuable feature. Unpleasant misunderstandings and lengthy lawsuits may be avoided when concise records have been kept.

Recording information accurately and promptly is also essential to an effective record system. On a small operation, for example, one person should be responsible for observing and recording teasing responses. This person could use symbols, colors, or numerical codes on a calendar to relay consistent messages to the stallion manager, veterinarian, mare owner, etc. (On some farms, several teasing attendants may be needed to insure that all the mares are teased regularly.) Information should be noted immediately and transferred to the proper file as soon as possible. At the close of each day, all files should be up-to-date; nothing should be left to memory.

It is also important that breeding farm information be easy to locate and accessible to appropriate personnel. Whether the record system encompasses a large office or a small corner of the tack room, a carefully organized and well-defined filing pattern is imperative. Accurate information is useless if it is lost in the shuffle of forms, letters, receipts and addresses. Filing systems, such as subject and alphabetical files, can be adapted to fit individual needs. A comprehensive list of necessary headings should be made prior to arranging the folders. In some instances, a cross-reference system may be helpful. This system involves a file of index cards for all possible duplicate subject titles or names, explaining how they can be found in the permanent system. Also, it is very important that any material taken from the file be identified on "out" cards which are then placed in the file. The material taken, the date it was removed, who has taken it, and when it will be returned should all be noted. Ideally, the files should be located near the telephone so that it is easy to pull the records on a particular client's mare and make a concise, verbal report.

Efficiency on the breeding farm is greatly enhanced by a well-organized office with space for wall charts and files.

Another important guideline to remember is that both the farm records and the filing system should be complete but simple. On many operations, time available to record and file important information may be limited. In order to insure that information is not left to memory or lost before it reaches the file (especially on busy days), the record system should be as efficient as possible. When selecting or designing records for a particular operation, the breeder should keep in mind the number and type of horses involved and the time that can be allotted for paperwork. An understanding of the basic elements of a breeding farm record system allows the breeder to select only those forms that apply to his or her operation and, thereby, eliminate unnecessary paperwork.

Identification Program

Careful identification of each horse is of primary importance to an effective breeding farm record system. The program should allow the farm owner, manager, or stable hand to find any horse at a moment's notice. It should also insure against accidental exchange of foals, substitution of one mare for another in the breeding shed, and other embarrassing problems. Although many cases of mistaken identity have been documented (e.g., disputed parentage, breeding a mare to the wrong stallion, etc.), it is likely that many exchanges and disputes are caused by accidents that might have been prevented through the use of a detailed, but simplified, identification system. Because there are many acceptable methods for identifying horses and because each technique has various advantages and disadvantages, the breeder should study all possibilities and develop an identification program that best accommodates individual needs.

LOCATION MAPS

On many well-managed breeding farms, breeders use identification maps which literally pinpoint the exact location of each animal on the farm. These maps can be easily updated, in case of new arrivals or departures, by simply adding or removing a labeled pin. Color codes might be used to distinguish outside mares from farm mares, colts from fillies, mares to be bred to a particular stallion, etc. By recording the movements of horses from one area to another, the location map eliminates time-consuming searches for specific horses.

Color-coded location maps can aid farm managers and employees by pinpointing the location of any horse on the farm.

OUTSIDE MARES

It is very important that each outside mare have an information tag attached to her halter or neck strap upon arrival at the breeding farm. These tags provide the receiving manager with very important information: the mare's identification, information about her reproductive status, last vaccination and deworming dates, stallion to which she is to be bred, etc. Information tags help the manager make immediate decisions about each mare's welfare and minimize the possibility of mistaken identity or errors in the breeding shed.

It should be noted that horses are occasionally injured or killed in accidents involving halters. (Leather halters, however, are usually safer than those constructed of nylon or rope because leather breaks more easily when stressed.) Therefore, it is wise to remove halters from all horses as they arrive on the farm and are placed in stalls, paddocks, or pastures. Alternatively, break-away halters could be placed on hard-to-catch horses. For identification purposes, halters can be replaced with neckstraps or tags as discussed in the following section.

An identification shipping tag provides important information on incoming mares.

Although an identification tag can be designed to hold many types of information, this simple tag bears information suitable for a broodmare being shipped to a breeding farm.

FARM-OWNED HORSES

Even if the breeder is familiar with all of the farm-owned mares, stallions, and foals, the use of labeled neck straps or mane and tail tags may protect the farm from chaos in case of an emergency, change in personnel, death of the owner, etc. Newborn foals should be carefully identified to prevent exchanges between mares—an uncommon yet potentially serious problem. Although they are less secure than neckstraps, mane and tail tags are probably safer identification methods for youngsters. Color-coded tags can be used to distinguish colts from fillies, making separation of large numbers of pastured weanlings much easier. Neck straps and tags that are color-coded to specific areas on a location map are very helpful on large operations.

On many breeding farms, individual mares are identified with numbered neck straps. Whether band or link style, the neck strap should be breakable and should fit the mare's neck closely.

SIGNALMENT

Signalment is one of the oldest identification methods for horses. It is a system based on a simple drawing of the horse's markings on a body outline. It is sometimes combined with a written description of color, sex, and distinguishing features for more accurate identification. The body outline is commonly divided into sections for greater ease in designating the exact location of identification marks. Several outlines showing different views of the horse's body should be used to insure accurate representation. Signalment records should be on file for all farm-owned and visiting horses. Young suckling foals should also have signalment forms located in their dam's file. (The foal's markings should be drawn on its registration application forms soon after birth.) Prior to breeding and upon departure, all horses should be checked against their signalment to insure proper identification.

Courtesy of Mark Gratny Quarter Horses

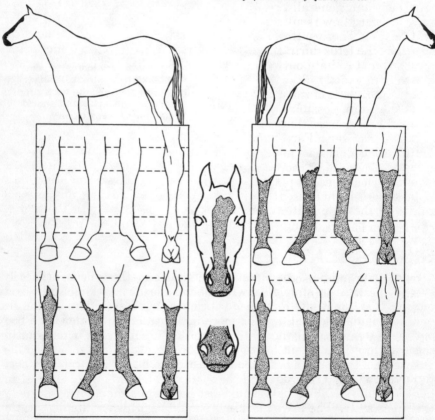

A signalment outline, such as this one of the Quarter Horse stallion shown above, is a commonly used method of horse identification.

CHESTNUTS

Chestnuts are the horny tissues located on the inside of the horse's front and, usually, hind limbs. Since these growths are unique for each horse, they are the rough equivalent of the human fingerprint. Tissue peeled from a horse's chestnuts provides an accurate and permanent means of identification. Close-up photographs of each chestnut can also be used to identify a specific horse. Computerized chestnut scan converters, which "fingerprint" the horse, convert the size and shape of each chestnut to a numerical value. This unique identification number is assigned to the horse and the scan reading is entered in computer storage. The computer's information can then be compared with the horse at some later date, if identity is uncertain or must be confirmed for sale or show. Close-up photographs of each chestnut can also be used for identification.

HAIR PATTERNS

With the exception of hair on the muzzle, around the eyes and in the mane, body hair generally grows in a tailward direction. Any disturbance in the tailward flow results in a unique hair pattern that can be used for identification. These patterns are formed in response to stresses placed on the skin of the fetus during gestation. Hairs originate from a center point, or vortex, and radiate outward in a variety of patterns. Because no two patterns are exactly alike and because the patterns remain unchanged throughout the life of the horse, they provide a reliable means of identification. The only possible way to change the hair pattern involves skin transplants, a process that leaves scars that can easily be detected.

The late Dr. Keith Farrell, an internationally known expert on equine identification, devised a method of recording the hair pattern on a horse's forehead. The area is thoroughly cleaned and moistened with a special solvent. A clear piece of plastic is placed on the forehead and pressure is applied. The solvent causes the plastic to melt, leaving a three dimensional imprint of the hair pattern. This imprint can be filed with the horse's registration papers, along with drawings of any brands or markings.

BRANDS

Hot iron brands, used religiously in the past, have been primarily replaced by less painful and more dependable forms of identification. Although hot iron branding leaves a permanent scar, hair may eventually grow over the mark, making accurate identification difficult. Also, many hot brands can be easily altered, another factor that affects their dependability.

FREEZE MARKING

Although a relatively new technique, freeze marking is recognized as a very important means of identifying horses. Because the freeze mark method can be used to destroy pigment-producing cells on a dark horse, or completely stop hair growth on grey or white horses, it leaves a permanent area of contrast (white hair on a dark horse and dark bald skin on a grey or white horse). On some large ranches, large numerals are freeze marked on all mares for permanent identification. A freeze mark system that was designed by Dr. Keith Farrell consists of a series of right angle symbols that usually represent the horse's registration number. The position of each symbol represents a numerical value, making it impossible for the brand to be altered to another number. (The number can be blotted out; however, this leaves a large area of contrast.)

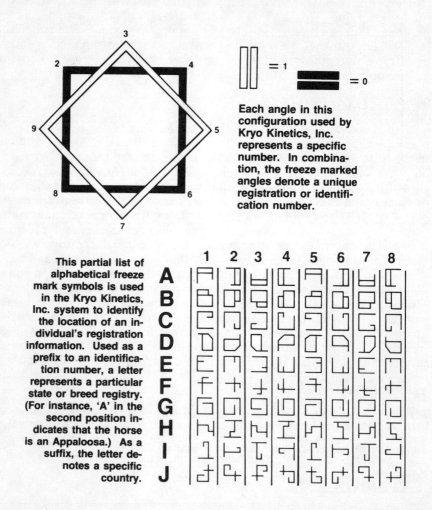

Each angle in this configuration used by Kryo Kinetics, Inc. represents a specific number. In combination, the freeze marked angles denote a unique registration or identification number.

This partial list of alphabetical freeze mark symbols is used in the Kryo Kinetics, Inc. system to identify the location of an individual's registration information. Used as a prefix to an identification number, a letter represents a particular state or breed registry. (For instance, 'A' in the second position indicates that the horse is an Appaloosa.) As a suffix, the letter denotes a specific country.

The mark is applied with an iron that has been submerged in a liquid nitrogen bath until cooled to -320°F, the temperature of liquid nitrogen. An area on the horse's neck is clipped and washed with alcohol while the iron is cooling. The iron is applied to the prepared area for 8-30 seconds, depending on coat color and age of the horse. The iron used on foals is smaller, and application time is shorter. Bald marks on grey horses require more time than those that merely destroy the color pigment. Initially, there is some swelling and, after a few days, a dry scab or scale forms. The hair grows back white, or growth ceases altogether, depending on application time. The easy and relatively painless freeze marking procedure was developed by Beverly P. Farrell and Michael Mucha. It leaves an unalterable message which can be used by anyone who is familiar with the system to identify a horse.

Courtesy of Texas A&M University

Shortly after an area is freeze-marked the tissue begins to swell. This is the only sign that is visible immediately after marking.

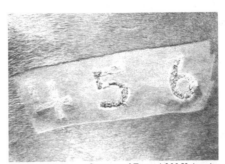

Courtesy of Texas A&M University

After a few days the new mark is covered by a dry scab which gradually peels away.

Courtesy of Texas A&M University

The shorter application times that are used on dark-colored horses allow depigmented hair to grow back, which then contrasts with the horse's hair color.

Courtesy of jbryn

On light-colored horses, such as this grey, hair follicles under the marked area are destroyed by the freeze mark, leaving bald skin.

TATTOOS

Lip tattoos have been adapted for use in race horses. Most tracks require that a horse be tattooed prior to racing as a means of "positive" identification. The use of lip tattoos has received some criticism because they can be altered. In addition, on breeding farms where long-term identification is necessary, lip tattoos are not the best means of identifying horses, since they fade with time, becoming difficult to read.

BLOOD TYPES

Massive blood typing programs, sponsored by various breed registries, have been initiated to supplement current identification methods. Laboratory analysis of prepared blood samples reveals certain inherited blood types, or markers. Because these factors are expressions of the horse's genetic make-up, they are established at conception and do not change with age. Since there are a number of blood type systems, and because there are many possible "blood types" within each system, the chances that two horses will carry identical markers are extremely small. With modern techniques for identifying inherited blood factors, blood typing is considered to be an excellent means of identifying horses.

IDENTIFICATION CARDS

Permanent identification (I.D.) cards are becoming an important tool throughout the horse industry. Designed to accompany horses in transit, these cards also have an important place on the breeding farm. Although most owners, managers, and stable hands can identify their own horses, problems often arise when horses are shipped from farm to farm for breeding or training purposes, or when mares and foals are turned out to pasture. The I.D. card should accompany the horse to all farms, shows, or races and should provide at least one clear photograph of the named horse. Two photographs (front and side view) or three photographs (front and both sides) are preferable, however. The card should state the horse's name, registration number, color, foaling date, sex, and markings. Information with respect to blood type, freeze brands, chestnuts or other identification marks are sometimes included.

Courtesy of Kryo Kinetics, Inc.

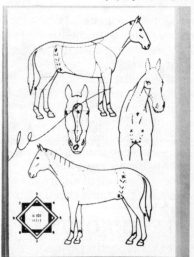

Kryo Kinetics, Inc. issues this card to owners of horses that are registered with its freeze marking identification system. Instead of a photograph, a signalment form showing the location of hair patterns and white markings appears on the back of the card.

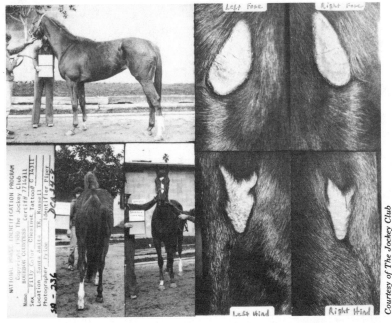

The Jockey Club currently uses three photographs of the horse and photographs of the horse's four chestnuts for identification purposes. This composite is kept on file with the registry and a miniature composite is issued to remain with the horse.

Daily Records

Any details or general observations concerning the breeding operation should be recorded immediately to minimize the chances of error. Naturally, this procedure is especially important on large farms, where several managers supervise many employees, and the responsibilities of one department affect efficiency in another area. This is not to say, however, that these records are unnecessary on a small operation. The efficiency of any operation is enhanced by its daily records system. Analyzing the farm's routine procedures and organizing a records system that reflects each of these procedures provide a strong management foundation for the operation. The breeder should note on a daily report or calendar the important events that occur each day (e.g., teasing responses, breeding dates, vaccination dates, accident information, etc). Later, this information should be filed so that the facts can be found without confusion.

DAILY REPORTS

The exact reporting system set up by the individual breeder depends on the size of the farm and on the services performed by the operation. The breeder with only two or three mares, for example, might record daily events on a calendar. On larger operations, the breeder might use daily or weekly stall (or pasture) cards for each horse or a daily report of notable activities kept by each supervisor. Medical reports that give the name and registration number of each horse examined by the veterinarian, as well as the details of any treatment given, are extremely important to the daily records system. This information, along with details on deworming, vaccinations, or hoof care might be recorded by the attendant on a general health care calendar or chart and later transferred to the individual's permanent file. (Special space-saving codes may be helpful when many horses are involved.)

GENERAL HEALTH CARE REPORT FOR _3/15/81_			
NECKTAG NUMBER	HORSE'S NAME	CONDITION	TREATMENT GIVEN
42	Lucky Day	Cut on right front shoulder from kick received in pasture	four stitches, penicillin (10 cc morning, night), Ta
543	Fine Time		Tr
63	Wanderlust Lady	two months prior to foaling	T, W (dichlorvos)
89	Persistence	thrush (mild)	Tr
300	Split Second	quarter crack	Tr

T–	tetanus toxoid	R–	Rhinopneumonitis
Ta–	tetanus antitoxin	I–	Influenza
EEE–	Eastern Equine Encephalomyelitis	W–	deworming; indicate product used
WEE–	Western Equine Encephalomyelitis	S–	shod
VEE–	Venezuelan Equine Encephalomyelitis	Tr–	trimmed

Perhaps the breeder's most essential daily reports are those that deal with the actual reproductive examination, teasing, and breeding procedures. Examples of these and other important reports are provided, but it should be remembered that these records will be easier to understand after reading subsequent chapters (e.g., "Heat Detection," "Broodmare Management," "Breeding Methods and Procedures," etc.). For this reason, it might be very helpful if, prior to organizing a daily records system, the breeder has read the entire text with possible daily record forms in mind.

DAILY REMINDERS

If set up appropriately, a record system can also provide daily reminders for teasing, breeding, foaling, vaccination, deworming, and hoof care dates. An example is the routine health care chart that lists every horse on the farm and all required vaccinations, dewormings, etc. Holes in these charts show, at a glance, which horses have not received certain treatments. Bulletin boards and chalk boards are also used to remind personnel of procedures or precautions to be taken on a particular day.

TEASING AND BREEDING CHART

Month _____ 19____

HORSE	1	2	3	4	5	6	7	8	9	10	11	12	13	14	15	16	17	18	19	20	21	22	23	24	25	26	27	28	29	30	31

I	in heat	F	foaled
Oh	out of heat	We	weaned
B	bred	Ft	feet trimmed
O	open	s	shod
P	pregnancy test	V	vet check
X	pregnant	Vt	veterinary treatment
M	medication*		

RECORD PERMANENT ENTRIES IN INK AND
PROJECTED REMINDERS IN PENCIL

*see individual records for details

MONTHLY PLANNER

FARM _____

MONTH_____

Horse–Reg. Number	Tet. Tox.	Flu	Booster	Rhino	Booster	EEE/WEE	VEE	Deworm	Teeth	Feet	Pregnancy Test	Wean

Permanent Records

Grouping an individual animal's records into one, easily accessible, well-organized folder simplifies the breeding farm filing system. To reduce the chances of loss or error, temporary forms (e.g., stall records, manager's daily reports, teasing calendars, service reports, etc.) should be used to update the permanent system. This update system limits the movement of an individual's permanent folder, thereby reducing the possibility of losing valuable information and insuring that this information is readily available at all times.

Individual folders for each farm-owned horse should include purchase papers, identification information, registration certificates, pedigree, any advertising copy, and a list of habits and vices. It is important to carefully record all veterinary examinations and treatments. The health report can provide a detailed history of an individual's physical status: routine checkups, deworming dates, vaccination dates, illnesses, injuries, drugs administered, dental care, special feeding requirements, etc. It also serves as an immunization and deworming reminder.

STALLIONS

In addition to the aforementioned material, the stallion's individual folder should contain information on his reproductive status: semen evaluation, reproductive examinations, notes on sex drive, general attitude, etc. To insure optimum conception rates, the results of routine semen examinations can be recorded for future reference and comparison. The

SEMEN EVALUATION

STALLION:	REG. NO.:
Date	**Date**
_____ % motility	_____ % motility
_____ % live–dead count	_____ % live–dead count
_____ morphology	_____ morphology
_____ sperm concentration	_____ sperm concentration
_____ pH	_____ pH
_____ bacteriological culture results	_____ bacteriological culture results
_____ total volume	_____ total volume
Evaluator's Signature	Evaluator's Signature
Date	**Date**
_____ % motility	_____ % motility
_____ % live–dead count	_____ % live–dead count
_____ morphology	_____ morphology
_____ sperm concentration	_____ sperm concentration
_____ pH	_____ pH
_____ bacteriological culture results	_____ bacteriological culture results
_____ total volume	_____ total volume
Evaluator's Signature	Evaluator's Signature

stallion's folder should also include a service reservation sheet, which is simply a list of mares to be bred by that stallion during a specified breeding season. Many breeders limit the number of mares on this reservation sheet in case a scheduling problem arises. If problems do not develop, several last minute additions can be made to the stallion's breeding list. The stallion's breeding report, discussed more thoroughly later in this chapter under 'Registry Reports and Forms,' provides the names, registration numbers, first and last breeding dates for each mare serviced by the named stallion, and a list of mares serviced each day by the stallion. Because most registries require that this list be on file prior to registering the resulting foals, the breeder must keep accurate records and submit a complete report at the end of every breeding season. Copies of each report should also be kept in the stallion's permanent file.

STALLION RESERVATIONS FOR ___FALL IS FAVORITE___

APRIL	Horse/Owner	Arrival Date	Comments
3	Freya's Rose / Armstrong Ranch	3/2	Maiden Mare
	Flory's Folly / A.A. Suzanne	2/15	Maiden Mare
	Grevey Dun / D. Weldon	3/6	Foal at side; born 2/25
4	Shani Kibibi / S.E. James	2/25	Foal at side born 1/9
	Dee Dreams / Wanderlust Farms	2/18	Twins; born 1/15
5	Melody's Majic / Hilltop Quarter Horses	2/8	
	Annie Bee / Armstrong Ranch	3/2	Foal at side
	Persistence / D. Copeland		
	Fancy One / D.K. Watson		

FARM-OWNED MARES

Folders for farm-owned mares should contain detailed information on current reproductive status and past reproductive performance. A complete report includes information on reproductive abnormalities, fertility status, injuries, veterinary examinations, treatments, breeding dates, etc. An account of each foaling should also be included in the mare's permanent file. In addition, information from teasing and breeding calendars and health care charts should be transferred to the permanent file at the end of each day. Teasing and breeding information is extremely important since it would be difficult for the handler to remember each mare's estrous cycle patterns, behavioral problems, and fertility status.

INDIVIDUAL TEASING & BREEDING CHART

Date	Teas.	RECTAL EXAMINATION						SPECULUM				TREATMENT/COMMENTS
		Cerv.	Ut.	LO	RO	CI	B/O/P	Cer.	Color	Muc.	Cult.	

BROODMARE PRODUCTION RECORD

Name _____ Year foaled _____

Reg. No. _____

Color and Markings _____

Sire _____ Dam _____

Breeder's Name _____

Address _____

Purchased from _____ Price _____

Remarks:

Sire			
	Paternal Grand sire	
	Paternal Granddam	

Dam			
	Maternal Grandsire	
	Maternal Grand dam	

	Dates Bred	Bred to	Date Foaled	Complications of Gestation or Foaling	Cond of Foal	# Days Gestation	Foal Color	Foal Sex	Foal Name	Date Weaned	Buyer	Price	Date
1													
2													
3													
4													
5													

OUTSIDE MARES

As mentioned earlier, the use of neck straps and tags on outside mares can save guesswork and confusion. It is equally important that each outside mare be accompanied by some type of identification as well as appropriate health and breeding soundness certificates. These records and the mare's arrival and departure sheet are important additions to each outside mare's file. The arrival and departure sheet should include the date of the mare's arrival, special notes about the mare upon arrival, description of the mare's exact location on the breeding farm, date of her departure, and special notes about the mare upon departure. When a breeding contract is signed, many farms request that the mare owner complete a special information sheet with respect to the mare to be bred. These sheets give a brief reproductive history of the mare, explain dietary and exercise requirements, and list any significant habits or vices that she might display. This information sheet should also be kept in the outside mare's permanent file.

FOALS

Each suckling foal (farm-owned or outside) should also have a folder included in his dam's permanent file. Information such as the foaling date and the foal's sire and dam should be included. A description of the foal and notes on its birth should be recorded. Vaccination, deworming, and detailed medical records should also be included in the foal's permanent file. Notes on nursing patterns, bowel movements, and sleeping patterns might also be included. Upon weaning, this information can be incorporated into the weanling's permanent folder.

INFORMATION SHEET FOR OUTSIDE MARE

Owner's Name_____ Phone No._____

Address_____

Mare Information:

1. Name_____ Reg. No. _____

   Color _____   Age_____

2. Innoculations:

Tetanus	Yes _____	No _____	Date _____
Flu – Vac	Yes _____	No _____	Date _____
WEE & EEE	Yes _____	No _____	Date _____
VEE	Yes _____	No _____	Date _____
Rhinopneumonitis	Yes _____	No _____	Date _____
Strangles	Yes _____	No _____	Date _____

3. Allergic sensitivities_____

   _____

4. Most recent deworming date_____dewormer name_____

5. Usual feeding procedure (amount, grain type, protein content, type of hay, and
   number of daily feedings)_____

   _____

   _____

6. Previous breeding and foaling history:

   Date of last foaling _____

   Has this mare ever been treated for
   genital infection?   Yes ____   No ____

   Does she show obvious, visible signs of heat?   Yes ____   No ____

   Must she be rectally examined to detect
   her heat cycle?   Yes ____   No ____

   Does she foal every other year although
   bred yearly?   Yes ____   No ____

   Has she ever slipped a diagnosed pregnancy?   Yes ____   No ____

   Has she ever received progesterone for
   pregnancy maintenance?   Yes ____   No ____

   Has she ever aborted?   Yes ____   No ____

   Does she have a history of retaining afterbirth?   Yes ____   No ____

7. Please give the name and phone number of a veterinarian who knows this mare.

   _____

   _____

PLEASE NOTE:

All mares should be accompanied by proof of negative Coggins taken within one
   year of arrival date.

Remove hind shoes before shipping.

FOAL RECORD

Name of Dam _____
Name of Sire _____

Name of Foal_____ Reg. No._____
Date Foaled_____ Date Weaned _____ Sex_____
Color and Markings_____

	Weight	Height at Withers	Girth (in)	Comments
1st Year				
2nd Year				
3rd Year				

Treatment

Date	Deworming	Vaccinations	Farrier	Injuries

COMMENTS (birth, health, etc.):

Contracts

As a business establishment, the breeding farm frequently participates in buy-sell or leasing transactions. Although most negotiations are completed without difficulty, legally binding contracts are the best insurance against unpleasant misunderstandings and potential lawsuits. Although a lawyer's services are indispensable when drafting important or complicated documents, the breeder should also be aware of the basic principles behind each type of business transaction and should understand the importance of keeping records of these agreements on file.

BILL OF SALE

The Uniform Commercial Code is the basic law for the sale of goods in most states. The code varies slightly between states but is a dependable guideline for most transactions. Any buy-sell agreement, or bill of sale, should give the names and addresses of both parties, a description of the horse (or horses) sold, and the date of the transaction. The price, the method of payment, and any allowable recourse for either party must be stated in the agreement. The date and place of delivery, responsibility for the animal before and after delivery, and any warranties, disclaimers, or insurance provisions should also be established.

LEASE AGREEMENTS

Leasing agreements should be carefully outlined so that any dispute concerning liability, ownership of subsequent offspring, or restrictions on the use of the stallion or mare can be readily solved. All terms of the agreement should be completely understood by both parties: the commencement and termination dates of the agreement should be set in advance; any provisions concerning the care of the leased animal should be established; and the lessee and lessor should decide who will be responsible for loss or damages during the lease agreement. (Refer to the discussion on leases within "Breeding Farm Economics.")

BREEDING CONTRACTS

The breeding contract is a written agreement between two parties stating that a certain stallion owned by the first party will be bred to a certain mare owned by the second party during a specified breeding season. Because there are many problems that can occur during a breeding transaction, breeding contracts should cover all contingencies. The following is a list of example elements that might be included in a breeding contract:
1. names and addresses of both the stallion owner and the mare owner;
2. services provided by the stallion owner and corresponding fees;
3. method and date (dates) of payment for board bill, booking fee, and stallion fee,
   —recourse for stallion owner in the event of nonpayment;

4. identification of the stallion and mare to be bred (name, registration number, description, age, parentage, etc.);
5. designated breeding season (e.g., no earlier than February and no later than June of a specified year);
6. statement of guarantees, conditional guarantees, or absence of guarantees;
7. insurance coverage;
8. payment of attorney's fees in case of breach of contract;
9. feed, care, and exercise requirements for the mare;
10. stallion owner's right to have the mare treated or dewormed by the farm veterinarian and to charge subsequent fees and other health-related expenses to the mare owner;
11. recourse for either party:
    —if the stallion is unavailable (e.g., due to sale or illness),
    —if the mare is barren,
    —if the foal is stillborn;
12. breeding method and location;
13. stallion owner's responsibilities:
    —to provide suitable care for the mare,
    —to see that the mare is bred to the named stallion,
    —to send a stallion breeding report to the breed registry,
    —to provide a valid breeder's certificate upon full payment of the breeding fee;
14. mare owner's responsibilities:
    —to deliver the mare with necessary health certificates,
    —to provide negative Coggins test (in some states),
    —to see that the mare is broken to halter,
    —to provide information on any health or soundness problems,
    —to provide information on any unusual habits or vices,
    —to send notice to the stallion owner in case of abortion or stillbirth;
15. disclaimers:
    —liability of each party in case injury or damage is inflicted by his horse or on his establishment,
    —responsibility for mare's veterinary fees;
16. special conditions:
    —maximum number of times the mare will be bred,
    —contract invalid unless completed in full,
    —contract nontransferable.

SYNDICATION CONTRACTS

The use of equine syndication has increased rapidly over the past few years. The benefits and responsibilities of such group participation are described in "Breeding Farm Economics." Again, it should be emphasized that the formal syndication agreement is a complex legal document that can have substantial financial and tax implications for each shareholder and for the original owner. For this reason, it is extremely important that a knowledgeable lawyer draft the contract and oversee each transaction.

Each member should have a complete copy of the agreement and should understand the importance of keeping this contract filed in a safe, yet accessible, location. (For a description of the basic points covered in a syndication agreement, refer to the discussion on stallion syndication within "Breeding Farm Economics.")

Registry Forms and Reports

To guarantee the registration of each foal and to insure the receipt of transferred registration papers, the breeder must be acquainted with, and adhere to, certain rules and standards set by the breed registry. For example, rules concerning artificial insemination, embryo transplant, blood typing, and transfer of ownership are determined by the breed association. These guidelines are established to protect the breeder and to insure breed purity.

REGISTRATION APPLICATIONS

The registration application is a permanent record used by the breed association to identify each horse. This important document should be completed in careful detail. The horseowner is usually defined as the owner of the horse's dam at time of foaling (unless a transfer of ownership is requested upon registration or a leasing agreement was in effect at the time of foaling). Most registries request several name choices in case duplicated or unqualified names are submitted. A description of the horse (sex, color, markings, etc.), its place of foaling, foaling date, and descriptions of its sire and dam (registration numbers, colors, markings, etc.) are also important. Some breed associations register horses of unknown or incompletely known parentage but usually require that each horse meet rigid breed standards (e.g., height, color, etc.). If a registry accepts horses whose sire and/or dam are not registered with that particular association, the owner is usually required to provide adequate proof of the horse's parentage before the registry will certify its pedigree.

Most associations require the signatures of both the breeder and the stallion owner at time of service to verify the horse's parentage. This statement may be in the form of a supplemental breeder's certificate or a clause within the registration application. A breeder's certificate states that a stallion and mare were bred on a particular date (or dates) or, in the case of pasture breeding, that the mare was exposed to a certain stallion within a designated time period.

BREEDING REPORT

A stallion's breeding report is a list of mares that he has bred along with corresponding breeding dates for a specified breeding season. This list must be submitted to the appropriate registry prior to the registration of any subsequent foals. Most registries require that the forms be returned before the end of the year and that they be signed by both the attending manager and the owner of the stallion at time of service. (Late fees may be charged for failure to submit these reports by a specified date.) The stallion's name, registration number, and location should be identified. The name, the registration number, and the owner (at time of service) of each mare bred by the stallion should also be included in the report. Dates of service for each mare should be noted on the report according to registry requirements. If the mare was pasture bred, the breeding date should be designated as the period of exposure to the stallion. Some registries also request (but do not require) that the breeder submit a list of mares that have not been bred during a particular season if those mares have foaled previously.

Courtesy of American Quarter Horse Association

CERTIFICATE OF REGISTRATION

A certificate of registration is issued to the owner of a registered horse. Reasons for delayed registration include failure to submit a stallion report; failure of the breeding date to correspond to dates on the stallion report; failure to describe or draw the horse's color, markings, sex, etc.; incorrect information on the sire or dam; incorrect signature; and an inconsistent or extremely unusual gestation period. In some registries (e.g., Thoroughbred), both the sire and dam (with certain qualifications) must have their blood types on file with the registry before their foals can be registered. Each registry has a special procedure that must be followed to change the recorded ownership of a horse. Many registries require that a transfer report be completed by the original owner or, in some instances, by the buyer before a new certificate of registration is issued or the original certificate is stamped and altered. These reports should be completed carefully; the buyer's name should appear as it is to be on the new certificate. The date sold, the horse's registration number, and the location of the animal are all important details.

TRANSFER REPORT
THE AMERICAN QUARTER HORSE ASSOCIATION
AMARILLO, TEXAS 79168

INSTRUCTIONS: (Print or type all requested information)
- **CAUTION:** The color and markings of the horse should be checked against the registration certificate.
- Registration certificate and transfer fee must accompany this form.
- All blanks must be completed, including written signature of seller. Seller's signature must conform with last recorded owner. **Any erasure or alterations of this form will necessitate verification.**

TRANSFER FEE $7.50. U. S. FUNDS ONLY.

I/We hereby authorize the AQHA to record the transfer of ownership on the horse.

NAME OF HORSE

AQHA REG. NO. OR APPENDIX CODE

IF NOT YET REGISTERED (list year foaled and number of sire and dam)

Year foaled   Sire   Dam

DATE OF SALE ▶   19 ◀ **IMPORTANT**
(list month, day and year horse actually changed ownership)

TO: BUYER
Print Buyer's name which must not exceed 30 characters (Letters, spaces and marks of punctuation).

BUYER'S AQHA ID NUMBER

MAILING ADDRESS OF BUYER

Street or Box Number   City   State   Zip Code

Buyer's Phone: Area Code   No.

I/WE further certify that the horse sold is the horse registered with the Association as described in the Certificate of Registration delivered to AQHA.

Written Signature of Seller X

Printed Name of Seller

SELLER'S AQHA ID NUMBER

MAILING ADDRESS OF SELLER

Street or Box Number   City   State   Zip Code

REQUIRED: IF CONSIGNED TO AUCTION SALE PLEASE GIVE NAME, DATE AND MAILING ADDRESS OF SALE CO.

OFFICE USE ONLY

Courtesy of American Quarter Horse Association

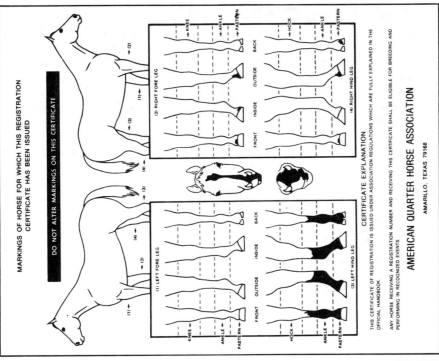

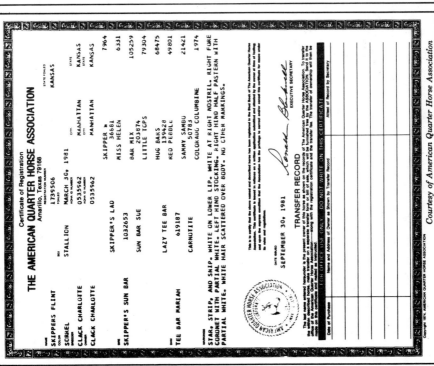

Courtesy of American Quarter Horse Association

OTHER IMPORTANT FORMS

Other forms issued by the breed registries include castration reports, leasing agreements, and death reports. The castration report identifies horses that have been castrated or spayed, lists the corresponding registration numbers, and gives the dates for each alteration. Many registries require that the certificate of registration accompany the castration report. After appropriate changes have been made on the certificate, it is returned to the owner. Most registries must also be notified of any mare or stallion leasing agreement. Breed registry leasing agreements identify the horse to be leased, the date the lease begins, and the date the lease terminates. The signatures of both the lessor and the lessee are required by most registries, and special arrangements must be made to insure that the lessee can register any foals that are rightfully his, according to the terms of the lease. Death reports give the horse's registered name and number, the date of death, and the owner's signature. The registration certificate is returned with the death report for cancellation. Upon request, some registries return cancelled certificates to the former owners.

GELDING REPORT

REGISTRY OFFICE,
The Jockey Club
380 Madison Avenue
New York, New York 10017

Date ....

Owner of animal ....   Address ....

NAME	Year Foaled	SIRE	DAM	DATE CASTRATED

IT IS *NOT* NECESSARY TO FORWARD REGISTRATION CERTIFICATES WITH THIS REPORT.

Signature ....

Use ZIP code ...., Address ....

Courtesy of The Jockey Club

AQHA
LEASE AUTHORIZATION

Alterations or added conditions shall make this form unacceptable.

The horse, _____

Registration Number _____

has been leased from _____
(Owner)

to _____
(Lessee)

_____
(Mailing Address)

_____
(City, State, & Zip Code)

for the period of time starting with _____
(Month, Day, Year)

and ending _____
(Month, Day, Year) or (If indefinite, so indicate)

and the lessee is authorized to sign all pertinent documents pertaining to this horse under the rules and regulations of the American Quarter Horse Association during this period, at the expiration of which period, the lessee's authority is terminated.

SIGNATURE OF LESSEE _____

SIGNATURE OF OWNER _____
(If individual owner, sign here.)

— OR —

If Ranch, Partnership, Corporation, etc., complete this portion:

_____
(Name of Ranch, Partnership, Corporation, etc.)

By: _____

Title: _____
(Owner, Partner, Officer, etc.)

A $15 FEE IS REQUIRED TO RECORD EACH LEASE.

Courtesy of American Quarter Horse Association

Form No. 1B

REGISTRY OFFICE

The Jockey Club
380 Madison Avenue
New York, New York 10017

Date ....

REPORT OF DEATHS

Property of ....   *Address* ....

NAME OF HORSE	Year Foaled	PEDIGREE		DATE OF DEATH
		SIRE	DAM	

P L E A S E   R E T U R N   C E R T I F I C A T E S   F O R   A L L   D E A D   A N I M A L S

Name of Farm ....

Signature of Owner or
Authorized Representative ....

Address ....   *Address* ....

The Gelding, Death or Change of Ownership of all Animals Must be Reported.
The Breeding Status of all Mares Must be Reported Annually.

Courtesy of The Jockey Club

Special Information File

    In addition to supplying important facts about the actual management of breeding stock, the farm records should provide helpful information on other important business aspects. Again, emphasis placed on various operational procedures varies from farm to farm, so that special information required or used by one breeder may be unnecessary on another farm. Each breeder should organize these files according to his individual needs. The following topics are included as examples of items that might be found in a special information section.

PASTURE RECORDS

High-quality pasture can be an important element of a successful breeding operation. It should provide excellent nutrition and allow horses to exercise freely and safely. In comparison to the benefits gained, the time and attention that are required to maintain high-quality pasture are insignificant. To insure that this maintenance program is followed, accurate pasture records should be kept. The pasture maintenance file should contain information on soil composition, changes in soil characteristics, fertilization dates and agents, and any problems with a specific area (e.g., poor drainage, overgrazing, disease, insect damage, etc.). Frequent observation of the pasture and careful notation of its condition are extremely important to a complete maintenance program. Because climatic conditions are closely correlated with pasture conditions, a record of weather extremes should be compared with changes in pasture quality when studying problem areas or damaged pasture. Pasture records should also contain notes on seeding, pasture rotation, irrigation, harrowing (spreading of manure), and mowing. Information obtained from local extension agents on the stages of plant development and approximate protein content at various stages might also be helpful. The breeder should also keep a list of harmful plants and a general description of each. (Refer to the discussion on pasture management within the Appendix.)

SHIPPING REQUIREMENTS

The breeder that ships or receives mares from other states must be aware of interstate health requirements. Because the regulations concerning required vaccinations, tests, and certificates are always subject to change, the horseman should keep a list of sources for current health and transportation requirements. In some states, for example, identification verification certificates or shipping permits may be required. Laws concerning the actual transport of animals (e.g., space requirements, maximum time between rest stops, etc.) should be a part of the breeder's transportation information file. Occasionally, a horse must be transported to another country for breeding, training, or sales purposes. In these instances, the proper authorities (USDA Animal and Plant Health Inspection Service) should be contacted for information on import and export regulations.

EMERGENCY NUMBERS AND PROCEDURES

A readily accessible file for emergency numbers and procedures is an extremely valuable addition to the special information file. Telephone numbers for several veterinarians, emergency veterinary service, insurance companies, farm managers, farm owner, owners of outside horses, the fire department, and the police department should be on hand. In addition, procedures to be followed if a horse dies, when stock or equipment are stolen, or when an animal is sick should be outlined in case someone who is not familiar with these policies must take a position of responsibility. The breeder should use forethought when analyzing the details of any emergency that might occur.

A clearly defined plan should also be established to be followed in the event of the owner's death. This plan provides for the succession of management, care of the horses, and any other actions that must be taken if the owner dies. Although estate planning may appear, at first, to be of low priority, it is actually an extremely important consideration that can have far-reaching effects on the breeding operation.

FINANCIAL RECORDS

Everyone who is subject to United States tax regulations is required by law to keep records of income and deductible expenses. To insure that the Internal Revenue Service considers the operation a business, the breeder must keep accurate documentation of all business proceedings, especially financial transactions. By special tax law, horse breeders are allowed to use the "cash" method of accounting. "Cash" method simply means that the

FINANCIAL STATEMENT FOR YEAR
BALANCE SHEET AT END OF YEAR

CURRENT ASSETS			CURRENT LIABILITIES		
Cash		$ 162,000	Accounts Payable		$ 270,000
Accounts Receivable		486,000	Accrued Interest		117,000
Inventory		702,000	Income Tax Payable		30,000
Prepaid Expenses		90,000	Notes Payable		220,000
Total Current Assets		$ 1,440,000	Total Current Liabilities		$ 637,000
			LONG – TERM NOTES PAYABLE		$ 300,000
FIXED ASSETS			STOCKHOLDER'S EQUITY (for corporations)		
Machinery, Equipment,			Paid – In Capital (for which stock		
Furniture and Fixtures	$ 696,000		shares are issued)	$ 933,000	
Accumulated Depreciation	116,000		Retained Earnings	150,000	
Book Value		598,000	Total Stockholder's Equity		$ 1,083,000
TOTAL ASSETS		$ 2,020,000	TOTAL LIABILITIES AND OWNER'S EQUITY		$ 2,020,000

INCOME STATEMENT FOR YEAR			NET WORTH STATEMENT FOR YEAR	
Sales Revenue		$ 4,212,000	Assets	$ 2,020,000
Costs of Goods Sold		2,808,000	Liabilities	937,000
Gross Profit		$ 1,404,000		$ 1,083,000
Operating Expenses	$ 936,000			
Depreciation Expense	116,000	1,052,000		
Operating Profit		$ 352,000		
Interest Expense		52,000		
Profit before Income Tax		$ 300,000		
Income Tax Expense		150,000		
Net Income		$ 150,000		

breeder reports income in the tax year that it is received and reports allowable deductions as they are paid. This method is much simpler than the accounting structure required for many businesses (accrual method) but still requires careful notation and organization of receipts, bills of sale, etc. (Note: The horse breeder can use the accrual method, but once either method is selected, those rules must be adhered to for the life of the business—unless special permission from the IRS to change the accounting method is obtained.)

The breeder should keep detailed records on income and expenses and, with the help of an accountant, should draft periodic financial statements. These financial statements can be used to analyze the annual profit or loss, evaluate past expenses, and form the coming year's budget. They also provide valuable information if the business status of the operation is challenged by the IRS. Careful notation of the dates for each purchase and sale provides proof of any holding period required for capital gains deductions. (Refer to "Breeding Farm Economics.") These lists also help the breeder itemize losses when making an insurance claim.

MARKET OUTLETS

Records pertaining to possible market outlets should be up-to-date and thorough. This information might include a list of former clients, a calendar of upcoming sales or auctions and their nomination deadlines, and a list of possible advertising outlets. Any response from advertisements and any inquiries about the operation should be carefully noted. It is also good policy to make sure all clients are satisfied with their purchase, lease, or choice for stallion service. An occasional telephone call or postcard to check on a former buyer's satisfaction and to inquire about the health and performance of the purchase is an example of good public relations—the key to keeping past market outlets open.

PROMOTIONAL MATERIAL

In addition to a special file for marketing outlets, the serious market breeder should place copies of pedigrees, production records, and performance records for all farm-owned horses in a special promotional file. Any information that might interest a prospective client should be kept in a readily accessible area. Such information could include a history of the farm, an explanation of management policies, a description of procedures used during the breeding season, maps of the farm, brief descriptions of each farm employee, and sales lists. In addition, clients will be interested in the pedigrees and accomplishments of notable horses on the farm. Rather than keeping all material in a file, the breeder might prefer to keep current farm literature in an open area. Direct, but tasteful, promotion of the breeding operation contributes to the farm's professional appearance and enhances public relations. ♞

BREEDING FARM ECONOMICS

The horse industry has evolved from a strictly agricultural-related enterprise into a dynamic industry. Consequently, the financial and legal aspects involved in a breeding operation have become vast and complex. It is very important, therefore, that the serious breeder have at least basic business knowledge—if only to keep adequate financial records, to use foresight when planning or expanding the operation, and to understand the principles behind sophisticated business concepts. Advice from lawyers, accountants, and business consultants is extremely important and most horsemen seek counsel on specific matters, depending on advisors to update their knowledge of current tax legislation and business trends pertaining to their operations. A complete analysis of equine economics is beyond the scope of this text, as an entire book could easily be written on the subject. Instead, this chapter should be regarded as a general introduction to basic concepts, contracts, and tax planning methods currently used by breeders.

Profitability and Economic Risk

An equine breeding operation often represents a significant investment in time and finances. In order to realize predetermined goals (e.g., profit, the production of a particular breed type, etc.), the breeder must concentrate not only on selection and husbandry techniques but also on financial concepts that help define the program's opportunities and risks. It is necessary to understand these concepts and use them to evaluate the operation's economic status in order to establish a successful breeding operation.

FINANCIAL COMMITMENT

The first few years of operation are generally the most costly to the horse breeder. Initial investments may include acquisition of breeding stock, investments in equipment, and purchase or renovation of facilities. The breeder must also consider maintenance costs (e.g., feed, bedding, labor, and insurance premiums) and special services from veterinarians, farriers, trainers, transporters, lawyers, and financial advisors. Because a number of variables affect initial costs (e.g., location of land, quality of breeding stock purchased or leased, number of acres required, and management practices), all aspects of the operation should be considered carefully before making the initial investment.

OPERATING COSTS

1. land investment (or rent)
2. site development costs
3. costs of establishing and maintaining pastures
4. irrigation and water access costs
5. interest on loans
6. cost of establishing or renovating facilities (barns, fences, etc.)
7. equipment (vehicles, tools, tack, etc.)
8. purchase of mares (or leasing fees)
9. stud fees (or purchase of stallion)
10. depreciation
11. taxes
12. management expenses:
    a. labor costs
    b. feed
    c. utilities
    d. veterinary costs
    e. farrier costs
    f. transportation costs
    g. incidental repairs
    h. vehicle operating expenses
    i. personnel training programs
13. losses from uninsured casualty or theft
14. insurance premiums
15. dues to organizations
16. travel expenses
17. marketing expenses
    a. entertainment expenses and facilities
    b. consignment fees
    c. advertising
18. continuing education to keep abreast of latest horse-related information

Due to the lengthy equine production period (i.e., the time between a foal's conception and useful or saleable age), the breeder may have to wait several years before realizing a return on the initial investment. For this reason, the market should be carefully studied before any financial commitment is made. Although marketing is the last stage of the production cycle, it is the key to breeding for profit and is, therefore, a very important factor in determining how much money can be safely invested in a breeding facility. A thorough understanding of the current and projected markets for a particular breed or breed type, a knowledge of effective marketing techniques, and careful investment and cost analysis increase the breeder's chances of establishing a significant profit margin in several years.

In order to produce profitable horses, the demand for the breeding program's product must be projected several years in advance of sale.

As a standard for culling and a measure of success, goals are an integral and exciting part of a breeding program. Carefully planned goals and responsible selection to attain those goals can make any breeding program more effective. By increasing the overall quality of the breeding herd, the breeder can actually increase the operation's profit margin, even if breeding stock numbers are reduced significantly. Further, by keeping and breeding only the best (as defined by predetermined goals), the breeder can increase the rate of herd improvement and reduce total maintenance expenses. However, producing horses haphazardly, without a selection program to maintain or improve the herd's quality, can be economically disastrous. To realize a profit, the breeder must consistently produce marketable or successful foals, which can only be achieved through careful planning and by culling undesirable breeding stock.

By culling, the breeder recognizes that not all the horses produced in a particular program are equally useful. However, it does not imply that the entire breeding program is below par or inadequate. Effective culling requires accurate evaluation of each horse with respect to overall breed type or function and should not be ruled by emotional involvement with individual horses. Selection criteria based on the individual breeder's goals and financial circumstances can be used to design a selection index, which is a list of desirable traits weighted according to their importance. The index may be used as a guide for judging and culling horses. Any overall standard for quality can be an effective selection tool, especially if it also allows the breeder to measure the rate and degree of improvement within the breeding program. (For a detailed description of selection techniques, refer to the complete text on this subject, EQUINE GENETICS & SELECTION PROCEDURES, Equine Research, Inc., 1978).

PRODUCTION RATE

Another important factor that affects the breeder's ability to establish a profit is the production rate, which is the number of foals born each reproductive year in relation to the number of mares bred. When projecting potential profits, the breeder should remember that production rates are seldom 100 percent and that the foal crop may contain some "unmarketable" foals. Although some registries report production rates of up to 60 percent, the national average production rate is probably only 55 percent. On individual farms, production efficiency can often be improved by proper breeding management as illustrated throughout this text. With proper management and culling, for example, a conception rate of 80 to 90 percent is not unusual.

INSURANCE

Horse life insurance protects an owner's investment by reducing the economic risk involved in purchasing a horse or paying an expensive stud fee. Once a horse produces a return on the breeder's investment or after the horse's value has depreciated significantly, the breeder may choose to lower the amount of insurance coverage or discontinue the policy entirely. Thus, to guarantee adequate coverage without unnecessary expense, the farm's insurance needs should be re-evaluated each year. Because the insurance needs of each farm vary significantly, the services of an insurance agent who is also familiar with the horse industry may be extremely helpful.

It is very important to note that paying insurance premiums does not guarantee that the owner will be able to collect insurance claims. When choosing an insurance company, the breeder should study service methods, rates, claims procedures, and requirements for several companies. After carefully selecting a company, the breeder should then make every effort to understand the policyholder's responsibilities and coverage and should check the policy very carefully for errors. The owner and everyone responsible for care of the insured horse should be familiar with the claims procedures and requirements. For example, some insurance companies require immediate notification before general anesthesia or extensive surgery. In addition, many companies require that the owner notify them before euthanizing an insured animal. Most insurance companies require immediate notice of the horse's death, and some require that the animal be autopsied at a specified clinic or by a particular veterinarian. Official reports from police, fire authority, owner, and witnesses may also be required. To limit the time between loss and payment and to avoid claim denial, it is important that the proper claims procedure be carried out promptly.

Other important protection measures include liability coverage for accidents to visitors, to employees, or to outside (i.e., visiting) horses. Insurance against loss of income due to abortion or stallion infertility might also be considered. In some instances, special insurance policies for employees (e.g., life, medical, etc.) are appropriate. Based on the number of employees, workmen's compensation is mandatory in some states. Also, insurance against loss of facilities or equipment (e.g., due to fire, theft, or natural disaster) is usually a wise investment, especially when large sums have been invested in their purchase, construction, or repair.

A breeding operation can be destroyed by uninsured losses of facilities, equipment, or breeding stock. An insurance policy covering valuable items that are essential to the farm's operation is a wise investment.

TAX ASPECTS

The subject of taxation as specifically applied to breeding farm economics must proceed from an understanding of the general view the Internal Revenue Service takes of breeding farm operations. The breeder must determine and substantiate the nature of the enterprise in which he or she is engaged—that is, whether it is profit-motivated or a hobby; and he or she must recognize the fact that in most cases IRS regulations treat the breeder as a farmer, with the consequent tax advantages.

While the breeder may also be subject to various state and local taxes (e.g., property tax, sales tax, state income tax, unemployment insurance tax, etc.) this discussion is concerned only with various aspects of federal income tax. The 1981 Tax Act provides many new tax benefits advantageous to the horse breeder; but since most are not automatic, the breeder must be aware and prepared to take advantage of them. It is important in this regard that the breeder consult both an accountant and an attorney to determine which laws apply to the breeding operation and to insure that all requirements are being met. This discussion deals with federal taxation laws in effect in 1982. However, because adjustments are made to the tax laws each year, the horse breeder must be prepared to stay abreast of changes.

BUSINESS VS. HOBBY

The IRS will allow the breeder to deduct breeding farm expenses from total income (i.e., income generated by the horses and by other unrelated activity) only if the operation is not considered a hobby. If the operation is considered a hobby, or nonbusiness activity, the breeder can deduct expenses up to the amount of profit derived from the horses only. To illustrate this "hobby loss" provision, assume that in one year a breeder spends $1200 on the purchase and care of a horse and then sells the horse for $800. Because the breeder's activity is not considered a business by the IRS, the $400 horse-related loss ($1200 − $800 = $400) cannot be deducted from other income. The hobby breeder deducts up to the amount of horse-related income ($800) and, at best, pays no tax on the income from the sale of the horse. If the owner wishes to deduct horse-related expenses beyond gross income produced by the horses, he or she must prove that the operation is an actual business venture. (Exception: Specific items, such as taxes, interest, and charitable contributions that are allowed as deductions under the tax law can be deducted regardless of hobby loss limitations.)

Because horse breeding is a long-term venture, usually requiring several years to show a profit, the IRS has set up guidelines used in determining whether a profit motive exists. To insure that the IRS will consider the operation a business, therefore permitting business deductions, the breeder should try to adhere to as many of these nine characteristics as possible:

1. Manner of Carrying Out the Activity: By operating in a business-like manner, keeping accurate records, and maintaining separate records for different activities, the owner indicates a profit motive. Also, in attempting to improve the operation by doing away with

methods proven to be unprofitable and in planning expenditures and projecting profits, the owner indicates that he or she has more than a hobby in mind. This category of operating the business is perhaps the most important of the nine in proving profit motive.

Accurate record-keeping is one of the criteria used by the Internal Revenue Service to determine whether the breeder is profit-motivated.

2. Expertise: Knowing the business cannot be overemphasized as a criterion for determining profit motive. Expertise is the key word here—concentrated effort toward improving knowledge or skill in such areas as farm management, reproductive physiology, marketing, etc. Consulting expert advisors and utilizing their input indicates a profit motive.
3. Time and Effort Expended: The amount of nonrecreational time a breeder spends with his operation is looked upon as a significant factor by the IRS. Curtailing or abandoning involvement with another occupation in order to devote more time to the breeding farm may be evidence of a profit motive. However, lack of time spent with the operation does not indicate lack of profit motive if persons knowledgeable in the field of horse breeding are employed.

4. Appreciation of Assets: Profit, according to IRS regulations, includes appreciation in the value of assets used in the activity. Thus, even if the breeding operation itself is operating at a loss, the overall picture may be one of profit. Land (unless purchased primarily for its appreciation), equipment, facilities, and horses may be considered by the IRS in a total evaluation of the breeding activity.

5. Success in Other Activities: If the breeder's past record is one of turning unprofitable operations into profitable ones, this factor may be taken into account by the IRS in its evaluation.

6. History of Loss or Income: A series of years of profit is considered by the IRS to be strong evidence of a profit motive. In the horse breeding business, however, losses during the first few years are not unusual, and long-term losses are even acceptable to the IRS if the breeder can prove that the losses were due to external (unexpected and uncontrollable) conditions. In this case, the breeder must be able to produce records showing that a profit could have been made if certain events had not occurred. A gradual decline in losses or a series of profit years after the initial loss years are good indications of profit motive.

7. Amount of Profit in Relation to Losses: Occasional, small profits from an operation in which the owner has made a substantial investment or which has a history of large losses are factors not generally favorable in determining a profit motive. However, the IRS does look favorably in this regard on an operation which turns a substantial profit if only on an occasional basis.

8. Financial Status: Lack of substantial income from sources other than the horse operation is a generally favorable situation in determining a profit motive. Conversely, if losses from the breeding farm constitute a substantial tax benefit with respect to other income, the IRS may be suspicious of any true profit motive.

9. Elements of Recreation: The fact that a business venture yields a certain amount of personal pleasure or recreation to the owner does not in itself indicate that the venture is a hobby. A hobby is indicated, however, when no obvious business motives exist.

TWO-OUT-OF-SEVEN-YEAR PRESUMPTION

A horse breeder will be presumed to be engaged in an activity for profit if the operation has two profit years in a defined seven year period. If the breeder achieves two profit years in the specified period, the burden of proving the operation is not engaged in for profit is shifted to the IRS. If the breeder does not show two profit years out of seven, there is no negative presumption, but it will be the breeder's obligation to demonstrate that the operation is a business activity engaged in for profit.

It is important to note that the two-out-of-seven presumption protects only the second profit year and the remaining years of the seven year period. For example, if an operation shows a profit in 1982 and again in 1985, the presumption covers the period from 1985 through 1988:

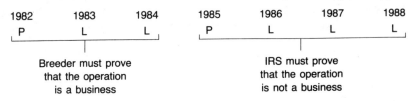

Deduction of losses prior to 1985 would be subject to challenge by the IRS unless those years were covered by a separate earlier presumption:

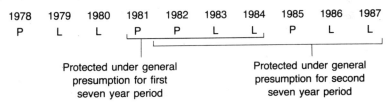

Important points to be noted are: 1) presumptions are not complete assurance that the IRS will consider an activity a business, 2) there is no negative presumption arising from a failure to achieve two profit years during a seven year period, 3) presumptions are possible only if the profit years involve the same activity as the loss years (without an obvious attempt to load certain years for profit).

The IRS has made provision for a "special" presumption for new horse breeding operations. This presumption covers the initial seven year period rather than just the period beginning with the second profit year. The IRS, in effect, is granting the new horse breeder a full seven year period to produce two profit years. A "special" presumption can be used only once and must be applied for within three years after the due date for the first year's tax return. If, within that three-year period, the IRS serves notice that it intends to disallow deductions for losses on the basis that the activity is a hobby rather than a business, the breeder has 60 days to request an appeal and apply for a "special" presumption. At the end of the seven-year period the breeder must be prepared to defend the activity as a profit venture rather than a hobby. Due to possible financial and legal complications, and because drawbacks may outweigh benefits, the breeder should consult a tax advisor before applying for a "special" presumption. ("Special" presumptions may be elected by filing IRS form 5213.)

TAX BENEFITS

Several tax benefits are available to the serious breeder. If the breeding operation qualifies for business deductions, an obvious advantage is the opportunity to make deductions against income. As previously stated, in most cases the breeder is considered a farmer by the IRS, so he or she may use most of the tax advantages associated with farming. Deductions are allowed for such items as labor, tack, tools, veterinary fees, breeding fees, farm-related educational expenses, feed (or production cost if feed is grown on the farm), etc. Capital expenditures (investments made to expand or improve a business operation, excluding operating expenses) are not deductible. However, most capital expenditures (e.g., machinery, buildings, horses, and fences) can be depreciated.

Depreciation

Depreciation is a method of measuring the use of, or wear and tear on, an item designated as a nondeductible capital asset. As the property's value diminishes, the owner is allowed annual depreciation deductions against income until the item's basis is depreciated to zero. Some items, such as land and goodwill, have no useful life and, therefore, cannot be depreciated. Livestock produced on the farm also cannot be depreciated, because production costs are deducted each year. Additionally, horses held primarily for sale cannot be depreciated.

Economic Recovery Tax Act

The Economic Recovery Tax Act of 1981 has put into effect a dramatic change in methods of depreciation. Prior to enactment of ERTA, depreciation was largely a matter of the owner's estimation of the useful life of his property. Using the straight-line method of depreciation, the cost of the property, less its salvage value (the estimated value of property at the time of sale or other disposition when property is no longer to be used by owner), is deducted in equal amounts over the estimated useful life of the property. Under the new legislation, depreciable property put in service after December 31, 1980 is categorized, and the owner must follow specific guidelines in recovering the cost of his or her property. There are four recovery categories, but only two are directly applicable to the horse breeder:

1. Three-year property, which includes property such as cars and light trucks, previously eligible to be depreciated in four years or less. Race horses more than two years of age and other horses more than 12 years of age are included in this category.

2. Five-year property, which includes all horses not covered in the previous category, single-purpose agricultural structures and most other equipment. (Barns and other farm buildings or depreciable real property placed into service after December 31, 1980 are depreciated over a 15-year period.)

The following tables, delineated in the Internal Revenue Code, are applicable in determining cost recovery deductions for property which falls within the three-year and five-year recovery categories:

THREE-YEAR PROPERTY

Year Placed Into Service — Amount of Cost

1st Year .... .25 %
2nd Year .... .38 %
3rd Year .... .37 %

100 %

FIVE-YEAR PROPERTY

Year Placed Into Service — Amount of Cost

1st Year .... .15 %
2nd Year .... .22 %
3rd Year .... .21 %
4th Year .... .21 %
5th Year .... .21 %

100 %

Each percentage applies to the total original cost of the property. It is important to note that under this system no salvage value is given. Write-off is for the total cost of the property. (In exception to this rule, if an investment tax credit is taken on an item acquired after 1982, the item's depreciable basis must be reduced by one-half of the amount of the investment tax credit.)

For example, assuming that a 14-year-old broodmare (three-year property) was purchased in October of 1981 for $20,000, cost recovery deductions would be:

1981	25 % x $ 20,000 =	$ 5,000
1982	38 % x 20,000 =	7,600
1983	37 % x 20,000 =	7,400
		$ 20,000

Note that the entire first-year cost recovery percentage applied even though the horse was not purchased until October, 1981.

New breeding operations which acquire property in a short tax year (a year in which the new corporation or partnership files a tax return which reflects less than 12 months of operation) must modify the cost recovery schedules. First year depreciation will equal the normal ACRS percentage multiplied by a fraction, in which the numerator is the number of months in the short tax year and the denominator is 12. For example, if purchased by a new breeding farm starting operations on October 1, first year depreciation on the broodmare from the previous example would be figured in the following manner:

$$25\% \ \times \ 3/12 = 6.25\%$$

Because substantial losses may occur early in the life of an asset due to the new cost recovery regulations, an alternative method of depreciation is allowed under the Internal Revenue Code. This method allows the breeder to take straight-line depreciation for each category of property, allowing write-off over a longer recovery period. For three-year category property, the breeder may elect to take straight-line depreciation over 3 years, 5 years, or 12 years. For property in the five-year cost recovery category, the breeder may elect to take the straight-line depreciation over 5 years, 12 years or 25 years.

If the breeder elects to take the straight-line method of depreciation in either category, he or she must apply it to all of the property in the category. For instance, if the breeder elects straight-line depreciation over five years for an item classified in the three-year category, all three-year property must be depreciated over five years by the straight-line method. Should the breeder elect to use straight-line depreciation for a category of property (other than real property), he or she must use the half year convention regarding the property. The half year convention provides that all of this property is considered to have been placed into service on July 1, and the breeder claims one-half of the depreciation for the full year in the year the property is placed into service.

Farm buildings and other depreciable real property are depreciated over a 15-year period if they were placed into service after December 31, 1980. Such property may be written off by using the straight-line method at the rate of 6.67% annually (6.67% x 15 years = 100%) or by using accelerated depreciation at the rate of 175% of the straight-line rate annually. If the latter method is chosen, all depreciation must be treated as ordinary income (recaptured) to the extent of gain when the item is sold. However, if the straight-line method is used, all gain is considered capital gain, with the resultant tax advantages. The bottom line consideration of the breeder must be whether he or she would be better served by the rapidity of deductions offered by the accelerated depreciation (wherein all depreciation is treated as ordinary income recaptured to the extent of gain at the time the building is sold); or by taking advantage of the fact that in straight-line depreciation all gain is considered capital gain, which is taxed at a much lower rate. Thus, if a breeder buys property which may later be sold at a profit, the straight-line method should be considered, since the accelerated depreciation method will result in higher tax when the property is sold.

The following table has been formulated by the IRS for use by taxpayers choosing the 175% x straight-line rate:

ALL REAL ESTATE (Except Low-Income Housing)
The applicable percentage is:

YEAR	MONTH											
	1	2	3	4	5	6	7	8	9	10	11	12
1	12	11	10	9	8	7	6	5	4	3	2	1
2	10	10	11	11	11	11	11	11	11	11	11	12
3	9	9	9	9	10	10	10	10	10	10	10	10
4	8	8	8	8	8	8	9	9	9	9	9	9
5	7	7	7	7	7	7	8	8	8	8	8	8
6	6	6	6	6	7	7	7	7	7	7	7	7
7	6	6	6	6	6	6	6	6	6	6	6	6
8	6	6	6	6	6	6	5	6	6	6	6	6
9	6	6	6	6	5	6	5	5	5	6	6	6
10	5	6	5	6	5	5	5	5	5	5	6	5
11	5	5	5	5	5	5	5	5	5	5	5	5
12	5	5	5	5	5	5	5	5	5	5	5	5
13	5	5	5	5	5	5	5	5	5	5	5	5
14	5	5	5	5	5	5	5	5	5	5	5	5
15	5	5	5	5	5	5	5	5	5	5	5	5
16	-	-	1	1	2	2	3	3	4	4	4	5

The table is used in this manner:

If you erected a qualifying depreciable structure at a cost of $30,000 on your property in August of 1982, you would thenceforth use the percentage rates in column 8 (i.e., 5 percent in 1982, 11 percent in 1983, 10 percent in 1984, etc.)

The 15-year write-off period does not apply to property acquired from relatives or to property acquired in a tax-free transaction. Nor is use of the 15-year write-off period mandatory. The breeder may choose a 35-year or 45-year write-off period, but in such case he or she must use the straight-line method of depreciation.

The Economic Recovery Tax Act of 1981 provides that improvements made on leased buildings must be written off over a period of 15 years or, if it is a shorter period, over the term of the lease. Improvements on leased equipment are written off over a period of five years or over the term of the lease, whichever is shorter.

Beginning in 1982, breeders may elect to expense a portion of the cost of qualified depreciable property placed into service in that year. To qualify, the property must be acquired by purchase, and must be eligible for the investment credit. (The expensing election is not applicable to horses or buildings.) If the expensing election is used, neither cost recovery deductions nor investment credit can be taken on the amount that is expensed. There are also special rules governing partnerships and certain limitations on expensing. Therefore it is advisable to consult an accountant before electing this method.

Depreciation Prior to 1981

The Asset Depreciation Range System was enacted by Congress in 1971 and adopted by the IRS as a guideline for determining useful life values for a number of farm-related items and for depreciating farm property and equipment. Use of the ADR system is optional, but it is wise to employ useful life periods that are acceptable by the IRS, if the three- and five-year recovery periods of the ACRS are not employed.

ASSET DEPRECIATION GUIDELINES
FOR FARM PROPERTY
(ADR)

machinery and equipment, but no other land improvement . . . . .10 years

horses: breeding or draft . . . . .10 years

farm buildings . . . . .25 years

land improvements (roads, landscaping, etc.) . . . . .20 years

Since the ADR table does not recognize the acquisition of previously owned breeding stock, and because preownership transactions are commonplace in the horse industry, the American Horse Council has adopted its own set of guidelines in this regard. Based on statistical studies of Thoroughbreds, Standardbreds, and Quarter Horses, this useful life table for stallions and mares has been established. Other tables have been developed by several breed organizations concerning useful life values for horses of various ages, types and services.

AMERICAN HORSE COUNCIL
USEFUL LIFE GUIDELINES
FOR BREEDING HORSES

Age When Placed in Service*	Years of Remaining Useful Life
6 years and under	10 years
7	9
8	8
9	7
10	6
11	5
12	4
13	3
14 years and over	2

*Assuming that the horse was placed into service on January 1.

Capital Gains Deductions

Profit resulting from the sale of a capital asset may qualify for capital gain treatment. If the capital asset has been held for at least 12 months (or at least 24 months in the case of breeding, draft, or sporting horses), the capital gain is referred to as long-term and can be reduced by 60 percent before being added to the taxpayer's total taxable income. (In some instances, long-term capital gain is subject to an alternative minimum tax established by the IRS.) As a case in point, a long-term capital gain of $100,000 is reduced by the 60 percent capital gain deduction to $40,000 (i.e., $100,000 − $60,000 = $40,000). The $40,000 is added to other income, and the sum is reduced by itemized deductions to determine the individual's total taxable income. Under the Economic Recovery Tax Act of 1981, the maximum tax rate to be paid on capital gains is reduced from 28 percent to 20 percent (40% capital gains rate x 50% maximum tax rate = 20%).

Involuntary Conversion Deduction

Involuntary conversion is any loss of property (e.g., horses, equipment, buildings, etc.) due to theft, natural disaster, fire, disease, or other unforeseen circumstances. Losses due to involuntary conversion can be deducted against ordinary income (or against capital gains, if applicable). The insurance proceeds and salvage value are subtracted from the tax basis of the lost or damaged property to determine the actual loss. The tax basis is the original investment or expense, minus any deductions that have been made with respect to that item over the years. A horse's tax basis, for example, is its purchase price plus any additional costs involved in the animal's maintenance or training minus tax deductions that have been made with respect to those expenses and depreciation deductions since the horse's purchase:

$$\text{Purchase Price} + \text{Maintenance, Training, etc.} - \text{Tax deductions made since purchase} = \text{Tax Basis}$$

$$\text{Tax Basis} - \left[\text{Insurance Proceeds} + \text{Salvage Value} \right] = \text{Actual Loss}$$

Due to depreciation deductions and business expense deductions, it is not unusual for a horse's tax basis to be zero. The tax basis of a horse produced and raised on the farm is also zero, since the production costs are deducted each year. If a profit results from involuntary conversion of a horse (e.g., due to a tax basis of zero or high insurance proceeds) tax liability can be limited by investing the money in an animal "similar or related in service or use" within two years of the loss.

It should be noted that horse business losses can be deducted from total income (both horse-related and from other business activities or investments) only up to the total amount "at risk" with respect to the horse business. The amount "at risk" is the total amount invested in the business minus all the tax deductions that have been made over the years. If the amount "at risk" in a horse business is zero, losses cannot be deducted until more money is invested in the operation. Any loss which is not deducted because of the "at risk" limitation, can be carried back or forward to another tax year.

$$\text{Amount At Risk} = \frac{\text{Total Investment in}}{\text{the Breeding Farm}} - \frac{\text{Total Deduction for}}{\text{Breeding Farm Expenses}}$$

Tax Credits

In addition to tax deductions, the breeder is also allowed various tax credits. Investment tax credits, for example, help to reduce the owner's tax liability (total tax due after deductions and exemptions have been made). If new or used depreciable property is purchased for business use, up to 10 percent of the property's cost can be subtracted from the owner's tax liability. However, there are certain limitations to this allowance:

1. The value of used depreciable property in excess of $125,000 does not qualify for investment tax credit. This applies through 1984. After 1984 the limit will be increased from $125,000 to $150,000.
2. If eligible property is in the three-year cost recovery category, 60 percent of the owner's investment is eligible for the investment tax credit. The credit is 10 percent of the eligible cost or 6 percent of the total cost (60% of cost x 10% credit = 6%).
3. If eligible property is in the five-, ten-, or fifteen-year categories, 100 percent of the owner's investment qualifies for the investment tax credit. The credit is 10 percent of the cost.
4. The maximum amount of investment tax credit is $25,000 plus 90 percent of tax liability over $25,000 (if income tax liability exceeds $25,000).
5. Horses and general purpose farm buildings are not eligible for the credit.

The following examples illustrate the proper method of computing the amount of tax credit:

1. Assume you purchase a new light-duty truck in 1982 for $10,000. This purchase, under the cost recovery rules, is in the three-year recovery category. The amount of the tax credit is computed:

$$60\% \times \$\ 10,000 \times 10\% = \$\ 600$$

2. Assume that in 1982 you also purchase used equipment for $4,000 and your total purchases of used property in 1982 are less than $125,000. This purchase falls in the five-year recovery category. The amount of the tax credit is computed:

$$100\% \times \$\,4,000 \times 10\% = \$\,400$$

The Tax Act of 1981 limits the amount of tax credit that can be claimed to the amount of the tax basis of the property for which the taxpayer is at risk at the close of the tax year. This limitation specifically concerns property placed in service after February 18, 1981.

Broodmare Economics

The mare's genetic and maternal contributions to her offspring are of particular importance on the breeding farm since they may determine the breeding farm's economic success.

1. The mare's ability to consistently produce marketable or useful foals can affect farm income significantly.
2. The mare's ability to conceive affects breeding costs, production rates, and efficient use of the stallion.
3. The mare's ability to carry a foal to term is extremely important. Abortion is costly not only because of the lost foal but also because of wasted time, breeding fees, and other pregnancy-related expenses.
4. The mare's influence on her suckling foal is also important. Maternal influence has been shown to affect the foal's growth rate, tractability, and susceptibility to vices—factors that affect the marketability and usefulness of the foal.

Because the broodmare band is the financial foundation of the producer's breeding operation, each mare should be carefully evaluated before purchase for breeding soundness, genetic potential, and quality with respect to breed type or function. (Refer to the discussion on breeding soundness examinations within "Broodmare Management.") A mare that has never been bred (i.e., maiden mare) may cost less, but the breeder should remember that a maiden mare is not necessarily an easy breeder or a good mother, and she may not produce high-quality foals. Reproductive examinations and pedigree studies are very important prerequisites for selecting maiden mares. Purchasing a mare that has been certified "in foal" by a veterinarian limits the risks involved in buying a maiden mare and enables the breeder to realize a return on the investment earlier than if an open mare had been purchased.

When buying a mare in foal the purchaser should make sure that the stud fee has been paid. If the fee was not paid by the previous owner, the stallion owner may refuse to sign the necessary foal registration papers. Also, any guarantees or stipulations within the breeding contract should be understood. The buyer should note that, in many instances, the sale of a mare voids her breeding contract. The buyer should also note that the fact that the mare is pregnant does not insure that she will carry the fetus to term. Due to abnormal (and sometimes uncorrectable) conditions within the uterus, some mares consistently abort during certain stages of pregnancy. (Refer to "Abortion.")

Perhaps the safest broodmare investment is in the so-called proven mare. A proven mare is valuable because she has produced quality offspring (i.e., foals that have competed or produced successfully and, consequently, enhanced their dam's value). Because a proven mare is likely to be aged and very expensive, the purchase price should be considered with respect to her remaining reproductive years. Such a mare should be given a thorough reproductive examination to determine whether she will be able to pay back the breeder's investment by producing several foals. Although

Courtesy of The Thoroughbred Record

Although they are usually very expensive, proven mares can provide excellent opportunities for upgrading the breeding herd's quality. A proven mare should have produced several high-quality offspring and should be in good breeding condition. The Thoroughbred mare, Exclusive, is a notable example of a proven mare, having produced 9 stakes winners from 16 foals.

the reproductive career of an older mare is limited, she may be an excellent investment if she has been an outstanding producer. Thus, the chances of obtaining one last outstanding foal from an older mare should be weighed against the possibility of age-related pregnancy and foaling complications.

MARE OWNERSHIP

Long-term ownership of a broodmare implies that the owner hopes to profit from the sale of her offspring. For the owner to profit from the investment, the price of each foal must be greater than that foal's production cost, which includes a percentage of the mare's purchase price. (Refer to 'Cost of Raising a Foal' within this chapter.) The breeder should recover an investment in a young mare through the sale of about four or five foals, and unless the mare proves to be an exceptional producer, it may be economically beneficial to sell her while she is still fertile and insurable. With increasing age, the mare's reproductive capacity often diminishes, her insurance premiums increase, and her market value decreases (unless her foals are in demand). Under certain circumstances, short-term ownership may provide the owner with a rapid return on the investment. For example, some breeders find it profitable to buy open mares, breed them to popular stallions, and sell them (in foal) for a substantial profit. The risks involved with such a practice (e.g., failure to conceive) should be carefully considered. Other breeders may purchase a young, high-quality broodmare, breed her to comparable stallions for several years, and sell her for a slight profit, keeping her best filly as a breeding replacement. A mare that is purchased to be sold after producing several foals will be easier to market if she meets the following qualifications: 1) successful show or race record, 2) promising or proven pedigree, 3) good conformation, 4) reputation for producing marketable or successful foals, and 5) good breeding condition.

MARE LEASE

A mare lease is a contract that states a mare owner's agreement to give another party the use and possession of a mare for a certain length of time. Mare lease provisions may vary but are generally conducted on either a cash or a foal basis. In a cash lease, the lessee (i.e., person to whom the property is leased) pays the lessor (i.e., owner) money for the use of the mare. (Cash paid for leasing broodmares is usually deductible.) In a foal lease, the mare usually produces two foals at the lessee's expense, one of which is given to the lessor. A combination of both the cash and foal basis may also be used in a mare lease. In many instances, a lease gives the breeder with limited cash a chance to use a quality mare for several breeding seasons and, hopefully, to obtain a foal from that mare. Alternatively, the mare owner who is unable to afford an expensive stud fee might lease a mare to the stallion owner on a foal basis.

Misunderstandings between lessor and lessee can be avoided by executing a contract before the onset of the lease. For example, the lessor may require that the mare be insured by the lessee for the duration of the lease with a loss payable clause in favor of the lessor. It is also important that the breed registry be notified of a lease agreement and that the lessee be given permission to register specified foals by power of attorney. Several items should be included in a lease contract:

1. names, addresses, and capacities of the parties involved;
2. identification of the leased horse by name, breed, registration number, sire, dam, foaling date, color, and markings;
3. date the contract becomes effective and length of lease;
4. warranties, if any;
5. the type of care that the horse is to receive;
6. the terms of payment;
7. provisions for the registration of lessee's foal (or foals);
8. the owner of the foal, if mare is pregnant when the contract is signed;
9. estimated value of the leased horse;
10. provisions for responsiblity, including who is to assume the cost of injury to the horse, damages caused by the horse, or death of the horse;
11. party responsible for insurance;
12. provisions for default (breach of contract), including specific remedies to be used in case of default.

Stallion Services

Although the mare plays a very important role in determining overall production rates, the stallion is often crucial in determining foal market values. His popularity, pedigree, stud fee, performance record (i.e., race, show, etc.), and production record are factors that influence the value of his foals. The reason for the stallion's importance on the breeding farm is the stallion's ability to sire many foals per reproductive year and, therefore, his potential ability to influence the breed as a whole. Because of these influences, and because fewer stallions are required in a breeding operation, the importance of careful stallion selection for service or purchase cannot be overemphasized.

FARM-OWNED STALLION

A farm-owned stallion can have a great impact on farm management. As a case in point, mares can be bred at home, thus eliminating outside stud fees and minimizing the chances of transportation-related injuries to the mare or her foal. Stallions that are successful at stud and that have been carefully promoted can also provide an important source of income through stud fees. In addition to the obvious benefits of owning a stallion, there are several disadvantages that should be considered.

1. The stallion's purchase and promotion may represent a substantial investment which may not be repaid through foal production or by stud fees.
2. Farm-owned stallions require special facilities and expert handling. The stallion should have secure and separate (not necessarily isolated) housing with a pasture or paddock of sufficient size to accomodate free exercise.
3. The stallion should be managed by a qualified horseman that is experienced in stallion handling and breeding methods.
4. Extra labor costs involved in breeding, teasing, and palpating mares on the farm should be considered prior to purchasing a breeding stallion.
5. If outside mares are to be bred, necessary facilities, additional land, and extra labor may be costly. The mare care fee, paid by the mare owner to cover the expense of feed and care, may not be large enough to cover all the expenses encountered in connection with outside mares.

If a breeding stallion is kept on the farm, allowances must be made to provide a sturdy, spacious stall that gives the stallion a view of farm activities.

LEASED STALLION

By leasing a stallion, the breeder may be able to acquire a better stallion than he or she could afford to purchase and, consequently, may be able to introduce outstanding bloodlines to the breeding program. The concerns of the stallion owner, such as housing, care, and additional expenses are also relevant to the stallion lessee. In addition, the interests of both lessor and lessee should be protected by a contract. A stallion lease agreement should include most of the items listed in the mare lease agreement.

OUTSIDE STALLION

The use of outside stallions (i.e., stallion standing at another farm) allows the breeder to select a stallion that will enhance each mare's strengths and minimize her weaknesses. If the breeder has only a few mares, the cost of outside stallion service is usually less than the cost of buying, maintaining, breeding, and promoting a stallion. Costs that the mare owner should consider include transportation costs, mare care fees, breeding expenses (palpations, treatments, etc.), and stud fees. If the stallion is a proven sire, the mare owner's investment will probably be returned in the form of a marketable foal. However, because the dam also influences the foal's quality and marketability, the mare should be of similar quality to the stallion.

It is essential that the mare owner understand any guarantees provided in the breeding contract, which should include a detailed description of the guarantee. If the contract includes a live foal guarantee, for example, the stallion owner's definition of live foal and a description of the procedures to be followed if a live foal is not produced should also be included in the guarantee. In some instances, the mare may be rebred until she produces a live foal by that stallion. Other contracts may only guarantee that the mare will be rebred during one specified breeding season without guaranteeing the mare owner a live foal. In many instances, the stallion owner guarantees the return of all or part of the stud fee if the mare owner does not obtain a live foal. Special color guarantees are sometimes made by owners of Paint, Pinto, Appaloosa, and Palomino stallions. The mare owner should study the benefits and limitations of any breeding guarantee when selecting an outside stallion.

STALLION SYNDICATION

A mare owner may decide to buy interest (i.e., shares) in a syndicated stallion as an alternative to purchasing a stallion or buying annual services. Syndication spreads among several people the economic risks involved in owning a valuable stallion. In addition to the purchase price, the shareholder pays for a portion of the stallion's upkeep and insurance. In return, the shareholder is guaranteed a certain number of breedings to the stallion and usually has special voting rights concerning how the syndicate should be managed.

If the entire cost of a share is paid, title to that share (or shares) is received by means of a bill of sale. If payments are made in installments, on the other hand, a bill of sale and security agreement provide for a transfer of title when the principal and interest are paid as agreed. The terms for installment payments (e.g., amount owed, interest rate, due date of payments, action taken upon failure to make payments, etc.) are usually outlined in a security agreement that accompanies a promissory note. Although each shareholder may have the right to sell shares or to sell breeding rights for a given year, it is customary to require that other shareholders be given the "right of first refusal" before a share is sold to an outside party.

In many syndicates, the number of shares sold corresponds with the number of mares that the stallion will breed in a given year (excluding breedings that may be granted to the stallion manager in exchange for specified duties). If a designated number of shares are not sold, the stallion owner (i.e., seller) may want to cancel the syndication. If the seller elects this option, an escrow account should be established. The stallion's registration certificate and any funds collected from the initial purchasers are placed in the escrow account (e.g., at a bank or with a special escrow agent) until all, or a specified number, of the shares are sold. The original owner may also elect to retain a security or lien in undivided interests until full payment by the shareholders has been made. If the shareholders' obligation to pay the seller is joint, each syndication member could be held liable for other unpaid shares. Provisions for escrow accounts, retained liens, and transfer of shares should be incorporated into the syndication agreement. Because this agreement is a sophisticated legal document and because it can have substantial financial and tax consequences for each member, a knowledgeable lawyer should be obtained to draft the document. Four important aspects of the stallion syndication should be dealt with in the agreement: 1) provisions for the purchase of shares, 2) rules under which the syndicate will operate, 3) tax aspects of the syndicate, and 4) state and federal laws governing securities.

Courtesy of The Thoroughbred Record

Many successful performance stallions are syndicated before they begin their breeding careers. Spectacular Bid, shown here as a yearling, was syndicated for breeding before being retired from racing.

PROVISIONS FOR PURCHASE

The type of syndication and the shareholders' rights within the syndicate should be designated in the agreement. For example, shareholders within a type "A" syndication are entitled to a designated number of stallion services per year, and the shareholder's right to receive these services is insurable. Although additional breedings are not available to the general public, individual members can sell their breeding privileges to outsiders. (By limiting the number of breedings, the syndicate may enhance the value of the stallion's offspring.) The cost of maintaining the stallion is shared by the syndicate members, but the costs of maintaining each mare at the stud farm are paid by the individual mare owners (e.g., shareholders). In a type "B" syndicate, each purchaser receives a share of the stallion's income (after expenses have been deducted). Breeding privileges are available to anyone for a designated fee. Members of this type of syndicate should take special care to comply with state and federal laws governing securities. (Refer to the discussion on securities within this chapter.) The type "C" syndication is a combination of types "A" and "B." Shareholders pay for a portion of the stallion's upkeep and are entitled to a designated number of breedings. Fees from additional services, which are available to the shareholders and to the general public, are distributed among syndicate members. The type "C" syndication is advantageous because shareholders are assured a designated number of breedings, receive income to defray the cost of their share, and can purchase additional breedings.

MANAGEMENT OF THE SYNDICATED STALLION

In most instances, the syndicated stallion is handled by an appointed manager who follows guidelines set forth in the syndicate agreement. The manager supervises the stallion's daily care, handling, and breeding and is usually required to keep a complete account of the stallion's health, breedings, income, and expenses. The manager's report should also show the names of all mares bred to the stallion, the names of all mare owners, and the number of times each mare was bred. The manager may also be required to furnish special items, such as halters, blankets, special feeds, veterinary services, payment of telephone bills, transportation costs, medicines, advertising, etc. In exchange for these services, the manager receives a handling fee for the care and board of the stallion in addition to a specified number of free breeding rights. If advised by a veterinarian, the manager has the right to limit or increase the stallion's breedings. Provisions for such an event should be included in the syndication agreement. For example, shareholders may be required to draw for breedings if the stallion's book is reduced. The members that miss a year's service due to a reduced book are usually given priority in subsequent years. The manager also has the right to refuse to breed the stallion to a particular mare if, according to the attending veterinarian, it would be injurious to the stallion's health or fertility.

It should be noted that each shareholder can insure the right to a specified number of stallion services. This so-called fertility insurance specifies a certain standard for fertility (e.g., a 60 percent conception rate and/or the ability to pass a thorough semen examination) and provides compensation for missed services due to infertility. If the stallion is found to be sterile, the syndication agreement should provide that shareholders receive a refund of the initial purchase price and that any shareholder whose mare (or mares) did conceive pay a designated stud fee. Although most syndication agreements include a veterinary health certificate, which warranties the stallion's general health, each shareholder should arrange to have individual interest in the stallion covered by full mortality and fertility insurance effective when the syndication agreement is signed.

TAX ASPECTS

Depending on the syndicate agreement and the surrounding circumstances, a stallion syndicate is usually treated as either a co-ownership or a partnership for income tax purposes. Co-ownership implies joint ownership and maintenance of property. Services provided or property produced by the co-owners must be sold individually. A partnership, on the other hand, involves a common trade or business from which the profits are divided. Co-owners are allowed to file separate income tax returns, while partners must file a special partnership information tax return. Failure to submit this form can result in a stiff penalty for the manager.

There are special income tax considerations that the seller and shareholders should understand. For example, if the syndicated stallion has been used in connection with the original owner's horse business for at least 24 months, the owner may qualify for capital gains treatment when the stallion is syndicated. If the stallion has not been held primarily as "inventory," any profit that the seller derives from the stallion's syndication can be treated as long-term capital gain, receiving a 60 percent capital gains deduction.

If the shareholder's horse operation is considered to be a business by the IRS, the shareholder can deduct expenses such as depreciation, handling fee, insurance, etc. Again, it is very important to remember that deductions are limited to the amount at risk, not including any money borrowed from persons with an interest in the business (e.g., the seller).

Another important consideration concerns the tax on the stallion manager's breeding rights. If the syndicate agreement does not specify that free breeding rights are given annually (i.e., not for the lifetime of the stallion), the IRS might attempt to tax the manager for the entire value of all possible breedings as income received in one year.

SALE OF SECURITIES

The investor in a security relies on another's efforts to make the investment profitable. If the Security and Exchange Commission (SEC) considers the sale of interest in a stallion to be the sale of securities and if the securities (i.e., stallion shares) are not exempt from registration, the SEC may require that the promoter (stallion owner) cancel the transaction and return the money plus interest to the participants within one year. To avoid this, the syndicate can be registered on form S-1. However, SEC registration requirements are expensive, complicated, and time-consuming. If required for all horse syndications, many would be impractical. For this reason, a syndication agreement should specify that the purpose for selling shares is to insure each shareholder of the right to breed a certain number of mares to the stallion. To avoid legal complications, shares should not be sold to nonbreeders, syndicates should not be represented as investments, and potential financial gains should not be advertised. A "no action" letter, issued by the SEC in July of 1977, describes the terms under which a stallion syndication would not be considered a security. Syndication agreements which do not follow the eight listed provisions, or which do not conform to the factual pattern of the SEC "no action" letter will not necessarily be considered as securities. The SEC letter merely defines a "safe harbor" area.

1. The cost of the stallion's upkeep is to be paid by the shareholders in proportion to the number of fractional interests each owns. For example, if a shareholder owns 1/34 of the horse, 1/34 of the expenses are paid by that shareholder.
2. The shareholders may annul the transaction if a specified level of fertility is not achieved at the end of the first breeding season.
3. The stallion must stand at a location designated in the syndicate agreement, and a syndicate manager must act as agent, performing custodial functions. The agreement must specify a percentage of votes required to move the stallion or to replace the manager.
4. The manager's compensation should be the right to a specified number of stallion services for each year the position of manager is fulfilled.
5. Any extra breedings should be drawn for by lot among the syndicate's shareholders.
6. Each shareholder must have an insurable interest in the stallion.
7. Shareholders may sell their shares to nonsyndicate members only after extending the "right of first refusal" to the other syndicate shareholders.
8. A shareholder may sell breedings to another breeder, and that breeder does not necessarily have to be a syndicate member. The shareholders may obtain a list of interested breeders from the syndicate manager, but the shareholder must not ask the syndicate manager to sell breedings.

In addition, the Securities Act of 1933 offers other guidelines:

1. Syndicates should sell shares only to those who have sufficient knowledge, information, and experience to evaluate the merits and risks of the purchase, or the purchaser must have a knowledgeable advisor.

2. The purchaser must be able to assume the risk of a share of the necessary syndicate expenses, bear the economic risk of the purchase, and bear the loss of the entire investment.

Cost of Raising a Foal

It is important to analyze the costs of horse production before committing a large investment to a breeding enterprise, since only extremely efficient operations manage to show a significant return on their investment within several years. Breeding farms that show continual long-term losses are often viewed suspiciously by the IRS, especially if the owner seems to be using the business as a tax shelter for other income. Written cost studies are important tools for proving that the operation is being conducted with a profit motive. Regardless of whether the operation is conducted as a business or a hobby, however, the breeder needs to evaluate expenditures periodically and should allocate the annual production costs to the annual foal crop.

Several charts are provided in this section to assist the breeder in determining the cost of raising a foal from conception to weaning or sale age. Because exact costs vary with the farm's location and size, these charts should be used as guidelines to estimate future expenditures. (If faithful cost accounting is conducted each year, average expenditures will gradually become easier to estimate.)

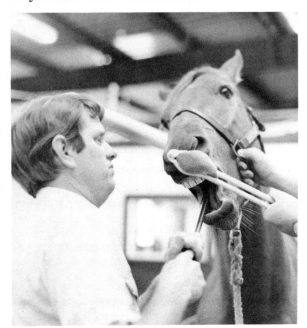

Routine items such as dental care must be included as part of the cost of raising a foal.

Because cost accounting need not be as precise for the hobby breeder as for the profit-motivated breeder, this discussion treats the cost of raising one foal separately. The following "hobby expense" table assumes that the mare is boarded at a professional stable that supplies labor, stall, and pasture. (The mare may spend several months at the breeding farm each year, during which her board is paid in the form of mare care.) A detailed and critical comparison of boarding versus home stabling costs may prove that boarding is an economical proposition for the hobby breeder, especially if additional facilities required for the production of one foal are unnecessary for any other purpose. A possible increase in boarding fees for the foal is not reflected in this table, since this value varies greatly among commercial stables. Feed is figured as a separate cost because broodmare nutrient requirements increase during the last trimester of pregnancy and again during lactation. Because the mare is not considered an investment in this example, depreciation expenses are not reflected as part of the foal cost. (A detailed cost analysis supplement for these tables is included in the Appendix.)

COST FOR ONE FOAL: nonbusiness enterprise *

Feed (mare)	$ 389.81
Feed (foal)	109.05
Board	1,500.00
Mare Care	240.00
Transport to breeding farm	150.00
Farrier Service (mare)	96.00
Farrier Service (foal)	48.00
Breeding Fee	1,500.00
Veterinary Service (mare)	227.00
Veterinary Service (foal)	184.00
TOTAL COST OF HOBBY FOAL	**$ 4,443.86**

* It is assumed that the mare produced a live foal. Costs increase when the mare fails to conceive or aborts.

A second example, designed to illustrate the cost of producing several foals on a professional basis, involves the following hypothetical situation. A breeder owns 50 mares, 40 of which produced foals—now weanlings. All of these mares were bred to a farm-owned stallion, valued at $100,000. In the chart 'Average Cost Per Foal,' the expenses for maintaining the stallion, 10 barren mares, and facilities are divided among the 40 offspring. The chart 'Cost For One Foal' represents a more accurate estimate of the cost for a particular individual, since it reflects the true value of its dam, rather than an average value for the entire broodmare band. These are only two examples of the many accounting methods that can be used. The breeder should study several techniques and adopt the most suitable method. Expenses not mentioned on the cost charts include entertainment, telephone, rent, showing, utilities, and tack. To effectively evaluate the operation's financial status from year to year, the breeder should use the same accounting method each season. When accurately noted and carefully organized, these cost analysis sheets provide the breeder with helpful information and a basis for making important management decisions.

AVERAGE COST PER FOAL: business enterprise*

Expenses:

Feed	$ 31,077.58
Farrier Services	6,816.00
Veterinary Services	18,925.00
Pasture Maintenance	3,550.00
Sales Commissions, Consignment Fees	6,000.00
Stallion Depreciation	10,000.00
Mare Depreciation	50,000.00
Facilities Depreciation	1,840.00
Stallion Insurance	5,000.00
Labor	25,600.00
Cost to raise 40 foals	$158,808.58
COST PER FOAL	**$ 3,970.21**

* 50 farm-owned mares, 1 stallion and 3 employees

COST FOR ONE FOAL: business enterprise*

Expenses:

Mare		
Feed	$ 525.42	
Farrier Services	96.00	
Veterinary Services	227.00	
Depreciation	1,000.00	
Total Mare Expense		$1,848.42
Stallion		
Feed	444.58	
Farrier Services	96.00	
Veterinary Services	215.00	
Depreciation	10,000.00	
Advertising	6,000.00	
Insurance	5,000.00	
Total Stallion Expense	$21,755.58	
Divided by 40 Foals		543.89

Foal

Feed through 6 months of age	109.05	
Farrier Services	48.00	
Veterinary Services	184.00	
Total Foal Expense		$ 341.05

Facilities Depreciation	46.00
($1,840.00 divided by 40 foals)	
Pasture Maintenance	88.75
($3,550.00 divided by 40 foals)	
Farm Labor	640.00
($25,600.00 divided by 40 foals)	
Maintenance for 10 barren mares	462.11
($18,484.20 divided by 40 foals)	
TOTAL COST PER FOAL	**$3,970.22**

Cost is figured for one foal produced from a $10,000 mare, assuming an 80% foal crop at weaning, and maintaining a band of 50 mares (10 barren mares).

Estate Planning

Estate planning allows the breeder to provide for continued maintenance or dispersal of his breeding stock, farm equipment, and land after death. First and foremost, such a plan must insure the availability of sufficient cash, labor, and expert guidance to maintain and protect the property until other arrangements can be made. These arrangements should be specified by the owner in a legally binding document, or will. When drawing up a will, the breeder should choose an attorney who specializes in estate planning, preferably one who is also familiar with the horse industry. Estate planning can also reduce the burden of estate and inheritance taxation by making special provisions for the payment of such taxes and by reducing the taxable estate. The estate plan should be structured so that the farm is managed and death taxes (e.g., federal estate tax, state inheritance tax, etc.) are paid with minimal confusion. It is extremely important that both an accountant and an attorney be consulted when determining the best methods for accomplishing these goals.

WILLS

A will is a legal document that takes effect upon the death of an individual to dictate specific property distribution, fair estate distribution, the specific exclusion or inclusion of friends or relatives, etc. When properly drafted, the will becomes a means by which the deceased may supervise property disposal. In the absence of a will, the state provides for the distribution of the deceased's property, according to that particular state's succession laws. Unfortunately, property distribution under these laws may not result in distribution that would have been desired by the deceased and may even cause confusion and bitterness between family members and friends. To avoid such conflicts the will should be drafted by an attorney in accordance with state laws. Once the will has been drafted, it may still be changed or revoked anytime before death, as long as the benefactor is believed to be of sound mind. (If a new will is drafted, the previous will should be destroyed.) Although the drafting of a new will may be necessary if substantial changes must be made, minor modifications may be accomplished through a "codicil." It is important to note that these additions must be drafted by a lawyer with the same formalities as the original document.

TRUSTS

A trust is an arrangement that allows a property interest to be held and managed by a trustee (e.g., bank official, appointed friend, or relative) for a designated beneficiary or beneficiaries. If the trustor completely relinquishes control over the property prior to death, it is referred to as an irrevocable trust. This type of trust provides significant tax benefits because it reduces the taxable estate. (The taxable estate is the property which is given to the beneficiary after the owner's death. The value of the property determines how much inheritance tax must be paid.) A revocable lifetime trust, on the other hand, is not a tax-saving technique but is very helpful in relieving beneficiaries of the details of estate management. Special trusts, designed to keep the estate together as an income-producing whole, might also be considered by the breeder. Trust arrangements may also be used to meet special needs, such as the care of the breeder's horses or the continuation of a breeding program, following the breeder's death. As with the drafting of wills, trusts should be set up by an attorney, since the language used must meet both legal and tax regulation standards.

CHOOSING AN EXECUTOR OR TRUSTEE

The breeder's wishes for the estate, as dictated in the will, should include the appointment of persons to carry out certain duties. For example, every will must have a designated executor (or executrix, if female) to oversee the estate settlement and make sure that the wishes of the deceased are honored. (In some cases, more than one executor may be named.) If an executor is not named in the will, the probate judge must appoint one. To insure that all property is managed properly, the breeder should appoint a trustworthy individual who is familiar with the business (e.g., friend, relative, attorney, or accountant) to execute the provisions of the will. Several alternative executors should be appointed in case the original choice is unable or unwilling to accept this responsibility. In some cases, it is recommended that the trust department of a financial institution be named co-executor to avoid problems that might result if the named executor predeceases the breeder.

Any trust set up by the breeder must be supervised by a trustee who takes responsibility for the property, making investments or in some way managing the estate's assets. (In some instances, the donor may serve as an investment advisor to the trustee for lifetime trusts.) Because the trustee has important legal and fiduciary obligations to manage the trust property and any investments with utmost care, this appointment merits very careful consideration. The individual's frugality, knowledge of the horse business, understanding of economic concepts, and access to accounting and investment advice should be studied closely.

The breeder should also appoint someone to manage the breeding operation until the estate is settled. In this instance, humane treatment of breeding stock should be the utmost consideration. For this reason, someone familiar with equine management should be selected to carry out or oversee daily farm routine. This specification might be stated in a "Letter of Last Instruction," along with details on burial requests, location of all estate property (e.g., savings bonds, bank accounts, stocks, etc.), names of lawyers, accountants, or consultants who are familiar with the estate, and location of the will.

The breeder should make arrangements for daily farm management in the event of his or her death to insure that animal care will continue uninterrupted.

ESTATE AND INHERITANCE TAXES

The federal estate tax is imposed on the transfer of an individual's property at death. This tax is levied on the fair market value (as defined by the federal government) for all property within the estate and is paid with assets from the estate. Taxable property includes the following:

1. property owned by the deceased,
2. property jointly owned by another party (or parties) in addition to the deceased,
3. revocable gifts (or trusts),
4. gifts that the donor still had control over at time of death,
5. certain powers of appointment (e.g., the power to name benificiaries of a trust),
6. some life insurance proceeds,
7. certain annuity contracts (investment that yields a fixed sum of money each year).

The Economic Recovery Tax Act of 1981 provides for tax rate schedules which will reduce the maximum estate tax rate from 70 percent to 50 percent over a four-year period beginning in 1982. Another benefit of the 1981 Tax Act is the removal of all limits on the marital deductions for estate tax purposes. If the estate passes directly to the decedent's current spouse, its value is 100 percent deductible if the decedent dies after December 31, 1981.

Most states have either an estate tax or an inheritance tax. Unlike estate taxes, inheritance taxes are paid by the individuals that receive a share of the estate. Death taxes vary considerably from state to state and are frequently subject to change. For this reason, a discussion on state death taxes is not included in this text. ◤

4

THE STALLION REPRODUCTIVE SYSTEM

The stallion's importance on the breeding farm is highlighted by his ability to produce many foals per reproductive year. His role on the farm is simple, but his effect on the stallion-oriented operation can be very complex. The mere presence of a stallion, for example, dictates a need for special facilities, experienced handlers, and carefully planned procedures. The stallion's fertility status may have a very significant effect on the farm's annual profit-loss figure. In addition, the quality of the stallion's offspring directly influences the farm's income and reputation.

There are many contributions and complications that a stallion's residence incurs, but the most important characteristic is that he requires special management and a knowledgeable handler or owner — one that can

identify potential fertility or behavioral problems before they affect an entire foal crop. The breeder should remember that the stallion's physical ability and psychological willingness to cover a mare are just as important as the usefulness or marketability of his offspring. A basic understanding of the stallion's reproductive organs and normal sexual responses is extremely important to the stallion owner that hopes to manage his charge for optimum productivity. The following discussion provides background information for subsequent chapters on management and infertility. Without this working knowledge, important (but somewhat technical) aspects within later chapters may be difficult to understand. This chapter introduces useful anatomical and physiological terms and explains the relationship between reproductive structure and function, information that is especially important background material for the chapter "Stallion Infertility and Impotency."

Reproductive Anatomy And Physiology

The term *anatomy* refers to the study of an animal's physical make-up, or anatomical structures. Its importance stems from the need to identify and refer to various body parts when discussing development, function, disease processes, injuries, etc. Stallion reproductive anatomy identifies the structures that produce, nourish, store, and transmit the spermatozoa. Other important structures include tissues that produce reproductive hormones, glands that secrete seminal fluid, and organs that deposit the stallion's semen into the mare's reproductive tract. Reproductive physiology is the study of how these structures carry out their normal processes.

TESTES

The sperm-producing, hormone-producing organs of the stallion's reproductive tract are the testes, or testicles. During fetal development, the testes are located against the upper abdominal wall just above and behind the kidneys. Prior to birth, the strong colorless membrane that lines the abdominal cavity moves through an opening in the lower abdominal wall. This opening, which is referred to as the internal inguinal (or abdominal) ring, leads to the inguinal canal — a narrow passageway between the abdomen and the future testicular sac, or scrotum. The abdominal membrane forms a pouch (processus vaginalis) that extends downward through the inguinal canal, through the external inguinal (or vaginal) ring, and into the scrotum. During the later stages of gestation or soon after birth, the testes follow the processus vaginalis to the scrotum and become partially enclosed by the abdominal membrane. This membrane serves as a protective scrotal lining, or tunica vaginalis. As the testes migrate to the scrotum, their blood vessels and nerves are relocated along the inguinal canal. These events normally occur prior to birth, but temporary or permanent retention of one or both testes is not unusual. (Refer to the discussion on cryptorchidism within "Stallion Infertility and Impotency.") The exact physiological events that cause testicular descent are not clear at this time, but the following factors may be involved:

1. The testes may be pulled downward by the shortening of a fetal ligament attached to the bottom of the scrotum and the lower end of the epididymis (sperm storage vessel attached to the testes);
2. Increased intra-abdominal pressure due to organ growth and development may cause migration of the testes;
3. The testes may follow a path of least resistance to the scrotum;
4. The actions of fetal hormones probably play an important role in testicular descent.

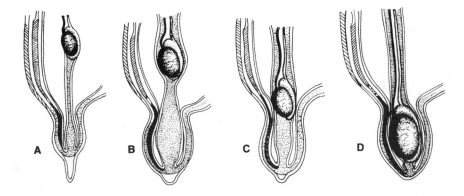

Testicular descent: (A) the abdominal testes enter the inguinal canal via the internal inguinal ring; (B) as the testes pass through the inguinal canal, the spermatic vessels are repositioned along the canal; (C) the testes pass through the external inguinal ring; and (D) the testes are positioned in the scrotal sac.

The normal adult stallion has two testes located in the prepubic region, enclosed in the scrotum. The testes are suspended from the abdomen by connective tissue and muscle which form the spermatic cord. This cord also contains important blood vessels and nerves which nourish and support these important male gonads. Within the cord, an intricate network of blood vessels allows arterial blood to be cooled by returning venous blood, keeping the testicles at an optimum temperature. The muscles within the spermatic cord (cremaster muscles) and within the scrotum (tunica dartos muscles) also control testicular temperature by positioning the testes close to or away from the body. (Because sperm production is inhibited by an increase or extreme decrease in temperature, these mechanisms are very important to the stallion's fertility.)

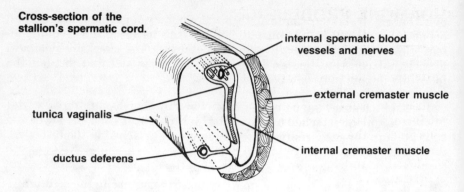

Cross-section of the stallion's spermatic cord.

internal spermatic blood vessels and nerves

external cremaster muscle

tunica vaginalis

internal cremaster muscle

ductus deferens

Testicular size, an inherited characteristic, varies significantly between stallions. Size may decrease with age (as will be explained later), but on the average, a testicle weighs about 8-10 ounces and measures about 5x3x2 inches. The long axis of the testis is usually horizontal within the scrotum, but may be vertical when the testicle is drawn up next to the body. Although both testes are located in a single external sac, they are separated by connective tissue and muscle (septum scroti) which form two protective pouches within the scrotum. Beneath the tunica vaginalis (abdominal lining within the scrotum), each testis is surrounded by a strong fibrous capsule, or tunica albuginea. Sheets of this fibrous tissue enter the body of the testis from various directions, dividing the organ into small lobules (Lobuli testis). Within each lobule is a reddish-grey mass of small intercoiling seminiferous tubules. The interior of each tubule contains primary germ cells, which are capable of dividing and eventually forming viable spermatozoa. (Refer to the discussion on sperm production within this chapter.)

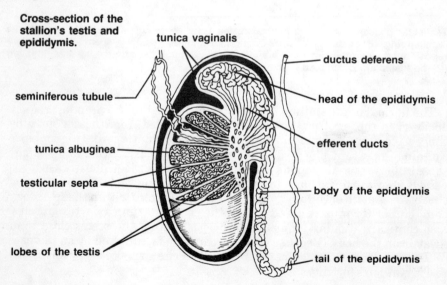

Cross-section of the stallion's testis and epididymis.

tunica vaginalis

ductus deferens

seminiferous tubule

head of the epididymis

efferent ducts

tunica albuginea

testicular septa

body of the epididymis

lobes of the testis

tail of the epididymis

HORMONE PRODUCTION

Special "nutrient" cells are also found within the seminiferous tubules. Scientists believe that these cells, sometimes referred to as Sertoli cells, provide nutrients for the developing sperm and that they may be responsible for the production of estrogen. (Although estrogen is a female reproductive hormone, large quantities are found in the stallion's urine.) The connective tissue between the seminiferous tubules also contains hormone-producing cells, referred to as interstitial cells or the cells of Leydig. These cells produce the male reproductive hormones, or androgens. Testosterone, for example, is an androgen that is responsible for the initiation of sperm production and the development and maintenance of masculine characteristics.

Reproductive hormone production is controlled by a special center in the brain, referred to as the hypothalamus. In response to certain stimuli, such

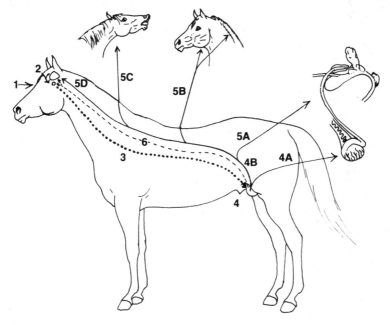

A simplified diagram of the basic hormonal mechanisms that control reproduction in the stallion:
1. Photo stimulation and other external stimuli affect the hypothalamus.
2. The hypothalamus produces and releases gonadotropin releasing hormone (GnRH).
3. GnRH stimulates the anterior pituitary gland which releases gonadotropic hormones, FSH and LH (. . .).
4. Gonadotropic hormones stimulate the production of (4A) spermatozoa and (4B) androgens.
5. The androgen, testosterone, (5A) maintains the accessory sex glands, (5B) stimulates the development of masculine characteristics, (5C) stimulates libido, and (5D) affects behavioral patterns.
6. Testosterone also controls the release of GnRH from the hypothalamus by a negative feedback system (– – –).

as daylight length and hormone levels, the hypothalamus produces the gonadotropin releasing hormone (GnRH) which, in turn, stimulates the production of gonadotropic hormones from the anterior pituitary gland (a small gland that is suspended from the base of the brain). Although the precise role that these hormones play is not understood, it is believed that they somehow work together to stimulate the production of androgens (e.g., testosterone) and spermatozoa. Scientists do know that the production of gonadotropin by the pituitary is inhibited by high levels of testosterone. This phenomenon indicates that a negative feedback mechanism keeps a check on the production of all reproductive hormones. For example, when testosterone reaches a certain level, gonadotropin production decreases. When the testosterone levels drop in response to reduced gonadotropin, the hypothalamus is stimulated to produce more gonadotropin releasing hormone. As a result, increased gonadotropin causes increased testosterone production, and the cycle continues. (Note: Some sources argue that estrogen, also produced by the testes, affects the regulatory feedback system. The relationship is not fully understood at this time, however.)

SPERM PRODUCTION

As mentioned earlier, germ cells within the seminiferous tubules eventually give rise to the stallion's spermatozoa. The cellular division and maturation processes, which take place over about a 60 day period, are referred to collectively as spermatogenesis. In order to explain how these primitive germ cells (spermatogonium) divide to form sperm cells, a review of basic cellular properties is in order. (The following is an elementary review of spermatogenesis. For a detailed discussion, the interested student should refer to the text EQUINE GENETICS & SELECTION PROCEDURES.)

The cell is the smallest living unit within the body. An outer membrane encloses the cellular gel-like fluid, or cytoplasm. Located within this cytoplasm are small bodies, or organelles, each with a specific life-sustaining function. Together these organelles maintain the cell's integrity and insure that cell reproduction is carried out normally. The cytoplasm also contains a membrane-enclosed entity, or nucleus, that carries the genetic messages which dictate how and when each cell will divide or function. These messages, or genes, are the same for every cell within a particular individual, but for unknown reasons, a specific cell obeys only a few selected messages. The genes are organized on long protein strands, or chromosomes, and each chromosome is paired with another of similar size, shape, and genetic content.

The pairing of chromosomes is an extremely important concept to understand when studying spermatogenesis. When the primary germ cells divide to form the sperm (or the ovum in the mare), the number of chromosomes must be cut in half so that each parent gives its offspring 50 percent of the necessary chromosomes. The sperm cells carry one chromosome from each chromosome pair and, therefore, have only half the normal chromosome number. When the sperm and the ovum unite, has the normal number (not twice the number) of chromosomes for that particular species is restored.

As a case in point, the mature horse has 64 chromosomes (32 pairs) within every body cell. The stallion's sperm and the mare's ovum, on the other hand, have only 32 chromosomes (one from each pair). When these cells unite during fertilization, their chromosomes combine to produce the normal pattern of 64 chromosomes.

During the first few cell divisions, the chromosomes undergo complex chemical changes that allow them to replicate (duplicate their structures). As a result, the two daughter cells formed by the division of the parent cell carry exactly the same genetic information. However, when a primary spermatocyte divides to form two secondary spermatocytes, the information carried in the two daughter cells is not necessarily the same. The reason for this change is very complicated and will be examined in detail within the Appendix. At this point, however, it is important to note that the purpose of the unusual replication/division process is to insure that each secondary spermatocyte receives only one chromosome from each chromosome pair.

As these complex replication/division processes occur, each new daughter cell is located further away from the basal membrane of the seminiferous tubule. For this reason, layers of primitive sperm cells at various stages of development line the seminiferous tubules. (In other words, the primitive germ cells form the deepest layers, while the spermatids are located near and released into the tubule's cavity.) This variation along the tubule results in the so-called "spermatogenic waves" associated with continuous sperm production. Even though the formation of several spermatids from a primary germ cell takes about 60 days, the seminiferous tubules release immature sperm cells in a continuous wave-like pattern.

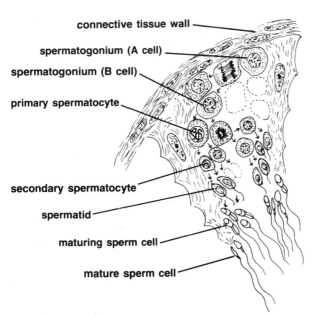

connective tissue wall

spermatogonium (A cell)

spermatogonium (B cell)

primary spermatocyte

secondary spermatocyte

spermatid

maturing sperm cell

mature sperm cell

A single microscopic "pie-like" section of the seminiferous tubule shows the gradual cell division processes that result in the production of several sperm cells from a single primitive germ cell. This process occurs continuously along the tubule, resulting in "spermatogenic waves."

After the spermatid is released, it is pushed along the coiling seminiferous tubule by rhythmic surface contractions and secreted fluids. As the immature sperm cell migrates through the coiling seminiferous tubule and into a less tortuous "straight" tubule, it is transformed into a mature sperm cell. By the time the sperm cell is released into one of several efferent (exiting) ducts, it has usually lost an extensive amount of cytoplasm and formed a tail, or flagellum. Although the cell has acquired its mature form, it is not yet able to fertilize the mare's ovum, and its tail is not yet motile. The final maturation process is completed within an accessory structure, referred to as the epididymis.

EPIDIDYMIS

The epididymis is a long U-shaped structure attached to the top surface (long axis) of its respective testis. Like the testis, its tubular system is enclosed by a tunica vaginalis and a tunica albuginea. Anatomically, the epididymis is divided into a head, body, and tail. The head consists of several intercoiling ducts which are continuous with the efferent ducts on one end and, on the opposite end, unite to form a single epididymal duct within the body of the epididymis. The epididymal duct is extremely long and tortuous, and its inside surface area is increased by the presence of many projections, or microvilli. Scientists believe that this additional surface area aids in the absorption of epididymal fluids. Although the cells of the epididymal duct secrete fluids that nourish the sperm and aid in their transport, one of the most important functions of the epididymis is to reabsorb fluids and, consequently, to concentrate the sperm and facilitate maximum storage.

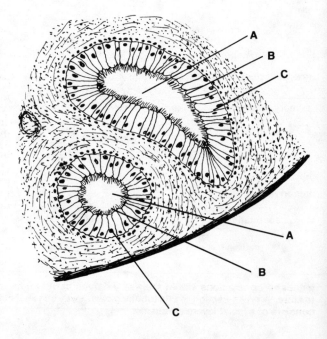

Microscopic anatomy of the epididymis showing the close association between (A) the intercoiling ducts and showing (B) the tiny projections, or microvilli, on (C) the epithelial cells.

The sperm cells move through the epididymis due to smooth muscle contractions along the walls of the epididymal duct. As they reach the tail of the epididymis, the cells have usually reached full maturity and are normally capable of fertilizing the mare's ovum. However, ionic concentrations (pH) within this area inhibit movement of the flagella, thereby conserving the sperm cells' energy stores until they are ejected into the mare's reproductive tract. The sperm cells can be stored in the tail of the epididymis for several days, but they do eventually die. Dead, dying, and abnormal cells are normally broken down and reabsorbed by the epididymis, a process that keeps the sperm reserves healthy and the stallion fertile.

VAS DEFERENS

The vas deferens or, as it is sometimes called, ductus deferens is a muscular tube that transmits mature sperm and their surrounding fluids from the epididymis to the urethra. As it leaves the tail of the epididymis, the vas deferens is about ¼ inch in diameter. The duct's muscular coat gives it a firm, thick wall and a small lumen (i.e., cavity). Like the epididymal duct, the vas deferens has many folds along its inner wall to increase surface area and to maximize storage space. Passing through the

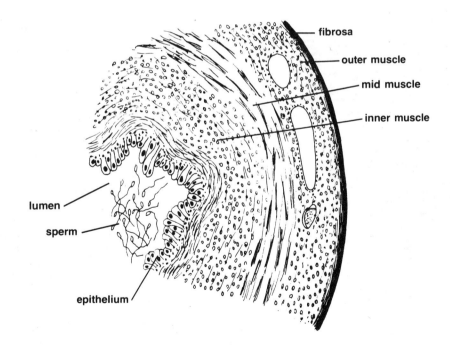

Muscular contractions within the vas deferens carry sperm from the epididymis to the urethra. A cross-section of this tubular passageway shows that the wall of the vas deferens consists of several layers of muscle.

inguinal canal as part of the spermatic cord, the vas deferens separates from the other spermatic vessels and enters the pelvic cavity. As it reaches the vicinity of the bladder, the walls of the duct enlarge to form the ampulla. This section of the duct, which is about an inch in diameter, is characterized by the presence of many glands and by the secretion of a seminal fluid. (Refer to the discussion on accessory sex glands within this chapter.) After about 6-8 inches of enlarged glandular tissue, the vas deferens decreases in size, unites with the urethra, and enters the body of the penis. The urethra receives urine from the bladder, sperm from the vas deferens, and seminal fluid from the accessory sex glands and, therefore, serves as a combination excretory and ejaculatory duct.

ACCESSORY SEX GLANDS

During ejaculation, the accessory sex glands secrete about 60-90 percent of the total seminal fluid volume. Although scientists believe that these secretions stimulate sperm motility, they emphasize that most studies have been made on test tube samples and that the effects of these secretions may be very different within the mare's reproductive tract. Nevertheless, it is likely that accessory sex gland secretions improve the fertilizing capacity of sperm by providing various energy sources and several protective buffers. Ergothioneine, for example, is a seminal component that protects the sperm from the toxic by-products of certain chemical reactions that occur within the semen. Energy sources (e.g., sugar compounds) within the semen may prolong the sperm's life but are not believed to be essential for normal sperm function.

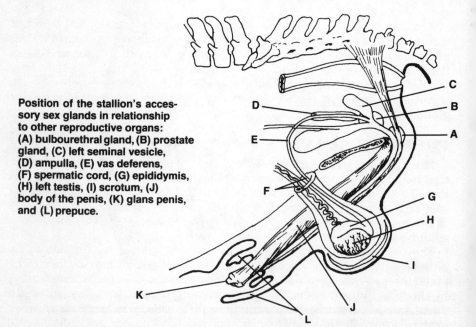

Position of the stallion's accessory sex glands in relationship to other reproductive organs: (A) bulbourethral gland, (B) prostate gland, (C) left seminal vesicle, (D) ampulla, (E) vas deferens, (F) spermatic cord, (G) epididymis, (H) left testis, (I) scrotum, (J) body of the penis, (K) glans penis, and (L) prepuce.

Size and activity of the reproductive glands are controlled by testosterone levels. For this reason, prepuberal (before puberty) castrates have under- developed sex glands, and postpuberal castrates have atrophied sex glands. It is important to note that testosterone injections improve glandular function only temporarily. Continued testosterone therapy results in a negative feedback response to the hypothalamus, causing decreased gonad- otropin production and testicular atrophy. Consequently, testosterone pro- duction by the testes is depressed, and normal stimulation of the accessory sex glands is inhibited.

SEMINAL VESICLES

The seminal vesicles produce a large portion of the seminal fluid. Al- though the name implies that the organs are storage vessels for sperm, they are actually true glands. (Some sources now refer to these bodies as the vesicular glands.) Vesicular secretions are slightly acidic and contain a high concentration of potassium and protein. Anatomically, the seminal vesicles are two elongated sacs which lie on either side of the bladder. The glands are characterized by a rounded, blind end; a narrow body; and a constriction or neck. Measuring approximately 6-8 inches in length and, at the greatest part, two inches in diameter, these vesicles are positioned with their long axes parallel to the vas deferens. Each gland is divided into lobules by sheets of connective tissue, and each lobule has one main duct with several branches. Smooth muscle contractions within the seminal vesicle walls empty the glandular secretions into the urethra during ejaculation.

AMPULLA

As mentioned earlier, the ampulla is the enlarged glandular portion of the vas deferens. Although the vas deferens produces small amounts of fluid along its entire length, the ampulla produces a much greater volume of fluid, and its secretions are more concentrated. The glands and the secre- tions of the ampulla are very similar to those of the seminal vesicles. In fact, scientists believe that the ampulla is actually a primitive form of vesicular gland.

PROSTATE GLAND

The prostate gland secretes a fluid that contains proteins, compounds that help to break up proteins, citric acid, and a very high concentration of zinc. Zinc levels are directly related to androgen production, but the role that zinc plays as a seminal fluid constituent is not known. The alkalinity of the prostatic secretions (pH 7.8) helps to neutralize the acidity of the epididymal secretions.

The prostate gland actually consists of two lateral bodies located on either side of the urethra and connected by a bridge of glandular tissue. The gland is surrounded by fibrous tissue and muscle and is divided into lobules by sheets of connective tissue. Each lobule has a main duct, and each main duct has many branches with numerous dilated sac-like areas. As in other accessory sex glands, muscle contractions deposit prostatic secretions into the vas deferens during ejaculation.

BULBOURETHRAL GLANDS

The clear fluid that is released from the stallion's penis during sexual stimulation is a product of the bulbourethral glands. This fluid flushes the urethra of urine, bacteria, and debris prior to ejaculation. In addition, the secretion lubricates the urethra for sperm passage and insures a more suitable environment for sperm longevity and motility.

The bulbourethral glands are two ovoid bodies, approximately two inches long and one inch wide, that lie on either side of the pelvic urethra. Although the bulbourethrals contain fewer muscle fibers and less distinctive lobules, they are for the most part very similar to the prostate gland. The bulbourethrals are compound glands with many side branches and saclike dilations. Several ducts from each body open into the urethra.

PENIS

The male copulatory organ is the penis, a cylindrical structure which is compressed on the sides and located between the thighs, extending forward to the umbilical region. In its quiescent state, the penis is about 20 inches long. Although most of this length is continuous with the protective skin covering, 6-8 inches lie free within the sheath. The sheath, or prepuce, is actually a double fold of skin that begins near the scrotum and makes its first fold in the umbilical region. Up to this point, the external skin of the prepuce is very similar to that of the scrotum. The first invagination is reflected back about 6-8 inches, forming an opening called the preputial orifice. The skin then turns forward and is folded back once again as it reaches a point just behind the first fold. This second fold forms another

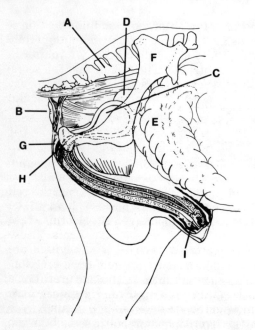

Position of the stallion's penis in relation to other anatomical features: (A) vertebra, (B) anus, (C) bladder, (D) colon, (E) cecum, (F) pelvic bone (cut through midline), (G) root of the penis, (H) pelvic urethra, and (I) urethral orifice (opening).

opening referred to as the preputial ring. The skin between the preputial orifice and the preputial ring is almost hairless and has irregular pigmentation. Large sebaceous (i.e., oil) glands within the area secrete a fatty, cheese-like matter, referred to as smegma. This dark, smelly glandular secretion tends to collect in the folds of the prepuce and, if it is not removed periodically, may become a source of extreme irritation to the stallion. The skin that is reflected backward from the preputial ring of the preputial-penis attachment contains no sebaceous glands but does tend to collect smegma within its cavity.

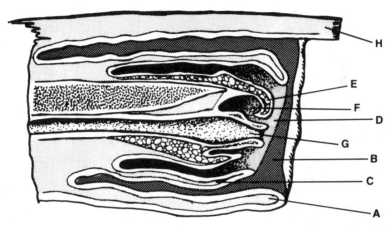

A sagittal (lengthwise) section of the stallion's penis shows that the prepuce consists of several folds of skin. (A) The first fold forms (B) the preputial orifice and (C) the second fold forms (D) the preputial ring. Other notable features include: (E) the glans penis, (F) the urethral diverticulum, (G) the urethral orifice, and (H) the abdominal wall.

The penis is attached to the lower pelvic bone by two branches which are extensions of the penile body. These continuations are referred to collectively as the root of the penis. The body of the penis begins as the two branches join together. At its origin, the body is suspended by a ligament that attaches to the pelvis. Upon closer examination, there are actually two tubular bodies within the penis. The dorsal (top) part of the the organ is the corpus cavernosum penis, a narrow tube that carries the nerves and blood vessels and contains most of the erectile tissue. This dorsal body is enclosed in a thick fibrous elastic capsule and makes up the greatest bulk of the penis. A network of trabeculae (i.e., sheets of connective tissue) forms areas of enclosed erectile tissue within the corpus cavernosum penis. These sections contain muscle tissue and scattered cavities. The cavities are specialized capillaries which fill with blood and enlarge during erection. The ventral (lower) part of the penis is the corpus cavernosum urethrae, a body that forms a tube around the urethra. The urethra originates at the neck of the bladder, receives sperm from the vas deferens and seminal fluid from the accessory sex glands, passes over the lower pelvic bone between the branches at the root of the penis, and enters the corpus cavernosum urethrae. The ventral body is similar to the corpus cavernosum penis, but has a less extensive trabecular network and larger intra-trabecular spaces.

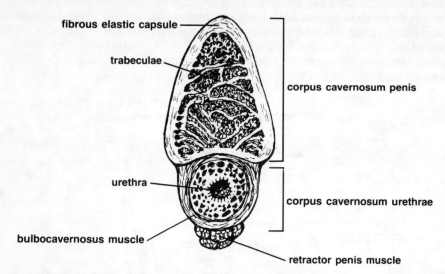

A cross-sectional view of the stallion's penis reveals that the body of the penis is characterized by several tubular passageways.

The enlarged free end of the penis is the glans. This portion of the penis is continuous with the corpus cavernosum urethrae and carries the external urethral opening. The urethra extends past the glans for about one inch as a free tube and is, at that point, referred to as the urethral process. The cavity (urethral sinus) which is situated around the urethral process sometimes fills with smegma and should be cleaned routinely. (Refer to "Breeding Methods and Procedures.") Like the corpus cavernosum urethrae, the glans penis has a trabecular network. The trabeculae of the glans, however, are extremely elastic, a characteristic which allows the glans to "flower" or "bell" during ejaculation. The skin that covers the free end of the penis has no glands but is well-supplied with nerves. The following discussion on erection and ejaculation explains the importance of these nerve endings.

ERECTION

To deposit the semen in the female reproductive tract, the penis must become rigid. The mechanism by which this so-called erection occurs involves psychic stimulation of the central nervous system which results in circulatory changes within the penis. Nerve impulses from the cerebral cortex of the brain (initiated by the presence of a mare in heat or some other sexual stimulation) induces muscles at the root of the penis to contract, pulling the body of the penis up against the pelvic bone. The resulting pressure on the penis cuts off its venous outflow. At the same time, nerves from the central nervous system and sensory nerves within the penis stimulate dilation of arteries so that the sinuses within the corpus cavernosum penis and the corpus cavernosum urethrae become engorged with blood. The fibrous elastic capsules that surround these two bodies allow the penis to become rigid without a great increase in diameter. The length of the penis, on the other hand, increases by about 50 percent.

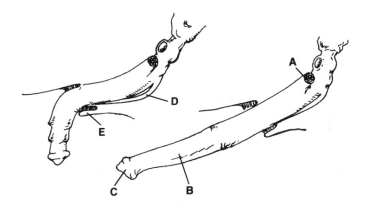

A comparison of the non-erect and erect conditions shows that erection in the stallion involves an increase in penile length rather than diameter. In response to psychic stimulation of the nervous system, contractions of (A) the ischiocavernosus muscles diminish the flow of blood from the penis, and dilation of the penile arteries causes (B) the body of the penis and (C) the glans penis to fill with blood. (D) The retractor penis muscle helps to keep the penis positioned within (E) the prepuce when sexual stimulation is not present.

EJACULATION

Ejaculation, or expulsion of semen from the penis, is stimulated by nerves within the base of the penis and by special stimuli from the brain. In response to nerve impulses, muscles within the walls of the accessory sex glands and the epididymis contract to expel sperm and fluid into the urethra. This movement of semen from the duct systems to the urethra is referred to as emission. Ejaculation is the contraction of muscles within the penis and urethra which causes the actual release of semen from the male copulatory organ during the breeding process. Ejaculation consists of about 10 pulsations that occur at approximately one second intervals. About 75 percent of the total sperm is present in the first three to four spurts of fluid.

Puberty

The development of the colt's reproductive organs is a gradual process that usually corresponds with overall body growth. This initial maturing process begins at birth and is caused by the low-level release of gonadotropin releasing hormone from the hypothalamus which causes the low-level release of special hormones from the anterior pituitary gland. The pituitary hormones cause the testes to secrete small amounts of male reproductive hormone which, in turn, stimulate gradual growth of the stallion's genitals. At puberty there is a marked increase in gonadotropin production. Subsequently, accessory sex glands develop, and masculine traits appear. The mechanism that triggers this sudden surge of reproductive development is not clear at this time, but scientists believe that changes in the central nervous system and changes in reproductive tissue sensitivity to androgens may be very important factors.

Courtesy of Stubblefield Farm

The young horse above has not yet begun to display the secondary sex characteristics of a stallion, while the mature stallion, Laddy's Image (below), shows distinctly masculine traits (e.g., heavy muscling, pronounced neck and jaw development).

Courtesy of Gene Young Quarter Horses
Photo by Gary Hamilton

Most colts reach puberty at about 18 months of age, but individual ages range from 12 to 24 months. Factors such as management, breed type, inherent growth rate, and presence or absence of disease are believed to affect the rate of reproductive development. Underfeeding, for example, can delay puberty by retarding growth of the testes and, if left uncorrected throughout puberty, may cause permanent damage to the testes. Overfeeding, on the other hand, encourages early sperm production but is not recommended due to close association between overfeeding and the development of leg and foot weaknesses in young horses.

Although puberty is characterized by the presence of mature sperm and aggressive sex drive, it should be noted that these traits do not indicate complete reproductive maturity. The stallion usually acquires full reproductive capacity at about four years of age and is generally not bred prior to his third year. Scientists who are familiar with stallion physiology and behavior emphasize the importance of gradually introducing the young stallion to his breeding career. A two or three-year-old stallion may be physically capable of covering and settling mares, but the youngster is also highly susceptible to psychological problems that may result in poor breeding behavior. In other words, behavioral maturity does not always parallel physical maturity. (Refer to "Stallion Management.")

The physical changes that occur during puberty are directly related to increased androgen production; testosterone is an especially significant hormone throughout this period. If the colt's testosterone supply is eliminated by castration, normal male characteristics do not develop. These testosterone-induced changes that occur in the normal stallion include:

1. increased muscling in the neck, head, shoulder, and back;
2. early closure of long bone growth areas (shorter stature than geldings with similar genes for size);
3. onset and maintenance of normal sex drive;
4. development and normal function of the accessory sex glands;
5. initiation of sperm production and gradual increase in sperm output.

Behavioral changes that occur during puberty are also effected by changes in hormone levels. Sex drive, or libido, is the stallion's instinctive reaction to sexual stimulation. He expresses his interest in the mare by approaching cautiously, snorting periodically, sniffing at her flanks, and nipping her on the croup and neck. When the stallion smells the genitals or urine of a mare in heat, his interest is intensified. The resulting Flehman response is characterized by upward extension of the neck and curling of the upper lip. Sex drive is an inherited trait that determines how quickly the stallion learns to mount and breed a mare. In contrast to this instinctive sex drive, the act of breeding is a trial and error learning process for the young stallion. The next chapter emphasizes how man's interaction in the breeding shed procedures affects the stallion's normal learning process. At this point, it is important to note that sex drive is an inherited response that determines how easily the stallion can be trained for breeding shed procedures and also determines how stable his behavior will be after training.

Semen Characteristics

Semen is the sperm cell/accessory fluid mixture that is expelled into the female genital tract during ejaculation. As mentioned earlier, the sperm cell carries the stallion's genetic contribution to his offspring and is, therefore, essential for propagation of the species. The seminal fluid is the sperm cell transport medium secreted by the complex reproductive duct system. Because semen quality is a valuable measure of stallion fertility, an ability to recognize abnormal characteristics is an important management tool. The steps involved in detailed semen evaluation are presented in the chapter, "Semen Evaluation," while abnormal semen characteristics are described in "Stallion Infertility and Impotency." Before these aspects are studied, however, it is important that the breeder have a basic knowlege of normal semen parameters.

THE MATURE SPERM CELL

During the maturation process, certain organelles within the spermatid undergo complex physiological changes that initiate the formation of a tail and an interconnecting neck. For example, most of the fluid, or cytoplasm, within the cell is lost as a droplet that migrates along the newly formed tail, leaving the cell's chromosome-filled nucleus to form the main body (head) of the mature sperm. The sperm head is about 6x3x1 microns (1 micron = 1/1,000,000 meter) and appears ovate (i.e., has a flattened egg shape). The

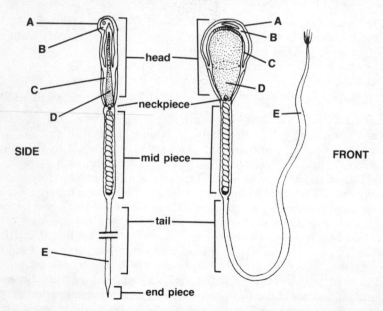

A normal, mature sperm cell is characterized by (A) the cell membrane, (B) an acrosomal cap, (C) a nuclear membrane, (D) a nucleus that forms most of the sperm head, and (E) a motile tail. Side and front views show that the sperm is flat and ovate.

tip of the head is covered by a double membrane cap containing a special protein that, upon contact with the ovum, breaks down the protective membranes of both the ovum and the sperm. This process allows the sperm to penetrate the mare's ovum and deliver the stallion's chromosomes during fertilization.

The sperm's neck piece serves as a connection plate between the head and tail. Although off-center placement of the tail is not considered abnormal, the long axis of the tail should be parallel with the long axis of the head to insure that movement is not impeded. Scientists believe that contractions of special muscle-like fibers within the neck result in the tail's whip-like movement. One whip of the tail causes the sperm head to rotate on its long axis from left to right and move forward in a slight wave-like pattern.

SEMINAL FLUID

Only about 3.5 percent of the stallion's total ejaculate consists of sperm cells. The remaining 96.5 percent is a thick mixture of fluids secreted by the accessory sex glands, the vas deferens, and the epididymis. The most important function of this fluid is to serve as a transport medium for the sperm cells. In addition, the seminal fluid provides nutrients and protective buffers that improve each cell's chances of survival within the mare's genital tract. The quality of a semen sample and, therefore, the status of the stallion's reproductive organs can be measured by qualified technicians during a study of the various seminal components (e.g., proteins, buffers, minerals, carbohydrates, etc.). Many seminal fluid constituents have a wide range of "normal" values. Total seminal volume, for example, varies between breeds and changes with the season (e.g., decreases in November and December in the northern hemisphere). Sperm per ejaculate also decreases during the winter months. The breeder should note that collection and evaluation techniques may have a very important effect on semen parameters. (Refer to "Semen Evaluation.")

SEMEN FRACTIONS

The stallion's ejaculate is characterized by four continuous, but recognizable parts: presperm fraction, sperm-rich fraction, postsperm fraction, and tail-end fraction. Because these fractions originate in different areas of the reproductive tract, they sometimes reflect accessory sex gland and epididymal activity.

PRESPERM FRACTION

The clear fluid dribbled from the stallion's penis during sexual stimulation is referred to as the presperm fraction. The secretion is usually sperm-free and averages about 10 cubic centimeters (cc). This initial fraction cleans and lubricates the urethra prior to ejaculation. By removing urine and debris and by adjusting the pH (urine is very alkaline), this presperm fraction provides a more suitable environment for the passage of sperm.

Because this fraction contains the greatest concentration of bacteria, it should not be collected for artificial insemination purposes. Studies indicate that this fraction is secreted by the bulbourethral glands, but researchers suspect that the prostate and vesicular glands are also involved.

SPERM-RICH FRACTION

When the glans penis swells to seal off the vagina, a sperm-rich fraction is discharged into the mare's cervix. This secretion, which averages about 30-75 cc total volume, contains about 80-90 percent of the total number of sperm. Most of the fluid within this fraction is secreted by the ampulla. The sperm-rich fluid also contains a high concentration of ergothioneine, a chemical agent that protects the sperm from the detrimental effects of a chemical reaction (oxidation) that occurs within the semen. In other species, sperm cells utilize glucose and fructose, found in the sperm-rich fraction, as an energy source. These carbohydrates are not as significant in stallion semen, but researchers believe that similar energy-conversion reactions occur in the stallion's epididymis. They also suspect that inositol and sorbitol, two sugars that do occur in the stallion's sperm-rich semen fraction, may supplement the sperm's indwelling energy supply. The sperm cells do not seem to depend on these sugars, however.

POSTSPERM FRACTION

The postsperm fraction is a sticky gel secreted by the seminal vesicles as the penis collapses after ejaculation or, occasionally, after the stallion dismounts. This fraction is extremely variable in volume, ranging anywhere from 8 ml to 85 ml, and may even be nonexistent or, in some samples, may be so extensive that it forms over 40 percent of the total ejaculate. The amount of postsperm fraction produced by the stallion is determined by several factors. First, some breeds tend to be characterized by greater gel production than others. Second, individuals with excellent sex drive tend to produce greater amounts of gel. Season is a third influence on gel production; studies indicate that gel occurs in about 50 percent of the samples collected in April to July, but occurs in 0-20 percent of the samples collected in September to February. Also, increasing the amount of sexual stimulation prior to breeding tends to increase the amount of gel in the ejaculate. Finally, when the stallion is bred twice in one day, the gel fraction in the second ejaculate is often halved, indicating that the seminal vesicles cannot completely replace their secretion capacity in less than one day.

The stallion's gel fraction is usually slightly more alkaline than the sperm-rich fraction, and the gel fraction alkalinity increases when he is bred more than once per day. The postsperm fraction also contains high levels of citric acid, a component that is closely related to testicular endocrine function and, therefore, is an indirect measure of androgen secretion. The gel fraction contains about 12-20 percent of the total sperm cells, a factor that should be considered when the gel is filtered out for semen evaluation or insemination purposes.

The seminal gel is not believed to have any detrimental effect on fertility when the mare is bred by natural cover, but there are several reasons for removing the gel when the stallion's semen is collected in an artificial vagina. Because the gel has a higher alkalinity than the sperm-rich fraction, many gel sperm migrate to the more suitable environment of the sperm-rich fraction. When the gel is mixed with the sperm-rich fraction, sperm concentration is not continuous throughout the sample. For this reason, sperm-rich/gel mixtures collected for semen evaluation may present a very unreliable estimate of sperm cells/ml, one of the most important measurements made during semen evaluation. (Refer to "Semen Evaluation.") Similarly, samples drawn from unfiltered ejaculates for insemination purposes may contain very few sperm cells if a large amount of gel is obtained. Also, the breeder should note that sperm survival is lower when gel/sperm-rich mixtures are frozen than when gel-free samples are frozen for storage.

TAIL-END FRACTION

The tail-end fraction is an accessory sex gland secretion that contains very little gel and few if any sperm, but is unfortunately the semen fraction used by some breeders to evaluate semen quality, since it is easiest to collect. (This practice will be discussed in the chapter "Semen Evaluation.") Although reasons are not yet clear, research studies indicate that, when semen is to be frozen and stored for future artificial insemination purposes, removal of this final fraction improves pregnancy rates. ♘

5

STALLION MANAGEMENT

The stallion manager is concerned primarily with maintaining the stallion's health and maximizing his reproductive capacity. In order to use the stallion effectively, the manager must attend to the animal's basic needs, such as proper nutrition and parasite control, and must understand the stallion's behavioral patterns and reproductive limitations. The importance of careful stallion evaluation, semen analysis, and other special management techniques cannot be overemphasized. The stallion's physical and mental status are also important to the breeding operation because they have a direct bearing on conception rates. One low-performing mare is unfortunate, but one poorly managed stallion can have a disastrous effect on the breeding program. The stallion owner, manager, and handler should keep abreast of the constantly expanding collection of facts known about stallion psychology, reproductive physiology, and animal husbandry. The following detailed study is presented to provide the reader with a compilation of the most current stallion management concepts and techniques which can be incorporated into the breeding farm's stallion management program.

Stallion Evaluation

There are several important characteristics that influence the stallion's value as a sire. Genetic potential, for example, is an extremely important aspect, since the stallion is normally able to sire many foals each reproductive year. An understanding of genetics (the transmission of traits from parent to offspring) may be very helpful to the breeder. The successful breeder selects animals according to their potential ability to improve or maintain the quality of a breeding herd. If the stallion passes desirable traits to a large percentage of his offspring, he fulfills very important economic and genetic roles on the breeding farm. To select intelligently, the breeder must evaluate his own goals, determine the inherent strengths and weaknesses of each animal, and cull animals which will not help achieve those goals. (A complete analysis of the heritability of specific traits and methods used to select for or against them can be found in EQUINE GENETICS & SELECTION PROCEDURES.)

The stallion's willingness to cover mares and his physical ability to settle them are also important factors to consider. When selecting a stallion, the breeder should weigh the stallion's merits against those traits that obviously impose serious limitations on his reproductive capacity. Ultimately, stallion selection depends on the individual breeder's goals and circumstances, but the breeder should remember that intelligent stallion evaluation—both before selection and before each breeding season—is the best insurance for a successful breeding program. A stallion reproductive evaluation entails thorough analysis of the stallion's reproductive status, physical condition, medical history, and temperament.

Photo by Don Shugart

Lady Bug's Moon, Champion Quarter Running Horse and a leading sire of race horses, has proven his ability to pass desirable traits to a large percentage of his offspring.

REPRODUCTIVE EXAMINATION

The stallion's ability to perform on the breeding farm is affected by the condition of his reproductive organs. Obviously, the testes must function normally to provide testosterone and viable sperm. The accessory sex glands must secrete the fluids that are essential for sperm transport, nutrition, and maturation. In addition, the condition of the penis affects the stallion's ability or willingness to breed. To insure that the stallion's reproductive tract is healthy in all respects, a thorough examination should be performed by a veterinarian.

The stallion's reproductive examination includes close evaluation of both the external genitals and internal reproductive organs. Because detection and diagnosis of physical reproductive abnormalities require special knowledge of reproductive anatomy and physiology, as well as a certain degree of experience, a thorough reproductive examination is best performed by an equine practitioner. The size, shape, texture, symmetry, and pliability of the reproductive organs can be readily evaluated by an experienced practitioner who, by virtue of this experience, has a basis for comparison and can easily detect abnormalities. As a case in point, it is not abnormal for the stallion's right testicle to be slightly smaller and carried slightly higher than the left, but a stallion with extremely small or malpositioned testicles is a likely candidate for infertility.

The penis and prepuce can be examined when they are washed prior to semen collection. (Semen evaluation, which is a very important part of the stallion's reproductive examination, is discussed in the following chapter.) The stallion's testes, spermatic cord, inguinal rings, inguinal canal, and accessory sex glands are usually palpated (either through the scrotum or the rectum) after semen collection. After collection, the scrotum should be more relaxed and less sensitive and the stallion is usually relaxed and easier to handle. Because palpation of an infected reproductive structure often causes an increase in the number of white blood cells within the semen, the examiner can detect inflammatory changes by measuring seminal white blood cells in samples collected before and after palpation.

The following summary of observations commonly made by the veterinarian during the stallion's reproductive examination is provided to help familiarize the breeder with the procedure.

1. The penis and prepuce should be examined for injury, squamous cell carcinoma, summer sores, sarcoids (warts), paralysis, infection, etc.

2. The scrotum should be examined for evidence of dermatitis and excessive fat, both of which can result in increased testicular temperature. (Refer to "Stallion Infertility and Impotency.") Excessive fat in the scrotum is also an indication of poor conditioning.

3. Each testis should slip up and down easily within its surrounding tunic. Adhesions of the testicle to the scrotum indicate a past injury and, possibly, defective sperm production. The testes should be characterized by uniform texture and should have a smooth surface except for small wave-like shapes caused by veins within the tunic. Although cancerous tissues and abscesses are unusual, the testes should be checked carefully for such abnormalities.

4. The thumb and index finger should be used to transpose scrotal width, height, and length upon a ruler. (If special calipers with pads

instead of points are used to measure the scrotum, extreme care should be taken to avoid injuring the stallion.) Scrotal size can be used to determine testicular size which is compared with average sizes for that specific breed, age group, etc.

5. The position of the epididymis helps the examiner identify torsion (twisting) of the testes. (Refer to "Stallion Infertility and Impotency.") As with the testes, the size and texture of the epididymis reflects overall health and ability to function normally.

6. The spermatic cords should be similar in size and uniform in diameter (about 2.5 cm in diameter) along their entire length. The spermatic cords normally disappear at their respective external inguinal rings.

7. The pulse of the spermatic artery should also be checked. Any deviations from normal pressure may indicate hemorrhage, an obstruction (clot, tumor, etc.), or an infection (the release of fluid from the blood into the infected area changes blood volume and alters blood pressure).

8. The condition of the accessory sex glands can be evaluated by studying the size, shape, and texture of these organs via rectal palpation and by performing a detailed semen exam. (Refer to "Semen Evaluation.")

9. The inguinal canal and inguinal rings should be checked for the presence of adhesions and hernias. Normally, these areas are palpated via the rectum and scrotum.

When evaluating the stallion's general physical status, close attention should be given to his physical ability to cover a mare. The stallion should be checked very carefully for any indication of lameness or any source of pain that might hinder his ability to mount or stand during breeding (e.g., hind leg weakness, arthritis, spinal injury, wobbler syndrome, or severe laminitis). A thorough soundness examination is often performed by a veterinarian prior to the purchase or syndication of a stallion, but should also be performed before each breeding season. Proper diagnosis of lameness and good judgement in determining the significance of a particular problem require extensive knowledge of equine anatomy and physiology. The limbs must be observed and carefully palpated before and after exercise. The stallion's movement is also studied from the side, front, and rear for deviations in stride. Weaknesses that might disqualify the stallion for athletic purposes do not necessarily reflect an inability to breed, especially if a phantom (dummy) mare is used to collect semen. It is important, however, to carefully consider the possibility of a trait or weakness being transmitted from parent to offspring.

Because a large percentage of equine lameness involves or originates in the hoof, routine trimming (e.g., every 5-6 weeks), frequent examination of the feet, and proper hygiene are very important aspects of stallion husbandry. (Refer to 'Efficient Use of the Stallion' within this chapter.) Rear leg lameness may cause the stallion to mount several times before ejaculating, or may severely restrict his ability to mount. In some instances, the lame stallion responds well to surgery, corrective trimming, or other treatments, and performs successfully on the breeding farm.

GENERAL HEALTH

Physical fitness, an important requirement for the breeding stallion, is the result of well-managed exercise, feeding, and health programs. The stallion's general condition is reflected in the texture and shine of his coat, by his attitude, by his weight, and by his heart and respiratory rates before and after exercise. To insure that the stallion is in top physical form for breeding activities, an evaluation of his general condition should be made prior to the breeding season.

Laboratory analysis of the stallion's blood helps the veterinarian detect problems such as low-grade infection, blood loss, cancer, nutritional deficiencies, blood-sucking parasites, etc. A low red blood cell count indicates that the horse is anemic, and additional tests identify the type of anemia. A high white blood cell count indicates the presence of an infection, while a low total protein index is often associated with starvation, anemia, blood loss, gastrointestinal disease, liver disease, or kidney disease. Although blood studies are complicated and difficult for most laymen to interpret, they can be extremely valuable to the veterinarian performing a complete physical examination. The following table presents normal equine blood values.

Characteristic	Normal Value*
red blood cells (RBC)	8.0-10.5 million per mm^3
white blood cells (WBC)	6.1-12.5 thousand per mm^3
packed cell volume (PCV)	38-51%
hemoglobin	12.5-17.0 gm%
band neutrophils	0-2%
segmented neutrophils	30-65%
lymphocytes	25-55%
eosinophils	0-4%
monocytes	0-3%
basophils	0-2%
mean corpuscular volume	35-38 g/dl
mean corpuscular hemoglobin	10-18 g/dl
mean corpuscular hemoglobin concentration	37-39%

These values vary depending on age, sex, conditioning, and breed of the individual.

MEDICAL HISTORY

Stallion evaluation is not complete without a careful analysis of past illnesses, injuries, and treatments. If accurate and complete records have been kept, a detailed medical history can be used to identify problems that may have affected the stallion's reproductive abilities in the past or problems that could cause infertility in the future.

Even a brief illness, if accompanied by fever, can destroy developing sperm cells and interrupt spermatogenesis, causing temporary infertility. Improvement in sperm quality may not be observed for several months due to the long sperm production cycle (approximately 60 days). Systemic infection (e.g., strangles, rhinopneumonitis, influenza) can cause hidden inflammation of the testes and, depending on the extent of tissue degeneration, the stallion may be left subfertile or permanently sterile. Testicular degeneration and the formation of tough fibrous areas within the testes can be caused by severe injury to the testes, another circumstance that should be recorded in the stallion's health records. If the damaged testicle must be removed, a record of the event will assure the prospective buyer that the stallion is not a unilateral cryptorchid. (Refer to "Stallion Infertility and Impotency.") Occasionally, injury to the sensitive genitals during breeding causes long-term behavioral problems. Any injury or illness should be considered carefully with respect to the stallion's future libido and fertility.

The stallion's medical history should also provide a complete description of medications administered over a period of several years. The effects of systemic drugs (i.e., drugs that affect the body as a whole) on stallion fertility has become the subject of much concern and study. For example, many researchers suspect that anabolic steroids, substances which are sometimes used to increase muscling in the young athlete, can be detrimental to the stallion's reproductive processes. When administered at relatively high dosages for extended periods, these steroids may reduce testicular size and lower sperm output. It should also be noted that any drug that has possible side effects, such as diarrhea or loss of appetite, may cause a temporary decrease in libido.

Medical records also supply information concerning the stallion's past vaccination and deworming schedules. All necessary vaccinations and deworming should be up-to-date prior to the breeding season. The stallion's role as a potential carrier of disease between mares emphasizes his need for adequate immunity. The close association between parasite infestation and poor condition explains his need for adequate parasite control. Each breeder should consult a veterinarian and establish a vaccination and deworming schedule that is suitable for the stallion's individual circumstances. (Refer to the Appendix for a calendar of suggested stallion health management procedures.)

DISPOSITION

Another important factor to consider during the evaluation process is the stallion's tractability, or ease in handling. Each handler's requirements for stallion disposition differ, depending on experience, personal preference, facilities, etc. Naturally, less selection emphasis is placed on disposition than characteristics such as pedigree, athletic ability, and conformation. Libido, however, is one aspect of the the stallion's temperament that should always be checked prior to purchase and before each breeding season. Preliminary teasing and, if possible, breeding can supply the horseman with valuable information about the stallion's sex drive and breeding manners. To allow sufficient time for correction, breeding behavior problems should be detected well in advance of the breeding season.

Another important aspect of the stallion's disposition is the presence and extent of any undesirable habits such as cribbing, kicking, self-mutilation, masturbation, etc. Such vices may not be apparent to the prospective buyer, but careful observation may reveal important clues about the stallion's undesirable habits.

Handling the Stallion

Stallion handling is a special skill that requires several important characteristics: physical strength, self-confidence, an ability to anticipate the stallion's behavior and make responsible decisions quickly, and an ability to communicate effectively with the stallion. It is very important to remember that even the most well-trained and even-tempered stallions can be dangerous. Stallions that become distracted or angry usually focus on the object of their concern and, thus, the handler's ability to keep the stallion's attention and to maintain physical control may prevent unnecessary accidents.

The tack used to control the stallion should be selected carefully and used wisely. Methods of restraint vary with the individual stallion's age, size, temperament, and with the handler's experience, facilities, and personal handling techniques. Although methods necessary to control different stallions may vary in severity, it is very important that any form of restraint not be abused. Severe and lengthy punishment or cruel handling may cause the stallion to become even more unruly. If the stallion is constantly corrected by ineffective means, he may resent his handlers, resist any form of restraint, or associate pain with expressed sex drive. It is often helpful to use different tack for breeding purposes, so that the stallion knows when he is allowed to express libido and when he is expected to mind his manners.

The stallion's halter (e.g., strong leather or webbed nylon halter) should be examined carefully prior to each use. A properly fitted halter applies pressure to the poll and both mandibles (lower jaws) only when the stallion pulls back. The halter should not be tight, nor should it be so loose that the stallion could catch a hoof in the slack. In case the stallion rears or strikes, the lead shank should be long enough to allow the handler to step back safely out of the way without turning his charge loose. The shank should always be held carefully to avoid catching an arm or hand in the slack (i.e., do not hold it in coils). It should also have a heavy-duty snap, which should be examined before each use.

Some stallions are bred with a headstall and snaffle bit. The headstall should not pull on the stallion's ears or slide back along his neck and the bit should not press tightly against the corners of the stallion's mouth. A lead strap with a chain shank and strong halter are often used to insure maximum control over the hard-to-manage stallion. Using the halter rings as guides, the chain is placed in various positions through the stallion's mouth, over his nose, or under his jaw. The severity of the chain depends on its location. For example, placement of the chain through the left halter ring, across the nose, through the right halter ring, and snapped to the end of the chain shank is probably the least severe method. The chain might also be passed through the left ring, under and then over the halter's nose band, through the right ring, and snapped to the ring below the ear. When pressure is applied to the shank the chain tightens, but wrapping the chain

Courtesy of Benedict's Funny Farm

Effective but only moderately severe restraint.

Courtesy of Benedict's Funny Farm

Pressure on a chain shank positioned on the lower jaw can encourage the stallion to rear.

once around the nose band reduces the severity of this method significantly. On the other hand, placement of the chain through the left halter ring, between the upper lip and upper gum, through the right halter ring, and to the ring below the right ear is an extremely severe method of restraint. (Pressure applied to the gum in the incisor area is very painful.) The most severe method, referred to as a gag bit, is made by running the chain from the left ring, through the mouth and the right ring, and to the ring at the base of the ear. The chain is sometimes run below the mandibles in a similar manner. It should be noted, however, that pressure applied to the lower jaw may cause the stallion to rear or strike. Each of these methods should be used very cautiously. For example, tying a horse with the chain applied to the nose, mouth, or jaw area can severely injure the animal if he pulls back suddenly.

A very important aspect of maintaining physical control is the handler's understanding of stallion psychology and his ability to apply that knowledge in making decisions concerning his own protection, the welfare of the stallion, and the safety of other horses, bystanders, or co-workers in the immediate vicinity. This understanding, which also helps to prevent the onset of behavioral problems, is the result of careful study and experience—aspects that cannot be overemphasized. The handler should be aware of the stallion's individual behavior and should be quick to recognize a potentially dangerous situation. Good judgement can prevent fights between stallions, self-inflicted injuries to the stallion, unplanned matings, and other potentially serious problems.

It is known that animals can sense insecurity and apprehension in others. The handler that has little confidence in his or her ability to control the stallion will find that the stallion senses this uncertainty and, in turn, places little confidence in his handler. The stallion's lack of trust can become a lack of respect, which may result in aggressive behavior. By attempting to hide fear of the stallion with a loud voice and rough treatment, the handler can cause irreparable damage to the stallion's attitude. It follows that a necessary characteristic for stallion handling is self-confidence. It is important, however, that carelessness not be substituted for self-confidence. The handler should balance self-assurance with an awareness of the stallion's unpredictable moods and reactions.

The handler's ability to communicate requests to the stallion is a talent that enhances the animal's overall manageability. The skilled attendant directs the stallion with a simple word or hand cue. Control over the stallion reflects an understanding between horse and handler. The well-trained stallion that trusts and respects his handler usually requires little discipline. When discipline is necessary, it should be delivered promptly and briefly in a calm, quiet manner with sufficient severity but without anger, and without inflicting extreme pain. Once the behavior has been corrected, the handler should treat the stallion with due respect and continue as if nothing had happened; unremitting, open hostility toward the stallion is counterproductive. The stallion should understand why he is being punished and should be confident that, by correcting the undesirable behavior, his handler will be satisfied.

Training the Stallion to Breed

Recent studies indicate that poor breeding behavior in the stallion is often caused by rough handling and improper training. As mentioned earlier, incompetent human intervention in the equine breeding process can significantly affect the stallion's attitude. If he is not treated with a careful combination of respect and quiet discipline, the stallion may develop undesirable breeding habits or resentment toward his handler. The stallion's courage and pride are two of his most striking attributes, but these same characteristics increase his need for intelligent and confident handling during the sensitive breeding training period.

The handler should be well acquainted with the stallion prior to training and must be familiar with the farm's established breeding procedures, have two or three competent assistants, and remember that breeding behavior patterns developed during the early years will probably last throughout the individual's career. Bad habits formed at this stage will be difficult, if not impossible, to correct later. The basis of any equine training program is to establish a procedure that is thoroughly understood by the student. In training the young or inexperienced stallion, a step-by-step breeding routine should be adopted, and the procedure should be carefully followed to avoid misunderstandings between stallion and handler. A change in schedule, a different handler, or the presence of strangers in the breeding shed may confuse the stallion and, consequently, disrupt his normal breeding pattern.

LET-DOWN PERIOD

If the stallion has been used in an athletic capacity, or if he has been subjected to a rigid show schedule or exercise regimen, a period of physical let-down should precede training for the breeding season. Emphasis on this let-down period does not imply that the stallion should not be in good condition for his new role on the breeding farm—only that there is a significant difference between breeding condition and top athletic form. Breeding condition is characterized by good health, good flesh (not excess weight), and a relaxed, alert attitude. This breeding attitude is basically a sense of security, enhanced by an established routine of exercise and feeding. The stallion should always know what is expected of him.

The let-down period allows the stallion several months to readjust to a new diet, different surroundings, a new handler, and possibly a less extensive exercise program. The stallion's new exercise program should be well established six weeks prior to breeding. Although stringent workouts are not usually necessary on the breeding farm, exercise should never be eliminated from the healthy breeding stallion's routine. Any injuries or problems that limit the stallion's movement (especially in the hind limbs or back) should be dealt with during this period. To insure that the stallion does not associate pain with the act of breeding, any required surgery or treatment (and healing) should be completed before the stallion's training period begins.

Depending on his new exercise program and the demands of his breeding schedule, the stallion's energy requirement may be significantly lower on the breeding farm and his diet may need to be adjusted to compensate for any change. When the stallion's optimum condition is reached, his weight should be noted. This weight should be maintained throughout the breeding season, since any significant change in weight can be detrimental to fertility. (Refer to 'Efficient Use of the Stallion' within this chapter.)

The let-down period also provides the handler with time to learn what to expect from the stallion and provides the stallion with time to become acquainted with his handler. It is very important that the handler and the stallion develop mutual respect, and equally important that the stallion trust his handler. The stallion's attitude may be enhanced by allowing him to become acquainted with (by distant observation) other farm-owned horses during the readjustment period. Stallions that are not isolated from other horses are often less nervous and seem to perform more consistently in the breeding shed. (Note: The sight of strange horses, mares in heat, or breeding shed activities may upset even the most well-adjusted stallion.)

TEASING

The period of sexual stimulation that precedes copulation is also an important part of the stallion's breeding routine. As discussed earlier, the stallion's reaction to a mare in heat is instinctive. He normally shows strong interest in the mare, prancing around her carefully with his head held in an alert manner. The courting period is characterized by prancing, snorting, smelling the mare's genitals, and nipping her about the croup and neck. As he smells the mare, the stallion displays the Flehman response, and the erectile tissue of the penis slowly becomes engorged with blood. Although the intensity of a stallion's sex drive is believed to be inherited, reaction time (period between initial stimulation and normal erection) depends on a variety of factors: age, experience, health, handling techniques, frequency of breeding, etc. For example, erection time may vary from 3 minutes in young or inexperienced stallions to 15 seconds in mature stallions.

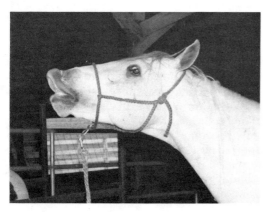

A stallion frequently shows the Flehman response (curling the upper lip) after smelling a mare in estrus.

Prior to his first cover, the stallion should be allowed to tease several mares that are known to be in heat. This practice helps the stallion to develop his natural responses and allows the handler to observe how the stallion will probably behave in the breeding shed. If the stallion associates interest in sex with punishment, he may be slow to express any interest in the mare. If the shy stallion is encouraged with repeated teasing and given some direction, however, he will usually develop the necessary enthusiasm.

THE STALLION'S FIRST COVER

Whether the stallion is to be used in a pasture, hand, or artificial breeding program, his first few covers should be by hand and under the close supervision of an experienced stallion handler. This procedure helps to insure that a valuable stallion is not injured during the sensitive and somewhat difficult learning period. The stallion should be taught to back readily on command before he ever enters the breeding area. In case of an emergency, this response may help prevent injury to both stallion and handler. Although proper discipline is very important in a handbreeding situation, excessive fear of the handler can distract the stallion to the point that he pays little attention to the mare or the job at hand.

The stallion's first attempts at covering a mare may be frustrating for both stallion and handler. The key, at this point, is patience. It should be noted that allowing young stallions to watch breeding shed activities seems to ease their trial and error learning period. There are also several contributions that the handler can make to help the stallion through this transition. The handler should see that the breeding area is free from potential hazards and that the floor provides safe footing and a soft landing. (It is not unusual for inexperienced stallions to fall off their mares.) The handler should also make sure that the stallion is slightly larger than his first mare. The mare should be gentle, definitely in heat, well-liked by the stallion, and willing to stand still when mounted. A maiden mare may be too nervous or unpredictable for a young stallion, and both animals may come away from the experience with strong foundations for poor breeding habits. Mares with a history of unpredictability in the breeding shed should not be bred by an inexperienced stallion.

An unpleasant incident during this sensitive training period may cause a sexual aversion to certain mares; the stallion seems to associate the experience with mares of the same color, size, type, etc. Occasionally, a stallion refuses to acknowledge the presence of a mare in heat. To limit problems of this nature, it is important that the young stallion be carefully introduced to each of his mares. The first few mares should be very cooperative and understanding of the youngster's limited talent. If the young stallion dislikes a certain mare, he should not be forced to breed her.

After the mare is prepared for breeding, the stallion is washed and/or rinsed and allowed an adequate period of stimulation. (Washing procedures for the mare and stallion are discussed in "BREEDING METHODS AND PROCEDURES.") The handler should guide the stallion calmly, but firmly, to the mare's left side. The stallion should be allowed to touch the mare, but should never be permitted to bite or strike. If the period of stimulation is too short, the stallion may not be ready to mount the mare. If the period is too long, the glans penis may become too large for normal penetration, or the stallion may eventually lose interest in the mare. When he appears to be fully extended and ready to mount, the stallion should be backed to a position behind the mare and allowed to approach her slowly. During his first few covers, the stallion may accidently mount the mare sideways (across her back) or may even attempt to mount her neck. The handler should try to avoid these awkward positions by keeping the stallion under control and in proper approach to the mare. It is also important that the mare's head be turned, so that she can see the stallion's rear-end approach. (After the stallion learns the basic procedure, he should be trained to mount from the mare's left side.) The mare may take one or two steps forward as the stallion mounts. Because this movement helps the stallion position himself, it should not (in most instances) be restrained. The handler's technique and attitude are very important at this point. The stallion's first attempt to cover a mare may be very unsuccessful, but the handler should remember that this lack of expertise is not at all unusual or abnormal. If the stallion loses interest, falls, or is injured, the handler should be prepared to deal with the situation calmly, in order to prevent further injury and minimize the stallion's anxiety or frustration. For this reason, an experienced stallion handler is an invaluable part of the training program.

Although it is very important to teach the stallion good breeding manners, training during the stallion's first few covers should focus on simply allowing him to figure out exactly what to do and how to do it. Biting, or any other form of roughness, should always be discouraged but detailed etiquette should be dealt with during subsequent covers. A young or inexperienced stallion may take longer to ejaculate, but repeated covers (more than 3-4) to accomplish ejaculation should not be allowed. It is important to remember that the average for trained stallions is about 1½ mounts per ejaculate and that 2-3 mounts per ejaculate is not unusual during the first cover. (Longer teasing periods may discourage repeated covers.)

ARTIFICIAL BREEDING AIDS

Stallions with good sex drive can usually be trained to mount a phantom mare (i.e., an imitation mare, sometimes referred to as a dummy mare) and ejaculate into an artificial vagina (i.e., a semen collection apparatus). Use of a phantom mare minimizes the possibility of injury to the stallion and allows stallions with aversions to certain mares to be collected for artificial insemination. The phantom should be of appropriate size for the stallion and should be padded for his comfort and safety. Because many stallions like to grasp the mare's neck during the breeding process, the phantom might also be equipped with something that the stallion can bite (i.e., hold on to).

The individual's inherent sex drive determines the ease or difficulty in training him to mount a dummy mare. Some breeders sprinkle the phantom with urine from a mare in heat to help stimulate the stallion's normal olfactory (sense of smell) responses. Most breeder's start out with a mare in heat standing on the phantom's right side. The stallion is allowed to approach the phantom's left side, smell the live mare, and jump the dummy mare. After a few collections using this procedure, the mare is moved away from the phantom. When the stallion becomes accustomed to mounting the phantom, the presence of a mare in heat is not usually necessary to stimulate sexual interest.

Use of an artificial vagina for semen collection is extremely helpful for semen evaluation and is an important tool in artificial insemination programs. Temperature and pressure within the artificial vagina are important controls for eliciting normal ejaculation. The use of both the artificial vagina and the phantom are examined more closely within "Breeding Methods and Procedures."

Stallion Vices

A vice is any habit that is detrimental to the stallion's health, usefulness, or tractability. Although inherent temperament and endocrine balance certainly affect the stallion's attitude, many vices result from improper handling. Again, it must be emphasized that stallion handling requires patience, firmness, and an extensive knowledge of stallion behavior. The insecure or inexperienced handler may attempt to hide his or her fear with harsh treatment or with constant, timid discipline. Either extreme encourages nippy, aggressive, and unruly behavior in the stallion. Boredom is also an important cause of vices in stalled stallions. Breeding stallions are often fed high-energy diets, kept isolated from other horses, and exercised irregularly. Under these conditions, the stallion becomes frustrated and bored. To limit behavioral problems associated with boredom, the stallion should be allowed to observe other horses or farm animals from a distance. A tetherball or plastic milk jug hung in the stall may also provide some diversion for a bored stallion. In addition, controlled exercise for at least 30 minutes each day is essential.

Many vices are very difficult to eradicate once a behavior pattern becomes firmly established. If possible, the causative factor (or factors) of any vice should be removed or corrected before the behavior becomes a habit. The stallion handler should recognize potential problems and use good judgement in correcting the behavior or, preferably, the circumstance. The causes of several stallion vices are presented in the following discussion. These vices can be organized into three categories: vices that are dangerous to the handler, those that are dangerous to the stallion, and those that are detrimental to fertility or libido.

AGGRESSIVE BEHAVIOR

A stallion that attacks his handler or other horses is extremely dangerous to everyone around him. The behavior should be studied closely to identify its cause so that corrective measures may be taken. For example, does the stallion react violently around one particular person? Does he resist restraint when a particular piece of tack is used? Does he attack, nonspecifically, every mare that he services? If he savages mares or becomes extremely aggressive during a certain time of the year, his behavior may be caused by seasonal endocrine changes. In most instances, however, aggressive or resistant behavior is due to mismanagement, and careful handling and retraining are the only cures. Some breeders advocate turning a rough stallion out with a large group of healthy, domineering mares to dampen his aggressiveness. Although the results of this corrective measure may be very impressive, the possibility of the stallion or the mares being injured must be carefully considered.

If the root problems cannot be identified or corrected and if the behavior ruins the stallion's usefulness, gelding may be indicated. (Note: There is a 4 percent chance that gelding will have no effect on the horse's behavior.) If, for economic reasons, the savage stallion must be kept entire, special handling precautions should be utilized. The bull pole is sometimes used to prevent the stallion from bearing in on his handler. This safety device is a strong metal pole, approximately six feet long. One end of the bull pole fastens to the stallion's halter; the opposite end is held by the handler. In addition, the savage stallion should wear a heavy halter and muzzle when he is handled. If a stallion savages his mares and if the undesirable behavior cannot be corrected, he should wear a muzzle during the breeding process and each of his mares should wear a special pad which protects her neck and shoulders. An alternative would be to use the stallion in an artificial insemination program (if the breed registry allows), so that he mounts only a phantom mare.

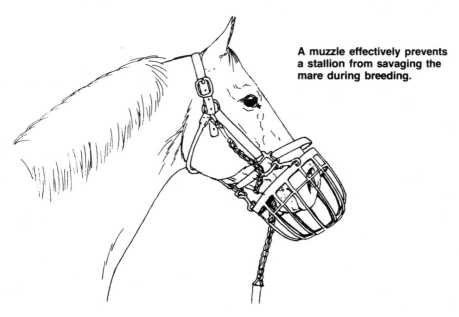

A muzzle effectively prevents a stallion from savaging the mare during breeding.

BITING

Although biting is common behavior in most horses, prompt correction usually prevents habitual biting. Due to his naturally aggressive nature, the stallion is more inclined to bite than is the gelding or mare and a biting stallion can be very dangerous to those who must handle him. Like most vices, biting is much easier to prevent than it is to cure. To avoid situations that encourage biting, the stallion should not be disturbed while he is eating, should not be hand-fed, should not be handled excessively about his head, and should never be allowed to nuzzle his handler. The most effective corrective measure is probably a sharp jerk on the lead shank. Another effective means to stop biting is to punch or strike the muzzle very hard. Slapping the face should be avoided because the stallion may make a game of dodging the flying hand or he may become head shy. It is important not to assume that a stallion's biting habit has been cured. Given the opportunity, the corrected stallion may quickly renew his old habit.

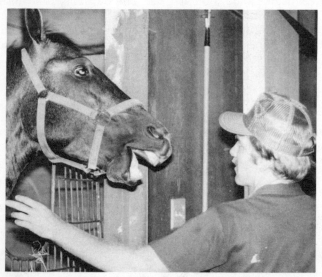

Biting and nipping should be controlled at an early age. The playful nip of a young colt may develop into a dangerous vice that is difficult to correct.

REARING AND STRIKING

Rearing and striking are also natural behavioral responses for most horses. For example, the stallion may rear slightly and strike when teasing a mare. Handling the stallion's face, eyes, ears, mouth, neck, or chest may also cause him to strike. For the handler's safety, this behavior must be corrected. The stallion that rears or strikes should be restrained with a long lead shank and a strong halter with a metal nose band or chain. (Note: Because positioning the chain below the lower jaws may encourage rearing, the chain is best placed over the stallion's nose when correcting this particular vice.) The handler should always stand to the side of the stallion—never in front of him. When the stallion attempts to rear, or after he completes a rear, his lead shank should be jerked in sharp disapproval. Jerking the lead shank when the stallion is standing on his hind legs may cause him to fall backwards if he throws his head back and loses his balance. Backing the stallion rapidly for a few steps immediately after correction discourages a second attempt at rearing.

A stallion who rears is a threat to himself and to his handler. If the lead shank is jerked while the stallion is on his hindlegs, he may throw his head backward and fall.

SELF-MUTILATION

Self-mutilation is a very unusual form of savaging sometimes found in extremely bored horses. A horse with this vice bites his own legs, shoulders, and sides. Increased exercise or companionship with another horse or farm animal may solve the problem. A neck cradle or muzzle may prevent the horse from inflicting self-injury but will not cure the root problem. If the stallion savages himself only after breeding a mare, the problem may be alleviated by washing his chest, forelegs and belly (to remove the mare's scent) and by tying him for an hour or two after breeding.

CRIBBING AND WINDSUCKING

Boredom also is believed to be the primary cause of cribbing and, once this habit is initiated, it is very difficult (if not impossible) to correct. A cribber habitually chews wood (e.g., fences, mangers, stalls). From cribbing, the horse may develop the habit of windsucking in which he sets his upper teeth into a piece of wood, arches his neck, and gulps air. Chronic cribbing or windsucking results in a tendency toward colic, reduced appetite, poor condition, and excessive wear on the upper incisors. Coating wooden surfaces with creosote may discourage cribbing. Cribbing may also be checked with a muzzle or cribbing strap. The cribbing strap is a specially constructed leather strap that fastens snugly around the horse's throat-latch to prevent the horse from tensing certain neck muscles which are used to arch the neck. In some cases, the condition has been corrected by severing neck muscles that attach to the hyoid bone (situated at the base of the tongue). About 80 percent of these cases show complete correction, and an additional 10 percent show significant improvement. A newer treatment involves severing the spinal nerves to these muscles. This surgical procedure causes less trauma to the animal, and the response is much better.

WEAVING AND STALL WALKING

A vice which has only been reported in stalled horses is weaving. The weaver stands in one place, swinging his head and shifting his weight from side to side, apparently because of extreme boredom. The compulsive weaver may ignore his feed and, subsequently, become thin and hard to condition. The chronic weaver may damage his legs and tire himself to the point of exhaustion, a condition that indirectly affects fertility. Occasional weaving may be discouraged by the presence of chains, or other obstructions, hanging from the ceiling so that the horse hits one and then another as he weaves. Chronic weaving may be reduced by allowing the stallion free access to a large paddock, although the weaver frequently reacquires his habit when confined to his stall.

In moderation, stall walking and tail chasing are more frustrating to the handler than they are harmful to the horse. In extreme cases, however, displaced bedding, ruts in the stall floor, and clouds of dust cause an unhealthy environment for the offender. Although laying tires on the stall floor or hanging chains or hot wires (carrying a very low electrical charge) from the ceiling may discourage stall walking, the only way to cure this vice is to eliminate its cause—in most cases, boredom due to stall confinement.

SLOW BREEDER

Slow breeding is characterized by the need for extended stimulation prior to breeding or by repeated mounts to ejaculate. This behavior may be the result of a hormone imbalance, improper management, or inexperience. The young or inexperienced stallion may require several mounts or even several hours to cover his first mare. Although, in this instance, slow breeding is not abnormal, it should be strictly discouraged as the stallion becomes more experienced. The best way to discourage slow breeding is to limit distractions in the breeding shed and to avoid overcorrection.

In the past, slow breeders were sometimes treated with injections of testosterone or HCG (human chorionic gonadotropin). Testosterone is necessary for the development and maintenance of male sex characteristics, including the desire to breed. HCG stimulates the testosterone-producing cells (Leydig cells) within the testes and, thereby, causes an increase in the stallion's testosterone levels. However, recent research shows that hormone therapy does not cause a noticeable improvement in libido, suggesting that hormone imbalance is not a relatively significant cause of slow breeding. In addition, indiscriminate or improper use of hormones may result in testicular degeneration and serious fertility problems. For these reasons, the use of hormones to treat a slow-breeding stallion is discouraged.

Slow breeding is usually due to management problems, such as an unsuccessful breeding training period, extremely rough handling, or overuse in the breeding shed. An unpleasant training period may cause the stallion to fear his handler, the breeding process, or one particular mare (or type of mare). He may resent restraint during breeding or may become completely frustrated with the entire breeding process. Forcing the stallion to breed usually exaggerates the problem. Frequently, this type of slow breeder is encouraged when allowed to tease and breed several gentle mares in strong heat, especially if he is allowed to breed without restraint in a corral or pasture. This procedure has been very successful in helping the shy breeder gain self-confidence and, as a result, become a more aggressive breeder.

Extremely rough or cruel handling at any point in the stallion's life may cause irreparable damage to his general attitude and to his response to handling during the breeding process. The stallion that fears or resents his handler may pay little attention to the mare—concentrating instead on his handler's every movement and expression. Because correction of this type of behavior problem is time-consuming, if not impossible, emphasis must be placed on prevention. Stallion handling is an important management aspect, one that affects every element of the stallion's productivity. Stallions are individuals, and they should be treated as such. If a stallion has idiosyncrasies, they should be dealt with carefully. Patience, understanding, and an ability to discipline effectively without anger or hostility are the key factors in preventing attitude problems that affect the stallion's performance in the breeding shed.

Overuse of the stallion is also an important management problem—not only with respect to consequently lowered sex drive but also lowered semen quality. (Refer to "Stallion Infertility and Impotency.") The stallion's inherent sex drive and his ability to produce viable spermatozoa determine how extensive his breeding schedule should be. This is an extremely important management consideration—one that can determine the success or failure of a stallion's breeding season. Stallions that are teased or bred excessively may become discouraged with the breeding process. In addition, the sperm production process may not be able to keep up with the heavy breeding schedule and, consequently, sperm reserves may be depleted. Sexual rest is the best treatment for slow breeding or lowered semen quality caused by overuse in the breeding shed. These problems can be prevented through controlled collections (e.g., 2-3 per week) in an artificial insemination program, if the registry allows artificial insemination.

MASTURBATION

Masturbation, or rubbing the erect penis against the belly, is a relatively common vice among stalled stallions. Although erection occurs, most stallions do not actually ejaculate and, therefore, fertility is not usually affected by the vice. In rare instances, excessive masturbation does lead to ejaculation and consequent breeding problems (e.g., lowered sex drive and reduced semen quality). Stallions that masturbate are difficult to identify, since the vice is usually effected at night or when the offender is alone. One definite indication of ejaculation during masturbation is the presence of dried semen on the belly and backs of the forelegs. This clue is not always present, however.

Boredom is probably the most frequent cause of masturbation. As with most vices, adequate space for free exercise, some forced exercise, and a clear view of other animals and farm activities are the best corrective measures. In some cases, an accumulation of smegma may cause masturbation. (The stallion masturbates to relieve the irritation caused by smegma.) Treatment and prevention of this problem entail routine cleansing of the stallion's prepuce.

When the cause of masturbation cannot be identified or corrected, the vice can be dissuaded by fitting the stallion with a plastic ring that fits snugly around the penis just behind the glans to prevent erection. It is important to note that stallion rings have been associated with problems such as penile lesions, soreness and swelling of the penis, the presence of blood in the semen, and erection past the ring causing an inability to retract the penis. Hence, the effectiveness and safety of stallion rings are questionable, and it is recommended that they be used only upon the advice of and under the direction of a veterinarian. The ring should be removed every day to routinely examine and cleanse the penis, and the ring must be removed before breeding. An alternative to the stallion ring is a wire brush attached to a band encircling the stallion's barrel so that the brush lies against the stallion's belly directly in front of the preputial opening. The bristles irritate the stallion's penis and, as a result, discourage masturbation.

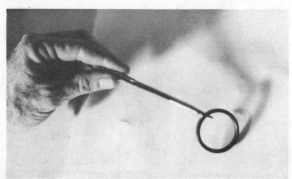

Stallion rings are sometimes used to prevent stallions from masturbating even though some authorities believe that this problem is actually caused by boredom or smegma accumulation in the prepuce.

Efficient Use of the Stallion

Efficient stallion management should also encompass a broad range of common-sense husbandry practices. These practices emphasize the importance of preventing, rather than correcting, illnesses, accidents, parasite infestations, vices, etc. Each breeder should evaluate his operation carefully and use forethought when planning suitable procedures for each aspect of the stallion's care and handling. Although it is impractical to suggest that one set of rules could govern all stallion programs, the following guidelines may be helpful.

NUTRITION

The importance of feeding a properly balanced diet to improve and maintain a horse's condition is understood by most horsemen. In fact, nutrition is believed to be the most important factor affecting all facets of the horse's well-being, including fertility. Many years of study have enabled researchers to establish equine nutrient requirements with respect to age, size, and level of activity. (Refer to the Appendix for more information on nutrient requirements.)

Because the stallion's size, condition, activity, and temperament affect his dietary needs, he should be observed closely and fed on an individual basis. The stallion's nutritional requirements are very simple as compared with those of the lactating mare or growing foal, but they are very important to his fertility and attitude. Although proper nutrition is especially important during periods of stress (e.g., breeding, teasing), the stallion's optimum feeding program should be maintained throughout the year. Generally, the stallion can be fed the same ration during the breeding season and the off-season; slight alterations in energy content might be made, depending on activity, weight loss, or weight gain.

The healthy stallion normally consumes about 2-3 percent of his body weight daily. At least half of that intake should be roughage. Plenty of quality pasture or hay supplies this roughage and, in addition, provides essential protein, vitamins, and minerals—except possibly calcium if a grass hay is fed. Adequate vitamin A (abundant in leafy green forage) is necessary for healthy sperm-producing tissue (germinal epithelium) and is, therefore, especially important during the breeding season. In most instances, a 10 percent protein ration is adequate for the mature stallion, but the requirements for immature stallions are higher due to growth. (A 12-14 percent protein ration is recommended for young stallions.) If overall feed consumption decreases, due to fatigue or nervousness, the need for a greater percentage of dietary energy and protein may be indicated. The individual's level of activity will dictate any need for a supplemental energy source (e.g., crushed corn, crimped oats). The amount of grain fed should be carefully monitored so that the stallion's optimum weight is maintained.

Many horsemen provide free-choice salt and free-choice mineral supplements for their stallions. Regardless of whether these micro-nutrients are provided in the diet or as a free-choice supplement, the breeder should study the horse's mineral needs carefully and make sure that all requirements are met. Clean, fresh water should be available at all times. It is very important that water troughs or buckets be cleaned thoroughly every day. Stock ponds, streams, and other natural water sources are not recommended for use on the breeding farm. If sources of this type must be used, it is suggested that water anaylsis by a qualified agency or university be made to determine whether the water is fit for livestock consumption. Refer to the Appendix and to the equine nutrition text, FEEDING TO WIN, for additional information on horse nutrition.

Equine nutritionists stress that stallions are frequently overfed and oversupplemented. There are no known dietary supplements that improve sperm concentration or sex drive. Contrary to popular belief, the stallion's protein requirements do not increase during the breeding season. (Increased sperm production does not require a significant increase in dietary protein.) The only dietary requirement for efficient sperm production and good breeding performance is a balanced diet that maintains the stallion at his optimum weight. Further, the presence of excess fat is closely associated with decreased sex drive and increased breeding time (i.e., time from sexual stimulation to ejaculation). It should be noted that weight loss during the breeding season can also reduce fertility and is, therefore, just as undesirable as obesity in the breeding stallion. Weight loss or gain can be corrected by adjusting the diet's energy content. To stop weight changes before sex drive is affected, the stallion's weight should be recorded monthly. If weight scales are not available, special measuring tapes may provide a fairly accurate weight estimate. These tapes (which can be purchased through most feed stores) are used to measure the girth. The girth measurement correlates with an approximate weight imprinted on the tape.

An alternative method of weight estimation employs measurements (in inches) of the girth and of body length (from point of shoulder to point of croup). The following formula is then used to convert the two measurements (in inches) to body weight (in pounds):

$$\frac{\text{girth x girth x body length}}{300} + 50 \text{ lb} = \text{body weight}$$

For example, if the girth is 75 inches and body length is 72 inches, the following calculations are made:

$$\frac{75 \times 75 \times 72}{300} + 50 \text{ lb} = \frac{405,000 + 50 \text{ lbs}}{300}$$
$$= 1350 + 50 \text{ lbs}$$
$$= 1400 \text{ lbs}$$

EXERCISE

A well-planned exercise program promotes healthy bone and good muscle tone, improves circulation, and increases lung capacity. The effects of exercise on cardiovascular condition are especially important to the breeding stallion, who must undergo short periods of extreme physical exertion. Although an uncommon occurrence, overweight and poorly conditioned stallions have been known to die of aortic rupture during the breeding process. Insufficient exercise also predisposes the stallion to health problems such as azoturia (i.e., "tying-up" syndrome caused by lactic acid build-up in the muscles.) A poorly conditioned horse is also more susceptible to a variety of infectious diseases that can affect his overall productivity. Thus, the importance of a carefully planned exercise program as a management tool on the breeding farm cannot be overemphasized.

Proper exercise complements stallion nutrition by maintaining normal digestion and a healthy appetite. Insufficient exercise contributes to poor muscle tone within the digestive tract, increasing the chances of digestive disorders. On the other hand, excessive exercise is detrimental to the stallion's digestion if it causes fatigue. (Fatigue interferes with digestion and increases the horse's need for frequent, but very small, quantities of easily digested feed.)

Courtesy of Paradise Farms

Free exercise in a large paddock is physically and mentally beneficial to the stallion and may prevent vices that are caused by boredom.

Photo by Alix Coleman

Courtesy of Iron Spring Farm

Forced exercise programs should be designed to meet the individual stallion's needs and, ideally, should occur at the same time each day.

Stallions that do not receive adequate exercise are more susceptible to vices such as cribbing, stall weaving, biting, rearing, and masturbation. Properly exercised stallions are generally less nervous, more sensible around mares, and more consistent in the breeding shed. In addition, exercise enhances fertility by developing muscle tone, improving sex drive, and preventing obesity. It is important to remember, however, that too much exercise can cause fatigue which may actually reduce sex drive. The proper balance of exercise and relaxation can influence the success or failure of any stallion's breeding career.

Health and weather permitting, the stallion should be allowed or forced to exercise regularly. The amount and type of exercise depend on the stallion's temperament and physical condition. When adequate space (e.g., two acres of safely fenced pasture) is available, for example, many stallions exercise freely and require little or no forced exercise. However, if a stallion's nervous temperament encourages him to pace the fence line until he actually runs off weight, exercise may have to be limited by allowing only a short period of freedom each day. (Weight loss can be just as detrimental to the stallion's fertility as obesity.) If pasture space is not available or if the stallion is passive and refuses to exercise on his own, forced exercise may be required to keep him in optimum breeding condition.

Forced exercise programs should not be so demanding that the stallion requires a lengthy cooling out period, but should be designed to meet the individual's needs. Exercise methods sometimes used for the stallion include riding, longeing, swimming, and walking on a treadmill or mechanical walker. Swimming is especially beneficial in strengthening the cardiovascular system and is particularly useful for exercising lame stallions. Walking, on the other hand, is not the best form of forced exercise, and special precautions should be taken to insure the stallion's safety if a mechanical walker or treadmill is used.

Regardless of the method used, forced exercise should occur at the same time each day. Ideally, daily exercise is divided into morning and afternoon sessions. In any event, a regular exercise schedule contributes to the stallion's psychological well-being by providing a certain element of security—security that evolves from knowing what to expect each day.

HEALTH CARE

Husbandry practices that minimize the chances of injury, infection, parasite infestation, and disease transmission are also essential to efficient equine production. The need for optimum conception and foaling rates to obtain a maximum profit margin reflects the importance of maintaining healthy breeding stock. The stallion's health is especially important, since any reduction in his productivity may cause a significant reduction in conception rates and, consequently, a large decrease in annual breeding farm profits. Each breeder must realize the importance of preventing health problems and plan a detailed health care program that fits the needs of that particular operation. Naturally, specific procedures will vary from farm to farm, but the following guidelines can be applied to most operations and are equally important for mares and young stock, as will be illustrated in later chapters.

HOUSING

The horse's natural environment, one of spacious prairies and lush vegetation, has been sacrificed to a great extent by domestication. When putting breeding hygiene into perspective, it is important to remember that the wild horse's natural habitat enabled him to migrate steadily, constantly in search of fresh vegetation and, therefore, always surrounded by a clean, relatively uninfected environment. It follows that the best possible situation for the stallion is a large open pasture, but this option is usually impractical or impossible.

Stalls and paddocks may be unnatural according to evolutionary standards, but when cleaned at frequent intervals and kept free of flies, excess dust, and humidity, they provide a suitable environment for the stallion—assuming that adequate exercise and proper nutrition are also provided. In most instances, a two-acre pasture with adjoining stall or shelter is a practical and suitable substitution for the stallion's natural habitat. However, the benefits of a stallion pasture can be completely overridden if proper maintenance is ignored. To minimize fly and parasite problems, manure should be picked up at least once daily. Because the stallion normally defecates in only one or two areas, keeping a two-acre paddock reasonably free of manure is not an endless task. The stall or shelter should also be kept as clean as possible, and the water and feed troughs should be scrubbed and disinfected routinely. Regardless of the stallion's allotted space, the primary health care objective is to maintain a clean, pleasant, disease-free environment. To succeed, the breeder must use both common-sense husbandry practices and a touch of empathy.

SAFETY

Special measures should also be taken to insure the stallion's safety. Handling techniques that minimize the chances of injury to the stallion, the handler, bystanders, and horses in the immediate area have been emphasized earlier in this chapter. All aspects of the stallion operation should be considered with respect to safety. For example, paddocks should be free of hazardous plants, farm equipment, debris, and pot-holes. Fences should be at least five feet high, of sound construction, and easily visible. To minimize fighting, the stallion should be separated from other horses by double fencing and solid stall partitions. Any roads through the pasture should be unpaved, and low hanging branches or electric lines should either be eliminated or fenced off. Leaving these obstacles intact or accessible can cause serious economic loss if the stallion is injured or killed.

These examples illustrate only a few of the measures that can be taken to insure the stallion's safety. Careful evaluation of the stallion's temperament when planning facilities and breeding farm procedures can prevent accidents, fights, and potential disasters. The stallion's aggressive nature is an integral part of healthy breeding performance and the breeder should make special allowances for this behavior—taking precautions that insure safety rather than suppress normal behavior. Although the main safety objective is to prevent injuries that might affect the stallion's fertility or libido, his behavior should never be allowed to jeopardize the safety of others.

Courtesy of The Quarter Horse Journal

Alleyways between pastures, paddocks, and the barn provide safe routes along which to lead the stallion.

HOOF CARE

The hoof is a complex and delicate structure that is extremely susceptible to a variety of problems, most of which are detrimental to soundness. Lameness in the stallion's back feet may result in reluctance to mount, often observed as a decrease in libido. Any resulting decrease in conception rates is costly to the breeder in terms of both money and time. For this reason, regular trimming (most stallions are not shod during the breeding season) and proper hoof care can be extremely important to the stallion's breeding career.

To prevent serious injury to the mare, the stallion should not be shod, especially on the front feet. Exceptions to this rule are, of course, cases where the stallion is ridden extensively over rough ground, where the stallion's hooves chip and split excessively when left unshod, and where

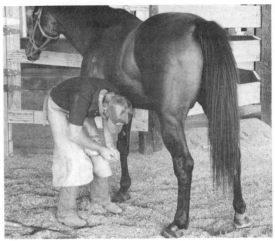

Courtesy of Paradise Farms

The stallion's hooves should be trimmed regularly in order to prevent foot and leg unsoundness which may impair his ability to mount the mare.

shoeing is necessary to correct some lameness or disorder. If the stallion must be shod, a special neckpad can be used to protect the mare's neck and shoulders from the stallion's front feet during breeding.

DENTAL CARE

The stallion's mouth should be checked regularly for lesions and excessive or uneven dental growth. Equine teeth are continual erupters, meaning that they grow steadily throughout the horse's life. If the grinding action of the lower teeth against the upper ones does not wear the teeth evenly, sharp points may form. These points can cause severe lacerations

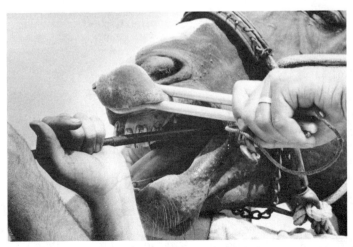

Regular examination and care can prevent dental problems from developing in the stallion.

or, if they interfere with normal chewing, may result in poor condition and weight loss. To prevent damage caused by these sharp points, the teeth should be checked routinely and floated (i.e., rasped) by a veterinarian as needed. It is not unusual for these points to form throughout the horse's lifetime.

PREVENTIVE MEDICINE

Illness threatens not only the stallion's fertility, but also the health and reproductive status of any mare that he contacts. For this reason, preventive health care should be emphasized on the breeding farm. Preventive health care involves disease control through prevention rather than treatment. Because these measures usually cost less in time and money, as compared to veterinary treatment and long-term recuperation, preventive medicine is an economically sound practice for the breeder.

The stallion acts as a potential carrier of disease between mares, and if outside mares are bred, the disease transmission problem is even more significant. It is, therefore, extremely important that the stallion's vaccinations be up-to-date prior to the breeding season. The breeder should, with the help of his veterinarian, outline a stallion vaccination schedule that is suitable for his location and his individual needs. (For a description of possible vaccination schedules, refer to the Appendix.)

Parasite control is a constant, year-round battle; nearly all horses are infested with parasites to some degree. The purpose of parasite control is to keep the degree of parasite infestation as low as possible. Heavy infestation can cause health problems, including weight loss, anemia, lethargy, diarrhea, lowered disease resistance, and reduced fertility. For this reason, the breeder should keep all breeding stock on a year-round deworming program. Some sources suggest that it is best not to deworm the stallion during the height of his breeding season. Horses may be listless for a few days following deworming and, if libido is affected, the stallion's breeding schedule may be disrupted. (For a description of specific deworming agents and schedules, refer to the Appendix.)

STATISTICS

Throughout this chapter, management practices that improve or protect the stallion's productivity have been described. The use of reproductive examination procedures to estimate the stallion's fertility potential were also examined. Although a complete reproductive examination is a very important management tool, the best indication of a stallion's fertility is the percentage of mares that he has settled at the end of each breeding season. In addition to pregnancy rate, there are other important values that can help the breeder evaluate the stallion's breeding performance. By determining each of these values and comparing the averages for each breeding season, the breeder can detect changes in the stallion's fertility and sex drive as well as weaknesses in the stallion management program. Careful observation and accurate notation of the following characteristics provide an excellent overview of the operation and indicate any need for corrective measures or management changes for the next breeding season.

1. number of mounts per ejaculate
2. average time to mount
3. average time to ejaculate
4. breeding time (from onset of sexual stimulation to completion of coitus)
5. number of covers per season
6. percentage of mares pregnant
7. percentage of mares actually foaling
8. total sperm production per week

Semen evaluation is discussed in the following chapter. For a more detailed discussion on the factors that control these characteristics, refer to "Stallion Infertility and Impotency."

The ability of a stallion manager to use these statistics and estimate the greatest number of mares that a stallion can breed efficiently during one complete breeding season is a very valuable skill. The number of mares that a stallion can breed without losing sex drive or semen quality varies greatly between individuals. When planning a breeding schedule, the breeder should remember that several factors affect the stallion's reproductive ability: testicular size, inherent sex drive, breed type, age, condition, health, acquired behavioral problems, and season. The chapter "Stallion Infertility and Impotency" examines these factors more closely. At this point, it is important to remember that the stallion's reproductive capacity is a fluctuating characteristic, a trait that should be closely monitored to prevent overuse and economic waste. Because unforeseen illness or injury can reduce the stallion's breeding potential, many breeders leave several openings in the service calendar and make last-minute additions if all goes well. If the stallion is well-managed, if his reproductive tract is normal and healthy in all respects, and if he has no significant breeding behavior problems, he should be able to meet the following demands in a live-cover situation:

Description	Estimated No. of Mares*	Comments (live cover)
3-year-old	10-15	The 3-year-old's covers should be scattered throughout the breeding season. He should not be required to give more than one service per day.
Mature	30-60	The ability of the mature stallion to breed several times a day depends on the individual. Generally, the mature stallion should not be bred more than twice a day. If the schedule involves two or more covers per day he should be given 1-2 days of

sexual rest once a week. Extremely fertile stallions with good sex drive may be able to handle much heavier breeding schedules. Mature stallions have been known to breed up to 100 mares (live cover) in a single season.

Over 20 years old	20-30	General health, overall breeding condition, and soundness determine the extent of an efficient breeding schedule for the older stallion.
Mature stallion with one testis removed (first year)	10-20	If severe injury or infection necessitates the removal of one testicle, the stallion's breeding schedule should be reduced accordingly. (Semen evaluation should dictate the extent of this reduction.)
Mature stallion with one testis (2nd year)	30-40	Eventually, the remaining testicle will compensate for the loss by increasing its sperm output. Most unilateral castrates reach over 90 percent of their original productivity.

*assuming he is not limited by seasonal subfertility.

LIGHT PROGRAMS

Although the bulk of equine reproductive research has focused on the mare, the stallion's contribution to poor conception rates and the effects of proper stallion management have received more attention in recent years. The effects of season (i.e., length of daylight) on the mare's estrous cycle are well known; simulating the photoperiod of the mare's natural breeding season by using artificial lights has improved conception rates on many breeding farms. (Refer to "Broodmare Management.") Recently, researchers performed studies to determine whether the stallion was also sensitive to such changes in daylight length. As a result of those studies, researchers concluded that, unlike the mare, the stallion's fertility is less responsive to increased photoperiods. Two semen characteristics are affected by light programs, however: both gel-free semen and gel fraction increase in volume as daylight length increases. Sources also indicate that libido is enhanced during the winter months by placing the stallion under a lighting system similar to that used on mares. Unfortunately, sperm-per-ejaculate, the factor most closely related to the stallion's fertility status, is not affected by these programs. ♞

6

SEMEN EVALUATION

Although the best overall indication of a stallion's fertility is the percentage of mares in foal at the season's end, many fertility problems can and should be detected before the extensive demands of a new season begin. By simply studying several semen characteristics, the breeder may be forewarned of abnormalities within the reproductive tract. The presence of blood in the semen, for example, indicates a potential fertility problem which will be examined in "Stallion Infertility and Impotency." A short survival period for the stallion's sperm cells suggests that he may have to be bred to each mare more than once, unless ovulation is detected by rectal palpation. (Refer to "Heat Detection.") A thorough evaluation of the stallion's semen aids in the diagnosis of infertility by revealing how well the testes, the epididymis, and the accessory sex glands are functioning. It is important to note, however, that a normal semen sample does not necessarily guarantee fertility, since the stallion's ability to settle mares is controlled by many variables.

When To Evaluate

A study of semen quality prior to the breeding season enables the breeder to predict the number of mares that a stallion will be able to cover efficiently. This capacity is directly related to two factors: 1) the stallion's ability to produce a sufficient number of normal sperm that move rapidly in a straightforward direction and 2) the stallion's ability to produce quality semen and maintain aggressive sex drive when collected daily for seven days. A very fertile stallion produces exceptionally concentrated semen throughout the testing period. His breeding schedule is usually limited by sex drive rather than semen quality. On the other hand, the subfertile stallion's sperm concentration decreases rapidly when he is collected daily for several days. This type of stallion should be bred to a minimal number of mares, and his semen should be checked at frequent intervals throughout the breeding season for changes in sperm concentration, motility, and morphology.

It is also important to monitor the stallion's fertility throughout the breeding season. Detecting problems as they occur minimizes any periods of lowered productivity and, therefore, limits economic loss due to lowered conception rates. Routine semen examinations also indicate when the stallion's breeding schedule is excessive. (Overuse is often reflected by reduced sperm per ejaculate.) To protect the stallion's breeding potential and the farm's breeding schedule, routine semen examinations should be made at least every six weeks. If the stallion has a heavy schedule, the intervals should be every 7-14 days.

Regardless of how extensive or impressive a stallion's breeding record is, the potential buyer should examine (or have a veterinarian examine) the stallion's semen prior to purchase. Any irregularities should be carefully considered with respect to the stallion's medical history. For example, a high fever or an injury to the testicles may result in temporary failure to produce normal sperm cells. The possibility of complete sterility or even an inability to produce a sufficient quantity of viable sperm cells are very important considerations. (Refer to "Stallion Infertility and Impotency.")

Many breeders test breed a young stallion to one or several fertile mares. This test of the young stallion's breeding potential is important, but a thorough evaluation of his fertility should also include a complete reproductive examination and a detailed semen study. It is important to note that, although sperm quality before and after several daily breedings is a good indication of reproductive maturity, the two-year-old stallion is, in most instances, too young for this type of critical semen evaluation.

Collection Guidelines

Many cases of stallion infertility can be corrected, but the breeder should remember that, due to the lengthy sperm development period, an improvement in semen quality may not be evident for 60 days. (Treatment may be in progress, or completed, but the stallion may not become fertile for several weeks.) For this reason, semen samples should be checked several months prior to the onset of the breeding season. Collecting the stallion two or three times before collecting a sample for evaluation helps to remove "stagnant" sperm from the reproductive tract. These preliminary breedings also help the breeder assess the stallion's libido and allow behavior problems (e.g., failure to ejaculate) to be dealt with early. After the preliminary breedings and a 4-7 day rest, the stallion should be collected twice; the second sample is collected at least one hour after the first. The most accurate evaluation of the stallion's ability to produce viable sperm is made by examining samples collected daily for seven days, although such frequent collection is impractical on many farms.

There are several methods that can be used to collect semen from the stallion. The use of a tampon, inserted in the mare's vagina before breeding, is the least desirable method. The tampon is removed after service and squeezed to recover the semen but, because excessive handling damages the sperm cells and because a representative sample is not usually collected, this method is not recommended. Contamination of the sample with microorganisms from the mare's vagina is another objectionable factor. When using the tampon as a means of collecting semen, the presence of live sperm is the only characteristic that can be determined accurately.

A stallion condom is sometimes used to collect semen for evaluation. The condom is fitted over the stallion's erect and thoroughly cleansed penis. To allow room for the semen, a small air space should be left at the end of the condom. If the condom is loose, the stallion can easily dislodge it during service. A rubber band may be used to tighten the condom if pressure is not excessive. (A tight condom may be very painful and, as in other cases of painful breeding, may even prevent ejaculation.) To insure that the sample is not lost, the condom should be removed immediately after dismount. The breeder's bag, another method used to collect semen, is similar to the condom except that it is made of heavier rubber and can be disinfected and reused several times.

The equine artificial vagina (AV) provides the best possible environment for semen collection. Temperature control and sanitary conditions minimize trauma to the sperm cells; the shape, warmth, lubrication, and water pressure of the artificial vagina stimulate normal ejaculation. One drawback, however, is that the stallion must be trained to use the artificial vagina. Although some stallions refuse the apparatus, most stallions accept it, and some even seem to prefer it over mares. (For a more detailed description of the artificial vagina and its use, refer to "Breeding Methods and Procedures.")

Analysis of the dismount sample (i.e., tail-end fraction) is not acceptable for fertility assessment. The sample usually contains very few sperm cells and does not reflect the quality of the accessory sex gland secretions. Although there is no relationship between sperm cell concentration in the total ejaculate and in the final semen fraction, some sources believe that the dismount sample can provide a very rough estimate of semen quality. Therefore, if routine semen collections cannot be made, a routine check of the stallion's dismount sample might be considered. The breeder should remember, however, that this method is not suitable for the identification of specific fertility problems.

Evaluation Equipment

Although many large-scale breeding operations are equipped to perform complete semen examinations, many breeders prefer to send the stallion to a special laboratory or qualified university for detailed semen analysis. A complete laboratory evaluation performed before and after the breeding season should be supplemented by periodic (about once every 4-6 weeks) microscopic examinations on the farm. These examinations are usually made to simply check sperm concentration, motility, and morphology. An equipment and supply chart is provided to give the reader a better idea of what is required for various laboratory procedures.

SEMEN EVALUATION EQUIPMENT AND SUPPLIES*

I. Supplies for Care of Equipment

dust-free cabinet
constant electric source
Nolvasan solution (3 oz./1 gallon warm water)
rubber-coated test tube rack
several soft test tube brushes of different sizes
glassware brush
tap water
deionized or distilled water
wash water and mild soap (e.g., Ivory)
autoclave
supplies for sterilization process (bags, tape, etc.)
oven
aluminum foil
70% ethyl alcohol
lens paper (e.g., Kimwipes)
facial tissue

II. Collection

PREPARATION:
liquid Ivory soap, diluted 1:1 with water
sterilized cotton on roll
2 buckets of clean water (104-108°F)
sterile plastic bags to line wash and rinse buckets
disposable cup
paper towels
squeeze bottle for soap
disposable obstetrical sleeves

TEMPERATURE REGULATION:
microscope slide warmer
incubator (100°F)
incubator microscope stage
light bulb, wire frame and rheostat (to warm protective AV cone if Colorado AV is used)
freezer
refrigerator

---

*When organizing and acquiring laboratory equipment, use area veterinary, laboratory, and chemical supply dealers to locate necessary items. Local extension agents, agricultural university representatives, or large animal practitioners that advise clients on artificial insemination are also helpful contacts.

EQUIPMENT:
    sterile lubricant that has not been chemically sterilized (e.g., KY Jelly)
    dial thermometer
    warm water (120-140°F depending on type of AV used and environ-
        mental temperature)
    sterile obstetrical sleeve
    scale to weigh AV to insure proper amount of water is added

ARTIFICIAL VAGINA:
    Colorado AV (leather cover with handle, heavy plastic casing with
        valve assembly, rubber liners, collection bottle, and radiator clamp)
        or
    Nasco AV (leather carrying case, combination rubber inner liner and
        cone, collection bottle, and radiator clamp)
        or
    Other AV model and necessary equipment (see instruction booklet)

III. Examination

MOTILITY:

Method I
    incubator
    pipettes (various sizes)
    microscope (preferably phase-contrast)
    slides
    coverslips
    slide warmer
    incubator stage
    skim milk diluent: *
        graduated cylinder (100 ml)
        deionized (or distilled, autoclaved) water
        nonfortified skim milk
        glucose (powdered)
        penicillin
        streptomycin
        5-10 ml vials with caps
        balance (scales)
        freezer

Method II
    incubator
    pipettes (various sizes)
    microscope (preferably phase-contrast)
    slides
    coverslips
    slide warmer
    incubator stage
    vials (5-10 ml)

---

* *Diluent formulations are given later in this chapter.*

    cream-gelatin diluent:*
      Knox gelatin
      distilled water
      double boiler
      heat source (bunsen burner, stove, etc.)
      half & half cream
      spoon
      100 cc flask or beaker
      potassium penicillin
      streptomycin
      polymyxin
      dark-colored bottles (250 cc/rubber stoppers)
      balance (scales)
      freezer
  Method III
    incubator
    pipettes (various sizes)
    microscope (preferably phase-contrast)
    slides
    coverslips
    slide warmer
    incubator stage
    Ringer's saline
  Method IV
    incubator
    pipettes (various sizes)
    microscope (preferably phase-contrast)
    slides
    coverslips
    slide warmer
    incubator stage

MORPHOLOGY:

  Method I
    microscope (preferably phase-contrast)
    nigrosin dye
    eosin Y dye
    deionized water
    sampling pipettes
    slides
    slide warmer
    dropper bottle
    bunsen burner (optional)
    balance (scales)
  Method II
    microscope (preferably phase-contrast)
    slides

*Diluent formulations are given later in this chapter.

slide warmer
bunsen burner (optional)
high grade India Ink (Pelican Yellow Label)

Method III
see 'Live-Dead Percentages'

Method IV
see 'Motility Method IV'

Method V
microscope (preferably phase-contrast)
pipettes (various sizes)
slides
slide warmer
ethyl alcohol
ether (must be used with extreme caution)
5% eosin B solution
1% phenol solution
distilled water

LIVE-DEAD PERCENTAGES:

microscope
pipettes (various sizes)
slides
slide warmer
dropper bottle
sodium citrate dehydrate
distilled water
eosin B
nigrosin
balance (scales)

CONCENTRATION:

Method I
spectrophotometer (e.g., Bausch & Lomb Spectronic 20)
spectrophotometer conversion chart (%T-sperm/ml)
colorimeter tubes
soft test tube brush
rubber-coated test tube rack
lens paper
1000 ml flask
sodium chloride
deionized or distilled water
formalin (37% formaldehyde solution)
balance (scales)
measuring and sampling pipettes
storage bottle (1000 ml) with secure stopper

Method II
hemacytometer (Neubauer or other model)
hemacytometer cover slip

    disposable diluting pipette (Unopette)
    hot tap water
    microscope (low and high power required)
  Method III
    red blood cell pipette
    3% chlorazene solution
    Neubauer blood cell counting chamber
    coverslip
    microscope (low and high power required)

ACIDITY AND ALKALINITY:

  Method I
    litmus paper
  Method II
    pH meter
    electric outlet (unless battery-operated meter is used)
    deionized water
    beaker (glass jar)
    electrode tissues (Kimwipes, Kleenex, etc.)
    pH standards
    semen containers for insemination sample (if AI is planned)

WHITE BLOOD CELLS:

See 'Concentration Method II' and 'Concentration Method III'

RED BLOOD CELLS:

See 'Concentration Method II' and 'Concentration Method III'

LONGEVITY:

  microscope
  clock
  sterile container for semen (e.g., collection bottle)
  pipette (sampling).

The microsope is an essential breeding farm laboratory tool. Because of its cost and value to the breeding program, it should be used and cared for conscientiously. Each microscope is usually accompanied by an operator's manual; thus, operational instructions are not discussed in this text. Instead, the differences between two common types of microscopes and what the breeder might look for in a microscope are examined briefly.

With most modern microscopes, the viewer observes the specimen through two or more lenses or lens systems. The specimen is seen through an eyepiece (usually 10 times magnification) and an objective (10 times, 20 times, 45 times magnification, etc.). The total magnification equals that of the eyepiece multiplied by that of the objective (e.g., 10 x 20 = 200 times total magnification). The microscope usually has two or more objectives mounted on a rotating plate that is often designed so that additional objectives can be attached at a later date. A glass slide, upon which the specimen is placed, rests on the microscope stage and is held in place by special clips or by a mechanical clamp, which allows the slide to be moved slightly during observation simply by turning an adjustment knob.

One of the simplest and least expensive microscopes is the light microscope which uses brightfield illumination. With this type of illumination the specimen is seen against a brightly lit background. Brightfield illumination only has limited application in semen evaluation, however. Normally, areas of different density in the microscopic specimen give the image detail and contrast. Because there is little density variation in unstained, transparent cells, such as sperm, these cells are very difficult to observe under brightfield illumination. For this reason, morphology and motility details cannot be studied effectively with this type of microscope.

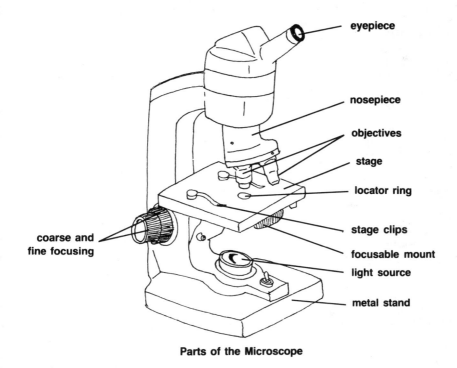

Parts of the Microscope

In order to avoid the problems of low contrast, a phase contrast microscope is often used to visualize detail in unstained sperm cells. This microscope operates on the principle that light is refracted (i.e., changes direction) when it passes through irregular areas of a transparent specimen. The remainder of the light passes through the specimen undiffracted. The phase contrast microscope separates the diffracted and undiffracted portions of the light. A special phase plate then reduces the intensity of the undiffracted light and causes a shift in its wave pattern. When the two portions of light are then recombined, the changes in the undiffracted light cause slight and otherwise invisible differences within the specimen to appear sharp and contrasting. This is the image which is observed through the eyepiece. Although the phase contrast microscope usually costs more than a simpler microscope, it is a wise purchase if the breeder intends to perform detailed semen analysis in the farm laboratory. It is particularly useful if more than one stallion is to be evaluated on a routine basis.

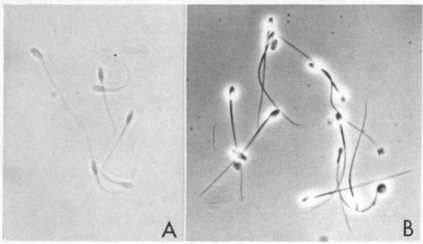

Courtesy of the Animal Reproduction Laboratory, Colorado State University

Photomicrographs of stallion spermatozoa as seen through (A) a brightfield microscope or (B) through a phase-contrast microscope. Magnification X325.

Other important laboratory instruments, such as the spectrophotometer, the hemacytometer, and the pH meter, are examined under the discussion on 'Examination Procedures' within this chapter.

Laboratory Design and Safety

A discussion on semen analysis would not be complete without mentioning the importance of planning a laboratory facility for efficiency and safety. When planning a laboratory, the breeder must consider lighting, temperature control, power source, water needs, furnishings, and equipment. The location of the lab and the layout of laboratory equipment should maximize efficiency. Locating the lab near the breeding shed, for example, minimizes the time between semen collection and incubation (or storage). The possibility of adding equipment or expanding the lab in the future should also be considered.

Safety considerations in any laboratory situation include careful handling and storage of glassware (e.g., pipettes, beakers, cylinders), special precautions and sanitary measures that minimize the transmission of infectious organisms, and the prevention of accidents and spills by carefully securing all tables and work benches to the floor. Electrical outlets should be sufficient in number and should not be located near a sink or basin. If an accident does occur, a well-marked cut-off switch for the main electrical supply and a fire extinguisher may be extremely valuable.

Handling the Semen Sample

Sperm cells are very sensitive to most changes in their environment. The cells are shocked by sunlight, water, soaps, antiseptic solutions, germicides (except antibiotics), and any sudden increase or decrease in temperature. Because of this extreme sensitivity to altered environmental conditions, special care should be taken when handling the semen sample to minimize changes in quality prior to examination. The following is a list of special precautions to be taken from time of collection to time of evaluation:

1. Prompt semen analysis insures a more representative evaluation since changes in sperm motility occur almost immediately after collection.
2. Semen collection bottles, and any other equipment that may contact the sample, should be prewarmed to 110-120 degrees Fahrenheit, allowing a 10-20 degree temperature decrease before the semen contacts the equipment. (The optimum temperature for sperm cells is 100°F; temperatures over 122°F kill them.) Semen containers and any equipment that touches the semen should not be set on cold surfaces. If the sample is not analyzed immediately, it should be placed in an incubator to maintain the proper temperature.
3. The semen sample should be protected from sunlight at all times. A prewarmed sleeve or cone, designed to cover the collection bottle, can be used to block sunlight and maintain temperature.
4. A filter (dairy milk line filter sewn together at one end or sterile gauze pads) should be used within the collection apparatus, to separate the gel fraction from the sperm-rich fraction. (The gel is removed to insure that sperm concentration is uniform throughout the sample.)
5. The collection apparatus should be sterile. Care should be taken to remove any soap, disinfectant, water, talc, etc. (This also applies to glassware, tubing, or storage containers used during the examination procedures.)
6. The stallion's penis should be washed, rinsed thoroughly, and dried carefully, because either residual soap or water alters sperm quality. The mare's hindquarters should also be washed, rinsed, and dried carefully (even if an artificial vagina is used to collect the semen).
7. If the stallion urinates in the sample, another sample should be

collected. Exercising the stallion prior to collection may eliminate this problem.

8. If the stallion's semen sample shows extremely poor motility or longevity, another sample should be evaluated and the laboratory technique carefully checked before ruling out the possibility of human error.

Examination Procedures

Prior to microscopic analysis, the total **gel-free** ejaculate is measured by pouring the filtered semen down the side of a pre-warmed graduated cylinder—very slowly. Extreme care is necessary to avoid damaging the sperm. Although gel-free volume is far less important than the total number of sperm cells per ejaculate, the volume is necessary to determine total sperm. In addition, a significant decrease in volume during the breeding season may indicate overuse of the stallion or malfunction of the accessory sex glands. It should be remembered, however, that total volume varies between breeds, between individuals, and with the season. A normal ejaculate for a light-breed stallion during the normal breeding season

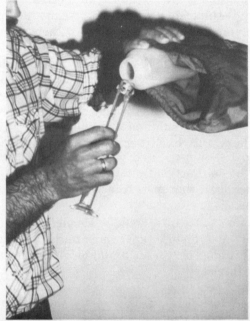

In order to determine sperm concentration in the sample, the gel-free volume must be measured. This is accomplished by carefully pouring filtered semen into a graduated cylinder, which is marked for measurement in milliliters.

varies from 50 to 150 milliliters (ml), depending on his inherent potential and on how often he is bred. If two samples are collected (approximately one hour apart), they should be nearly equal in volume. If the second sample is much smaller than the first, the epididymis and/or the accessory sex glands (which secrete most of the seminal fluid) may not be functioning properly. Reduced volume may also be the result of incomplete ejaculation, a problem that is also characterized by an alkaline shift (increase in pH).

After the volume is noted, semen color and density are studied. The sample should have an almost fluorescent "skim milk" appearance. It normally has a slightly blue or grey tint and has the same texture as light mineral oil. Unlike other domestic species, visual appraisal of the stallion's semen is not an accurate means of determining sperm concentration. Although a cloudy sample sometimes reflects a very high sperm concentration, it may also indicate the presence of debris, white blood cells, red blood cells, mucus, and/or pus. Clumps, clots, and large flakes within the sample are often caused by pus, usually from the accessory sex glands. If the stallion produces an extremely cloudy sample, or one with blood streamed throughout, a veterinarian should be consulted immediately. (Refer to "Stallion Infertility and Impotency.")

After examination is completed, the semen-filled cylinder is placed in an incubator or water bath to maintain normal physiological temperature (about 100°F) and thereby minimize changes in semen quality. The incubator is designed to maintain a specified temperature, but the water bath (a beaker of water over a hot plate or low flame) should be checked

An incubator regulates the temperature of semen samples as well as semen handling equipment such as pipettes and graduated cylinders.

Courtesy of Lab-Line Instruments, Inc.

Courtesy of National Appliance Company

A water bath can be used in place of an incubator to protect fresh semen samples from temperature variation.

repeatedly to insure that the desirable temperature is maintained. Temperature in excess of 122°F kills the sperm cells very quickly. If at all possible, the sample should be kept within a narrow range of its normal temperature to insure that it presents an accurate picture of what the stallion is capable of producing. (Semen can be frozen or refrigerated for future insemination, but is usually evaluated before storage.

To minimize the time spent on each laboratory procedure, all necessary supplies and equipment should be clean and ready to use **prior to collection**. Preliminary steps that can be taken to improve efficiency are included in the following discussion on various laboratory procedures. The precision of the laboratory technician can have a great effect on examination results. If extremely abnormal values are obtained, the procedure should be repeated carefully to identify any errors that might have affected semen quality (e.g., setting the collection bottle on a cold surface, using contaminated equipment, exposing the sample to sunlight).

MOTILITY

Motility is an extremely important measure of stallion fertility. Sperm should move forward rapidly, preferably in a straight line. Those that do not show progressive movement are unlikely candidates for the fertilization process and should not be counted as "motile" sperm. Although large, circular movements are not unusual, cells that move in tight circles are undesirable. Obviously, a high percentage of completely immotile cells suggests very poor breeding potential. The following guidelines for judging sperm motility may be helpful:

Motility Evaluation*

0 -immotile;

1 -stationary or weak rotary movements;

2 -backward and forward movement or rotary movement, but fewer than 50 percent of the cells are progressively motile and there are no waves or currents;

3 -progressive, rapid movement of sperm with slow currents, indicating that about 50-80 percent of the sperm are progressively motile;

4 -vigorous, progressive movement with rapid waves, indicating that about 90 percent of the sperm are progressively motile;

5 -very vigorous forward motion with strong, rapid currents, indicating that about 100 percent of the sperm are progressively motile.

according to Herman, Swanson, & Haq

A motility score of 0 or 1 reflects infertility; even a score of 2 suggests that the stallion is subfertile. If all other factors (e.g., sperm concentration, morphology, pH) are acceptable, a motility score of 3, 4, or 5 reflects good fertility, with a 5 being especially favorable.

Because progressive movement is altered significantly by time lapse, motility studies should be performed as soon after collection as possible. As with other studies, environmental control should be closely monitored. To accurately judge sperm movement, it is also important that the sample is not so concentrated that individual cells cannot be clearly seen. For this reason, dilution techniques mentioned later in this discussion should be carefully noted. These techniques also help to eliminate electrostatic attractions between sperm cells, allowing the observer to study the movements of individual cells one at a time.

To insure the greatest possible efficiency and accuracy during the evaluation procedures the following preliminary steps should be taken:

1. At least one hour prior to collection,
   a. remove vials of skim milk diluent from freezer (Method I); **or**
   b. remove vials of cream gelatin diluent from freezer (Method II);
2. At least 30 minutes prior to collection,
   a. preheat thawed extender to 38°C or 100°F (Methods I and II); **or**
   b. preheat physiological saline solution (Method III); **and**
   c. preheat all necessary equipment (stage warmer, slide warmer, slides, cover slips, pipettes, etc.).

3. About 10 minutes prior to collection,
   a. check incubator (or water bath) temperature; it should remain at about 100°F;
   b. if the Colorado AV is used, heat AV protector cone and fill AV with 134-140°F water; after temperature stabilization, check internal temperature using a dial thermometer (temperature stabilization may take longer than 10 min.);
   c. if the Nasco AV model (modified Missouri model) is used, fill combination liner cone with about 120°F water and allow to stabilize between 105 and 112°F (some breeders fill part of the Nasco AV with water and adjust the pressure by adding air);
   d. if the Japanese model is used, fill with 118-120°F water; the AV will stabilize and hold its temperature for at least 30 minutes.

Either of the following formulations can be used for motility evaluation.

Dried Skim Milk Diluent*

1. In a graduated cylinder, add 50 ml deionized (or distilled autoclaved) water, 5 grams nonfortified, dried skim milk, and 5 grams glucose.
2. Add 5 ml deionized water and 1,000,000 units penicillin to a vial.
3. Add 5 ml deionized water and 1.0 gram streptomycin to a vial.
4. Allow the antibiotics to dissolve and gently mix the streptomycin and penicillin solutions together.
5. Add 0.5 ml of the antibiotic mixture to the graduated cylinder.
6. Fill the cylinder to the 100 ml mark with distilled water.
7. The extender should be divided into 4.75 ml portions, stored in capped vials, and frozen. (Unfrozen or unrefrigerated diluent should be used as soon as possible, e.g., within 12 hours.)

(Note: This solution should not be used for insemination.)
*B. W. Pickett & D. G. Back, 1973

Cream Gelatin Diluent**

1. Dissolve 1.3 grams Knox gelatin in 10 cc distilled water; heat gently if necessary.
2. In the top of a double boiler, heat half & half cream to 95°C (203°F) for 5 minutes. Then remove any scum from the top.
3. Mix 10 cc gelatin solution with 90 cc half & half cream and cool the mixture's container in cold water.
4. Add the following:
   100,000 units potassium penicillin,
   0.1 gram streptomycin,
   20,000 units polymyxin.
5. Store unused mixture by freezing in dark-colored bottles (e.g., small, dropper bottles with a plain cap).

(Note: This solution should not be used for insemination.)
**B. W. Pickett, 1973

Place the semen samples on a slide using one of the following methods, and immediately return the semen-filled cylinder to the incubator or water bath.

MOTILITY METHOD I

Using a 0.5 ml prewarmed pipette, place 0.25 ml raw semen into 4.75 ml skim milk extender (100°F) and mix gently. Then pipette a drop of this extended semen onto each end of a prewarmed slide and cover carefully with prewarmed coverslips.

MOTILITY METHOD II

Dilute raw semen by placing about 5 drops prewarmed cream gelatin diluent in a prewarmed vial or on a prewarmed slide. Using a prewarmed pipette, add 1 drop raw semen to the extender and mix very gently. Place a drop of this mixture on each end of a prewarmed slide and cover with prewarmed coverslips.

MOTILITY METHOD III

Place a drop of raw semen on each end of a prewarmed slide. Add a drop of prewarmed physiological saline (0.9 percent sodium chloride) solution to each end of the slide. Mix the semen/saline drops with a prewarmed glass rod and cover each end carefully with prewarmed coverslips.

MOTILITY METHOD IV

Place a drop of raw semen on each end of a prewarmed slide and cover carefully with prewarmed coverslips. (Note: This method is not recommended for concentrated samples due to the difficulty usually encountered when trying to view individual cells within a concentrated area.)

Place the specimen slide in an incubator stage which maintains a desired temperature while keeping the slide positioned under the microscope lens. The cells should be viewed under 200x magnification, preferably with a phase-contrast microscope. Motility is determined by estimating the percentage of **physically normal** sperm that are moving in a fairly straight pattern. For example, each area viewed can be estimated as "one out of two moving," "four out of five moving," etc. If a sample contains less than 40 percent actively motile sperm in both drops, it is considered potentially subfertile. If two samples are collected one hour apart, the second sample should show equal or better motility than the first, since the first sample usually contains a larger percentage of "stagnant" sperm. If undesirable results are obtained, repeat the procedures to insure that human error has not affected motility and observe the following precautions.

1. Make sure that air bubbles are not trapped beneath the coverslips.
2. Observe several fields (areas) within each drop and average the percentage of motile sperm in each; if the difference in motility between the two drops is greater than 10 percent, the procedure should be repeated.

3. Adjust the lens up and down while counting in one field. This allows the viewer to see several levels, or planes, of spermatozoa within the same field.
4. Do not press the lens against the coverslip when adjusting the plane of focus.
5. Be careful not to measure motility within the edges of either drop, since drying may have occurred in these areas.

MORPHOLOGY

'The Stallion Reproductive System" included a description of the sperm production process and normal sperm cell structure. Without this knowledge, the viewer is poorly equipped to differentiate desirable and undesirable cell structure. Drawings of various sperm abnormalities are included in "Stallion Infertility and Impotency" to give the reader a better idea of what to look for during the examination process. The sample should contain at least 65 percent normal sperm. Although a high percentage of abnormal sperm indicates a malfunction in the sperm production process and suggests that the stallion is at least temporarily subfertile, the viewer should not rule out the possibility of secondary abnormalities caused by human error.

Several other points to note prior to the morphology exam are:

1. Detailed structure is best observed under a phase-contrast microscope. If brightfield illumination is used, a staining technique should also be employed to increase contrast.
2. The sperm concentration should be regulated so that individual cells (at least 100) can be viewed easily. Highly concentrated samples can be diluted with physiological saline to a more desirable concentration (about 200,000/mm^3).
3. If a staining technique is used, the gel fraction should be filtered from the sample during or immediately after collection. (Gel interferes with the staining process.)
4. If a stain is to be prepared in the laboratory, sufficient time should be allowed for preparation. For example, preparation of Wright's stain may take 2-4 weeks.

MORPHOLOGY METHOD I

Prepare the Hancock stain by mixing 6.0 grams Nigrosin in 60 ml deionized water. After the dye has dissolved, add 1.0 gram Eosin Y to the Nigrosin solution and allow it to dissolve. The resulting stain can be used at room temperature, but should be stored at 5°C (41°F). Using a pipette, place one drop of semen on a prewarmed slide. Using a dropper bottle, add a drop of stain to the semen, taking care not to contaminate the dropper. Mix the two drops gently with the corner of a second slide. To smear the stained semen, hold the second slide at a 45° angle to the first and pull it slowly across the drop with short zig-zag motions. The smeared slide can then be dried by passing it over a low flame or placing it in a slide warmer. (Allowing the sample to cool prior to drying may cause secondary sperm abnormalities.) For best results, this preparation should be viewed under a phase-contrast microscope, and the viewer should be able to count and study at least 100 spermatozoa. If the concentration is too heavy or too thin, a second slide should be prepared. The following factors control sperm cell concentration on the slide:

1. size of the semen and stain drops,
2. speed and pressure of the pull when smear is made, and
3. the angle at which the second slide is held against the first.

MORPHOLOGY METHOD II

The easiest method for preparing a morphology slide is to simply stain the sample with a high grade India ink. The stain provides contrast so that the sperm outline can be seen clearly against a dark background. For example, 5 parts Pelican Yellow Label India Ink are mixed gently with 1 part semen (i.e., 5 drops ink and 1 drop semen on a prewarmed slide). One drop of this mixture is applied to a second prewarmed slide and smeared by gently pulling the edge of another slide or a glass rod across the drop. The smear should then be air dried in a slide warmer or dried slowly over a low flame. After drying, the preparation should be examined under a microscope as described under Method I.

MORPHOLOGY METHOD III

Use the eosin-nigrosin stained slide prepared for the Live-Dead Percentage study. (Refer to 'Live-Dead Percentages' within this chapter.)

MORPHOLOGY METHOD IV

After the cells have slowed down somewhat, use the motility preparation to study morphology. Because the sample is not stained, a phase-contrast microscope is essential for an accurate evaluation of sperm cell structure.

MORPHOLOGY METHOD V

Prepare an air-dried semen smear by spreading a drop of semen carefully across a prewarmed slide and allowing it to dry in a slide warmer. Then immerse the slide in a 1:1 mixture of ethyl alcohol and ether for 3 minutes. (Note: ether is extremely flammable and should be used with extreme caution.) Allow the slide to dry and then immerse it in a previously prepared Casarett's stain (30 ml of 5% eosin B solution and 15 ml of 1% phenol solution). Rinse the stained smear in distilled water and allow to air dry prior to microscopic examination.

LIVE-DEAD PERCENTAGES

The percentage of dead sperm cells within the sample measures the ability of the testes to produce viable sperm and the ability of the epididymis to absorb dead cells. Most sources agree that a semen sample should contain at least 60 percent live cells. The reader should note, however, that the presence of live cells is not necessarily an indication of fertility. The cells may be characterized by undesirable movement or shape. For this reason, all semen characteristics should be observed and compared before an evaluation of fertility can be made.

If the breeder intends to study live-dead percentages, the following eosin-nigrosin stain should be prepared prior to semen collection:

1. Dissolve 3 grams sodium citrate dehydrate in 100 cc distilled water.
2. Add 5 grams nigrosin and, after the dye has dissolved, add 1 gram eosin.
3. Store the prepared stain at 41°F (5°C) but bring it to room temperature before use.

Because this mixture stains only the dead cells, dead sperm are easily differentiated from the viable ones. The stain may kill a few cells, but the live-dead count is reasonably accurate and provides a good basis for comparison throughout the season if the same procedure is used for each examination. The procedure involves:

1. placing a drop of semen and a drop of stain on a prewarmed slide;
2. sandwiching this mixture, by covering with another slide;
3. allowing the smear to dry prior to microscopic examination; and
4. counting the ratio of live to dead cells in a manner similar to that used for motile and immotile cells (refer to 'Motility').

SPERM CONCENTRATION

The normal range for sperm concentration in stallion semen is 30 million to 600 million sperm per milliliter of semen. If two samples are collected (1 hour apart), the second sample should show approximately half the concentration of the first. If the sperm concentration decrease is greater than 50 percent, or if concentration is dimished by several daily collections, the stallion should be examined by a veterinarian.

The total number of sperm within the ejaculate is the single most important factor affecting fertility. Researchers have determined that a total of 100 million progressively motile sperm are necessary for conception and that numbers greater than 500 million no longer improve fertility. To determine the total number of sperm per ejaculate, multiply the total gel-free volume by the number of sperm per ml:

ml/ejaculate x number of sperm/ml = number of sperm/ejaculate

Measuring the gel-free volume is the first step involved in a detailed semen examination. When using this value to determine sperm/ejaculate, it is very important 1) that the semen is measured in a graduated cylinder, 2) that the gel fraction is removed, and 3) that the volume is carefully recorded for future reference. The number of sperm/ml cannot be measured as precisely as volume but can be estimated to a relatively accurate value. Methods for estimating sperm concentration are presented in the following discussion.

CONCENTRATION METHOD I

The spectrophotometer is a sensitive instrument sometimes used to determine sperm concentration. (Unless several stallions are evaluated, this instrument is probably not worth the expense.) The spectrophotometer measures semen density by passing light through the sample and indicating, in percent transmittance (%T), how much light has passed through. Using a special chart, %T can be converted to an estimate of sperm concentration. It is important to note that the spectrophotometer must be calibrated for equine semen evaluation and a special conversion chart must be designed for that particular instrument. If the manufacturer does not provide this service, an accredited school of veterinary medicine might be consulted. If the breeder is familiar with the use of logarithms or if the spectrophotometer also reads "absorbency" on the meter, a chart can be made in about 3-4 hours on the farm. (Refer to the spectrophotometer conversion chart instructions within this chapter.)

The instrument recommended for use on the breeding farm is the Bausch and Lomb Spectronic 20 (VWR Scientific). The purchase of any spectrophotometer is a substantial investment but, with proper care and usage, the instrument should provide years of excellent service. It is extremely important that the spectrophotometer be placed on a dedicated circuit with a constant power supply. The colorimeter tubes (cuvettes) that hold the semen samples should be handled carefully; scratches, fingerprints, and dust can cause a significant error in the %T reading. These tubes should be stored in rubber-coated test tube racks and should be washed only with a soft test tube brush and wiped only with soft tissue paper (e.g., Kimwipes or Kleenex).

A spectrophotometer measures the amount of light which is transmitted through a liquid sample in order to accurately determine concentration.

Because the most accurate estimations of sperm concentration are obtained when the %T is between 20 and 80 percent, the sample usually needs to be diluted. For this purpose, the following formalin solution should be prepared prior to semen collection.

Formalin Solution*

Using a 1000 ml flask, dissolve 9.0 grams sodium chloride in 500 ml of deionized or distilled water. Then add 100 ml formalin (37 percent formaldehyde solution) and mix thoroughly. After the solution has been mixed carefully, fill the flask to the 1000 ml mark with deionized or distilled water, transfer the solution to a storage bottle, and seal carefully to avoid evaporation.

*B. W. Pickett, 10% formalin/0.9% NaCl

Depending on the initial semen concentration, either a 1:20 or 1:40 dilution of semen to formalin solution should be used. With practice, the breeder should be able to choose the correct ratio by simply viewing the sample.

For a 1:20 dilution ratio, pipette 7.6 ml of the concentrated formalin solution into a spectrophotometer tube. If a 1:40 dilution is needed, 7.8 ml of the solution must be used. Stopper the formalin tube securely to prevent evaporation. It is also very important that the tube be clean and unscratched. Before the semen is added to the tube, the spectrophotometer must be standardized so that the formalin diluent does not interfere with the %T of the semen. To accomplish this, the instrument must be zeroed.

1. Set the wavelength at 550 mμ (millimicrons).
2. Turn the "zero" knob until the indicator is at 0.
3. Place the formalin tube (without its stopper) in the spectrophotometer, making sure that the line on the tube matches the line on the instrument.
4. Adjust the right knob until the indicator reads 100%T.
5. Remove the tube and replace stopper.
6. Adjust the left knob, until the indicator reads 0.
7. Repeat steps 3-6 until the spectrophotometer reads 0 without the tube and 100% T with the tube in position.

When adding semen to the formalin solution, a prewarmed pipette should be used to measure the **exact** amount: 0.4 ml semen for a 1:20 dilution or 0.2 ml semen for a 1:40 dilution. Adding too much or too little semen is the most common source of error during the entire spectrophotometer procedure.

After the semen is added, the tube is stoppered and turned gently upside down and back ten times. (Vigorous shaking introduces air bubbles which cause inaccurate readings.) Clean the tube with lens paper and remove the stopper. With the wavelength set at 550 mμ, insert the tube so that the line on the tube and the line on the instrument match. Carefully note and record the percent transmittance, and then use the conversion chart (prepared for that particular spectrophotometer) to determine sperm concentration.

WAVELENGTH=550   CALIBRATION OF SPEC 20   MAR-05-82

                           STALLION SEMEN 1-20

T	MIL/ML	T	MIL/ML	T	MIL/ML	T	MIL/ML
15.0	632.6	33.0	352.9	51.0	198.5	69.0	91.3
15.5	621.0	33.5	347.6	51.5	195.0	69.5	88.7
16.0	609.7	34.0	342.3	52.0	191.6	70.0	86.2
16.5	598.8	34.5	337.2	52.5	188.2	70.5	83.6
17.0	588.2	35.0	332.1	53.0	184.9	71.0	81.1
17.5	577.9	35.5	327.0	53.5	181.5	71.5	78.6
18.0	568.0	36.0	322.1	54.0	178.2	72.0	76.2
18.5	558.2	36.5	317.2	54.5	175.0	72.5	73.7
19.0	548.8	37.0	312.3	55.0	171.7	73.0	71.3
19.5	539.6	37.5	307.6	55.5	168.5	73.5	68.9
20.0	530.6	38.0	302.9	56.0	165.3	74.0	66.5
20.5	521.8	38.5	298.2	56.5	162.2	74.5	64.1
21.0	513.3	39.0	293.7	57.0	159.9	75.0	61.7
21.5	504.9	39.5	289.1	57.5	155.9	75.5	59.3
22.0	496.8	40.0	284.7	58.0	152.9	76.0	57.0
22.5	488.8	40.5	280.3	58.5	149.8	76.5	54.7
23.0	481.0	41.0	275.9	59.0	146.8	77.0	52.4
23.5	473.4	41.5	271.6	59.5	143.8	77.5	50.1
24.0	465.9	42.0	267.4	60.0	140.8	78.0	47.8
24.5	458.6	42.5	263.2	60.5	137.9	78.5	45.5
25.0	451.4	43.0	259.0	61.0	135.0	79.0	43.3
25.5	444.4	43.5	254.9	61.5	132.1	79.5	41.8
26.0	437.5	44.0	250.9	62.0	129.2	80.0	38.8
26.5	430.7	44.5	246.9	62.5	126.4	80.5	36.6
27.0	424.1	45.0	242.9	63.0	123.5	81.0	34.4
27.5	417.6	45.5	239.0	63.5	120.7	81.5	32.2
28.0	411.2	46.0	235.1	64.0	118.0	82.0	30.0
28.5	404.9	46.5	231.3	64.5	115.2	82.5	27.9
29.0	398.8	47.0	227.5	65.0	112.5	83.0	25.7
29.5	392.7	47.5	223.7	65.5	109.7	83.5	23.6
30.0	386.7	48.0	220.0	66.0	107.0	84.0	21.5
30.5	380.9	48.5	216.3	66.5	104.4	84.5	19.4
31.0	375.1	49.0	212.7	67.0	101.7	85.0	17.3
31.5	369.4	49.5	209.1	67.5	99.1		
32.0	363.8	50.0	205.5	68.0	96.4		
32.5	358.3	50.5	202.0	68.5	93.8		

Sample calibration chart for the Spectronic-20 spectrophotometer

SPECTROPHOTOMETER CONVERSION CHART INSTRUCTIONS

1. Determine a semen sample's concentration using the hemacytometer. (Refer to Concentration Method II or III.)
2. Measure the percent transmittance or, preferably, the absorbency, if such readings are possible. (Refer to Concentration Method I.)
3. Note these values and repeat the procedure for samples of varying concentrations. Be sure to keep the wavelength set at 550 mμ.

4. Prepare a table with the following headings: "concentration," "%T" and "absorbency." Fill in the known concentrations and their corresponding %T values.

5. Using the following formula, convert percent transmittance to absorbency:

$$A = \log {}^{1}\!/T$$

Where A is the absorbency and T is %T/100. (Logarithm values can be obtained from a slide rule, a table of logarithms, or a calculator with logarithmic function.)

6. Add the absorbency values for each concentration to the table. (Note: If the spectrophotometer reads absorbency, step 5 can be deleted.)

7. Plot concentration versus absorbency. The resulting graph should form a straight line.

8. This graph can be used to convert absorbency to sperm per milliliter only when the spectophotometer is set at 550 mμ. (The angle of the graph changes as the wavelength is altered.)

CONCENTRATION METHOD II

If the breeder chooses not to purchase a spectrophotometer, a fairly accurate estimate of sperm concentration can be made by simply counting the number of sperm seen through a microscope. Because it is essential that the cells are counted within a known volume, a special slide (hemacytometer) must be used. This slide is indented and calibrated with grids so that small, fluid-filled squares can be viewed under a microscope. A coverslip rests on the shoulders of the gridded area so that a fluid-filled depth of 0.1 mm is allowed for the semen. The total volume over any area on the slide can be determined by multiplying the grid depth by the area viewed. In the Neubauer hemacytometer, for example, each grid is made up of 9 large squares, each of which is outlined with triple lines. Because

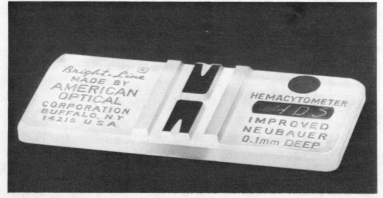

Courtesy of American Optical Corporation

Sperm concentration can be estimated by microscopically examining a small portion of the semen sample on a hemacytometer.

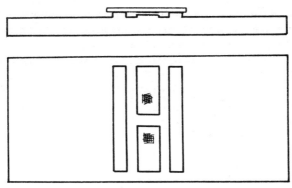

The hemacytometer coverslip rests on shoulders which flank the grid area. Thus, the coverslip is held at a precise height so that the semen sample covers the grid to a depth of 0.1 mm.

the area within each triple line is 1 square mm, the volume of fluid over one of these large squares is 0.1 cubic millimeter (mm^3).

depth x length x width = volume
0.1 mm x 1 mm x 1 mm = 0.1 mm^3

The central large square is divided into 25 smaller squares (0.2 x 0.2 mm), each reflecting a volume of 0.004 mm^3. These values are very important when cell counts per square are converted to sperm per milliliter.

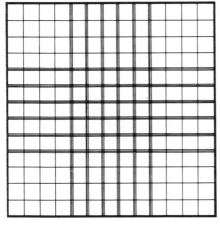

The hemacytometer grid is divided into nine large squares. Five of the nine squares are inscribed with triple lines so that 25 smaller squares are formed within the large central square.

The sperm cells in five of the small squares within the central square are counted. Because the volume of each square is already known, this information can then be used to determine sperm concentration.

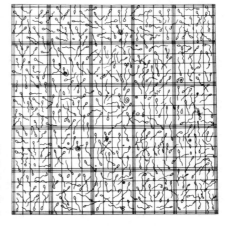

Because the sperm are difficult to count in concentrated samples, a diluting technique should be adopted. The use of a diluting pipette allows precise dilution for accurate concentration estimates. The disposable diluting pipette (Unopette) consists of a shield, a pipette, and a reservoir. This instrument contains a premeasured amount of diluent (for 1:200 dilution ratio) and can be used to accurately measure and dilute the necessary amount of semen.

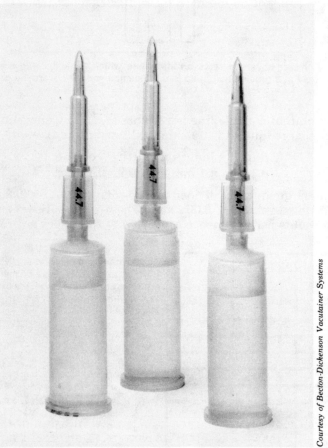

Courtesy of Becton-Dickenson Vacutainer Systems

Unopettes (diluting pipettes) are often used to precisely dilute semen samples before filling the hemacytometer grid.

Cover the hemacytometer grid with a coverslip and fill (from the side) with semen, taking care not to overfill. Next, position the hemacytometer under the microscope and, using a low magnification, locate a counting area. Moving diagonally from left to right, count the sperm cells within five 0.2 mm squares located in the large center square under high-power magnification. The volume for these 5 squares equals 1/50 mm^3, and the dilution ratio is 1:200. These values can then be used to convert the sperm count to sperm concentration.

$$\underset{\text{ratio}}{\text{dilution}} \quad \text{x} \quad \underset{\text{volume to 1 ml}}{\text{factor to change the}} \quad \text{x} \quad \underset{\text{counted}}{\text{number of sperm}} = \text{sperm/mm}^3$$

$$\textbf{200} \quad \text{x} \quad \textbf{50} \quad \text{x} \quad \textbf{?} \quad = \textbf{10,000 x (?)/mm}^3$$

Because 1 ml equals 1000 mm³, the above formula can be changed:

sperm concentration = 10,000,000 x (?)/ml

Therefore, the number of sperm cells counted within the five 0.2 mm hemacytometer squares is simply multiplied by 10,000,000 (add seven zeroes) to determine sperm/ml.

CONCENTRATION METHOD III

A third technique for determining sperm concentration is similar to Method II in that a special slide is used to estimate sperm numbers under a microscope. To dilute the sample, raw semen is drawn into a red blood cell pipette to the 0.5 mark. Then 3 percent chlorazene solution is added to the 1.01 mark and the mixture is shaken carefully. (The chlorazene dilutes the semen and kills the sperm.) After a few drops are released, shaking is continued. A few more drops are released and then a drop is placed on a Neubauer blood cell counting chamber and covered with a cover slip. As in method II, five 0.2 mm squares are counted in a diagonal direction, and the final count is multiplied by 10,000,000 to determine sperm/ml.

APPLYING CONCENTRATION VALUES

Once sperm concentration has been determined, the total number of sperm per ejaculate can be calculated using the following formula:

concentration x total volume = sperm number
sperm/ml x ml/ejaculate = sperm/ejaculate

To determine the number of motile sperm per ejaculate, multiply sperm per ejaculate by the percent motility value determined earlier:

sperm/ejaculate x % motile sperm = motile sperm per ejaculate

This final value is an extremely important measure of the stallion's ability to settle mares. As mentioned earlier, researchers have determined that 100-500 million progressively motile sperm (in a sample with normal values for morphology, longevity, etc.) are required per cover for efficient conception rates. In most instances, live cover provides adequate sperm numbers per mare; depending on the stallion and the circumstances (e.g., how often he has been bred, his general health), a normal ejaculate contains anywhere from 900 million to 150 billion sperm. When dividing a semen sample to inseminate two or more mares, the breeder should remember the 100-500 million rule. (Refer to the discussion on artificial insemination within "Breeding Methods and Procedures.")

Deciding how many mares to book to a given stallion is a serious consideration. Both overbooking and underbooking a stallion can have substantial economic impact on the operation. To make an intelligent decision, the breeder should study the following points carefully:
1. the presence of adequate numbers of progressively motile sperm in the stallion's ejaculate,
2. the stallion's ability to produce normal ejaculates when bred once or more per day for 4-7 days,
3. the stallion's sex drive.

A general guideline is that a normal, healthy, mature stallion, one that is capable of producing satisfactory semen samples, should be able to cover at least 30 mares per season (live cover). (Refer to the statistics discussion within "Stallion Management.") It is important, however, that the breeder use routine semen evaluation, libido studies, and good judgment to regulate the stallion's use throughout the season.

ACIDITY AND ALKALINITY

Thorough semen evaluation should also include a pH check—a measure of its acidity or alkalinity. Semen that is abnormally alkaline may have been contaminated by urine. Alkaline shifts may also indicate an infection in the reproductive tract, a malfunction of the accessory sex glands, or an incomplete ejaculation. Changes in pH alter the sperm cell's delicate environment and can, therefore, lower fertility. The normal pH range for stallion semen is 6.9 - 7.8. The number 7 is used to designate a neutral (i.e., neither acidic nor alkaline) substance. Increasing alkalinity corresponds with increasing numbers 7-13. Increasing acidity is designated by decreasing numbers 7-1.

1 2 3 4 5 6 7 8 9 10 11 12 13 14
← Increasing Acidity | Increasing Alkalinity →

pH METHOD I

Determining the exact pH of a sample involves complex chemical analysis, but a rough estimate of semen pH can be made on the farm by using a series of pH sensitive paper strips (litmus paper). Upon contact with a solution, a litmus strip changes colors if a certain pH level is reached. Each strip is sensitive to a specific pH level so that, by using paper that is sensitive between pH 7 and 8, a sample's pH can be placed within a narrow range. Because litmus paper cannot determine exact pH, this method does not indicate slight deviations from the normal pH range.

pH METHOD II

If the breeder wishes to perform accurate pH studies on the breeding farm, he should purchase a pH meter. This instrument measures ion concentration by means of sensitive electrodes. (Ions are electrically charged atoms; their concentration determines acidity or alkalinity.) The instrument converts ion concentration to acidity or alkalinity and indicates this value as pH on the instrument dial. Because there are many types of pH meters, it would be impractical to describe the use of each within this discussion. Instead, the breeder should study the manufacturer's instructions carefully. It is very important to note that, with any model, proper care of the electrodes is essential to accurate pH determinations. Prior to any pH study, the electrodes should be rinsed in **deionized** water and wiped dry with special tissues (e.g., Kimwipes, Kleenex). The operator should always check the calibration with pH standards (solutions with known pH) prior to use each day. (These pH standards are readily available from chemical reagent suppliers.) Immediately after use, the electrodes should be rinsed again in deionized water and stored with the tips immersed in deionized water.

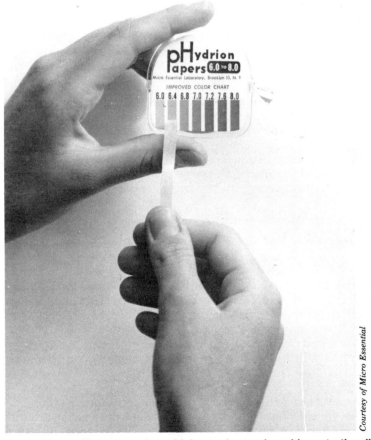

Courtesy of Micro Essential

Litmus paper with specific ranges of sensitivity can be purchased in protective dispensers for laboratory use. A sensitivity range from pH 6 to pH 9 is suitable for semen evaluation techniques.

WHITE BLOOD CELLS

If the pH of a semen sample is alkaline (exceeds pH 7.8), and if there is no indication that urine is present, a white blood cell count should be made to determine whether an infection is present in the stallion's reproductive tract. These counts are made in a manner similar to that used to count sperm cells/ml under a microscope. In many instances, the viewer counts white blood cells as a routine procedure during the semen concentration study. If more than 1500 white blood cells per cubic millimeter (mm^3) are counted, an infection is likely. These infections may originate in the testes, epididymis, accessory glands, vas deferens, urethra, prepuce, or penis. (Disease-causing organisms will be examined more closely in "Stallion Infertility and Impotency.") Isolation and identification of any pathogenic organism is a complex procedure that requires extensive laboratory facilities and a special knowledge of microbiology (i.e., the study of microscopic organisms). To determine which organisms are present, laboratory technicians perform various biological tests, propagate and isolate (culture) seminal bacteria on different growth mediums, and study the shape and behavior of the organisms under a microscope.

RED BLOOD CELLS

A few red blood cells are often seen in the semen during microscopic analysis. Unless more than 500 red blood cells/mm^3 are counted, there is no cause for alarm. Numbers exceeding this do not affect sperm concentration, motility, or morphology, but do (in some unknown fashion) cause infertility. The presence of blood in the semen, or hemospermia, is closely associated with excessive breeding schedules and with the improper use of stallion rings. Hemospermia may also be caused by lesions on or severe trauma to the stallion's reproductive tract, especially the prepuce, penis, or urethra.

LONGEVITY

Sperm cells with unusually short lifespans are unable to travel through the mare's reproductive tract and fertilize the ovum. For this reason, adequate sperm longevity is essential to the stallion's fertility and should be studied when the stallion's breeding potential is evaluated. Researchers are quick to point out, however, that sperm longevity in the test tube may be very different from sperm longevity in the mare's reproductive tract. They also admit that knowledge about sperm life in the mare's uterus is very limited. Until more is known about sperm in vivo, (i.e., in the living body) the following longevity study is, for all practical purposes, an effective test for that aspect of fertility.

Longevity is measured as the period of continued motility within an undiluted semen sample kept in a sterile container at room temperature. Every two hours, a drop of raw semen is taken from the sample and examined for motility. This procedure is continued for at least 8-10 hours. (Refer to Method IV under "Motility.") In an average sample, sperm motility can be observed for 8-10 hours, while an exceptional sample may show movement for as long as 24 hours at room temperature.

Normal Semen Parameters*

Volume	30-250 ml
Concentration	30-600 million/ml
pH	6.9-7.8
White Blood Cells	less than 1500 per mm³ (cubic millimeter)
Red Blood Cells	less than 500 per mm³
Morphology	at least 65% normal
Live/Dead Count	at least 65% live cells
Motility	at least 40% actively motile (i.e., moving in a straight line)
Longevity	at least 40-50% alive after 3 hours at room temperature; at least 10% alive after 8 hours at room temperature (with 2-5% motility)

according to S. J Burns

7

STALLION INFERTILITY AND IMPOTENCY

Because equine reproductive efficiency is extremely important to the success of a breeding program, it deserves special emphasis within a breeding management text. Future chapters emphasize the importance of female reproductive soundness and factors that cause infertility in the mare, topics that have been researched and studied for years due to their obvious economic importance. Stallion reproductive fitness, on the other hand, has only recently gained special recognition as a major contribution to reproductive efficiency. Difficulty in establishing fertility standards for the stallion, reluctance of the stallion owner to select for fertility, and failure to evaluate the stallion for reproductive soundness have contributed to low conception rates within the industry (currently estimated at 50-60 percent). Today, the serious breeder is beginning to focus on the stallion's contributions and limitations, realizing that there are a number of stallion-related management aspects that can make or break an entire breeding season. Therefore, more and more emphasis is being placed on stallion management and handling techniques.

Throughout this chapter, the term *infertility* is used to describe a **reduced** ability of the stallion's sperm to fertilize the mare's ovum or a **reduced** ability of the stallion to produce sperm. (*Subfertility* is sometimes used to indicate slight infertility.) *Sterility,* on the other hand, describes a **complete** and permanent loss of the sperm's fertilizing capacity or of the stallion's ability to produce sperm. Another term, *impotency*, is used to describe the stallion's reduced ability, complete inability, or lack of desire to perform the breeding act. These problems can be caused by physical or psychological abnormalities. Although stallion reproductive problems can be neatly divided into three basic categories, placement of a specific abnormality into one category is not always so simple. The causes of infertility are often complex and interrelated. Thus, a discussion of each individual abnormality would involve repetitious analysis of contributing physiological and psychological factors. For this reason, the following presentation emphasizes the basic causes of reproductive inefficiency and failure. Technical terms used throughout the discussion are defined as they are presented, and the reader should also note that a glossary and index are provided at the back of the book.

Management

The chapter on "Stallion Management" emphasized the importance of proper care and handling of the stallion, explaining how the stallion's inherent attitude and physical ability are constantly subjected to an array of environmental (e.g., management) influences. Therefore, several of the following environmental factors have been discussed previously from a management viewpoint. Depending on the stallion's resistance to stress and abuse, many of these factors can temporarily impede reproductive function or cause substantial damage to his reproductive tract.

DRUGS

Anabolic steroids are sometimes used in an attempt to improve muscle mass, growth rate, and performance ability in young horses. Some horse breeders have attempted to enhance the stallion's libido with anabolic steroids, but research studies indicate that such treatment has no significant effect on sex drive. Due to the detrimental effects of male steroids on reproduction in man and laboratory animals, these drugs have long been suspected as a possible cause of infertility in horses.

Recent studies indicate that the stallion's reproductive ability is significantly affected by certain anabolic steroids. Although libido was apparently unaffected, researchers observed the following physical changes in response to treatment* with two types of anabolic steriods:

1. The average testicular weight decreased to 40-60 percent of its original weight.
2. The average scrotal width decreased to 80-90 percent of its original size.
3. Fewer sperm per gram of testicular tissue were produced in treated stallions.
4. Motility decreased significantly, reducing the fertilizing capacity of the stallion's semen.

*These results were noted when higher than recommended dosages were administered.

It appears that reproductive sensitivity to anabolic steroids varies between stallions. In general, however, administration of recommended doses of two commonly used anabolic steroid products has been shown to cause significant changes in sperm quality and concentration in stallions. Until further research concludes that other anabolic steroids do not have similar effects, researchers recommend that these drugs not be administered to normal stallions that are used (or may be used) for breeding. (It should be noted that these anabolic steroids can be very useful therapeutic agents when used for other purposes.)

Studies have also shown that testosterone therapy adversely affects stallion fertility. In order to understand the effect of testosterone therapy, the actions of reproductive hormones in the normal stallion should be kept in mind.

1. Neural stimuli cause the hypothalamus to secrete gonadotropin releasing hormone into a capillary system that relays it to the pituitary.
2. The gonadotropin releasing hormone stimulates the anterior lobe of the pituitary to release the previously produced gonadotropins, follicle stimulating hormone (FSH) and luteinizing hormone (LH).
3. FSH and LH are responsible for normal spermatogenesis and testosterone production.

It is important to note that this system is equipped with a delicate negative feedback mechanism. The hypothalamus is sensitive to testosterone blood levels and regulates the production of releasing hormones accordingly. (Refer to "The Stallion Reproductive System.")

When testosterone is administered to the breeding stallion to improve libido, the release of gonadotropin releasing hormone from the hypothalamus may be inhibited. Consequently, FSH and LH production decrease and spermatogenesis is impeded. The possible effects on stallion fertility include decreased scrotal width, reduced sperm per ejaculate, and depressed motility. For these reasons, testosterone is not recommended for use in normal breeding stallions and should be administered only under the supervision of a veterinarian.

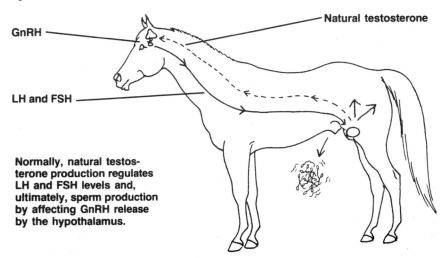

Natural testosterone

GnRH

LH and FSH

Normally, natural testosterone production regulates LH and FSH levels and, ultimately, sperm production by affecting GnRH release by the hypothalamus.

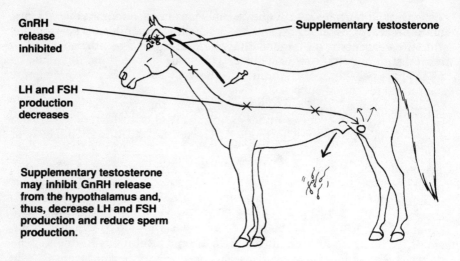

GnRH release inhibited

Supplementary testosterone

LH and FSH production decreases

Supplementary testosterone may inhibit GnRH release from the hypothalamus and, thus, decrease LH and FSH production and reduce sperm production.

Although studies have not been made, there are a number of drugs that are suspected of interfering with the stallion's reproductive processes. Because these allegations are not yet supported with sound research data, it would be inappropriate to mention product names. It should be noted, however, that any drug which causes temporary side effects (e.g., loss of appetite, diarrhea) can limit the stallion's immediate productivity. For this reason, treatments used during the breeding season should be considered carefully. When possible, detailed notes on the stallion's reaction to a particular drug should be studied before that drug is used during the breeding season.

NUTRITION

Proper nutrition is extremely important to normal sperm production and sex drive. Although there are no known dietary supplements that improve the stallion's reproductive ability, a severe nutritional deficiency can delay puberty, cause testicular atrophy (i.e., loss of testicular tissue), decrease sperm numbers per ejaculate, and reduce sex drive. Research on the effects of nutritional deficiencies on stallion reproductive physiology is limited, but severe vitamin A deficiencies have been associated with the cessation of spermatogenesis and reduced testicular weight in other species. Deficiencies of important minerals (e.g., cobalt, iron, and copper) may also affect fertility and potency by causing loss of appetite, reduced weight, and anemia.

Although the stallion requires a diet that contains sufficient protein, vitamins, minerals, and energy, the stallion manager should not confuse optimum diet with excessive feeding. Because obesity decreases sex drive and increases breeding time, it is a direct cause of impotency in the stallion. The presence of fat within the scrotum may increase scrotal temperature, causing degeneration of sperm-producing tissue. In rare instances, obesity may be caused by hypothyroidism, a condition that results in a deficiency of the thyroid hormone. A severe deficiency can cause reproductive failure, while less severe deficiencies are associated with delayed puberty, decreased testicular size, abnormal spermatogenesis, and poor sex drive.

OVERUSE

The key to maximum reproductive efficiency is in knowing how often a stallion can be bred without reducing the quality of his semen or lowering his sex drive. The number of sperm cells per ejaculate depends on the testes' ability to produce spermatids and the ability of the epididymis to support these immotile cells during their maturation and storage periods. Testicular size and, therefore, the amount of sperm-producing tissue has a pronounced effect on sperm output. Stallions with large testicles, for example, are usually able to produce greater than average numbers of sperm per week when collected daily for one week, while the production capacity of young stallions (2-3 years) or stallions with comparatively small testes is often limited. For this reason, optimum use of a stallion depends on his individual sperm-producing ability, a characteristic that should be determined for each stallion prior to the breeding season.

After daily use for one week, the average, mature stallion's sperm reserves are depleted by about 80-100 percent. After this point, he can then ejaculate only the sperm that reach maturity each day. When he is used every other day, he still gives the maximum total number of sperm cells that can be collected each week. However, the sperm concentration is higher than in samples taken from daily collections. If he is bred every day, the total number of sperm produced per week does not change but the sperm concentration per ejaculate decreases. It is important to note that daily use in the breeding shed or excessive use as a teaser (especially in the younger horse) may cause the stallion to lose interest in the breeding process. (e.g., He may be slow to breed or fail to ejaculate.) Although demands on the stallion may have to be increased in order to overcome scheduling problems, it is important to remember the potential consequences of stallion overuse — temporary infertility and impotency.

PAINFUL BREEDING

The importance of thorough physical examinations on a routine basis cannot be overemphasized. If, in the past, the stallion has experienced even minimal pain during the breeding process, he may hesitate to breed, refuse to breed, fail to ejaculate, or dismount from the mare rapidly. There are undoubtedly a number of reasons why the stallion would experience pain during breeding, but the most common man-made causes are as follows:

1. use of an artificial vagina that contains too much or not enough water or that is excessively hot or cold;
2. improper care of the feet, resulting in discomfort when standing or mounting a mare;
3. painful swelling or lesions on the penis due to improper use of a stallion ring;

   4. irritation due to the presence of smegma within the preputial ring
      or failure to rinse soap from the area;
   5. the association of pain with breeding due to rough handling, a
      breeding accident, or being forced to breed a mare against his will.
Again, emphasis on proper management is essential. The breeder must
exercise extreme care when breeding a mare naturally, when collecting
semen, or when using any type of breeding aid. (Refer to "Breeding Methods
and Procedures.") Proper handling techniques are important in all breeding stallions but are especially important when training the young stallion
to breed under supervision.

ISOLATION

The stallion's general attitude and willingness to carry out his breeding
duties consistently (an important element of potency) depend, to a great
extent, on the success of his breeding shed training period, his relationship
with his handler, and his awareness of farm activities. The importance of
this last factor is often overlooked. Many well-meaning breeders that
isolate their stallions for safety reasons often find that their stallions are
extremely excitable, very unpredictable around their handlers, and inconsistent in the breeding shed. A careful balance between safety and "stallion
participation" may provide the necessary stimuli for a healthy breeding
attitude, optimum potency, and increased breeding efficiency.

SEASON

Just as the mare's estrous cycle is affected by seasonal changes in
daylight length, the stallion's reproductive capacity also shows seasonal
variation. Studies have shown that, in the Northern Hemisphere, sperm
numbers per ejaculate are highest in May, June, and July and that numbers
slowly decline to a minimal production level during December and January.
A relationship between sexual behavior and season has also been noted:

   1. The time between initiation of sexual stimulation and the stallion's
      mount increases significantly during periods of seasonal subfertility
      (i.e., October through February in the Northern Hemisphere).
   2. The number of mounts required per ejaculation increases signifi-
      cantly during periods of seasonal subfertility.

In other words, research indicates that the stallion's libido decreases
substantially during the winter months. When the horse's evolutionary
development is considered, the reasons for this seasonal variation seem
obvious. Prior to domestication, equine reproductive patterns adapted for
survival purposes. Fertility and sex drive peaked during late spring and
early summer to insure that foals were born when warm days and optimum
nutrition were plentiful.

Today, the mare and stallion are still characterized by this period of
maximum fertility, a period which often conflicts with man-made breeding
seasons. In many instances, the busiest part of the man-made season
corresponds with a normal period of subfertility and reduced sex drive in

the stallion and erratic estrous patterns in the mare. If, during this time, the stallion is subjected to overuse, either in the breeding shed or as a teaser, he is far more likely to suffer from infertility or decreased sex drive. Depending on the individual stallion's inherent sex drive and sperm-producing capacity, overuse during periods of seasonal subfertility may be an important cause of impotency and reduced conception rates.

Age

Age is an important characteristic to consider when studying the stallion's reproductive abilities. The young stallion, for example, has limited sperm-producing ability and is very susceptible to sexual behavior abnormalities due to improper handling and poor training techniques. The youngster that has little experience in the breeding shed may have difficulty mounting the mare, may hesitate to breed her, or may fail to ejaculate. In other instances, he may ejaculate prematurely or may be unable to enter the mare due to premature enlargement of the glans penis.

The older stallion, on the other hand, frequently loses some sperm-producing ability due to gradual degeneration of testicular tissue, which is a normal aging process. Also, arthritis and other age-related physical problems may affect the older stallion's ability or willingness to mount a mare. This is not to imply that the older stallion should be retired from breeding. Many older stallions have above-average fertility, active sex drive, excellent breeding shed manners, established siring records, and are often an important part of the breeding program. The breeder should note, however, that the older stallion's sperm production capacity must be monitored closely and, if semen evaluation indicates a gradual decline in the number of progressively motile sperm per ejaculate, his breeding schedule should be decreased to compensate for the reduction in fertility. If the stallion has difficulty mounting a mare, special collection techniques (e.g., artificial vagina, short phantom mare, breeding platform) may be helpful. (Refer to "Breeding Methods and Procedures.")

Photo by Kay Coyte

With proper management, stallions may retain fertility and libido well into old age. The noted Thoroughbred stallion, Nashua, sired foals until 29 years of age.

Genetic Aspects

Since the horse's domestication nearly 4,000 years ago, selection criteria (characteristics that the breeder looks for in a horse) have emphasized the importance of conformation, trainability, athletic ability, and pedigree. Today, breeders continue to emphasize performance, appearance, and marketability. Consequently, fertility is usually a low-priority selection characteristic, and many inherited problems that contribute to infertility are propagated. Research studies indicate that, in other species, males with the highest androgen levels are socially dominant and able to breed a relatively larger percentage of the population in a natural situation. Prior to man's domestication of *Equus caballus*, subfertile and impotent stallions had little chance of producing offspring and, consequently, transmitted very few of their deficiencies to the population. Fertile, aggressive stallions, on the other hand, produced a large percentage of the subsequent generation.

Although it is impractical to suggest that breeders should select stock strictly on the basis of fertility, a knowledge of inherited reproductive abnormalities and a willingness to recognize and admit that inherent problems exist within a herd are the first steps to improving equine reproductive capacity. However, no amount of proper breeding management can overcome inherited reproductive limitations.

Although the following discussion includes a list of inherited abnormalities that may affect reproductive potential, a detailed explanation of how these abnormalities are inherited is beyond the scope of this text. The reader should note that a review of basic genetic concepts and terminology is included in the Appendix. For a more complete study of inheritance mechanisms, refer to EQUINE GENETICS & SELECTION PROCEDURES.

HORMONE PROBLEMS

Hormones are responsible for initiating and maintaining many biological systems. Of importance here is the extensive influence that reproductive hormones have upon the stallion's reproductive abilities. Without a proper balance of hormones from the hypothalamus, the testes, the pituitary, the adrenals, and the thyroid glands, normal spermatogenesis and active sex drive are impaired. Conclusive research on the heritability of hormone deficiencies and imbalance is not available at this time, but it is very probable that hormone systems, along with all other life-sustaining systems, are controlled by the coordinated action of many genetic messages, or genes. Over the years, failure to select stallions on the basis of sound hormone status has probably contributed to our present-day fertility problems. Obviously, selecting horses on the basis of hormone function is not realistic, although selecting for fertility and libido could (theoretically) result in a decrease of hormone-related abnormalities.

In some instances, it is difficult to pinpoint the relationship between genetics, hormone abnormalities, and reproductive failure. It is likely, however, that many reproductive abnormalities are related, either directly or indirectly, to malfunctions within the hormone synthesis, release or feedback mechanisms. Two disorders which illustrate and support this concept are testicular feminization and hypothyroidism.

TESTICULAR FEMINIZATION

Failure of the male embryo to respond to testosterone is a rare inherited abnormality that leads to a condition referred to as testicular feminization. Testicular feminization (TF) is characterized by the expression of female sex characteristics during fetal development. The resulting foal is born with underdeveloped testes (located within the abdominal cavity), male sex chromosomes, and female genitalia. These external sex characteristics are usually so well-defined that affected males are frequently registered as mares, although horses affected by TF are sterile and often exhibit stallion-like behavior. The dam rather than the sire carries the controlling gene and she is capable of producing similar offspring in the future.

HYPOTHYROIDISM

Hypothyroidism has been examined briefly with respect to associated obesity and reproductive failure. (Refer to the discussion on nutrition within this chapter.) The thyroid gland normally produces a number of important hormones that affect many different physiological systems (e.g., reproductive, digestive, nervous). In many instances, reduced ability of the thyroid to produce these hormones, or hypothyroidism, is believed to be inherited. The consequences of this disorder are widespread: thin hair, skin problems, excess fat deposition, decreased brain weight, muscle tissue disturbances, etc. As mentioned earlier, hypothyroidism is rare in horses but can be detrimental to fertility if it results in delayed puberty, obesity, decreased testicular size, reduced sperm concentration, or poor sex drive. If severe, hypothyroidism may cause complete reproductive failure. It has also been suggested that, in some species, reduced thyroid production can be caused by extremely hot weather and, therefore, could be a contributing cause of summer infertility in some climates. However, studies to prove that this occurs in horses have not been cited.

INHERITED ANATOMICAL ABNORMALITIES

Research on the heritability of specific anatomical abnormalities in the horse is limited in comparison to available data on humans, cattle, and rats. The importance and the need for such research is reflected in the fact that approximately one out of every five horses is born with some type of anatomical abnormality. These flaws may have little effect on the horse's usefulness, may reduce his market value significantly, or may be so severe that death ultimately results from malfunction of a life-sustaining process. The serious breeder should be familiar with all known inherited abnormalities and should carefully inspect every new breeding prospect for deviations from normal, or acceptable, anatomical structure. Although this chapter emphasizes the importance of anatomical problems that affect the stallion's fertility potential, the importance of other inherited anomalies should not be overlooked. (Refer to the text EQUINE GENETICS & SELECTION PROCEDURES.)

CRYPTORCHIDISM

During gestation, the gonads of the male fetus are located in the abdominal cavity. Just prior to birth, or within a few days after birth, each testis normally migrates from the abdominal cavity through a passageway (inguinal canal) to the scrotum, where optimum temperature for spermatogenesis can be maintained. As described in "The Stallion Reproductive System," thermoregulation encompasses the combined effects of several mechanisms:

1. The cremaster and tunica dartos muscles raise and lower the scrotum, positioning the testes closer to or away from the body and, thereby, increasing or decreasing testicular temperature.
2. When the testes are raised, the skin is wrinkled and overall surface area of the scrotum is reduced; this characteristic reduces heat loss from the scrotum, while lowering the testes creates the opposite effect.
3. The vascular system within the scrotal walls and the spermatic cord is intertwined, allowing warm arterial blood to be cooled by returning venous blood. As a result, scrotal temperature is lower than body temperature.

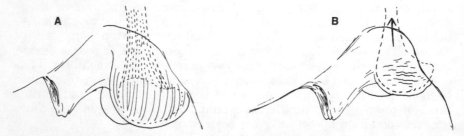

Thermoregulation within the testes is an important aspect of sperm production. As the temperature varies, the cremaster and tunica dartos muscles regulate the proximity of the testes to the body: (A) lower the testes during hot weather and (B) raise the testes during cold weather.

When the testes do not migrate to the scrotum, thermoregulatory failure results in high testicular temperature, gradual tissue degeneration and failure of the seminiferous tubules to produce sperm. The retention of both testes is referred to as *bilateral cryptorchidism*, while the retention of only one testis is called *unilateral cryptorchidism*. The term *monorchidism* has been used to describe the unilateral cryptorchid condition, but veterinarians currently use the term in reference to the complete absence of one or both testes. To avoid confusion, the terms *unilateral cryptorchidism* and *absence of testes* will be used in this text.

The term *abdominal cryptorchidism* indicates that one or both testes are located in the abdominal cavity. *Inguinal cryptorchidism,* on the other hand, is used to describe the retention of one or both testes within the inguinal canal. There is a 2:1 chance that left-sided retention will be abdominal, while most right-sided retentions are inguinal. (Veterinarians differentiate between these two forms by palpating the spermatic vessels via the rectum to determine whether they enter the inguinal canal.) In many instances, the inguinal testis (and occasionally the abdominal testis) descends without assistance before the stallion's second or, in rare instances, third year. Cases have been reported where the testes descend normally at birth but, as the foal grows, migrate back into the body cavity. Although the descent of cryptorchid testicles can sometimes be initiated with hormones, this type of therapy is usually unsuccessful, especially when it is performed after puberty or when the retained testes are located in the abdomen.

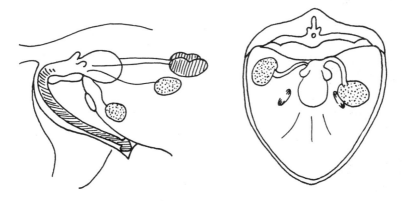

In the bilateral abdominal cryptorchid, both testes are retained in the abdomen.

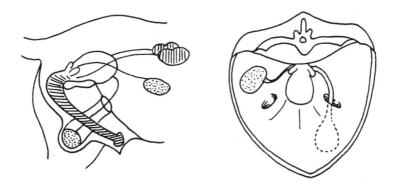

In the unilateral abdominal cryptorchid, one testis migrates to the scrotum, while the other testis is retained in the abdomen.

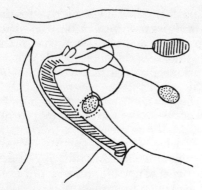
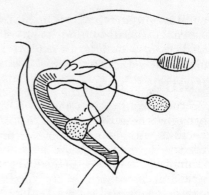

Inguinal cryptorchidism, in which one or both testes are retained within the inguinal canal, may be differentiated from abdominal cryptorchidism by palpating the spermatic vessels via the rectum.

Imperfect testicular descent is characterized by testes located high in the scrotum and near the external inguinal ring.

It is also important to note that testicles which remain in the abdominal cavity or inguinal canal for an extended period may be damaged permanently by extensive degeneration of the sperm-producing tissue (seminiferous tubules). Unlike the seminiferous tubules, the cells of Leydig, which produce testosterone, are not inhibited by body temperature and, consequently, the "corrected" bilateral cryptorchid may show very active sex drive and yet be infertile or sterile.

Another important argument against both the "correction" of cryptorchidism and the use of unilateral cryptorchids as breeding stock is based on the fact that the disorder is inherited. Bilateral cryptorchid stallions cannot propagate their defect since they cannot produce offspring. However, because this characteristic persists within the equine population, it is believed that the unilateral cryptorchid (who produces viable sperm within his descended testicle) carries the genetic information for the production of both bilateral and unilateral cryptorchid offspring. For this reason, many breeders, veterinarians, and registries argue that any colt with a retained testicle after one year should be castrated. Other important reasons for castrating the cryptorchid stallion are as follows:

1. Because testosterone production is not affected by body temperature, the bilateral cryptorchid retains normal stallion behavior without the benefits of normal stallion productivity. In fact, many sources report that sex drive in the cryptorchid stallion is exaggerated, making him even more difficult to handle than the average stallion.

2. Cryptorchid testes are believed to be more susceptible to neoplasms (tumors).

3. Torsion of the retained testicle may occur, causing severe abdominal pain and swelling.

(Note: Stallions may also be affected by imperfect testicular descent, a condition characterized by testes that are positioned near the body at the external inguinal ring. Although the testes are not actually located within the body cavity, thermoregulatory failure may cause degeneration of sperm-producing tissue and consequent infertility.)

SCROTAL HERNIA

Scrotal hernia is characterized by the displacement of a loop of intestine through the inguinal canal and into the scrotal sac. If the intestine enters the inguinal canal without descending to the scrotum, the disorder is referred to as an inguinal hernia. Scrotal hernias in newborn colts are common, since the opening between the abdominal cavity and the inguinal canal (internal inguinal ring) may not constrict for several months or, in some cases, several years after birth. Scrotal hernias in the mature stallion are usually caused by trauma and, probably, by an inherited predisposition. That is, the stallion inherits a characteristic (e.g., enlarged inguinal rings) that increases his susceptibility to the disorder when an injury occurs.

The signs of a scrotal or inguinal hernia include altered gait and abdominal pain. If severe, the scrotal hernia may cause obvious scrotal swelling and an increase in scrotal temperature. In some instances, a veterinarian can correct the scrotal or inguinal hernia by inserting a hand into the rectum and repositioning the intestine manually. (Note: This procedure requires special techniques and proper restraint and should only be performed by a veterinarian.) Other cases may require immediate surgical correction to prevent serious complications.

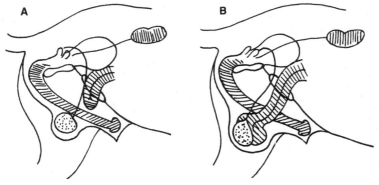

Hernias result when a portion of the intestine passes through the inguinal ring. A loop of intestine may enter only the inguinal canal, causing (A) an inguinal hernia, or may drop completely into the scrotum, causing (B) a scrotal hernia.

One of the most significant problems associated with scrotal and inguinal hernias is the possibility of twisting a section of intestine and drastically reducing its blood supply. Depending on the degree of bowel strangulation, the stallion may exhibit slight discomfort, altered gait, or extreme abdominal pain. If intestinal torsion is not diagnosed and corrected immediately, death may result. Another important problem associated with scrotal hernias is increased testicular temperature caused by the presence of abdominal tissue. As with cryptorchidism, an increase in testicular temperature causes degeneration of testicular tissue, cessation of sperm production, and a gradual decrease in future sperm production capabilities. Because the combined effects of pressure and temperature often cause extreme testicular degeneration, the veterinarian may suggest that one or both testicles be removed when the scrotal hernia is corrected. Castration is also recommended because it is difficult to prevent further herniation by surgical closure and still maintain an adequate opening for the spermatic vessels and because the tendency to develop scrotal and/or inguinal hernias is believed to be inherited.

TESTICULAR HYPOPLASIA

Testicular hypoplasia is an unusual abnormality of the male reproductive tract, characterized by small, flabby testes (each with a small, hard epididymis), by reduced sperm production, and by extremely low sperm reserves. Although testicular hypoplasia varies in severity, it is important to note that even slight hypoplasia seems to predispose the testes to tumors and further degeneration.

This testicular anomaly is believed to be caused by an inherent failure of the primitive sperm cells (spermatogonium) to appear and multiply during fetal development. Scientists admit, however, that germ cells may appear and then degenerate during a later stage of fetal development. The degree of infertility caused by this disorder is directly related to the number of functional spermatogonium found within the testes.

OTHER INHERITED PROBLEMS

Growth, development, and normal function of the reproductive tract is dictated by a number of genetic messages and influenced by environmental circumstances (nutrition, disease, etc.). If a mutation occurs so that these genetic messages are altered, normal mechanisms within the reproductive system may be inhibited. Due to the complexity of the reproductive system and its controlling genes, it is impossible to pinpoint every possible genetic alteration that might affect reproductive function in the stallion. Rather the reader should realize that, although such alterations are possible, the chances of seeing such anomalies are extremely remote. The following examples of rare anatomical abnormalities are included to illustrate the types of problems that can occur regardless of management skills or environmental circumstances.

Missing Testicles

Complete absence of one or both male gonads (sometimes referred to as monorchidism) has been reported in stallions. Other reports indicate that the abdominal cryptorchid testes are sometimes so small (due to tissue degeneration) that they are difficult, if not impossible, to find. Therefore, differentiating between the two abnormalities could present some difficulty.

Both Male and Female Sex Characteristics

When a foal is born with unusual sex chromosomes (normal being XX-female and XY-male), such as XXY or XXXY, he may develop both male and female sex characteristics. (Refer to the discussion on basic genetic concepts within the Appendix.) These intersex males often have both underdeveloped testes and ovaries, a clitoris that has developed into a small penis, an underdeveloped uterus, and they usually exhibit stallion-like behavior. Because the ovaries and testes are underdeveloped and the external genitalia so poorly defined, the animal is hopelessly sterile.

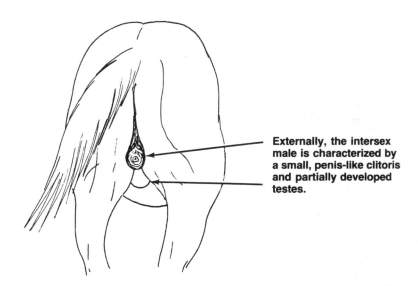

Externally, the intersex male is characterized by a small, penis-like clitoris and partially developed testes.

Three Testicles

Another unusual abnormality involves the presence of three testicles within the scrotum. This rare anomaly is not necessarily detrimental to fertility but can present problems if the animal is castrated and one testis is unknowingly left behind.

Defects of the Seminiferous Tubules

Inherent defects of the sperm-producing tissue, such as the previously described testicular hypoplasia, have also been reported. However, unless these abnormalities are severe and sperm production is inhibited significantly, the problem may be overlooked and the stallion's subfertility accepted.

Defects of the Efferent Ducts

As described earlier, the epididymis consists of many intercoiling tubes (efferent ducts) which serve as passageways for the maturing sperm as they travel slowly to the tail of the epididymis. Studies indicate that the presence of a number of blind (dead-end) ducts could be a significant cause of infertility in the stallion. Because affected stallions produce semen devoid of sperm cells, the condition is difficult to differentiate from that caused by a complete failure of the testes to produce sperm.

LETHAL GENES

Although studies are not yet conclusive, it appears that certain genes transmitted from the sire and/or dam to their offspring may affect reproductive efficiency by altering normal embryonic development. If these abnormal genetic messages cause failure of a life-support system, fetal death (e.g., abortion, mummification, or stillbirth) results. If death occurs during the first few weeks of gestation, the embryo may be aborted or resorbed, and the breeder may observe only the mare's apparent failure to conceive and overlook the possibility of an inherent problem. Although other problems may be indicated, conception failure that is exaggerated when crosses between two particular individuals are made should be viewed suspiciously. If this failure is characteristic of a certain stallion or mare, the breeder should consider the possible presence of lethal genes when fertility studies are performed. As a case in point, when the dominant roan coat color gene (R) is transmitted from both parents, the resulting combination (RR) is lethal to the developing embryo.

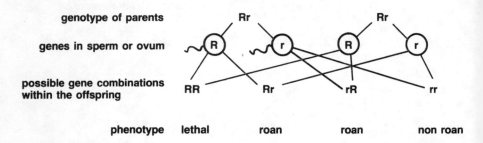

Due to the stallion's greater influence on the genetic pool (type of genes) within each new generation as compared to the mare's influence, there is a greater chance that any lethals he carries will be transmitted to future breeding stock. An explanation of why these lethals do not affect all embryos has been established but is beyond the scope of this text. (Refer to EQUINE GENETICS & SELECTION PROCEDURES.) Although information on specific lethals is still limited, careful observation and detailed breeding records may help reveal early embryonic death within a breeding herd.

Trauma

Because infertility and impotency are sometimes caused by physical injury to the stallion's genitals, legs, back or pelvis, breeding techniques and farm safety measures should be given special consideration by the stallion owner. Any injury that causes anxiety, severe pain, or even slight discomfort may result in hesitation or refusal to breed. Accidents that occur during breeding are especially detrimental to the stallion's libido, since horses have a long subconscious memory and may associate pain with breeding for many years. If an injury results in pain when the stallion walks, stands to mount a mare, obtains an erection, or copulates, impotency is likely. An accident may also cause infertility, or possibly sterility, if the testes are injured. Severe trauma within the scrotal area can lower the testes' resistance to infection and often results in inflammation, tissue degeneration, and the formation of fibrous scar-like tissue.

Perhaps the most common and the most serious injuries suffered by stallions are those caused by a mare's kick during breeding or teasing. (Refer to the discussion on restraint within "Breeding Methods and Procedures.") Besides the obvious psychological damage that this type of accident can instill, serious physical problems may result:

1. localized accumulation of blood (hematoma) in the penis, caused by the rupture of blood vessels within the organ;
2. forcible tearing (rupture) of the penis;
3. painful swelling of the glans penis and constriction of the penis by the smaller preputial opening which can result in gangrene if left uncorrected;
4. swelling of the penis within the prepuce and inability to extend beyond the preputial opening;
5. displacement of the intestine through the abdominal wall (ventral hernia), into the scrotum (scrotal hernia), or into the inguinal canal (inguinal hernia);
6. fracture of the hind limb;
7. inflammation of the testes, resulting from lowered disease resistance after severe trauma;
8. inflammation of the prepuce which may result in adhesions (type of scar tissue) between the penis and prepuce and prevent normal erection;
9. injury to the dorsal nerve of the penis or severe injury to the glans penis, resulting in loss of feeling in the glans penis and, consequently, reduced sex drive and/or failure to ejaculate.

Another important cause of infertility and reduced sex drive is a fall during the breeding process. Although floors that are designed for optimum footing may not prevent the inexperienced youngster from falling off a mare, proper flooring in the breeding shed reduces the probability of serious injury to both the stallion and the mare. Injuries acquired during a fall may include the following:

1. dislocation of the hip bone;
2. fracture of the hind limb, pelvis, or back;
3. penile paralysis resulting from a spinal injury;
4. strains and tears of muscles and tendons;
5. inflammation of the stifle joint.

Occasionally, the stallion suffers from lacerations of the glans penis when the mare's tail hair becomes caught in the vagina during the breeding process or when the mare is sutured with a breeding stitch of fine material, such as nylon or Vetafil®. This problem is easily avoided by wrapping the mare's tail prior to each service or by changing suture material. If the stallion's penis bleeds after copulation, it should be checked carefully for lacerations and tumors. (It should be noted that blood on the stallion's penis may be due to an injury within the mare's reproductive tract. Refer to "Mare Infertility.")

Penile edema and swelling are sometimes observed when the stallion is tranquilized and in association with certain diseases. (Refer to the discussion on disease-related conditions within this chapter.) The swelling that results from edema may cause an inability to draw the penis back into the prepuce. To prevent increased edema within the affected penis (due to gravity), the penis should be supported gently within a sling and veterinary assistance should be sought without delay. If swelling prevents replacement, cold packs may be helpful.

When dealing with any type of injury to the stallion's genitals, permanent damage to his fertility and/or sex drive can be minimized by **prompt** medical attention. For example, allowing the penis to remain outside the prepuce (as with penile paralysis) may necessitate penile amputation. Gangrene within a swollen penis and necrosis (cellular death) of testicular tissue are serious conditions that should receive immediate veterinary attention. (Note: Degeneration of testicular tissue is sometimes indicated by a decrease in testicular size.) If the testes become rotated within the scrotum, a condition that is especially common in race stallions, normal blood flow and temperature regulation are disturbed. If this painful torsion is not diagnosed and corrected promptly, tissue degeneration may cause permanent damage to the stallion's sperm-producing capabilities.

Reproductive Disease

Although disease-related conditions that affect the stallion's reproductive ability are usually diagnosed and treated by the veterinarian, the serious stallion handler should be aware of any changes in the stallion's health and should be especially attentive to those changes that may affect fertility status. The long-term implications of any disease process, whether it encompasses the entire body or is localized within the reproductive tract, emphasize the importance of careful observation and routine reproductive examinations described in "Stallion Management." To help bridge the literature gap between the veterinarian and the horseman, the following discussion includes a brief overview of the relationship between pathology and infertility.

The study of a cell's response to foreign compounds, living microorganisms (e.g., bacteria, viruses), or to a deficiency of blood, oxygen, or other life-sustaining material is referred to as pathology. Because cells have only a few ways of responding to a number of possible physical insults, the pathologist's job of identifying the cause (or causes) or cellular change may be quite laborious. Using a microscope, the pathologist relates changes in cellular size, structure, components, color, and strength with various stages of degeneration. Initially, injured tissue is characterized by redness, swelling, heat, and pain. This normal reaction to an injury, cellular deficiency, or foreign compound is referred to as inflammation, a necessary element of the healing process. Although inflammation is generally beneficial, an extreme reaction can be detrimental to the cell's integrity and, consequently, to the tissue's function. For example, if severe inflammation produces fibrin (a blood component) or pus, the development of hard fibrous tissue may decrease the tissue's strength and elasticity. A prolonged inflammatory response causes progressive cellular breakdown and, eventually, cellular death. The death of cells within a living body is usually referred to as necrosis, while the death of cells that are exposed to airborne bacteria is often differentiated by the term *gangrene*. Gangrene is characterized by tissue decay and ultimately death due to blood poisoning. Necrosis, on the other hand, is usually characterized by atrophy (reduced size) of the affected organ or tissue.

TESTES

Diseases that affect the testes range from tumors of the anterior pituitary gland in the brain to localized infections within the testes. In any event, damage to the sperm-producing or hormone-producing tissue is usually irreversible and can cause extensive, undesirable changes in the stallion's reproductive ability. As noted earlier, an injury to the stallion's testes can reduce the organ's resistance to disease, resulting in the same inflammatory and degenerative changes that occur with localized infections and systemic illnesses. As a case in point, testicular torsion (an example of trauma) reduces the organ's blood supply and, if severe and prolonged, may result in diseased testicular tissue.

The spermatogonium (primitive sperm cells) and the sperm-producing process are both very sensitive to *orchitis*, a term used to indicate inflammation of testicular tissue. If inflammation interferes with spermatogenesis, semen may lack viable sperm for several months. If a significant number of spermatogonium are damaged, orchitis results in a permanent decrease in the stallion's sperm-producing capabilities. Because the hormone-producing cells are more resistant to stress than the sperm-producing cells, reduced semen quality is not necessarily accompanied by decreased sex drive.

The signs of testicular inflammation vary with severity. For example, acute orchitis is usually characterized by firm swelling of the testes along with extreme sensitivity and heat. The onset of testicular degeneration is marked by soft, flabby testes and poor semen quality. Small, hard testes, on the other hand, may reflect the consequences of prolonged inflammation: degeneration, necrosis, atrophy, development of fibrous tissue, and deposition of calcium. The reduction in testicular size is directly related to the decrease in sperm-producing tissue and, therefore, is believed to be a rough measure of reduced fertility. (Note: Observation of testicular size should not replace a semen evaluation to determine the stallion's fertility status.)

The relationship between trauma and testicular damage was presented earlier in this chapter. Other important causes of testicular degeneration can be organized into the following categories: systemic illnesses, localized infections, and tumors. (The reader should note that trauma is the leading cause of testicular damage.) For example, a fever may completely interrupt spermatogenesis even though the testicles are not directly involved in the disease process. Depending on the severity and duration of the illness, testicular tissue may be permanently damaged, and the stallion's fertility affected accordingly. If the affected stallion stays in a recumbent (lying) position for an extended period, increased testicular temperature caused by the organ's close proximity to the body may contribute to the degeneration process. Localized infections and tumors within the testes are rare but have been reported on occasion. It is interesting to note that testicular degeneration resulting from an infection is usually bilateral (affecting both testes) due to the systemic nature of most infections, while degeneration caused by tumors is often unilateral.

Most of the organisms and neoplasms (e.g., tumors) involved in testicular degeneration have been identified, but information related to specific reproductive diseases in stallions is limited at this time. However, because a basic knowledge of the causes and effects of testicular abnormalities is important to the serious stallion owner or manager, a list of conditions that encourage or cause testicular disease is presented. This list is not complete but is provided to show the reader how sensitive the testes are to a wide range of physical problems.

TESTICULAR PATHOLOGY

Systemic Diseases

1. Equine Infectious Anemia
   Description:
   EIA is a viral infection which causes weight loss, periodic fever, edema (swelling due to fluid accumulation), and anemia. The virus is spread by bloodsucking flies and improper use of hypodermic needles.
   Effect on Testes:
   Cessation of spermatogenesis and degeneration of testicular tissue occur due to increased testicular temperature caused by fever. In weak horses, these problems may be aggravated by lying down for extended periods of time.
2. Strangles
   Description:
   Strangles is a bacterial *(Streptococcus equi)* infection, usually noted in young horses but also seen in older horses. The condition is characterized by high temperature, inflammation of the upper respiratory tract, reluctance to swallow, swollen lymph nodes, and pus-

filled drainage from the nostrils and from abscesses in the head and throat regions.

Effect on Testes:

Cessation of spermatogenesis and degeneration of testicular tissue may occur due to increased testicular temperature.

3. Pneumonia

Description:

Pneumonia is an inflammation of the lungs caused by a bacterial infection or by the presence of foreign particles in the lungs, and characterized by nasal discharge, coughing, depression, and fever.

Effect on Testes:

Cessation of spermatogenesis and degeneration of testicular tissue may occur due to increased testicular temperature.

4. Actinomycosis

Description:

Actinomycosis is a bacterial (*Actinomyces bovis*) infection commonly referred to as fistulous withers or poll evil. The disease is characterized by abscesses and drainage in the throat and neck regions, swollen lymph nodes and fever.

Effect on Testes:

Cessation of spermatogenesis and degeneration of testicular tissue may occur due to increased testicular temperature.

5. Chronic peritonitis

Description:

Chronic peritonitis is prolonged inflammation of the membrane that covers the abdominal cavity and most of the intestinal tract, caused by microorganisms or by chemical agents. The disease is characterized by severe abdominal pain, restlessness (lying down and getting up again), high fever, dehydration, weight loss, and possibly distention of the abdomen due to an accumulation of fluids.

Effect on Testes:

Temperature elevation is caused by fever and close proximity of the testes to the abdomen. (The abdomen stays tucked up and tense). Thrashing may cause testicular torsion and, subsequently, testicular strangulation and degeneration.

6. Parasitic Infestations

Description:

Depending on the severity and type of parasites present, these infestations are often characterized by weakness, weight loss, diarrhea, and anemia.

Effect on Testes:

In cases of severe parasitic infestation, weight loss may result in complete cessation of the sperm-producing process. The duration and extent of such weight loss is directly related to the degree of permanent fertility reduction. The migration of strongyle (blood parasite) larvae to the testicular artery is occasionally a direct cause of testicular inflammation and degeneration.

7. Colic
   Description:
   Colic is defined as abdominal pain caused by a number of possible conditions (e.g., twisted intestine, peritonitis, grain overload, parasites). The condition may be characterized by lack of appetite, restlessness, sweating, rolling, increased and uneven pulse, pale gums, and/or abnormal gut sounds.
   Effect on Testes:
   If severe, colic from any source can cause problems similar to those described previously under 'Chronic Peritonitis.'

8. Laminitis
   Description:
   Laminitis is characterized by inflammation of and possibly irreversible damage to the laminae of the foot, caused by a number of problems including overeating, overuse on hard surfaces, and obesity. The disorder may be indicated by an exaggerated pulse in the hoof, heat and pain at the coronary band, and an awkward stance to eliminate pressure from the affected feet.
   Effect on Testes:
   If (due to severe pain) the stallion remains in a recumbent position for an extended period, increased testicular temperature may result in tissue degeneration and loss of fertility.

9. Equine Viral Arteritis
   Description:
   Equine arteritis is a highly contagious viral infection characterized by fever, inflammation of mucous membranes and edema (sometimes within the prepuce and scrotum).
   Effect on Testes:
   Fever may elevate testicular temperature. Fluids accumulated within the scrotum may cause circulatory interference due to increased pressure on the testicles.

10. Equine Influenza
    Description:
    Influenza is a highly contagious respiratory infection caused by one of two viral strains (Equine Myxovirus A-equi-1 and Myxovirus A-equi-2), characterized by extremely high fever, coughing, nasal discharge, labored breathing, and weakness.
    Effect on Testes:
    The extreme elevation in body temperature causes an interruption of spermatogenesis. Unless the condition is complicated by secondary bacterial infections, the fever should last only 1-4 days.

11. Escherichia Coli
    Description:
    Escherichia coli is part of the normal intestinal flora but can cause serious problems if it gains access to other body tissues. If bacteria enter the circulatory system, septicemia may occur.

Effect on Testes:
Tissue response to increased testicular temperature and, if the bacteria gain access to the testicles via the urinary tract or circulatory system, severe orchitis (testicular inflammation) may result.

Localized Infections

1. Hemolytic streptococcus
Description:
Hemolytic streptococcus is a bacterium which is known to cause reproductive infections in mares. Usually the stallion serves only as a carrier of the organism, but the bacteria occasionally migrate to the testes via the urethra.
Effect on Testes:
Orchitis may occur.

2. Klebsiella pneumoniae
Description:
Klebsiella pneumoniae is a bacterium that occurs naturally within the intestinal tract. It is known to cause septicemia and is sometimes a source of reproductive infection in mares. The stallion often serves as carrier of the organism from mare to mare, and the bacteria have been known to migrate to the testes via the urethra.
Effect on Testes:
Orchitis or extreme elevation of testicular temperature may occur in cases of Klebsiella septicemia.

3. Pseudomonas aeruginosa
Description:
Pseudomonas is a bacterium that is known to cause problems such as pneumonia and reproductive infections. The organism is sometimes found in the urinary tract of the stallion, and can migrate to the testes.
Effect on Testes:
Orchitis may result.

Tumors

1. Dermoid Cysts
Description:
Dermoid cysts are rare, skin-like tumors that contain small cysts and sometimes hair, bone, and/or teeth. These growths are observed most frequently in cryptorchid testes.
Effect on Testes:
Degeneration of sperm-producing tissue may result.

2. Seminomas
Description:
Seminomas are malignant tumors that arise from the sex cells (spermatogonium) in young males. These tumors are seen more frequently than interstitial cell tumors or Sertoli cell tumors.
Effect on Testes:
The increased testicular temperature and inflammation associated with seminomas and the displacement of testicular tissue caused by tumor growth results in decreased sperm production.

3. Teratomas
   Description:
   Teratomas are tumors containing cells from other unrelated tissues. These growths are rare in the stallion but are seen most often in cryptorchid testes. Testicular teratomas are usually malignant and, thus, resistant to treatment and frequently fatal.
   Effect on Testes:
   Sperm-producing tissue degenerates.
4. Tumors of the Sertoli Cells
   Description:
   Sertoli tumors are rare testicular neoplasms located on the cells that produce estrogen.
   Effect on Testes:
   Excess estrogen production caused by these tumors inhibits the release of gonadotropin releasing hormone from the hypothalamus. Thus, follicle stimulating and luteinizing hormones (FSH & LH) are not released, and spermatogenesis ceases.
5. Tumors of the Interstitial Cells
   Description:
   Interstitial cell tumors are rare, testicular growths located on the cells that produce testosterone.
   Effect on Testes:
   Excess testosterone production caused by these tumors inhibits the release of gonadotropin releasing hormone from the hypothalamus, causing an effect similar to that described under Sertoli cell tumors.
6. Tumors of the Anterior Pituitary
   Description:
   Anterior pituitary tumors are rare neoplasms on the area of the brain that releases follicle stimulating and luteinizing hormones. As explained in "The Stallion Reproductive System," these substances are necessary for normal testicular function.
   Effect on Testes:
   Testes may fail to produce sperm or reproductive hormones.

PENIS

The stallion's penis is relatively resistant to infection but is an extremely important carrier of pathogenic (disease-causing) organisms between mares. Contamination of the stallion's prepuce and penis should be of primary concern to the serious breeder. This aspect was emphasized in "Stallion Management" and is reemphasized in "Breeding Methods and Procedures." A basic review of the contamination/infection processes helps to explain why hygiene plays an important stallion management role.

CONTAMINATION

The moist, warm preputial and penile surfaces provide an excellent environment for microbial growth. The presence of several types of bacteria and other microorganisms is not unusual and is not an important cause of reproductive infection. This normal flora is believed, by some sources, to inhibit the growth of certain pathogenic strains (e.g., the contagious equine metritis organism). However, when the stallion is contaminated with a significant number of pathogenic microbes or when normal disease resistance mechanisms are impaired, serious problems may be encountered. Contamination may be the result of improper hygiene, contact with an infected mare, or exposure to a contaminated environment. Recent reports indicate that excessive use of soaps or antibacterial agents may disrupt the stallion's normal flora, thereby removing an important inhibitory effect and allowing less competitive pathogens to increase in number. For this reason, mild soaps and lukewarm water are recommended for routine washing. Strong soaps, antiseptics, and antibiotics should be reserved for treatment of infections.

Because the stallion often shows no clinical signs of infection, the presence of these disease-causing organisms may not be acknowledged until a serious fertility problem exists on the farm. When a problem is recognized, detection of the carrier stallion and infected mares and isolation of the infectious organisms are of immediate importance. The signs of reproductive infections in mares are examined in "Mare Infertility," while culturing the mare to identify and treat the causative organisms is discussed in "Broodmare Management." Unlike the mare, the infected stallion cannot be identified on the basis of clinical signs. Rather, the incidence of infected return mares (those mares that must be rebred) along with positive cultures taken from the stallion's pre-ejaculatory fluid, from the urethra following ejaculation, from the semen, from the urethral sinus, and/or from the sheath indicate contamination. The value of routine cultures from these areas to warn the breeder of possible cross-contamination has been questioned. The most practical and efficient safety procedure is to prevent such problems through common-sense hygiene: washing the mare's perineal area prior to breeding, washing or rinsing the stallion's penis prior to breeding, washing the stallion again after breeding with a mild soap, use of artificial insemination (when applicable) to control problems that do occur, etc. (Refer to "Breeding Methods and Procedures.") If contamination is suspected or if the stallion has been treated recently for infection or contamination, the breeder may have the stallion cultured at regular intervals. The breeder should have a good working relationship with the veterinarian so that arrangements might be made to allow trained personnel to collect culture samples and have those samples sent to a special laboratory.

INFECTION

Although the stallion's penis is relatively resistant to infection, there are a few problems that may occur. For example, habronema larvae may invade the penis during the summer months, displacing tissue cells and causing irregular lesions that bleed when handled. These so-called summer sores often cause intense itching and may interfere with normal urination and ejaculation. The larvae of the screwworm fly may also cause lesions on the stallion's sheath and glans penis. These areas are characterized by bleeding, sensitivity, swelling, and an extremely offensive odor. Due to an intensive eradication program, the screwworm fly is no longer a serious problem in most parts of the United States. However, the breeder should realize that screwworm lesions are not only detrimental to the stallion's libido but may also result in his death if left untreated.

Another infectious organism, *Trypanosum equiperdum* (a one-celled animal, or protozoa) causes dourine, a highly contagious disease that is no longer seen in the United States. Dourine is characterized by inflammation and depigmentation of the prepuce and penis. Coital exanthema, an infectious disease of the stallion's external genitals that is occasionally seen in the U.S., is characterized by circular, merging lesions on the skin of the penis and by the affected stallion's refusal to copulate due to substantial discomfort. Coital exanthema is believed to be caused by a herpesvirus (not the rhinopneumonitis virus) and is spread by coitus. The rhinopneumonitis virus, equine herpesvirus I, is suspected of causing penile paralysis (an inability to retract the extended penis). If the penis is not supported within the prepuce by mechanical means, the organ may become swollen (due to fluid accumulation) and possibly strangulated by a restricting preputial opening. If this occurs, gangrene may result and penile amputation may be required to save the animal's life. (Such paralysis has also been associated with the use of certain promazine tranquilizers.)

CANCER

A different aspect of penile pathology involves the occasional occurrence of squamous cell carcinoma on the stallion's prepuce and penis. Squamous cell carcinoma usually occurs around body openings where the skin is unpigmented or on other body parts where the skin is lightly pigmented (e.g., white face markings). These growths are malignant, meaning that they frequently recur after removal, are resistant to treatment, and may eventually result in the animal's death. Grey horses may also have prob-

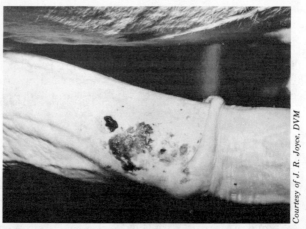

Courtesy of J. R. Joyce, DVM

Each time the stallion is washed he should be examined for developing abnormalities. This early stage of squamous cell carcinoma was found on a stallion's lightly pigmented penis.

lems with melanomas, a malignant cancer associated with grey pigmentation in horses. (Refer to "Mare Infertility.") Routine checks for cancerous growths should be made each time the stallion is washed and rinsed. An early diagnosis increases the chances of successful treatment.

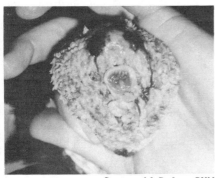

Courtesy of J. R. Joyce, DVM   *Photo by Jim Wright*

Squamous cell carcinoma of the glans penis (left) is readily noticeable, particularly when it is compared to a healthy glans penis (right).

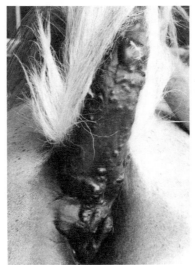

Melanomas are occasionally found in grey horses and appear as nodules located under the tail and around body openings such as the anus and the sheath.

EPIDIDYMIS

Tumors and cysts of the epididymis are very rare, but inflammation occasionally occurs in the duct of the epididymal tail, possibly due to the migration of various organisms through the vas deferens or through the blood and lymph vessels. Disease and resulting fibrous obstructions within the epididymal ducts interfere with the normal sperm support and storage processes. Sperm from an infected epididymis is often characterized by poor motility. There is no known cure for obstruction of the epididymal ducts. If the disease process is isolated in one testicle, however, removal of that testicle may protect the remaining testis from infection, degeneration, and obstruction. In time, the healthy testis will compensate for the lost organ by reaching nearly 90 percent of the previous sperm-producing capacity.

VAS DEFERENS

The sperm carrying duct that leads from the epididymis to the urethra is also subject to inflammation and degeneration. Infections of the vas deferens are indicated by a thick, firm enlargement of the ampulla (detected by a veterinarian during rectal palpation of the accessory sex glands). Occasionally, this enlargement is very painful. Sperm cells from the affected duct usually suffer from poor motility and reduced longevity.

SEMINAL VESICLES

The only other reproductive organs that seem to be susceptible to infection are the seminal vesicles. Infection of these glands (seminal vesiculitis) is characterized by pain, swelling, and the presence of pus within the vesicle. Palpation of infected vesicles causes an increase in the number of white blood cells within the ejaculate. For this reason, detailed semen evaluation may include collection of a sample, rectal palpation of the accessory sex glands, and collection of a second sample one hour later. (Refer to "Semen Evaluation.")

This table is a compilation of infectious organisms which can affect the major reproductive organs of stallions.

ORGAN	INFECTIOUS ORGANISMS
Penis	habronema larvae
	screwworm larvae
	Trypanosoma equiperdum (dourine)
	herpesvirus (coital exanthema or genital horse pox)
	herpesvirus I (penile paralysis)
Testes	*Streptococcus zooepidemicus*
	Klebsiella pneumoniae
	Pseudomonas aeruginosa
	strongyle larvae
Epididymis	*Streptococcus zooepidemicus*
	Corynebacterium pseudotuberculosis
	Actinobacillus
	Pseudomonas aeruginosa
Vas Deferens	*Streptococcus zooepidemicus*
	Corynebacterium pseudotuberculosis
	Actinobacillus
	Pseudomonas aeruginosa
Seminal Vesicles	*Corynebacterium pyogenes*
	Brucella abortus

Semen Abnormalities

Most of the conditions examined within this chapter have an effect, either direct or indirect, on semen quality. Many abnormalities and deleterious conditions can be detected through the detailed semen examination process described in "Semen Evaluation." To help the breeder relate the discussion on semen evaluation with a study of stallion infertility, several important semen abnormalities are described:

1. **Volume**: A low gel-free volume may be due to seasonal or inherent variation in semen production capabilities. Volume usually has little effect on fertility, since the number of viable sperm per ejaculate determines the likelihood of conception. Limited semen volume may cause handling difficulties if artificial insemination is used, but this problem is easily solved with the use of extenders.

2. **Color**: Cloudy semen may reflect the presence of pus, white blood cells, red blood cells, and/or miscellaneous debris. Sperm motility, live-dead percentages, and longevity are usually affected when the sample is visibly contaminated. Red streaks within the sample reveal the presence of blood. This blood usually originates from ruptured capillaries along the reproductive tract, sometimes in association with excessive breeding or the use of stallion rings. Hemospermia (i.e., blood in the semen) has no apparent effect on semen quality but does, in some unknown fashion, cause infertility.

3. **Density**: Thick and gelatinous or thin and watery semen indicates malfunction of the accessory sex glands. The corresponding effects on fertility are unknown, but the varied concentration of sperm cells throughout these samples makes an accurate determination of sperm concentration impossible. (Note: A normal ejaculate contains a gel fraction, but the entire ejaculate should not be heavily concentrated with gel.)

4. **Motility**: If a sample shows poor motility, the observer should analyze both his collection and handling techniques. If collection and handling errors are not present, reduced motility may indicate infection, hemospermia, or possibly testicular degeneration. In general, testicular degeneration is associated with the presence of 1/2 to 1/3 the normal number of progressively motile sperm cells. Fewer than 40 percent actively motile sperm causes a reduction in the stallion's conception rates. (Due to the presence of stagnant sperm within the epididymis, a sample collected one hour after the first should show equal or better motility.)

5. **Longevity**: Reduced longevity may be caused by the presence of blood, pus, or certain bacterial strains within the reproductive tract. Poor longevity reduces the sperm cell's chances of fertilizing the mare's ovum. As with motility, the observer should make sure

that laboratory techniques are not limiting sperm cell survival. (Refer to "Semen Evaluation.")

6. **Morphology**: The presence of abnormal cells may be caused by an interruption in the sperm production cycle (e.g., due to fever) or by testicular degeneration. Samples that contain less than 65 percent normal cells are directly related to reduced fertility. The reader should note, however, that the presence of abnormal cells has no adverse effect on normal cells.

SPERM ABNORMALITIES

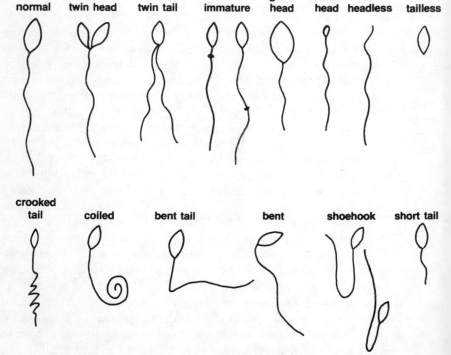

7. **Live-Dead Percentages:** If a semen sample contains less than 60 percent live cells, poor absorption within the epididymis or degeneration within the testicles may be indicated. If an increase in the number of live cells occurs when the stallion's semen is collected at frequent intervals, the epididymal environment is probably detrimental to sperm cell survival.

8. **Sperm Numbers**: A reduction in the number of viable sperm cells per ejaculate may be due to seasonal variation, overuse, testicular degeneration, or increased testicular temperature. Scientists have determined that sperm concentration in the stallion's

semen normally ranges from 30 million to 600 million sperm per milliliter, and that at least 100 million progressively motile sperm are needed for acceptable conception rates.

9. **pH**: An increase in the semen's pH (i.e., greater than the normal range of 6.9 - 7.8) may be caused by the presence of urine, by an infection within the reproductive tract, or by overuse in the breeding shed. A normal alkaline shift (increase in pH) is often seen when a second sample is collected, possibly due to a decrease in the amount of seminal vesicle fluid.

10. **White Blood Cells**: The presence of a large number of white blood cells within the semen sample (i.e., greater than 1500 WBC's per cubic millimeter) usually reflects the presence of an infection within the reproductive tract and, therefore, indicates reduced sperm longevity and motility. It should be noted, however, that the presence of pathogens within the stallion's tract or on his external genitals is not always accompanied by an increase in seminal white blood cells. ♞

8

THE MARE REPRODUCTIVE SYSTEM

Although the stallion's value is highlighted by his ability to sire many foals each year, the mare's role on the breeding farm encompasses a wider range of contributions. In addition to her genetic contribution, the mare protects and nourishes her offspring. Her ability to conceive and carry a foal to term influences the economic success and stability of the breeding operation, while her general health and disposition influence the potential quality and tractability of her foals.

The health and genetic constitution of the broodmare band as a whole are extremely important to the breeder's goals. With the recent increase in the number of horses bred each year, emphasis on careful selection and culling of broodmares has increased dramatically. Many serious breeders are demanding the same mark of excellence in their mares that has been traditionally expected in stallions. Unfortunately, the importance of inherent reproductive capacity is often overlooked due to the emphasis on "more economically desirable" traits. The profit-motivated breeder should realize the importance of inherent reproductive capacity and select stock only after carefully considering potential reproductive performance. This is not to suggest that breeders should select stock only on the basis of reproductive capacity. Obviously, this strict selection is impractical in many circumstances. It is important, however, that the breeder learn to recognize desirable and undesirable reproductive characteristics so that these traits can be considered when each mare's merits and handicaps are judged with respect to predetermined goals.

Subsequent chapters on infertility, breeding, pregnancy, and foaling underscore the importance of the mare's reproductive status and present important management concepts designed to enhance overall conception and foaling rates and to protect the mare's reproductive future. By describing the mare's physical needs, and by relating those needs to the complex irregularities that are potential hazards to her reproductive capacity, these chapters serve as a broodmare management "manual." Concepts and warnings presented throughout these discussions are the results of many years of research and deliberation. To bridge the literature gap between the researcher and the breeder, a presentation must first describe unavoidable technical terms and illustrate important scientific concepts. In this instance, a description of the mare's reproductive physiology and anatomy is important background information for effective utilization of future breeding management chapters.

Genital Tract

The mare's genital tract consists of the female gonads, or ovaries, and a tubular passageway suspended within the abdominal cavity by sheets of strong connective tissue. Although the ovaries and testes are corresponding reproductive organs (i.e., they are both gonads.), there are several important structural and functional differences that should be understood. The following introductory study of the mare's genital tract is included to provide important background material for future discussions on heat detection, infertility, gestation, etc.

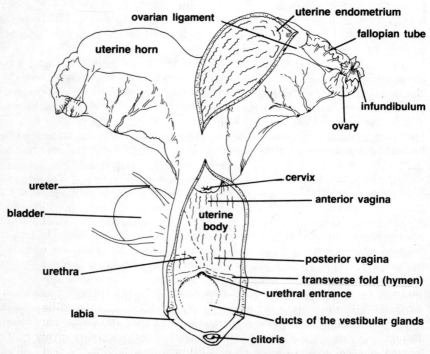

Dorsal view of the mare's reproductive tract.

OVARIES

Located at the most anterior (toward the head) point of the reproductive tract, between the last rib and the point of hip, the ovaries are responsible for production of the female gamete (egg, or ovum) and for the synthesis and release of essential reproductive hormones. Like other important organs, the ovaries are paired providing some natural insurance against complete sterility. For example, if one ovary is lost (due to disease or surgery) and the other remains sound, the drop in fertility, if any, should be minimal. Normally, there is not a decrease in fertility after loss of a single ovary.

Each ovary is kidney-shaped, measuring about 2 x 1.5 inches (5.1 x 3.8 centimeters) and weighing 1-2.5 ounces (28-71 grams). Located on the "pinched in" face of the ovary is the ovulation fossa, a unique anatomical feature of the equine ovary. The ovulation fossa is a narrow depression oriented toward the infundibulum (entrance to the tubular part of the mare's reproductive tract) that serves as the ovary's ovulation surface. In most mammals, ovulation occurs on the entire ovarian surface, but in the horse, a thin, tough layer of tissue (tunica albuginea) covers most of the organ's surface, limiting ovulation to a defined area.

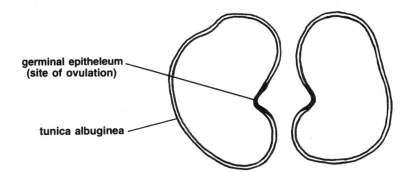

germinal epitheleum
(site of ovulation)

tunica albuginea

The ovulation fossa, located at the pinched-in face of the ovary is the site of ovulation in the mare. A membrane called the tunica albuginea covers the remainder of the ovary.

The ovary is composed of two basic cell types: interstitial and germinal. The interstitial cells provide structural support for the ovaries, while the germinal cells provide the mare's genetic contribution to her offspring. Production of the female gamete is somewhat similiar to that of the male sperm. (Refer to the discussion on sperm production within "The Stallion Reproductive System.") As in spermatogenesis, the division of cells to reduce the ovum's chromosome number is one of the most important aspects of oogenesis (i.e., egg production). Unlike the testes, however, the ovaries are supplied with partially developed ova (primary oocytes) at birth. The primary oocytes have the normal chromosome number (64), but when the mare reaches puberty, some will divide by a special chromosome reduction process (meiosis) to form secondary oocytes that contain half the normal chromosome number (32). While the primary oocyte is dividing, it is surrounded by a follicle (i.e., a fluid-filled sac). Follicles in various stages of development or degeneration can be found in an active, mature ovary at any time during the natural breeding season. (A more detailed description of follicle development, ovulation, and the natural breeding season is presented later in this chapter.) Meiosis involves two separate division processes. After the first division is completed, the follicle bursts, and the mare ovulates. The second meiotic division does not occur until the secondary oocyte (ovum) is fertilized.

Another important difference between gamete formation in the male and female is that the division-replication process of oogenesis results in the formation of only one ovum per primary oocyte. An explanation for this phenomenon requires a brief review of cell structure. As described in "The Stallion Reproductive System," the cell consists of a gel-like suspensory fluid, or cytoplasm, surrounded by cell membrane. Genetic information is distributed on several protein strands, or chromosomes, and enclosed within the cell's nucleus. (Normally, every cell within the living body contains the same set of chromosomes.) During any cell division process (e.g., for growth, repair), the chromosomes multiply and separate as the cytoplasm divides to form two separate cells, each with its own set of chromosomes. During oogenesis, however, one set of chromosomes retains most of the cytoplasm, while the other set is discarded as a polar body. (Rarely, polar bodies have been known to be fertilized, but they are not viable.) Occasionally, the first polar body may divide so that two ootids are formed.

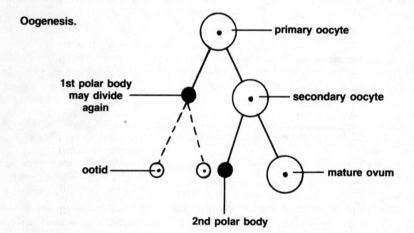

Oogenesis.

primary oocyte

1st polar body may divide again

secondary oocyte

ootid

mature ovum

2nd polar body

FALLOPIAN TUBES

After ovulation, the infundibulum (fringed structure which partially overlaps the ovulation fossa) traps the ovum and guides it to the fallopian tube, a twisting passageway that leads to its respective uterine horn. In the center of the umbrella-like infundibulum is a small opening, or ostium abdominale. Because this opening communicates with both the fallopian tube and the abdominal cavity, ova may be lost to the abdominal cavity on occasion.

With its length uncoiled, the fallopian tube (sometimes referred to as the oviduct) is approximately 8-12 inches long (20-30 centimeters). A sudden decrease in diameter forms a dividing point between the tube's ampulla and isthmus. The smaller and longer section, or isthmus, is the site of fertilization and early embryonic development (i.e., first 6 days post-conception). The ovum enters the isthmus through the ostium abdominale, while sperm cells enter the isthmus through the ostium uterinum(the fallopian entrance to the uterine horn). During estrus, the hormone estrogen stimulates the movement of tiny finger-like projections (cilia), located on the inner lining of the fallopian tubes. This beating action, coupled with the secretion of fluids within the tube, aids in the movement of sperm and ovum to their point of conception within the isthmus. If the ovum is fertilized, the resulting embryo is moved along the isthmus and into the uterine horn.

UTERUS

The uterus is a hollow, muscular organ that nourishes and protects the young embryo and developing fetus. Located between the fallopian tubes and the cervix, the uterus can be differentiated into two parts: the body and the horns. The right and left horns, each about 8 inches long and 3 inches wide (20 centimeters long and 8 centimeters wide) in an average size mare, are attached to the right and left fallopian tubes respectively. The horns then merge to form the uterine body, also about 8 inches long and 3 inches wide (20 centimeters long and 8 centimeters wide). (Measurements refer to the uterus of a healthy, non-pregnant mare.)

Both the uterine body and horns are suspended within the abdominal cavity by sheets of connective tissue, referred to as broad ligaments. These ligaments attach to the upper walls of the abdominal cavity, to the top of

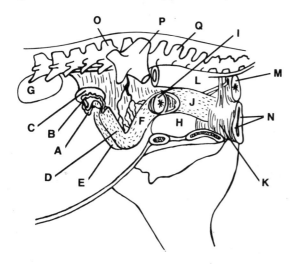

This view of the mare's reproductive tract shows the position of the uterus and other features of the tract in the abdominal cavity; (A) infundibulum, (B) fallopian tube, (C) ovary, (D) uterine horn (left), (E) uterine horn (right), (F) body of the uterus, (G) kidney, (H) bladder, (I) cervix, (J) vagina, (K) vulva, (L) rectum, (M) anus, (N) labia, (O) lumbar vertebra, (P) pelvis, (Q) sacral vertebra.

the uterine horns, and to the sides of the uterine body, supporting these important organs so that they do not contact the floor of their surrounding cavity. It should be noted for future reference that this suspensory mechanism is very important to reproductive health. Normally, the uterus is suspended in such a way that there is a natural drainage of fluids during each heat period. With age and repeated pregnancies, these ligaments gradually stretch, allowing the uterus to sag, preventing normal drainage, and predisposing the mare to reproductive infection. (Refer to "Mare Infertility.") In addition to their suspensory function, the broad ligaments contain blood vessels and nerves which supply both the uterus and the ovaries.

During a normal pregnancy, the fetus is located within the uterine body and one uterine horn. (Although the body/horn pregnancy is most common, other positions have been noted. Refer to "Abnormal Foaling.") The occupied horn becomes larger during gestation to accommodate fluid accumulation and fetal growth. Despite the organ's elastic nature, these size differences sometimes persist long after the mare foals. The changes that occur within the mare's reproductive tract during pregnancy will be analyzed more closely in "Gestation." Before these physiological changes are studied, however, the reader should have a basic understanding of uterine structure:

1. A protective outer coat, or serous membrane, surrounds the uterus.
2. Below it, a muscular layer (myometrium) contains blood vessels, nerves, and several layers of smooth muscle. During foaling, muscle contractions within the uterine and the abdominal walls result in waves of pressure which expel the fetus from the uterus.
3. The inner layer of the uterine wall, referred to as the endometrium, consists of a thin cellular (epithelial) lining, a glandular layer, and a layer of connective tissue. Uterine milk glands within the endometrium provide nourishment to the early embryo. Later, the placenta attaches to the endometrial surface, and the developing embryo gradually becomes positioned within the uterus around day 50 of pregnancy. Because failure to conceive, abortions, and reduced fetal growth are associated with infections or damage to this lining, a healthy endometrium is essential to fertility.

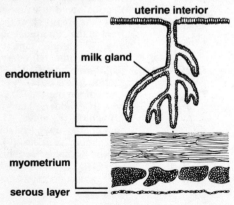

uterine interior

milk gland

endometrium

myometrium

serous layer

The uterine wall is distinguished by three major layers: the serous membrane, the myometrium, and the endometrium.

CERVIX

The uterus and vagina are separated by a muscular constriction referred to as the cervix. The uterine body tapers to form this 2-3 inch (5-7.6 centimeters) neck which projects into the vagina about 1.5 inches (4 centimeters). Consisting of a strong sphincter muscle and large mucous tissue folds, the cervix performs several important functions. For example, the cervix plays an extremely important role by protecting both the mare and the fetus from uterine infection. During pregnancy and between heat periods, the presence of the hormone progesterone causes the cervix to become pale and constricted. In addition to this cervical contraction, special secretory cells (goblet cells) within the cervix produce a thick mucus which forms a firm plug in the cervical opening during pregnancy, preventing the entrance or departure of disease-causing organisms, sperm, fluids, etc. (For information regarding the causes of uterine infections, refer to "Mare Infertility.") Hormone changes that occur during estrus or foaling cause the opposite effect: cervical dilation, an increase in the amount of blood within the cervical tissue, and relaxation of the cervical folds. This relaxation of the cervix allows the stallion's penis to deposit semen directly into the uterus during estrus and facilitates expulsion of the fetus during foaling.

VAGINA

The vagina is a passageway between the cervix and the external opening of the mare's reproductive tract. This area is 7-9 inches (18-23 centimeters) long and 4-5 inches (10-13 centimeters) in diameter. Serving as a birth canal and a reception vesicle for the stallion's penis, the vagina is capable of extreme dilation and is frequently subject to contamination and trauma. (Problems that occur in this area are examined later within "Mare Infertility.")

In the maiden mare, the vagina and vulva are usually separated by a thin fold of tissue, or hymen. Characteristics of this natural barrier vary greatly among mares. In some mares, for example, the tissue fold may be absent or partially open, while in others the tissue barrier is complete and so thick that surgical opening is required prior to breeding. The vaginal area posterior to the hymen, sometimes referred to as the vestibule, is common to both the reproductive and the urinary systems. The urethra leads from the bladder and enters the vagina through the vestibular floor. (Normally, the floor is positioned so that urine does not collect in the vagina.) Lubrication of the vagina and vulva is provided by the Bartholin's glands via special ducts that also enter through the vestibular floor.

EXTERNAL GENITALIA

The mare's reproductive tract ends at the vulva, a protective "door" positioned just below the anus. The lips of the vulva, or labia, are arranged vertically on either side of the vulval opening. Normally, these lips close tightly to prevent entry of feces and other contaminants into the reproductive tract. Proper muscle tension beneath the thin skin in the perineal area (area surrounding the anus and vulva) helps to keep the vulva and anus in vertical alignment and supports the vulval lips so that they form a firm seal. If the underlying musculature is damaged or if the anal area relaxes with age, the external genitalia may sag inward, preventing proper closure of the labia and predisposing the mare to reproductive infection. Even the passage of air between these lips can cause significant reproductive problems, as illustrated later in this text.

Between the vulval lips, located on the lower "V" of the vulva, lies the anatomical homologue of the male penis. This structure, referred to as the glans clitoris, is composed of erectile tissue. Upon sexual stimulation, contractions of the vulval and vestibular muscles cause the clitoris to protrude between the labia, resulting in the so-called "winking" associated with mares in estrus.

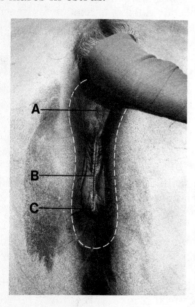

The external genitalia located below (A) the anus consist of (B) the labia on either side of the vulval opening, (C) the clitoris, and the perineal area (borders are defined by the broken line).

The mare's udder is located in the inguinal region and lies between her hindlegs. The anatomy of this important mammary structure is unique in the mare and will be described in detail in "Lactation."

Estrous Cycle

The mare's reproductive processes are further differentiated from the stallion's by her periodic production of gametes as compared to his continuous production of sperm. These female reproductive cycles, or estrous cycles, have been the subject of much research and discussion. Recently, the importance of understanding the equine estrous cycle has been highlighted by emphasis on improving conception rates through the use of such management tools as artificial insemination and heat synchronization. Because references to various characteristics of the mare's estrous cycle are made throughout this text, the following background information on ovarian cycles is presented. In general, this discussion explains how a delicate balance of hormones brings about special changes in the mare's behavior and reproductive tract—changes that allow conception and, ultimately, propagation of the species.

During the natural breeding season (e.g., May through August in most of the Northern Hemisphere), the production of gametes and the corresponding behavioral and physical changes usually occur in 21-23 day cycles. However, there are many factors that influence these estrous cycles, making possible a number of variations. (Abnormal and, for the most part, undesirable variations in the female cycle are examined under "Mare Infertility.")

FOLLICULAR STAGE

As mentioned earlier, the cellular changes that produce the ovum occur within a large, blister-like follicle. As the follicle develops, it secretes significant amounts of estrogen, a hormone that distinguishes this so-called follicular stage from the luteal stage of the estrous cycle. Although there is some confusion over the use of the terms *estrus* and *estrous*, in this text the two words are used in the following manner. The term *estrus* is used as a noun to identify only that portion of the mare's reproductive cycle during which she is sexually receptive to the stallion, or in heat. The term *estrous* is used as an adjective describing estrus. For instance, estrous behavior is the characteristic pattern of behavior shown by a mare in heat (estrus). In exception to the rule, however, *estrous cycle* is the term used to identify the 21-23 day reproductive cycle.

The early follicular stage is sometimes referred to as proestrus to distinguish it from the "classical estrus" more commonly associated with follicular development. Proestrus is a short period (about 2 days in most mares), characterized by a steady rise in estrogen levels and a gradual change in

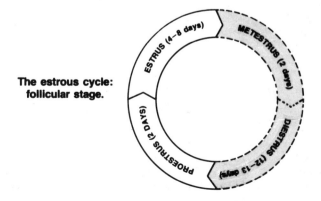

The estrous cycle: follicular stage.

the mare's attitude toward the teaser. Occasionally, mares show signs of receptivity during proestrus, but usually these signs are not evident until the onset of true estrus. (Note: The hormone changes that help to define each ovarian stage are described later in this chapter.)

The transition from proestrus to estrus is very gradual; the mare becomes more and more receptive to the stallion and, as she enters estrus, she normally shows a strong positive reaction (i.e., acceptance) to teasing. Estrogen produced by the maturing follicle is responsible for the behavioral and tissue changes that occur during estrus. For example, the mare is willing to accept (or is interested in) the stallion; she exhibits "winking" of the labia, squatting, and/or urinating; the vulva and cervix relax; cervical mucus is secreted; the blood supply to the uterus is maximized; the uterus becomes edematous (i.e., swollen with fluids); and secretions by the uterine glands increase. Although the strength of these responses varies greatly among mares, a characteristic pattern of increasing receptivity at the onset of estrus is often seen. (Refer to "Heat Detection.") The period of receptivity usually lasts about 4-8 days but varies with the individual, with the season, and with other factors such as nutrition and disease. (Refer to "Mare Infertility.")

During estrus, several follicles may develop, reaching a diameter of about 20 mm. By a special hormone-controlled mechanism, only one or sometimes two follicles reach maturity (35-55 mm in diameter). As the maturing follicle enlarges, its ovum gradually separates from the surrounding granulosa cells. Subsequently, a follicular cavity forms and fills with fluid, and the ovum is left resting on a mound of granulosa cells, referred to as the cumulus oophorus. The cells of the cumulus oophorus enclose the ovum and its two other protective layers: the corona radiata and the zona pellucida. (The sperm head is equipped with special enzymes that break down these protective layers, allowing penetration and fertilization of the ovum.) The wall of the developing follicle can be differentiated into two cellular layers: the theca interna and the theca externa. Researchers believe that the glandular theca interna produces the estrogen associated with behavioral and physical changes during estrus.

Cross section of a Graafian follicle.

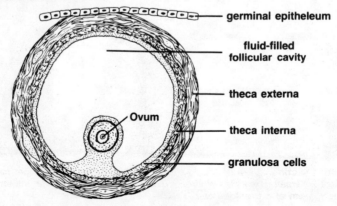

germinal epitheleum

fluid-filled
follicular cavity

theca externa

Ovum

theca interna

granulosa cells

Although the dividing line between the follicular and luteal phases of the estrous cycle is difficult to pinpoint, ovulation marks the beginning of this transition. Ovulation is a gradual process by which the ovum is expelled from a mature follicle (Graafian follicle) into the tubular portion of the mare's reproductive tract. The follicle ripens and prepares for ovulation while estrus is still in progress. It has been reported that 80 percent of all ovulations occur 24-48 hours before the end of estrus, with 75 percent occurring between 4:00 PM and 8:00 AM. Because ovulation is not marked by any obvious changes in the mare's physical or behavioral status, it is impossible to determine when the mare ovulates unless the ovary is palpated from the rectum. (Softening of the preovulatory follicle just prior to ovulation can be felt by palpation. However, this change is not always detected. Refer to "Heat Detection.")

Scientists have reported that, contrary to popular belief, the follicle does not necessarily "burst" and eject the ovum. Rather, they suggest that follicular degeneration allows the gradual release of accummulated fluids, thereby washing the ovum out of the follicle and into the infundibulum. Studies also indicate that concurrent maturation of two follicles and consequent release of two ova is not unusual in the mare. In fact, the occurrence of double ovulations has been estimated at as high as 25 percent. (i.e., One out of every four ovulations involves two ova.) Other extremes, such as triple ovulations and estrus without ovulations, have also been reported.

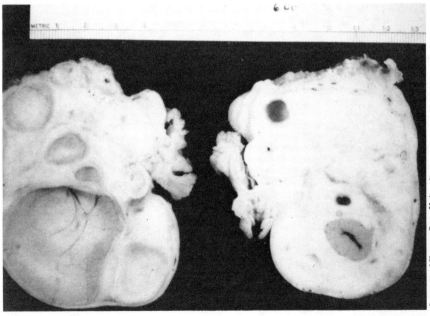

Photo by William C. Bergin

Courtesy of Kansas State University
Department of Surgery and Medicine

The cross-sectional view of the ovary on the left shows the follicular cavities of several developing Graafian follicles while the ovary on the right contains the small remnant of a corpus luteum of a previous estrus.

LUTEAL STAGE

About 14-30 hours after ovulation, the collapsed follicle fills with blood and forms a large clot (corpus hemorrhagicum). This blood clot, which is about 2/3 the size of the original follicle, forms the structural foundation for a very important hormone-producing body, the corpus luteum. The formation and degeneration of this temporary hormone source identify the luteal stage of the estrous cycle, a period that lasts approximately 15 days. (The reader should note that all time periods designated for certain stages of the estrous cycle are averages for normal, cycling mares. Extreme variation between mares is not at all unusual.)

The cells of the theca interna, described earlier, degenerate very quickly after ovulation. Within 10 hours post-ovulation, the granulosa cells begin to multiply rapidly. By a process known as luteinization, these proliferating granulosa cells are gradually transformed into hormone-producing luteal cells, forming a yellow mass of tissue. This yellow body, referred to as the corpus luteum, is the source of large amounts of progesterone, a hormone that distinguishes the luteal stage from the follicular stage of the estrous cycle. In response to progesterone, the mare rejects the teaser, the cervix becomes pale and constricted, and the uterus becomes firm, tubular, and less secretory.

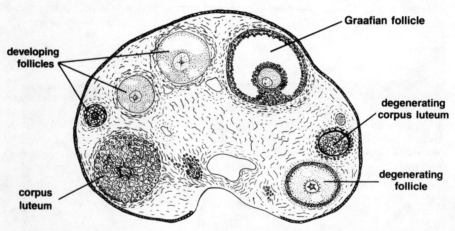

This cross-sectional view of an ovary shows a composite of follicular activity throughout the estrous cycle.

Like the follicular stage, the luteal stage can be differentiated into two parts. The early luteal period, or metestrus, is characterized by the formation of the corpus luteum and by increasing levels of progesterone in the blood. This period is not well-defined, but lasts approximately 2 days. Diestrus (the next luteal phase) is a 12-13 day period, characterized by a fully functional corpus luteum. If the mare's ovum is fertilized and if a viable embryo results, the uterus recognizes the pregnancy and forces the corpus luteum to remain functional · thereby supporting the pregnancy with adequate amounts of progesterone) until the placenta begins progesterone production. (Refer to "Gestation".) If the mare is not pregnant, the uterus produces and secretes a chemical agent, prostaglandin ($PGF_{2\alpha}$), which causes the luteal cells to stop producing progesterone and causes the corpus luteum to regress. As a result, progesterone production decreases, and diestrus is brought to a close as the ovaries continue to cycle. Although the uterus does not produce significant amounts of prostaglandin until almost two weeks after ovulation, the corpus luteum is sensitive to prostaglandin as early as 4 days postovulation. (Sensitivity is maximal at 6 days postovulation.) This early sensitivity allows estrous cycles to be hastened and/or synchronized by injecting prostaglandin 6 days postovulation. (Refer to "Broodmare Management.")

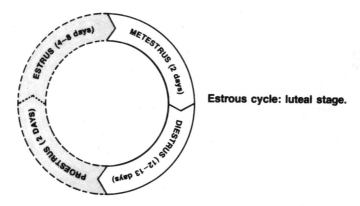

Estrous cycle: luteal stage.

Although the luteal phase is described as the period of corpus luteum formation, activity, and regression, it may also be characterized by follicular activity. In fact, diestrus ovulations are not unusual. However, any physical or behavioral effects that might be caused by estrogen from these maturing follicles are overridden by the high progesterone levels. In other words, progesterone produced by an active corpus luteum seems to suppress estrus.

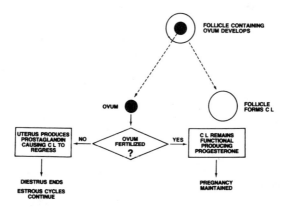

Negative feedback effect of prostaglandin on an active corpus luteum.

SEASONAL VARIATION

Although a few mares cycle throughout the year, at least 70 percent fail to show signs of estrus during fall and winter. This period of seasonal anestrus is a normal biological response to changes in the mare's environment. Scientists believe that seasonal anestrus is the result of decreasing daylight length. (Changes in environmental temperature and pasture quality may have a secondary influence on cyclic behavior, but research in this area is not yet conclusive.) A complex receptor mechanism relays this information to the brain, where hormone synthesis and release are altered accordingly. As mentioned in the chapter, "The Stallion Reproductive System," seasonal variation in reproductive capacity seems to be a protective process developed during the horse's evolution to insure that foaling occurred when weather was favorable for the newborn foal and when nutrition was plentiful for the lactating mare.

Associated with the shortest day of the year (e.g., December 21 in the Northern Hemisphere), "true" anestrus is characterized by complete absence of ovarian activity. (Any "absence of observable heat," regardless of cause, is referred to as anestrus.) As the mare enters and leaves this phase of reproductive inactivity, she usually experiences a transitional period of erratic estrous behavior: signs of heat without ovulation, split heat periods, prolonged estrus without ovulation, etc. The highest incidence of multiple ovulations also occurs during this period. Some mares exhibit these confusing patterns throughout the fall and winter, returning to normal cycles during the natural breeding season. This natural breeding season is a period of maximum fertility reached during May, June, and July (in most of the Northern Hemisphere) when estrous cycles are shorter and when ovulations are regular. In the Southern Hemisphere, this season of highest fertility occurs from November to January. For a more detailed description of anestrus and transitional receptivity, refer to the discussion on physiological aspects of infertility in "Mare Infertility."

CONTROLLING MECHANISMS

The intricate relationship between hormone levels and reproductive function is an extremely important concept to understand when evaluating and managing the reproductive status of breeding stock. Within the past 10 years, the available information on equine reproductive hormones has increased notably. However, many questions have yet to be answered, and many concepts are still based on supposition. For this reason, the following review of reproductive hormones and their various functions may seem abstract and incomplete. The reader should remember, however, that this review is presented during a period of extensive hormone research, a progressive and continuous period of investigation that gradually reveals the complex systems that control numerous physiological processes. As a review of the state of the art, the following discussion is provided to help both the interested student and the serious breeder correlate the mare's cyclic behavioral and physical changes with cyclic changes in hormone status.

In response to certain environmental stimuli, the hypothalamus sends gonadotropin releasing hormone to the anterior pituitary gland, located at the base of the brain. Consequently, the anterior pituitary releases follicle stimulating hormone (FSH) and luteinizing hormone (LH), the so-called gonadotropins that play very important roles within the ovarian cycle. (The reader should note that researchers believe two gonadotropin releasing hormones are secreted by the hypothalamus, but only one hormone has been identified, and explanations concerning the independently fluctuating levels of FSH and LH are not clear.)

The importance of both FSH and LH to a normal ovulatory cycle is well-established, but the exact functions of each are not so clear cut. It seems that the release of FSH at about 10 day intervals causes waves of follicular growth throughout each estrous cycle (i.e., except during "true" anestrus). These surges of FSH activity occur once during late estrus to early diestrus and again during mid-diestrus. It is believed that the mid-diestrus surge stimulates and prepares the follicle for subsequent production of a mature ovum.

LH levels also vary throughout the cycle, but a single outstanding surge of the hormone from about day 16 of diestrus to 1-2 days postovulation is believed to be responsible for the maturation of one or two follicles, for the increasing levels of estrogen (a hormone that stimulates the physical and behavioral characteristics associated with the mare's heat), and for the ultimate release of ova into the tubular portion of the mare's reproductive tract. (LH seems to have no effect on immature follicles.) The importance of this surge is illustrated by the failure of most diestrus follicles (i.e., follicles that are not stimulated by a prolonged increase in LH levels) to mature and ovulate. The formation and regression of the corpus luteum is also important to this surge of LH activity, since progesterone production (from the corpus luteum) seems to be inversely related to LH levels. In other words, as progesterone increases, LH decreases, and vice versa. It has been suggested that progesterone inhibits the LH surge and counteracts the effects of estrogen produced by diestrus follicles. As the corpus luteum regresses in response to prostaglandin release from the uterus, progesterone levels fall and LH levels increase, allowing the maturation of another follicle and the continuation of another ovarian cycle.

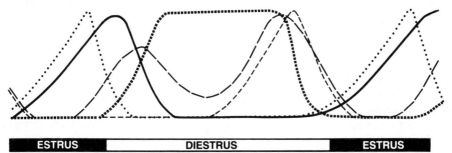

This graph shows the relative levels of five hormones involved in the estrous cycle: prostaglandin (- - -), luteinizing hormone (——), follicle stimulating hormone (— —), progesterone (••••••••), and estrogen (•••••).

Puberty

As the mare enters her second year, the hypothalamus and the anterior pituitary gland gradually become more active, resulting in an increase in gonadotropin production. During this period of adolescence, the ovaries become more sensitive to gonadotropin, and hormone levels prepare for the first ovulation. Because the mare does not have an active corpus luteum at this time, the proper interchange of reproductive hormones (i.e., decreasing progesterone and increasing estrogen) necessary for the expression of estrus is not present. For this reason, at least one ovulation must occur to prepare the nervous and endocrine (hormone) systems for estrous behavior. Normally, the first ovulation occurs at about 18 months of age, but extreme variation among mares is not unusual. As a case in point, the filly's date of birth has an important effect on her age at puberty. A filly born in January that cycles normally at about 18 months of age begins to show estrus in July or August. Alternately, a filly born in the summer should cycle in November, December, or January, but in the Northern Hemisphere the young mare's sexual receptivity would probably be delayed until the following spring, due to seasonal anestrus. Regardless of age or season, mares experiencing their first few cycles are notoriously erratic, making it difficult to describe a general pattern for puberty in mares.

It is extremely important to note that the onset of regular estrous cycles and the mare's ability to carry a foal to term successfully do not necessarily coincide. Because the mare's physical maturity is essential to her health during pregnancy and postfoaling and to the health of her newborn foal, it is best not to breed her before 3-4 years of age. Although a mare can be bred at 2 years of age (so that she foals at 3 years) without problem, the breeder should understand the physical stress that may result. Even when the mare delivers a normal, healthy foal, mineral drainage from her bones during gestation may have serious and permanent effects on the young dam's bone structure. When coupled with energy deficiencies (caused by the mare's generous support of her developing fetus), mineral drainage can prevent the mare from attaining her potential size. (The reader should note that the mare will sacrifice her own nutrition and health for that of her foal's only to a point where her own death is imminent. At that point, the mare will abort.)

Reproductive Lifespan

Theoretically, the mare continues to cycle and is capable of producing sound foals as she ages. However, damage to the reproductive tract, caused by problems such as difficult foaling or genital infection, can render the older mare (and, in some instances, the younger mare) incapable of conceiving, supporting, and/or delivering healthy offspring. This is not to suggest, however, that all aged mares should be culled from the farm's breeding program. Rather, the breeder should be aware of age-related reproductive problems and should understand that the future reproductive capacity of any mare depends on her reproductive history. A mare with good genital conformation and a history of proper breeding management may produce successfully for many years, while another mare may be significantly limited by the condition of her reproductive tract. (In the future, it may be possible to keep a valuable, but reproductively limited, mare in production through the use of embryo transfer. Refer to "Breeding Methods and Procedures.")

Discussions on important aspects of broodmare management and age-related infertility are presented later in this text. At this time, several exemplary factors that may affect the mare's reproductive lifespan are presented to emphasize the complexity of reproductive function and failure in mares.

- Repeated infections or foaling problems may render the uterus incapable of supporting a fetus.
- With age, relaxation of perineal muscles may prevent proper closure of the labia and allow contamination of the reproductive tract.
- With age and repeated pregnancies, the broad ligaments may lose their elasticity, allowing the uterus to sag and preventing normal drainage from the reproductive tract.
- It has also been found that aged mares have a higher incidence of twin ovulations than do younger mares, contributing to the reduction in conception and foaling rates that often accompanies age.

These and other fertility problems that may limit the mare's reproductive lifespan are examined more closely in "Mare Infertility."

9

MARE INFERTILITY

The mare's overall health, the condition of her reproductive organs, and her general attitude have an important influence on her reproductive capabilities. Although the stallion plays an important role in determining the quality and quantity of each year's foal crop, the mare is far more instrumental in determining the individual foal's health and tractability. In comparison to the stallion, the mare's complex reproductive role subjects her to greater physical stress and, consequently, to a longer list of potential fertility problems. Minimizing these problems through proper management is one of the most important concerns of equine production. Many regard broodmare management as a key factor in determining the success or failure of the breeding operation.

Management of the mare to insure that she reaches her optimum inherent fertility potential is an extensive subject that is examined closely in "Broodmare Management." Before these practices are considered, however, the breeder should have a basic knowledge of the factors that contribute to reproductive inefficiency and failure in the mare. Infertility, sterility, and failure to successfully carry a fetus to term are the visible results of a number of possible reproductive problems. In many instances, these problems are complex, interrelated, and disguised by the healthy outward appearance of the mare. The breeder that is unfamiliar with the causes of infertility or that does not understand the potential complications of a single reproductive abnormality may not seek veterinary assistance until damage to the mare's reproductive tract is irreversible. Early recognition of a reproductive problem followed by prompt diagnosis and treatment can be extremely important to the mare's future breeding soundness. The breeder that quickly recognizes changes in a mare's fertility status and that responds to those changes without delay may prevent substantial veterinary expense, breeding schedule delays, and subsequent economic loss.

An ability to recognize reproductive unsoundness or physical conditions that predispose the mare to infertility is very important when selecting breeding stock. Although a veterinarian cannot guarantee that a particular mare will produce successfully, a thorough preseason or prepurchase reproductive examination is a valuable precaution. The following discussion on reproductive problems in the mare is presented to increase the breeder's understanding of what the veterinarian looks for during a reproductive examination and to improve the breeder's ability to recognize fertility problems. The basic causes of infertility are divided into four categories: management, physiological, pathological, and anatomical aspects.

Management

It has been suggested that one of the best ways to improve conception rates is to eliminate man's interference with the breeding process and allow a more natural state through closed pasture breeding. (i.e., The herd is not open for exchanges or additions.) Increased fertility in a natural breeding situation may be the result of increased physical interaction between the mare and stallion. The pastured stallion is usually more persistent than man in his efforts to detect a mare in heat and often covers a mare several times during a single estrus period.

Unfortunately, pasture breeding is not practical or desirable on most breeding operations. When a large number of mares are booked to one stallion during a limited season, the stallion must be used as efficiently as possible. For this reason, hand breeding or artificial insemination are far more feasible than pasture breeding in most cases. More importantly, the transport of mares between farms and the resulting threat of contagious venereal disease demands that special precautions be taken to limit contamination and the spread of disease between animals. Pasture breeding may also result in serious injury to the stallion or mare and cause substantial economic loss if fertility or potency are affected.

Because controlled breeding is emphasized today, changes in the mare's lifestyle have significantly affected her reproductive potential. The serious breeder must understand why certain management practices limit the mare's inherent reproductive capacity and, as caretaker and decision-maker, must recognize his responsibility to protect the mare's health and reproductive soundness.

IMPOSED BREEDING SEASON

The evolutionary process (i.e., survival of the fittest) emphasized optimum reproductive efficiency by limiting the breeding season to a period of summer months so that the mare foaled when the weather was warm and the grass was nutritious. This seasonal control over fertility is still an important part of the equine reproductive process. Many breed registries have de-emphasized this natural breeding season by designating a universal birth date on January 1, so that all foals born in a certain year automatically become yearlings on January 1 of the following year. This designation, along with the competitive spirit of the horse industry, places more emphasis on early birth dates for larger, stronger two-year-olds and less emphasis on the importance of the physiological breeding season. In order to insure competitive foal crops, the imposed breeding season ends just as the mare's natural breeding potential reaches a peak. Also, the imposed breeding season is much shorter than the physiological season, making it even more difficult to overcome problems associated with breeding during a period of natural subfertility.

SELECTION

In addition to the fertility problems caused by an imposed breeding season, reduced fertility potential has resulted from years of selection on the basis of conformation and athletic ability rather than reproductive capacity. (Fertility has been described by geneticists as an inherited characteristic, one that is controlled by many different genes affecting many different physical processes.) In the wild, subfertile mares were bred, but due to their subfertility, their inherent reproductive problems were not propagated extensively. Fertile mares, on the other hand, produced a large percentage of the next generation's breeding stock, helping to insure survival of the species. Today, artificial breeding methods have become effective tools for getting subfertile mares in foal. As a result, the occurrence of inherited reproductive problems has been encouraged. Although fertility is an economically important aspect of the equine industry, conformation, speed, bloodlines, and other elements of marketability are far more important to most breeders.

NUTRITION

Since the mare's domestication, man's influence over her nutrition, exercise, and exposure to disease has also affected her reproductive status. Underfeeding during early development may delay puberty, while starvation may permanently damage normal development. Insufficient nu-

This starving mare is unlikely to conceive and carry a fetus and may even fail to show normal estrous cycles.

trients in the adult may inhibit follicular growth, alter normal reproductive behavior patterns and, possibly, result in deep anestrus. Insufficient nutrition may be caused by an inadequate or poorly balanced diet, parasite infestation, loss of appetite due to a systemic illness, lesions in the mouth, malalignment of the teeth, etc. The breeder's responsibilities include correcting parasite problems caused by confinement, protecting each individual from infections that spread rapidly through heavily populated areas, and seeing that all the mares' teeth are worn evenly.

Although a proper balance of vitamins and minerals is essential to the mare's overall health and ability to produce a live foal, the exact roles that these nutrients play in the reproductive process are not clearly understood. Vitamin B deficiencies may be indirectly related to lowered reproductive ability, resulting in signs similar to those caused by starvation. Absence of vitamin A (supplied naturally by leafy green plants) severely limits reproduction because it is essential to the health of the mucosal lining of the reproductive tract. Reports indicate that vitamin E deficiencies may limit the regeneration of reproductive tissue, but its exact reproductive function is unknown. Because vitamin D affects the body's utilization of calcium and phosphorus (two important minerals that affect the overall health of the mare and her foal), it is also essential to efficient reproduction. Workers have also suggested that manganese deficiencies are related to malfunctions of the ovaries in some species and that inadequate iodine can reduce fertility if it supresses thyroid function. (Suppressed thyroid function reduces gonadotropin secretions.) A severe salt deficiency results in anestrus and loss of body weight in cattle and may have similar effects on horses. The Appendix contains nutrition tables for easy reference. In addition, a more extensive discussion of nutritional requirements can be found in the text FEEDING TO WIN.

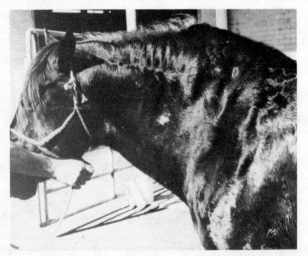

Hypothyroidism resulting from an iodine deficiency can depress fertility. The pronounced, cresty appearance of this mare's neck is characteristic of hypothyroidism.

FAILURE TO DETECT ESTRUS

Failure to detect estrus or failure to breed the mare within 48 hours prior to ovulation is an important cause of lowered conception rates. Overlooking the presence of estrus may be caused by:

1. failure to understand the mare's estrous cycle;
2. teasing during the late winter/early spring transitional period when the mare's behavior is usually erratic and not a reliable indicator of impending ovulation;
3. poor teasing records;
4. poor teasing techniques;
5. failure to detect a mare's fear or dislike of the teaser;
6. failure of the mare to show heat due to psychological reasons.

TIME OF BREEDING

The best time to breed the mare is just prior to ovulation. Normally, the mare ovulates 24-48 hours prior to the end of estrus. However, ovulation can vary anywhere from 48 hours before the end of estrus to 24 hours after. This depends, of course, on the individual mare and on whether she is being observed during the natural breeding season or during her transitional period of receptivity. During the transition from winter anestrus to the natural breeding season, she may have regular ovulatory cycles, extended periods of estrus without ovulation, or split estrus.

Once the mare begins to cycle regularly, good records on her behavior during past seasons help the teasing manager predict the length of estrus and breed the mare accordingly. If records are not available, or if the mare's behavior patterns are extremely unusual, rectal palpation may be required to determine the optimum breeding time. (Refer to "Heat Detection.") If the mare is not palpated and if the stallion is not being overused, the mare might be bred more than once (e.g., third day of estrus and every other day thereafter) to enhance the chances of conception.

MISUSE OF HORMONES AND DRUGS

Another management aspect that can lower conception rates is the extensive use of hormones and drugs. Mares that are raced or shown extensively may be subjected to a variety of drugs—drugs that inhibit estrus, improve muscling, reduce nervousness, etc. The effects of many of these drugs on the reproductive processes are still unclear, but it seems that the use of anabolic steroids to improve muscling can also cause masculinization and subsequent fertility problems in the mare. Similarly, drugs that mask or prevent the mare's estrus period can interfere with normal cycles when normal cycles are desired. It should be mentioned that improper use of hormones to stimulate or speed up the estrous cycle or to treat specific fertility problems, can cause temporary and sometimes even permanent damage to the normal reproductive processes. Until more is known about the mare's endocrine system and until more research data on the long-term effects of administered drugs or hormones is collected, use of such agents (e.g., progesterone, PMSG, estrogen, anabolic steroids) should be limited to that employed by, or under the direction of, a veterinarian.

The best treatment for most drug-related fertility problems in the mare is time. A mare that comes directly from the track or the show circuit may require several months to rest and "let down." She is given time to lose her lean, finely tuned performance condition or, in the case of some show mares, to lose excess weight. During this period, drugs are also metabolized (i.e., broken down) and eliminated from her system. After this period, the mare should be evaluated for fertility potential. (Refer to "Broodmare Management.")

MISUSE OF ARTIFICIAL LIGHTING

In "Broodmare Management," the use of artificial lights to hasten the mare's physiological breeding season is described. It should be mentioned in passing, however, that improper use of such light programs can actually be detrimental to fertility. For example, providing the mare with a special light program at home and then shipping her to a breeding farm where lights are not employed may cause the mare to have erratic cycles or revert to anestrus. Similarly, mares that are to be shipped to breeding farms that use lights to simulate daylight lengths of the natural breeding season conceive more readily if they are on a similar schedule at home.

BREEDING DURING FOAL HEAT

Breeding the mare during her first estrus period after foaling (i.e., foal heat) is a controversial practice, one that is performed to minimize the reproductive year but that, in some instances, lowers overall conception and foaling rates. While some studies have shown that reduced conception rates and increased abortions occur when mares are bred on foal heat, many sources suggest that mares without postfoaling problems can be bred safely and efficiently during foal heat. Prostaglandin is sometimes administered to shorten the first estrous cycle after foal heat in order to avoid breeding during foal heat. For a discussion of this therapy, refer to "Broodmare Management."

The single most important factor that determines whether the mare can be bred successfully during her foal heat is the condition of her reproductive tract after foaling. For example, if she has experienced a difficult foaling, if she retained the placenta for an extended period, or if involution of her uterus is delayed, the reproductive tract may not be capable of supporting a new pregnancy. On the other hand, if the mare passes a 7-day post-foaling examination, her chances of conceiving and maintaining pregnancy are the same as they would be if breeding were postponed until the next cycle. In other words, indiscriminate breeding during foal heat is a management practice that can reduce overall conception rates.

Physiology

As described earlier, the mare's estrous cycle is a complex physiological process, controlled by a delicate balance of reproductive hormones. Irregularities of this cycle have probably frustrated even the most experienced and competent breeders. To deal confidently with these problems, the breeder must be able to recognize different types of estrous irregularities and, in many cases, should be able to identify their respective causes and use appropriate management techniques to get the mare bred. Problems that are examined in this discussion include absence of estrus, prolonged estrus, split estrus, aggressive stallion-like behavior, failure to ovulate, failure to luteinize, and multiple ovulations.

ANESTRUS

Throughout this discussion, the term *anestrus* is used to describe any absence of observable heat. (Although, strictly defined, the term refers to a complete absence of ovarian activity associated with the winter months.) Because these periods of apparent sexual quiescence stem from a number of possible causes, ovarian activity may or may not be present. If ovulation does occur in the anestrus mare, chances of detecting the event are remote due to the mare's failure to exhibit signs of heat. Ovulation could possibly be detected by rectal palpation, and the mare force-bred or inseminated. When analyzing apparent anestrus in the mare, the breeder should consider the following possibilities.

PREGNANCY

A natural cause of anestrus in mares, and the first cause to consider, is pregnancy. Unlike the cycling mare, the pregnant mare usually has a tight cervix and a dry vaginal vault. She may or may not have ovarian activity. Treating the pregnant mare as though she were anestrus (e.g., allowing extensive rectal palpation, saline infusions of the uterus, hormone injections, etc.) will probably cause early embryonic death or abortion. For this reason, the possibility of pregnancy must be the first consideration when dealing with anestrus mares.

SEASONAL ANESTRUS

True anestrus is a natural response to decreasing daylight length and corresponding changes in climate and nutrition. It is a physiological phenomenon which developed during the horse's evolution to limit the foaling season, thereby protecting the newborn foal from extremely cold weather and providing extra nourishment for the mare when she needed it most. To insure this natural foaling season, most mares have no ovarian activity during the winter months. (i.e., Follicles do not develop, estrogen levels do not surge, and signs of estrus are not exhibited.) Many mares enter deep winter anestrus, characterized by complete absence of ovarian activity. The only cure for this normal nonovulatory period is time. Some mares have follicular activity during winter anestrus, but these developing follicles usually regress prior to ovulation. From all outward appearances, these mares are in true anestrus. Occasionally, a mare continues the erratic patterns of her transitional period throughout the winter. (The mare's transitional period is the natural period of erratic receptivity and reduced fertility from late fall to early spring. Refer to "The Mare Reproductive System.") Mares in transitional estrus and mares that have some follicular activity during winter anestrus may respond to certain stimuli (e.g., lights) and return to fairly regular cycles with ovulation.

PSYCHOLOGICAL ANESTRUS

Anestrus during the man-made breeding season is often caused by the seasonal patterns just described, but failure to show estrus during the natural physiological breeding season may be the result of more complex problems. Psychological anestrus, or "silent heat," is common in nervous mares, shy mares, and mares with foals. Researchers have suggested that many cases of psychological anestrus are caused by failure of the sex center within the hypothalamus to respond to normal signals. (Feedback signals are provided by levels of certain hormones in the blood.) Occasionally, a mare in estrus refuses to respond to a particular stallion but shows readily to another teaser. Some mares show only to other mares or in the privacy of their own stalls. This dislike of a certain stallion or fear of stallions in general, is a problem that requires diplomatic handling and special teasing techniques. The nervous mare that is concerned about her foal and that, consequently, refuses to show to the stallion must also be managed and observed carefully during teasing. (Refer to "Heat Detection.") Psychological anestrus in foaling mares should be distinguished from lactational anestrus, a period of complete anestrus that sometimes affects mares before or after their foal heat. Recent reports indicate that this form of anestrus is not caused by lactation, as suggested by the name, but by prolonged presence of the corpus luteum—a problem that is examined later in this chapter.

CHROMOSOME ERRORS

Anestrus may also result from defects in the mare's genetic constitution. The protein rods that carry each individual's genetic information within every living cell are referred to as chromosomes. Genetic messages on each of these chromosomes direct the development and maintenance of all life-sustaining and life-propagating systems. (Refer to the discussion on basic genetic concepts within the Appendix.) Each plant and animal species is characterized by a certain type and number of chromosomes, a trait that can now be studied by electron microscope photography. Both the mare and the stallion, for example, have 64 chromosomes (32 pairs). One special pair of chromosomes is associated with sex determination: the presence of two "X" chromosomes within the embyo's genetic make-up results in the development of a female fetus, while the presence of an "X" and a "Y" chromosome stimulates the development of male characteristics. When cell division abnormalities result in an embryo with only one "X" chromosome per cell, the resulting "female" is characterized by certain physical defects. These deviations include small stature, flaccid uterus, small inactive ovaries, complete absence of ovarian activity and, consequently, anestrus. (During winter anestrus, the immature mare is normally characterized by small inactive ovaries and a flaccid uterus.) Mares that are missing an "X" chromosome in every cell are referred to as 63XO aneuploid mares. (The O designates the missing "X" chromosome.) Mares that have normal "XX" chromosome pairs in some cells and missing "X" chromosomes in others are referred to as 64XX/63XO mosaics. These mares often have erratic estrous cycles without ovulation. Neither 63XO aneuploidy or 64XX/63XO mosaicism can be accurately diagnosed without a chromosome study, especially in young mares.

PERSISTENT CORPUS LUTEUM

One of the most important causes of anestrus during the physiological breeding season is prolonged functioning of the corpus luteum. As mentioned earlier, the corpus luteum is the progesterone-producing structure that forms in the follicular cavity after ovulation. This structure develops rapidly and has a normal life span of about 14 days (in the nonpregnant mare). If pregnancy is not established, the uterus normally produces and releases prostaglandin, which causes the corpus luteum to regress and blood progesterone concentration to decrease, allowing the mare to express estrus. In other words, the formation and regression of the corpora lutea correspond to the normal rise and fall of progesterone levels which, in turn, correspond to the normal pattern of the mare's estrous cycle.

If the uterus does not produce or release adequate amounts of prostaglandin or if the corpus luteum fails to respond to prostaglandin, the corpus luteum may persist for several months. This prolongation results in an absence of estrus behavior, even though follicles may be developing and ovulating. Although many cases of prolonged corpora lutea are spontaneous (i.e., related to no known cause), several physiological conditions have been associated with failure of the progesterone-producing structure to regress. For example, chronic uterine infections, accompanied by pus and fluid in the uterus, may destroy parts of the uterine lining and, consequently, interfere with the production and release of prostaglandin. The infected mare may appear normal except for slightly prolonged periods of non-receptivity between normal periods of estrus. Extensive damage to the uterine lining caused by the presence of excessive pus and fluid causes a substantial reduction in prostaglandin levels, possibly resulting in deep anestrus.

Another known cause of the persistent corpus luteum is embryonic death. After day 36 of pregnancy, endometrial cups form within the uterus and produce pregnant mare serum gonadotropin, a hormone that is believed to stimulate the development of numerous corpora lutea. (Refer to "Fetal Development.") If the embryo is resorbed or lost **before** day 36 of pregnancy, the original corpus luteum gradually regresses, and the mare returns to estrus. A mare that loses her pregnancy **after** day 36 or 38 will not usually return to heat until the endometrial cups cease to function, an event that occurs between days 120 and 130 postconception. (Early embryonic death is actually a form of abortion but, because it is often mistaken for failure to conceive, the condition is sometimes described as a form of infertility. Refer to "Abortion.")

Because follicular activity sometimes continues in the presence of a persistent corpus luteum, maturing follicles may be detected by rectal palpation. Although it is possible to inseminate these anestrus mares prior to (or immediately after) ovulation, this procedure is not always acceptable, and the chances of any resulting embryo surviving are minimal. A safer and more sensible approach involves correction of the persistent corpus luteum through prostaglandin therapy. When used properly, this luteolytic agent and its synthetic analogues produce very consistent results (90 percent effective) and promise to be extremely helpful in inducing luteal regression. Prostaglandin can be infused into the uterus or injected into the muscle or under the skin. It should be emphasized that prostaglandin is effective only when a mature corpus luteum is present (e.g., after day 5 postovulation). The agent ends diestrus by causing the corpus luteum to regress, not by inducing ovulation. Treatments that stimulate uterine production of prostaglandin (e.g., saline infusions) are unsuccessful when the uterus is physically unable to produce or release the agent. For this reason, prostaglandin treatment has been far more successful in treating prolongation of the corpus luteum.

One important point to remember about prostaglandin treatment is that the stage of follicular development at the time of treatment determines when the mare will ovulate. For example, mares with follicles that are smaller than 35mm (diameter) usually ovulate about 7 days posttreatment. Mares with follicles 45 mm or larger sometimes ovulate the large follicle by 5 days posttreatment. Occasionally, the large follicle regresses, and another ovulates about 9 days posttreatment. A large follicle that is preparing to ovulate in the presence of the persistent corpus luteum may ovulate 24-72 hours posttreatment. In these instances, the mare may not show heat (50 percent of the time) or shows for only one day, and owners often suspect that the treatment has failed. When follicular status is known, however, the results of prostaglandin treatment are usually very predictable.

It should also be noted that prostaglandin treatment has been associated with certain temporary side effects. Among these side effects, sweating and decreased rectal temperature have been observed most often. Other effects include incoordination, slight abdominal pain, and increased heart and respiratory rates.

GRANULOSA CELL TUMORS

Granulosa cell tumors are another important cause of anestrus in mares. Although ovarian tumors are unusual, the granulosa cell tumor is reported with greatest frequency. These tumors usually occur in mares between 5 and 7 years of age and are commonly located on only one ovary. The affected ovary may have many fluid-filled, blood-stained cavities or may carry only one, large, fluid-filled cyst. Because estrogen produced by the granulosa cell tumor exerts a negative feedback effect on the hypothalamus and the pituitary, follicular development and ovulation are inhibited in both ovaries. Although mares with granulosa cell tumors often fail to show estrus, they are sometimes characterized by other abnormal reproductive patterns. (Refer to the discussions titled 'Aggressive Behavior' and 'Prolonged Estrus' within this chapter.)

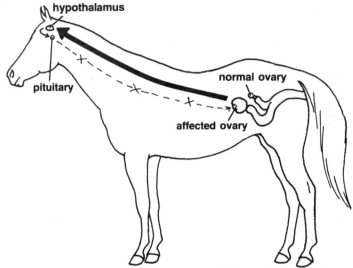

The estrogen produced by a granulosa cell tumor inhibits follicular development in the uninvolved ovary, because of the tumor's negative feedback affect on the hypothalamus and the pituitary.

OTHER CAUSES

Other significant causes of anestrus in mares include removal of both ovaries (bilateral ovariectomy), permanently underdeveloped ovaries (ovarian hypoplasia), and infantile reproductive tract. Each of these conditions is characterized by complete absence of ovarian activity and by the absence of normal hormone changes associated with the estrous cycle. In rare instances, tumors of the hypothalamus or anterior pituitary may cause endocrine imbalances and subsequent anestrus.

Because anestrus is a response to severe stress, it may be the result of long-term weight loss (e.g., parasitism, starvation), excessive work, or any drastic change in physical, mental, or environmental conditions. For example, mares that undergo surgery resulting in a long-term convalescent period often enter an anestrus state.

PROLONGED ESTRUS

Prolonged heat periods are occasionally seen in immature mares, maiden mares, and in mares of all ages during the spring. It seems that most cases of prolonged estrus occur between winter anestrus and the natural breeding season. It is not unusual for the maiden mare's first estrus to last 10-30 days without ovulation, and this condition is also seen in open and barren mares. After the first ovulation finally occurs, estrous cycles decrease in length until a normal cycle of about 21-23 days is reached, usually in April or May (in the Northern Hemisphere). Regular cycles can sometimes be hastened by stimulating the mare's reproductive system with hormones or lights. (Refer to "Broodmare Management.") A granulosa cell tumor may cause estrus to last as long as 120 days (with or without ovulation). In this instance, removal of the affected ovary usually corrects the abnormal behavior. (Refer to 'Pathological Aspects' within this chapter.)

SPLIT ESTRUS

Split estrus is an abnormality of the mare's estrous cycle, occurring in only about 5 percent of all estrous periods. The mare with this deviation shows to the stallion for a few days, goes out of heat for several days, and then returns to estrus and ovulates. Split estrus usually occurs during the mare's transition into or out of seasonal anestrus. This syndrome also seems to be closely associated with silent heat, making it important that the handler observe the mare carefully, especially during the second phase of estrus. Rectal palpation by an experienced palpator to detect follicular development and impending ovulation may be very helpful when dealing with these mares. Although the split estrus period usually lasts about 14 days, it may range from 4 days to 3 weeks. The cause (or causes) of split estrus is not known at this time.

AGGRESSIVE BEHAVIOR

Mares that attempt to breed other mares and mares that become extremely vicious during estrus or when handled about the hindquarters may suffer from chromosome errors, granulosa cell tumors, or inherent neural or hormonal disturbances. In rare instances, close observation of an aggressive mare reveals the presence of both male and female sex organs. These so-called "intersex" horses usually have extremely underdeveloped testi-

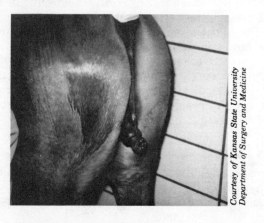

Courtesy of Kansas State University
Department of Surgery and Medicine

The overdeveloped clitoris seen in this horse is characteristic of the pseudo-hermaphrodite (intersex) condition.

cles within the abdominal cavity, in addition to secondary sex characteristics (i.e., outward appearance) of the female. The external genitals of the
intersex horse may appear to be those of a mare and may even be so
pronounced that the animal is registered as a mare. However, the external
genitals of the intersex horse are usually a compromise between the male
and female organs (e.g., an exaggerated clitoris with a small glans penis).
Intersex has been related to errors in the normal sex chromosome combinations (XX = female and XY = male). Reported deviations include the
following combinations: all XXY cells throughout the body, all XXXY cells,
XX cells with XY cells, and XY cells with XXY cells. For explanations
concerning these chromosome anomalies, the interested student should
refer to modern cytogenetics textbooks.

Another rare, but nevertheless important, cause of aggressive behavior
in the mare is a condition similar to intersex. Testicular feminization is
the failure of a genetic male to respond to testosterone during fetal
development. As a result, the animal develops female sex characteristics
which are sometimes so predominant that affected males are registered as
females. Testicular feminization is believed to be an inherited disturbance.

Nymphomania is a term often used to describe extreme excitement or
aggressive behavior in association with breeding. (The term *pseudonym-
phomania* is sometimes used to describe mares in prolonged estrus due to
seasonal influences.) Mild forms of nymphomania are characterized by
varying degrees of excitement during estrus. During diestrus and anestrus,
these mares usually behave normally. This form of nymphomania is believed to be caused by unusually high levels of estrogen. If the mare's
behavior detracts from her usefulness, the removal of one ovary may lessen
the problem. In extreme cases, removal of both ovaries may be considered,
although this will result in sterility. Mares with severe nymphomania are
characterized by continuous heat and extremely violent behavior when
presented to a stallion or handled about the hindquarters. These mares
are usually aggressive toward both people and horses and often present a
significant management problem. Severe nymphomania is believed to be
an inherent nervous disorder, possibly influenced or aggravated by an
inherited endocrine disturbance. There is no known cure for this form of
nymphomania.

FAILURE TO OVULATE

Absence of follicular activity and/or failure to ovulate have been closely
associated with conditions such as ovarian hypoplasia, granulosa cell
tumors, ovarian cysts, uterine infections, malnutrition, chromosomal errors, fetal developmental abnormalities, hypothyroidism, tumors of the
hypothalamus or anterior pituitary, severe nymphomania, prolonged estrus, and unfavorable climate. If the ovaries are defective, or if a hormone
disturbance is present, follicular development is usually inhibited. Seasonal anestrus is probably the most significant cause of failure to ovulate.
This failure is also a common occurrence during the mare's transitional
estrous period. During this period, a mature Graafian follicle may develop

and then suddenly regress without ovulating. As with other seasonal irregularities, time is usually the best treatment.

FAILURE TO LUTEINIZE

Failure of the follicular cavity to form the progesterone-producing corpus luteum is another important cause of infertility in the mare. For reasons not clearly understood at this time, the mare may enter estrus, ovulate at the proper time, but fail to produce a corpus luteum. Consequently, insufficient progesterone levels allow the formation and ovulation of a second follicle, and the handler observes prolonged estrus. If the follicular cavity fails to luteinize, progesterone levels are insufficient to maintain a suitable uterine environment for pregnancy. (Progesterone normally prepares the uterus for reception of the fertilized ovum.) Also, rising estrogen levels from the developing follicle interfere with ovum transport through the oviduct during the first 4-5 days following ovulation. Subsequently, death and resorption of the first embryo occurs. For this reason, the mare should be bred again just prior to the ovulation of the second mature follicle.

MULTIPLE OVULATIONS

Unlike other species which commonly produce more than one offspring per pregnancy, the mare's reproductive tract is not designed to properly maintain more than one fetus. Because the equine fetus must utilize the uterine body and a uterine horn to develop normally, twin pregnancy usually results in crowding, failure of one or both fetuses to receive sufficient nutrients, abortion of both fetuses, or the birth of one or two undersized, weak foals. (For an explanation of why this occurs, refer to "Fetal Development.") Although cases of healthy twin foals have been reported, twinning is generally considered to be undesirable due to increased chances of embryonic loss, fetal loss, difficult foaling, or the delivery of weak or dead foals. Because multiple conception contributes to numerous reproductive problems in the mare, it is considered a physiological aspect of infertility. When feasible, it is best to avoid breeding a mare if a good chance of twin conception exists.

At least 50 percent of all heat periods are characterized by the development of more than one follicle. Usually, a large percentage of the extra follicles regresses as one of the follicles becomes more prominent. In as high as 25 percent of all estrous periods, however, two follicles develop to maturity and present the possibility of multiple ovulation. It is difficult, if not impossible, to determine whether one follicle will regress after the other ovulates or whether both follicles will ovulate at some unknown interval. Even when the mare is carefully palpated for the presence of two mature follicles and declared "safe" to breed, twin ovulation may result from rapid development and ovulation of an unsuspected follicle, ovulation of a mature follicle that is positioned deeper than usual within the ovary, or ovulation of two closely positioned follicles that have been mistaken for one mature follicle. When dealing with the possibility of twin ovulations, it might be helpful to remember that a tendency toward twinning is believed to be an inherited trait. In other words, mares that have had twin ovulations or produced twins in the past are very likely candidates for twin ovulation in the future. Also, researchers have reported that a greater incidence of twin ovulations occurs in March, April, and May and that the frequency of twin ovulations increases with age.

When there is a chance of twin ovulation, the mare should be palpated daily and bred 48 hours after the first ovulation occurs—if 48 hours lapse between ovulations. If both ovulations occur on the same day or if only 24 hours lapse between ovulations, the possibility of twin conception is very high, and it is best to postpone breeding until the next cycle.

Reproductive Disease

Perhaps the most serious forms of infertility in the mare are the disease-related abnormalities. Damage caused by pathological processes within the reproductive tract is often irreversible. Even if an infection is cleared up or a tumor is removed, the mare's resistance to future problems may be seriously impaired. Before these disorders are examined, it may be helpful to review the definitions of important biological terms presented under 'Pathological Aspects' within "Stallion Infertility and Impotency."

INFECTIONS

The mare's reproductive tract is not designed for easy removal of infectious organisms and exudate (fluid formed in reaction to tissue injury or infection). In many species, including humans, the pull of gravity enhances the removal of purulent (i.e., pus-filled) fluid from the reproductive tract. The horizontal position of the mare's uterus and vagina, on the other hand, makes the battle against pathogens far more difficult. It is very important to remember that serious fertility problems arise when a mare sustains a chronic infection that goes unnoticed or uncorrected for a lengthy period. These infections may result from a number of man-made problems, including improper hygiene during breeding or foaling, improper sterilization of examination equipment, breeding the mare at the wrong point in the estrous cycle, etc. Resulting economic losses (e.g., money, time, and fertility) can have a serious impact on the breeding operation. For this reason, the breeder should understand how the mare's fertility is affected by infections, how these infections are recognized and, more importantly, how such problems can be prevented.

PREDISPOSING FACTORS

The presence of bacteria within the mare's reproductive tract is not an unusual or necessarily harmful condition. Organisms that cause infections are found throughout the mare's environment, but the pathogens that cause an infection in one mare may be harmless to another. For example, a natural cover (live cover) contaminates the uterus with organisms that have the potential to become pathogens. Development of a uterine infection depends on the mare's natural resistance and on the type, number, and virulence of the organism involved.

The mare's susceptibility to infection plays an important role in determining whether she will become infected, how serious the infection will be, and how easily the problem can be corrected. The young, healthy mare has natural mechanisms that resist and eliminate most pathogens. These resistant mares do not usually become infected unless they are subjected to extreme management errors or neglect. When a resistant mare is inoculated with infectious organisms (i.e., when pathogens are placed directly in the uterus), the uterus becomes very turgid (i.e., swollen) and tubular within a few hours. The uterus also produces large amounts of pus-filled fluid, and the ovaries produce large amounts of estrogen. These sudden changes in the reproductive tract are usually successful in eliminating the infectious organisms.

The reproductive tracts of susceptible mares, on the other hand, show little or no reaction to inoculation. Scientists believe that this failure may be due to scarring of the inner uterine lining, endocrine abnormalities, malfunction of the immune system, or an inability of the cells to produce bactericidal substances. These mares must be managed carefully to prevent contamination of their reproductive tracts. Mildly susceptible mares may eventually "clean up" on their own, but most mares within the susceptible range require special medical assistance. More susceptible mares become reinfected very easily after treatment. In mares that are highly susceptible to uterine infection, even the best therapy often fails and reinfection is almost inevitable. The chances of these mares ever conceiving are very poor.

Although the exact physiological changes that reduce the mare's ability to fight reproductive infections are not clear, it appears that each time an infection occurs her resistance to future problems decreases to some extent. Several environmental and physical problems that are known to increase the mare's chances of becoming infected have been identified, however. When studying a mare's fertility potential, the breeder should evaluate each of these factors carefully.

Perineal Conformation

One of the first characteristics that the breeder should evaluate is the conformation and condition of the mare's perineal area. The vulva normally protects the reproductive tract by sealing the vagina from the external environment. If lacerations, adhesions, scar tissue, or poor conformation interfere with proper vulval closure, the vagina may become contaminated with fecal matter and airborne bacteria. This inspiration of air and debris, which is referred to as pneumovagina, increases the mare's susceptibility to infection, especially when it causes ballooning of the vagina or uterus. Pneumovagina seems to be more prevalent in slim race mares, mares in poor condition, and mares with poor vulval tone due to injury, age, estrus, or foaling. In any instance, pneumovagina is the direct result of poor closure of the vulval lips.

Another abnormality of the perineal area is tipped vulva, a condition characterized by unfavorable positioning of the vulva in relation to the anus and the pelvic brim. Optimum conformation is characterized by minimal slant as a line is drawn from the anus to the vulva. A virtually straight line helps to maintain proper vulval closure and allows fecal matter to fall away from the vulval opening, but as the angle increases, the occurrence of pneumovagina and other related fertility problems also increases. (An increase in this angle is closely associated with aging and loss of condition.) Mares with high tail sets, concave vulvas, or sharply tipped vulvas tend to catch feces on the vulval shelf. These conditions, which are usually inherited characteristics, are frequently accompanied by poor vulval tone and consequent fecal contamination of the vagina.

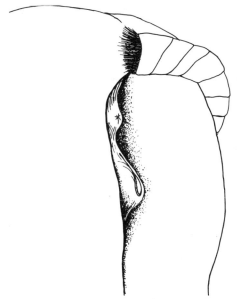

Poor muscular tone caused by a sunken anus prevents proper closure of the vulval lips.

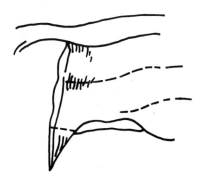

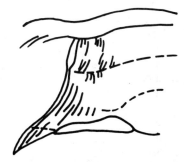

Ideal vulval conformation   **Sloping vulva caused by sunken anus**

A common procedure for preventing the inspiration of air and fecal matter into the vagina is referred to as Caslick's operation. This operation simply involves suturing the lips of the vulva together, preferably to a point below the pelvic brim or vulval shelf. Naturally, a small area must be left open so that the mare can urinate. (Refer to "Broodmare Management.")

Sunken Vagina

A vaginal floor that slants downward as it approaches the cervix is also detrimental to disease resistance if it causes an accumulation of urine within the vagina. This so-called "sunken vagina" is often seen in older mares and in mares with poor conformation. Poor muscle tone may also cause the vaginal floor to sink below the pelvic brim. This condition is frequently associated with pneumovagina and may be aggravated by a Caslick's operation, if urination is in any way impeded. The seriousness of the disorder is directly related to the degree of vaginal slope. Mild cases can be dealt with by wiping pooled urine out of the vagina before breeding and a small percentage of urine pooling cases can be corrected by surgically extending the urethral opening to the vulva. However, surgical correction is seldom successful when the vaginal slope is greater than 35 degrees.

Trauma

Trauma to the mare's reproductive tract stimulates the formation of scar tissue and adhesions and may lower overall resistance to pathogens. For example, improper use of iodine solutions within the mare's uterus has been associated with severe uterine adhesions and infertility. Other problems, such as foaling difficulties, may stress the mare's reproductive tract and contribute to a loss of disease resistance. Breeding or foaling injuries that prevent proper cervical closure allow continuous contamination of the uterus and increase the possibility of uterine infection significantly. Past infections, extensive uterine therapy, physical injury, excessive breeding by natural service, severe malnutrition, and the inevitable loss of muscle tone and strength that accompanies aging are also closely related to increased susceptibility to reproductive infections.

Inherent Susceptibility

The breeder should not rule out the possibility that certain mares are less resistant to infection due to their genetic constitution. The conformation defects mentioned earlier (poor vulval tone, tipped vulva, etc.) are good examples of inherited traits that predispose the mare to infection. In addition, inherited defects of the endocrine and immunologic systems could affect the normal infection-fighting mechanisms.

TYPES OF INFECTIONS

Although infections have been noted in all sections of the mare's reproductive tract, the ovaries seem to be very resistant to pathogens. Unlike the testes, the ovaries are usually unaffected by systemic diseases (i.e., diseases caused by microorganisms in the circulating blood). However, the small tubes that lead from the ovaries to the uterus are somewhat more susceptible to infection. Even slight inflammation of these fallopian tubes (mild salpingitis) can cause strictures or adhesions that prevent passage of semen through the tubes and passage of the ovum to the uterus. Complete adhesions of both oviducts, resulting in permanent sterility, is a rare occurrence in the equine, but has been observed in other species. Although most instances of salpingitis are probably caused by migrating uterine infections, organisms within the abdominal cavity and the circulating blood are also possible sources of infection. Unfortunately, there is not yet a practical method for diagnosing salpingitis in the mare.

Infections within the uterus are by far the most detrimental to the mare's fertility and the most frustrating for breeders and veterinarians. The presence of microorganisms in the uterus may result in mild inflammation of the endometrium or in the formation of pus-filled fluid and scar tissue. The severity of a uterine infection depends on 1) the mare's susceptibility, 2) the types of organisms present, 3) the number of organisms present, 4) how the mare was infected, 5) at what point during the estrous cycle she was contaminated, and 6) how long the infection has been present.

The general term used to describe inflammation of the uterus is metritis, and inflammation of only the endometrium is differentiated as endometritis. Endometritis is characterized by the presence of excess mucus, increased white blood cell count, and increased blood supply. Edema causes the endometritis uterus to feel enlarged and doughy when palpated. Besides failure to conceive, the only visible indication of endometritis may be a shortened estrous cycle. This change is caused by premature regression of the corpus luteum, a response that is triggered by the presence of inflammation and irritation within the uterus. The resulting rise in estrogen levels allows relaxation of the cervix and removal of any infectious exudate at about 8-10 days postovulation. (The exudate contains mucus, debris, bacteria, pus, etc.) If a mare drops from a normal 21-day cycle to one of 16 days or less, the breeder should suspect endometritis and seek veterinary assistance.

Acute metritis is characterized by areas of deep hemorrhage, missing endometrium, and granular tissue within the uterus. Other changes associated with this type of infection include hypertrophy, abscessed uterine glands, formation of pus-filled fluid, cellular degeneration, and cellular death. Acute metritis may be indicated by the presence of irregular estrus or by the expulsion of exudate through the mare's vulva during estrus. The exudate is usually a milky or creamy, white fluid that mats the mare's tail hairs and forms a crust on portions of her hindquarters.

Pyometra (chronic uterine infection) is not characterized by the normal inflammatory responses but by failure of the tract to fight an infection. This type of long-term and extremely damaging infection often occurs in the mare with a sagging uterus. It can also occur when cervical adhesions close off the cervix and prevent the normal release of infectious exudate. The pyometra uterus may be enlarged and pendulous with thin, flaccid walls or with very thick leathery walls in long-standing cases. The formation of tough, fibrous, nonfunctional areas on the uterine wall is not uncommon. Damage to the endometrium is often extensive during a chronic infection, and the mare's future breeding prospects are very poor. Perhaps the most important problem associated with pyometra is that these chronic infections are often disguised by the mare's healthy appearance and by the absence of exudate on her hindquarters. The breeder should note, however, that the distended, pus-filled uterus is usually unable to produce prostaglandin due to a damaged endometrium. For this reason, pyometra mares are often characterized by persistent corpora lutea and resulting anestrus.

Septic metritis is a severe, and often fatal, infection of the entire uterus caused by contamination of the mare's reproductive tract during or shortly after foaling. Fetal fluid and pieces of fetal membrane left in the uterus provide a perfect growth medium for microorganisms introduced into the area by hands used to assist during foaling or by air sucked into the uterus when the mare rises after foaling. Decomposition of the fetal debris, rapid spread of infection, and absorption of various by-products of tissue decomposition result in severe systemic illness and, in many instances, death of the mare. (Refer to the discussion on retained placenta within "Abnormal Foaling.")

The muscular body of the cervix is very resistant to infections, but the mucous membrane is susceptible to infections and irritation, much like the endometrium. Inflammation of the cervical mucosa (i.e., cervicitis) is characterized by a dark red or purple cervical opening. The opening may also be swollen, dilated, pendulous, or covered with thick mucus. When cervicitis is present, the handler may observe thick mucus on the mare's vulva or on the floor of her vagina. If the infection is severe, cysts may develop on the cervix, and adhesions may completely block the cervical opening, possibly resulting in pyometra. If scar tissue forms and prevents normal closure of the cervix, the mare's susceptibility to future uterine infections increases drastically. Although cervicitis certainly predisposes the mare to metritis or endometritis, the presence of cervical infection does not necessarily indicate uterine infection.

Because the vagina is frequently exposed to a variety of microorganisms, infections within the area are not uncommon. Mares with tipped vulvas or poor vulval seals are prone to vaginal infection (vaginitis) due to the presence of air, urine, and fecal matter. Infections originating in the vagina usually involve the cervix and can migrate to the uterus under certain conditions (e.g., low resistance, inability to cleanse the uterus due to high progesterone levels). Also, the migrations of purulent exudate from the uterus or cervix to the vulva may cause inflammation within the vagina. Fortunately, the vagina is more resistant than the cervix or uterus to inflammation-related tissue damage.

EFFECTS ON FERTILITY

Reproductive infections have three important effects on the mare's productivity: 1) they contribute to conception failures, 2) they reduce the mare's ability to carry a foal to term, and 3) they predispose the newborn foal to serious health problems that may even result in neonatal death. Fetal loss due to reproductive infections is examined in the chapter "Abortion," while related problems in the newborn foal are studied within "Neonatal Foals." Three very significant problems caused by reproductive infections are presented to emphasize the importance of preventing these disease processes in broodmares.

1. The pus-filled exudate present in an infected reproductive tract is a very effective spermicide (sperm-killing agent). For this reason, infections of the vagina and cervix can lower conception rates even though the uterus is clean. If the uterus is infected, the sperm's chances of survival are even more limited and, even if a sperm cell manages to reach the oviduct and fertilize the ovum, the embryo has little chance of survival when it migrates to the infected uterus. Therefore, any infection of the reproductive tract, whether inflammation of the vagina with slight increase in mucus or endometritis with large amounts of purulent fluid, affects the mare's fertility status. When pyometra or severe metritis is present, conception is impossible.

2. After an infection has been corrected, the presence of adhesions within the reproductive tract may also cause conception failure if the migration of sperm is impeded. Severe adhesions within the uterus, for example, may reduce the percentage of sperm that moves to the oviduct, causing a significant reduction in fertility. Adhesions of the cervix can prevent conception if they completely block the passage of sperm, and adhesions within the vagina may prevent normal coitus if they cause painful breeding and subsequent psychological associations between pain and breeding.

3. Another aspect is the inability (or reduced ability) of a fluid-filled uterus to produce sufficient luteolytic agent. This change in prostaglandin production causes prolongation of the corpus luteum which, in turn, causes anestrus. In addition, chronic infections may cause permanent damage to uterine tissue, thereby affecting future production of the luteolytic compound. Consequently, the mare may suffer from irregular cycles for the rest of her reproductive life.

INFECTIOUS ORGANISMS

In addition to the mare's susceptibility to infectious organisms, the type and concentration of these pathogens also determine the severity of any reproductive infection. Researchers have identified several organisms that are frequently involved in infections of the mare's reproductive tract. Identification of each pathogen can be extremely important to the eradication of an infection. (Refer to 'Diagnosis and Treatment' within this chapter.) To identify these microorganisms, samples of bacteria-filled fluid are taken from the vagina, cervix, uterus and/or clitoral fossa (depression within the clitoris), a process known as swabbing. (Refer to "Broodmare Management.") The swabs are then rubbed on special gel-like substances that provide the necessary nutrients for a specific organism or group of organisms to multiply. These cultures are then incubated for 12-48 hours. Depending on the type of medium that supports the organism and on the characteristic color and growth pattern, the organism can usually be identified—unless the sample was handled carelessly or the growth medium swabbed too heavily. Although a detailed study on how these cultures are made and how the pathogens are identified is beyond the scope of this discussion, a brief description of the more common pathogens is included to help the breeder understand what the veterinarian and the lab technician are dealing with. (Interested students should refer to a modern microbiology textbook.)

Streptococcus: Streptococci are spherical bacteria found in chains of various lengths. These bacteria are commonly found on skin, mucous membranes, and in the intestine. For identification purposes, these bacteria are divided into two main categories: alpha hemolytic streptococcus and beta hemolytic streptococcus. The alpha organisms are characterized by their ability to partially destroy red blood cells, while the beta group causes complete destruction of red blood cells (hemolysis). The beta hemolytic streptococcus organisms play a very important role in initiating reproductive infections in the mare and are usually indicated when a milky-white discharge is present. The most common beta streptococcus is *Strep. zooepidemicus,* a pathogen capable of causing severe cervicitis and metritis, possibly resulting in abortion, sterility, or death. As many as 75 percent of all uterine infections may be caused by beta hemolytic streptococcus organisms.

Staphylococcus: The staphylococcus bacteria have been isolated from the reproductive tracts of infected and noninfected mares. Under a microscope, these organisms are identified by their characteristic spherical or ovoid shape and by their grape-like clusters. Although there is no well-defined classification system for the staphylococcus organisms, the species can be differentiated by the type of toxin (i.e., poisonous by-product) produced in response to their invasion of living tissue. *Staphylococcus aureus* is a widely distributed species found within the normal bacterial flora on the skin and mucous membranes of both man and animals. This bacterium is considered an opportunist since, under suitable conditions, it will invade and destroy tissue. *Staphylococcus aureus* is frequently associated with endometritis in the mare.

Pseudomonas: Pseudomonas species are distributed throughout the world and are commonly found in fresh water, salt water, and soil. They produce a characteristic green, blue, or yellowish-green exudate. Only one species, *Pseudomonas aeruginosa,* has been associated with reproductive infections in the mare. This species is a slender rod with rounded ends and 1-3 flagella (i.e., propelling tails). It has been isolated in stallion semen and is believed to be transmitted by carrier stallions. (The presence of pseudomonas within the stallion's semen can be associated with reduced conception rates for that particular stallion, but many normal stallions produce a growth of pseudomonas from a semen or urethral culture.) Although pseudomonas cultures are characterized by the colorful exudate, pseudomonas infections are usually identified by a cloudy discharge that has flecks of pus distributed throughout.

Klebsiella: Klebsiella are encapsulated rod-shaped bacteria that are associated with chronic infections of the mare's reproductive tract. Although little is known about the distribution of the causative organism, Klebsiella infections are believed to be caused by carrier stallions and improper hygiene during breeding. There seem to be two Klebsiella species associated with reproductive infections in mares: *Klebsiella genitalium* and *Klebsiella pneumoniae*. However, these organisms are usually classified together. The presence of Klebsiella within the mare's uterus often causes a slimy, pus-filled exudate and irregular estrous cycles. It is also important to note that Klebsiella infections can be extremely difficult to eliminate. Because of this resistance and because Klebsiella infections are characterized by excessive exudate, the affected mare often develops cervical adhesions and pyometra.

Haemophilus equigenitalis: Due to the recent impact of *Haemophilus equigenitalis* on breeding farms in a number of countries, the bacterial species has gained wide-spread recognition as a serious source of equine reproductive infections (contagious equine metritis). The *H. equigenitalis* organism is a long, slender, rod-shaped bacterium that becomes more spherical with age. The organism cultures well on chocolate agar (growth medium), especially when its exposure to oxygen is reduced. Recent research and literature have focused on this organism and its effect on equine fertility. Although the stallion does not seem to be affected by the organism, he plays an important role as carrier of the bacterium from mare to mare. When the mare's reproductive tract is exposed to *H. equigenitalis*, she may develop an acute or subacute infection. Acute contagious equine metritis (CEM) is characterized by inflammation of uterine, cervical, and vaginal membranes and by the formation of a copious grey discharge within 3-5 days after exposure. In severe cases, this creamy exudate may be found on the mare's tail and hindquarters or in pools on the stall floor. Even when the clinical signs are less severe, as in subacute infections, degeneration and hyperplasia of uterine tissues are likely. The mare may also develop a chronic infection, or she may be asymptomatic (i.e., without clinical signs), contaminating healthy stallions who, in turn, infect healthy mares. Due to its highly contagious nature, strict breeding control and detailed codes of practice have been utilized in many affected countries. Special culture procedures and blood tests provide added assurance that infected animals are not bred. Identifying carrier animals and determining when formerly infected animals are safe to breed are significant problems, but researchers are actively seeking answers to these and other important questions about contagious equine metritis.

Escherichia: Escherichia organisms are commonly found in the intestinal tracts of man and animals and seem to be transmitted by flies or by the presence of fecal matter in food or water. Of particular interest to horsemen is the bacteria *Escherichia coli*, short rod-shaped bacteria that occur alone or in short chains. These bacteria can grow with or without oxygen and are capable of producing severe, and sometimes fatal, infections within the circulating blood. *Escherichia coli* is the second most common organism isolated from the mare's uterus and is believed to be an important cause of uterine infection.

Corynebacteria: Most of the corynebacteria are nonpathogenic, rod-shaped organisms that form clumps of irregular, club-shaped, or swollen bacteria when cultured. A member of this group, *Corynebacteria equi,* is associated with foal pneumonia and has been isolated from the mare's uterus. Although the organism has not yet been related to a specific type of infection, it is a suspected pathogen and is believed to cause abortion and prenatal (before birth) infections in foals.

Proteus: Proteus bacteria are described as multi-form rods with flagella distributed uniformly over the entire cell. These bacteria are commonly found in the intestinal tract of man and animals and, like many other pathogens isolated from the mare's reproductive tract, are opportunists. Although certain Proteus species have been isolated from the uterus of endometritis mares, little is known about the specific role that these organisms play in an infection.

Coliforms: Coliforms are any of the enterobacteria (e.g., Escherichia, Klebsiella) other than Salmonella, Shigella, and Proteus. A number of these infectious organisms have been isolated from the mare's reproductive tract.

Fungus: Fungi also contribute to reproductive infections and may become extensive when bacteria are eliminated by specific antibiotic treatments of the mare's uterus. *Candida albicans*, a yeast-like fungus that sometimes affects the skin and mucous membranes of animals and man, has been isolated as a cause of uterine infection in the mare. Strains of Actinomycetes, Penicillium, *Allescheria boydii*, and Aspergillus have also been isolated from the mare's reproductive tract.

Virus: Although coital vesicular exanthema (a viral genital disease commonly known as genital horse pox) is rare in the United States, it is still a significant problem in parts of the world. The disease is caused by a herpes virus which is transmitted by coitus, grooming, and association with infected horses. Coital exanthema is highly contagious, and the characteristic papules (circular elevated areas) on the vulva often appear suddenly. As the infection progresses, the vulva usually becomes red, tender, and swollen; special care should be taken to insure that secondary bacterial infections do not occur. The papules develop into scab-covered ulcers which, after healing, leave pigmented areas for several weeks. Although the disease is not detrimental to the mare's fertility, coital vesicular exanthema lesions on the stallion's penis can cause impotency. For this reason, affected mares should not be bred by natural cover until their lesions have healed.

Protozoa: Another venereal disease that occurs in the equine is dourine, an infection caused by the protozoa (one-celled animal) *Trypanosoma equiperdum.* Dourine has been eradicated in many countries (including the United States), but still occurs in some temperate areas, where the disease is usually chronic, and in some tropical areas, where its onset is usually acute. The dourine protozoa are transmitted by coitus. Initially, the infection is characterized by inflammation of the vagina and vulva and by a vaginal discharge. As the disease progresses, large circular areas of raised skin occur all over the body. Depigmentation of the genitals and periods of fever are also characteristic of the disease. About 50-75 percent of the reported cases have resulted in death.

DIAGNOSIS AND TREATMENT

Proper management can be extremely helpful in maintaining the health of the mare's reproductive tract. Proper sterilization of examination and artificial insemination equipment, treatment of the stallion's semen to reduce the number of potential pathogens, suturing the tipped vulva, correcting urine pooling, and using efficient breeding schedules are examples of important management practices that can be used to prevent reproductive infections. However, even the most careful and conscientious breeder occasionally suffers the frustration associated with infected mares. Because of this impending possibility, every breeder should remember the most important step in correcting reproductive infections: **early recognition that a problem exists.** The following list of important clues used to detect a reproductive infection may be very helpful:

1. irregular estrous cycles,
2. estrous cycles that are shorter than the normal 21-23 day period,
3. anestrus,
4. matted tail hair and crusty hair on the hindquarters,
5. failure to conceive,
6. visible slimy, milky, or creamy exudate on the vulva, tail, inner thighs, or vaginal floor,
7. slightly red or fiery red mucous membrane within the vagina,
8. dark red or purple cervical opening, and
9. excessive amounts of cervical mucus or exudate.

Identifying the Organism

The next step in the fight against reproductive disease is to have a veterinarian culture the mare and identify the causative organisms. It is important to note that vaginal and cervical cultures are not necessarily indicative of organisms present in the uterus. For this reason, vaginal and cervical cultures should be made in addition to, not in place of, uterine cultures. It is also important to run special tests on each organism to determine the most effective treatment for a particular infection. These so-called sensitivity tests are especially important when the mare is infected with antibiotic-resistant strains of bacteria. (It appears that many microorganisms are becoming resistant to the more commonly used antibiotics.) Although recently introduced broad-spectrum antibiotics are still useful, it may only be a matter of time before their effectiveness is lost.

Therapy

The importance of early recognition, diagnosis, and treatment of an infection cannot be overemphasized. Early therapy is important due to the possibility of long-term inflammation and cellular degeneration —circumstances that may eventually lead to cellular death, permanent tissue damage, and loss of tissue function. Exact treatments for reproductive infections vary with the infection's severity, the organisms present, the pathogen's drug resistance, and the veterinarian's past experiences and personal preferences. Usually, emphasis is placed on uterine infusion of antibiotics, disinfectants, steroids, enzymes, etc. Local (on the infected tissue) ointments and parenteral (injected into a vein, muscle, or under the skin) antibiotics are sometimes given in conjunction with medicated infusions of the reproductive tract. Parenteral antibiotics may be extremely important aids in the treatment of septicemia caused by (or in association with) severe uterine infection but, when used without the reinforcement of uterine infusion, they are not as effective in cleaning up the actual reproductive infection.

Uterine Infusions

To insure that the uterine cleansing process functions properly, it is important that the cervix be open when treatment is initiated. For this reason, many practitioners prefer to treat uterine infections when the mare is in estrus. However, this arrangement is not always possible due to the close association between reproductive infections and anestrus. Prostaglandin injections are sometimes given to decrease the time between heat periods and, therefore, to increase the number of treatments that can be given within a certain time span. If the mare has pyometra, anestrus is usually a problem, and the cervix and any cervical adhesions must be opened so that uterine fluids can be evacuated prior to treatment. Sterile gauze placed in the cervical opening may help to prevent the reformation of adhesions and allow the uterus to drain, but this is not a common practice. Some practitioners place sterile plastic tubes in the cervix to prevent reclosure and to allow the removal of exudate. The pyometra uterus may be flushed several times with about a gallon of mild disinfectant to help remove the purulent fluid. Because a fluid-filled, pendulous uterus does not drain properly without assistance, the disinfectant and exudate are usually siphoned out after each infusion. Drugs that stimulate involution (i.e., return to normal size) of the uterus, such as diethylstilbestrol, estradiol, and oxytocin, are sometimes given to hasten the removal of exudate and to prepare the uterus for antimicrobial therapy.

Uterine infusions are special mixtures of specific or broad-spectrum antibiotics and other agents that the veterinarian considers important (e.g., enzymes, steroids, estrogens, disinfectants) in about 500 ml of sterile saline, oil, or some other sterile base. For example, twice the oral or parenteral dose of a selected antibiotic might be mixed with 500 ml of sterile physiological saline solution and infused into the mare's uterus daily for 3-5 days while the mare is in heat. If additional treatments are needed, the veterinarian usually increases the dosage and extends the infusion schedule several days. For example, the antibiotic concentration may be increased by 25 percent and the mare infused daily for 8 days during her second treatment. If a third treatment is necessary, it may involve another increase in antibiotics and daily infusions for about 15 days.

Infusion solutions are frequently injected into the mare's uterus through a sterile plastic catheter, or tube, which has been inserted into the uterus via the cervix. The veterinarian then inserts a gloved arm into the mare's rectum and massages the uterus gently to insure that the solution reaches each uterine horn. Because reproductive infections are often difficult to correct and usually require several treatments, indwelling catheters have been used in an attempt to minimize trauma caused by passing and removing the infusion tubing. The indwelling catheter, which remains in the reproductive tract until the infection is corrected, has connective flexible tubing that extends through the vulva, allowing the veterinarian to infuse the uterus without re-entering the mare. However, because the indwelling catheter provides a direct path for contamination and because it is very irritating to the uterus and cervix, it is not used very often.

Courtesy of Department of Large Animal Medicine and Surgery Texas A&M University

Infusion solutions are usually introduced into the uterus through flexible, plastic tubing which is inserted through the cervix.

The indwelling catheter is inserted through the cervix and into the uterus. The small loop holds the catheter in place so that repeated infusions can be made easily.

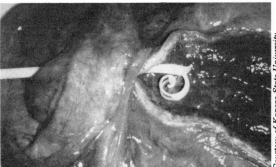

Courtesy of Kansas State University Department of Surgery and Medicine

Due to possible side effects, drug interactions, and drug resistance, the choice of drugs used to combat any infection should be dictated by a veterinarian. The veterinarian is qualified to determine the type, amount, and frequency of drug administration and is familiar with drugs that cannot be given simultaneously. The veterinarian also knows which drugs may not be effective due to the presence of antibiotic-resistant bacterial strains and understands the potential complications that may evolve in response to uterine antibiotic infusions. For example, certain drugs may alter the mare's estrous cycle or cause necrosis of the endometrium. If a drug is absorbed by the uterus, it may have certain effects on other parts of the body. Long-term use of uterine antibiotics can cause the development of fungal and yeast infections, especially when they are used with steroids. (Growth of the organisms becomes extensive when competition with bacteria is eliminated.) The veterinarian is aware of these and other important aspects of drug selection and knows how to deal with problems that may arise.

Prebreeding and Postbreeding Infusions

Uterine infusions performed immediately before or after breeding to control reproductive infections, have been used with varying degrees of success. The antibiotic semen extenders that are used in artificial insemination can also be used for uterine infusion before live cover in mares predisposed to uterine infections. Although some reports indicate that postbreeding infusions of the mare's uterus aid in the treatment of pseudomonas infections, other sources question the effectiveness of such treatment, stating that the infusion of certain drugs may actually be detrimental to conception. Their statistics also indicate that postbreeding infusions in normal mares are related to significant reduction in live-foal percentages.

Uterine Curettage

Uterine curettage is any mechanical or chemical irritation of uterine tissue, performed to stimulate nerve endings and to somehow initiate a cleansing process. The uterine curettage instrument has a plate or loop with a scraping edge, an extended flexible shaft, and a handle. The instrument's cutting end is introduced into the uterus very carefully, avoiding injury to the vagina and cervix. The operator then makes about 30 push-pull movements against the endometrium, rotating the instrument's position slightly with every 3-4 movements so that both uterine horns are treated uniformly.

This process has been virtually discredited in mares and is considered an unfortunate carryover from human practice where the procedure is often indicated but for totally different reasons. Uterine curettage has the potential to cause excessive scar tissue and adhesion formation in the mare's uterus, and its effectiveness in treating chronic infections is questionable.

REPRODUCTIVE TRACT ABNORMALITIES

Disease processes within the reproductive tract are not limited to tissue changes caused by infections. As a case in point, the mare may have inherited a physiological or anatomical abnormality that prevents conception or predisposes her reproductive tract to infection. In rare instances, the mare's reproductive tract may develop cancerous growths or cysts that interfere with normal reproductive function or actually threaten the mare's life. It should also be mentioned that trauma to the reproductive tract may cause tissue degeneration, hyperplasia, formation of scar tissue, and adhesions within the tract—responses that contribute significantly to reduced fertility. Due to the mare's extensive reproductive role, she is subject to significant physiological stress throughout her breeding farm career. It is not uncommon for a mare to begin her career with a healthy, disease-resistant reproductive tract but to develop a tract that is permanently damaged by years of foaling problems, breeding injuries, retained placentas, abortions, etc. For this reason, a study of mare infertility must include a review of noninfectious reproductive abnormalities with respect to the ovaries, fallopian tubes, uterus, cervix, vagina, and vulva.

OVARY

Although the ovaries are essentially unaffected by pathogens, they occasionally form neoplasms and cysts. The most common ovarian tumor is the granulosa cell tumor, mentioned earlier with respect to anestrus and aggressive behavior during estrus. Granulosa cell tumors are fluid-filled growths, originating from the cells that form the wall of an ovarian follicle. When palpated through the rectum, these tumors may feel firm and smooth or lobulated. Due to the tumor's tendency to enlarge gradually, its size is variable. This characteristic can be used to differentiate granulosa cell tumors from ovarian hematomas (blood-filled areas), which gradually decrease in size and cause no changes in estrous cycle patterns or behavior. Granulosa cell tumors do not seem to spread to other parts of the body and are usually found on only one ovary. However, because these tumors produce large amounts of hormones (estrogen and progesterone), follicular activity in the opposite ovary ceases as the organ gradually decreases in

Granulosa cell tumors of the ovary may cause stallion-like behavior in a mare.

size. For this reason, mares with granulosa cell tumors on one ovary are often characterized by the presence of one very large ovary and one small inactive ovary. The affected mare usually has a history of reproductive failure, irregular estrous patterns, and changes in temperament. In addition, she may have experienced occasional abdominal discomfort. As the tumor enlarges, the mare may show anestrus or continuous estrus, and she may occasionally develop aggressive, stallion-like behavior. Fortunately, the condition is easily corrected by removal of the affected ovary. Many mares return to normal reproductive function within a year after surgery.

Other ovarian neoplasms and cysts have been reported in the mare, but their occurrence is rare. Ovarian cysts do not seem to affect fertility except in unusual instances, when their position and size interfere with normal migration of the ovum through the oviduct. Malignant neoplasms may cause more significant problems, but information on these effects is limited. The following abnormalities are examples of unusual ovarian growths that have been reported in the mare:

Dysgerminomas are rare, malignant neoplasms of the ovary, characterized by firm greyish-yellow bodies with a center of blood and dead cells. Dysgerminomas are also associated with high estrogen levels and continuous estrous behavior. The growths seem to spread to other parts of the body through the lymph and blood vessels.

Teratomas are tumors that consist of many different types of tissue, including tissue that is not normally found in the ovary. Although these neoplasms are frequently malignant in the stallion's testicles, they are usually benign in the ovaries.

Parovarian cysts are relatively common, but they rarely cause fertility problems in the mare. These cysts are located around the ovary on non-functional tubules (remnants of a primitive excretory system referred to as the Wolffian body).

Paroophoron cysts are located behind and between the ovaries. Like the parovarian cysts, they are not usually a cause of infertility in the mare. These cysts originate from remnant tissue of the Wolffian body.

Rete cysts are derived from special cellular columns that are formed in the embryonic ovary to give rise to future follicles. These rare cysts are usually present on or near the concave side of the ovary.

Fimbrial cysts are thin-walled cysts that originate from the remnant tissue of the Wolffian body. They range from 2-3 millimeters in diameter.

Epithelial inclusion cysts are found in the ovaries of older mares, usually near the ovulation fossa. These rare cysts are sometimes found in large grape-like bunches, and their size may actually interfere with blood circulation to the rectum.

The congenital condition 63XO aneuploidy was described earlier with respect to its effect on estrous behavior. Affected mares are slightly smaller than the average for their breed and have small, hard, inactive ovaries but appear normal in all other respects. As mentioned previously, the condition is caused by an inherited chromosome anomaly, involving the absence of one sex chromosome. Because the condition cannot be corrected, affected mares are hopelessly sterile. It is important, however, to distinguish the "small, hard ovaries" of aneuploidy from the small, firm ovaries of young mares or mares that are slow to mature. For this reason, the condition should not be diagnosed without chromosome studies until the mare reaches reproductive maturity. (Refer to the discussion on chromosome errors within this chapter.)

Another abnormality of the ovaries, ovarian hypoplasia, is diagnosed after the mare reaches reproductive maturity. This condition is caused by failure of one or both ovaries to reach normal size during embryonic development. The ovaries of young anestrus mares may be mistakenly described as hypoplastic, but true ovarian hypoplasia is usually observed in only one ovary. In addition, the hypoplastic ovary maintains its small size throughout the year, while normal ovaries become larger during the breeding season.

Other unusual conditions of the ovary include the formation of fibrous strands, believed to be the result of bleeding during ovulation, and the migration of strongyles through the ovary causing the formation of scarred areas, or fibrin tags. If severe, these tags may interfere with normal ovarian function. Also, a localized area of blood (hematoma) within the ovulation fossa may be mistaken for a cyst or tumor. These hematomas usually cause no problems and normally regress within several days.

FALLOPIAN TUBES

As mentioned earlier, infection and related disease processes within the oviducts are uncommon. When inflammation does occur, stricture of the tube can reduce fertility by preventing the ovum from migrating to the uterus. Inflammation of the oviduct, caused by chemical irritation (e.g., due to uterine infusions or chemical curettage), is possible but unlikely. In rare instances, a large ovarian cyst or tumor could constrict that ovary's oviduct, limiting the passage of the ovum and, possibly, interfering with the oviduct's circulation. (Continued restriction of blood to any area can cause tissue degeneration and necrosis.) Tumors of the oviduct are extremely rare.

UTERUS

The presence or absence of uterine abnormalities can be very important in determining the mare's value as a producer. Natural tissue changes that accompany the aging process (e.g., fibrosis, cellular degeneration, cellular death, atrophy) often cause a gradual reduction in the mare's fertility and in her ability to maintain a pregnancy. Certain management practices and injuries may damage the uterus and speed up the natural, degenerative process. For example, failure to detect and correct poor vulval closure allows the inspiration of air and debris into the vagina, which may cause ballooning of the uterus. The presence of air within the uterus can irritate and infect the entire tract, possibly resulting in significant tissue damage and reduced fertility. Failure to tie the mare's afterbirth up off the ground can result in serious damage to the uterus if the mare steps on the placenta and tears it from the tract. Sudden removal of the afterbirth, whether by the mare's hindlegs or by the hands of an anxious attendant, may result in injury, bleeding, retention of pieces of placenta, infection, and formation of scar tissue. After reading the chapters on breeding and foaling, the breeder should be well aware of problems that may cause degeneration of uterine tissue and, consequently, a reduction in the mare's fertility. In addition, the breeder should be familiar with the following uterine abnormalities and their effects on fertility.

Uterine Dilation

Dilation of a portion of the uterus (usually between the uterine body and horn, along the bottom half of the uterus) is a problem seen in mares that have had several foals. While the dilated portion is characterized by flattened mucosal folds, endometrial atrophy, and lowered resistance to bacterial invasion, the rest of the uterus is usually normal. The exact cause of uterine dilation is not known, but researchers suspect that several pregnancies in one uterine horn may interfere with local blood supply, resulting in cellular degeneration and death within a particular area. The bulge of the dilated uterus is distinguished from the pregnant uterus by uterine tone; the dilated uterus is flaccid and limp, while the pregnant uterus has significant tone.

Uterine Asymmetry

When one horn of the uterus fails to involute (return to normal size) and becomes flaccid and unresponsive to hormones, the condition is referred to as uterine asymmetry. Reasons for this abnormality are unknown, but its effect on fertility has been documented. Because the enlarged uterine horn is flaccid and sagging, sperm motility through the area is reduced significantly. Researchers have estimated that fertility is reduced 45-55 percent when the right horn is involved and 55-65 percent when the left horn is involved.

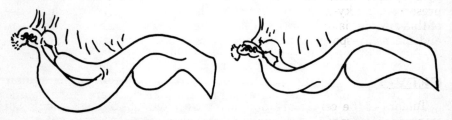

Uterine asymmetry   **Normally proportioned uterus**

Infantile Uterus

A small uterus is characteristic of the immature mare, but it is seen occasionally in older mares. This condition results in poor reproductive performance, usually observed as failure to conceive. If the infantile tract is not the result of a chromosome error (e.g., 63XO aneuploidy) or an inherent endocrine disturbance, time may help to correct the situation.

Lymph Stasis

When the microscopic study of a uterine tissue sample reveals the presence of many lymphocytes (disease-fighting blood cells), but uterine cultures show no indication of an infection, a condition referred to as lymph stasis may be present. In addition to normal cultures, the affected mare's speculum examination reveals no problems, and her estrous cycles appear normal. In spite of this absence of pathological processes, the mare exhibits poor fertility, frequent abortion, and a large spongy uterus. The cause of lymph stasis is not clear, and there is no known treatment.

Tumors

There are no frequently encountered uterine tumors, and those that do occur are usually benign. Leiomyomas and fibromas, for example, are benign neoplasms of the uterine smooth muscle and fibrous connective tissue respectively. Carcinomas, or malignant neoplasms of the cells covering most organ surfaces, are extremely rare in the equine uterus. Although most uterine tumors are benign, they can interfere with fertility if their size causes the uterus to sag or causes a mechanical barrier against normal sperm transport.

Cysts

Uterine cysts are also uncommon in the mare. They are usually seen in conjunction with an increase in intercellular fluid, the formation of small "lakes" within the tissue, and early embryonic death. One large endometrial cyst might be mistaken for a small fetus during rectal palpation, while numerous cysts located just below the endometrial surface give the uterus a soft, glistening appearance. The presence of many small uterine cysts is closely associated with mucometra, a condition that is characterized by a spongy, thick uterine wall; a pendulous, flaccid uterus; and the presence of milky-white mucus within the uterine cavity. The exact cause of this disorder is unknown, but it is believed to be somehow related to an endometrial response to abnormal estrogen and progesterone levels.

CERVIX

Tumors of the cervix are practically nonexistent, but adhesions and improper closure of this essential muscle are extremely important abnormalities. If the cervix is torn, or otherwise injured (e.g., during foaling, breeding, or in response to irritating agents, mechanical procedures, or cervicitis), its ability to open or close properly may be limited. Lacerations of the cervical os (opening), for example, may result in the formation of scar tissue and small adhesions in the area. These adhesions partially close off the opening and reduce the number of sperm that enter the uterus. Palpation reveals the presence of these adhesions, especially when the mare is in estrus. Minor adhesions can usually be corrected by severing the scar tissue and placing a sterile plastic tube or sterile gauze through the cervix to prevent the reformation of adhesions. (Note: This should be performed by, or under the direction of, a veterinarian.) Severe injury to the cervix, on the other hand, can result in failure of the cervix to close properly or in complete blockage of the cervical os. Improper cervical closure predisposes the mare to uterine infection, conception failure, and abortion. Extensive adhesions that completely block the cervix are equally detrimental since they prevent the passage of sperm into, or infectious exudate out of, the uterus. Correction of these extensive adhesions is usually unsuccessful.

VAGINA

Like the cervix, the vagina is more susceptible to injuries and adhesions than to neoplasms and cysts. During foaling, an abnormal fetal position, difficult delivery, or forced extraction of the fetus can cause lacerations of the vaginal mucosa (cellular lining), submucosa, or deep smooth muscle. Because resulting adhesions can cause the mare to refuse the stallion during estrus, routine speculum examinations should include a check for these vaginal strictures. (Refer to "Broodmare Management.") In unusual instances, the foal's hoof may enter the mare's rectum through the vagina, forming a rectovaginal fistula. If the foal's position is not corrected, the mare's extremely powerful uterine contractions may force the foal through both the rectum and the vagina, extending the rectovaginal fistula to the exterior. These vaginal injuries can cause extensive scarring and strictures of the canal. (Refer to "Abnormal Foaling.") Rupture of the vagina may also occur during breeding as a result of an accident, an overanxious stallion, or failure to use a breeding roll. (Refer to "Breeding Methods and Procedures.") If urine, air, and debris enter the abdominal cavity through these injured areas, severe peritonitis and death may result. Although vaginal tears can be corrected successfully, reformation of adhesions is not uncommon. (Corrective surgery cannot be performed until tissue edema and inflammation have subsided.)

In unusual cases, the hymen has no opening and drainage of mucus and debris from the uterus is prevented. When an unusually strong membrane persists into maturity, breeding can cause severe lacerations, formation of scar tissue, and a strong association between breeding and pain. For this reason, many veterinarians surgically rupture the persistent hymen several weeks prior to the maiden mare's first cover. Similarly, a persistent vaginal septum (membrane that follows the longitudinal plane of the vagina) should be corrected prior to breeding.

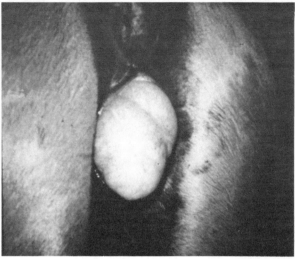

Courtesy of Kansas State University Department of Surgery and Medicine

This maiden mare had an unusual persistent hymen. Shown prior to surgical correction, abdominal pressure has forced the membrane to protrude from the vagina.

VULVA

The vulva is also subject to breeding and foaling injuries. Failure to open a sutured mare before foaling can cause extensive vulval lacerations and permanent damage to the area. Depending on how far the mare's vulva has been stitched, breeding may also necessitate opening the mare or, perhaps, reinforcing the sutured area with a special breeding stitch. Neoplasms of the vulva have been reported and seem to be limited to melanomas, carcinomas, and papillomas. Melanomas are malignant neoplasms originating from pigment-producing cells. These "pigment tumors" are most common in grey horses due to an accumulation of pigment that occurs in association with the greying process. Initially, melanomas are seen under the dock of the tail, around the anus, on the vulva, and below the ear. The tumors spread throughout the body, sometimes affecting life-sustaining organs such as the lungs, liver, brain, and lymph nodes. Carcinomas are malignant neoplasms that affect epithelial cells (cells that form the outer layer for most body and organ surfaces). Squamous cell carcinoma, a type of cancer that sometimes affects the stallion's prepuce, also occurs on the mare's vulva. Small benign vulva neoplasms, or papillomas, have also been reported. These growths are characterized by confined areas of many small tissue projections, or villi.

The presence of an enlarged clitoris in conjunction with outward stallion-like behavior may indicate an inherent sex chromosome anomaly. Abnormal combinations of the sex chromsomes (e.g., XXY, XXXY) may occur within an embryo due to abnormal spermatogenesis or oogenesis within the parents. These unusual chromosome combinations cause the resulting horse to have incompletely developed sex characteristics of both male and female. Intersex horses are sterile.

In conclusion, there are a number of fertility problems that may affect the mare's reproductive status. It is important to note, however, that many of these abnormalities are uncommon or extremely rare. Descriptions of both widespread and atypical fertility problems have been presented in this chapter to serve as a reference source and to emphasize the importance of protecting and enhancing the mare's reproductive status through proper management. ◀

10

BROODMARE MANAGEMENT

Many factors affect the mare's ability to conceive and carry a healthy foal to term. Although fertility is partly determined by inheritance, management practices that influence the mare's environment also significantly affect her overall productivity. By preconditioning each mare for the upcoming breeding season through skillful and prudent management, the breeder helps to insure optimum conception and foaling rates.

Preconditioning is actually a continuous process which requires conscientious husbandry practices and precise planning throughout the year. However, many aspects of preconditioning need to be considered only a few months or weeks before the breeding season to give the mare a better chance of conceiving and carrying the resulting foal successfully. Specific preconditioning procedures should be selected according to the type of breeding operation, the size of the broodmare band, and the mares' reproductive status (maiden, pregnant, or barren). However, many preconditioning practices such as breeding soundness examinations and nutritional management are applicable to all broodmare operations. For this reason, the following discussion describes currently used preconditioning techniques and explains their applications in broodmare management.

Breeding Soundness Examination

Before the mare can be conditioned to reach peak breeding form, she must be examined thoroughly to identify specific reproductive problems and to evaluate her reproductive status and fertility. A prebreeding examination should be carried out when the mare is purchased and several months prior to each breeding season. If the mare then fails to conceive or maintain pregnancy she might also require additional examinations during the breeding season. In fact, the best time to analyze the barren mare's reproductive status is during the season that her problems are first noticed. Some treatments should be done when the mare is in estrus, and since it may take more than a few months to correct a serious fertility problem, the breeder should take advantage of any time available to treat and condition the problem mare before the next breeding season.

BREEDING HISTORY

Complete records of each mare's reproductive history are indispensable to efficient breeding management and often aid in the diagnosis of fertility problems. A detailed breeding history includes a record of the mare's estrous behavior and past estrous patterns, a list of the foals that she has produced along with their dates of birth, notes on the health of each foal, information on past foaling or breeding difficulties, and an account of previous reproductive infections or injuries with their corresponding treatments. Ideally, this information should be organized on a chart and placed in the mare's permanent file. This file might also include general medical information detailing past vaccinations, dewormings, and any significant illnesses. (Refer to "Breeding Farm Records" for suggestions on record keeping.)

Many broodmares spend most of the year being boarded as outside mares and may be sold several times during their lifetimes. Consequently, those charged with handling a mare may be unfamiliar with her reproductive idiosyncrasies. In these cases, a detailed history is an important money-saving and time-saving tool for everyone interested in the broodmare's welfare. An accurate, written record of the mare's past reproductive performance is also a helpful reference during a breeding soundness examination, since it can provide the veterinarian with essential data for the diagnosis of fertility problems. These records can be used to compare the reproductive performance of individuals and to establish culling standards for the broodmare band. The breeding history should accompany the mare when she is sent to an outside stallion or when she is entrusted to anyone who is unfamiliar with her reproductive characteristics.

EXTERNAL EXAMINATION

The first step of a prebreeding examination is evaluation of the mare's apparent physical condition and external reproductive conformation. Do the mare's hair coat and attitude reflect good overall health? Will she require extensive conditioning before she could be considered an addition to the breeding herd? Lift the tail, and inspect her external genitalia and hindquarters. Does the mare have desirable pelvic and vulval conformation? Is she sutured? The labia should be positioned so that they form a tight vertical seal which protects the genital tract. If the labia and anus sink inward, feces cannot fall freely past the vulva. If the relaxed muscles also allow the labia to gape, air and feces may enter the tract, predisposing the mare to reproductive infections. Are exudates or signs of past drainage present on the tail or hindquarters? Creamy or milky discharges indicate reproductive infections. Is the external area surrounding the anus and vulva free of scars and puckers? External scars may be important clues of internal damages that cannot be detected during this preliminary examination. The answers to those questions help to pinpoint obvious problems and may hint at what will be found during the next stages of the genital examination. (It is also important to answer these basic questions before purchasing a prospective broodmare.)

Because the mare's overall health affects reproductive performance, her general condition should be considered during the external soundness examination.

RECTAL PALPATION

Because the condition of the ovaries, fallopian tubes, and uterus is so important to fertility, it is fortunate that the reproductive tract lies conveniently below the rectum. By inserting a well-lubricated, gloved hand through the anus and into the rectum and by gently feeling the reproductive tract and picking it up through the rectal wall, an experienced palpator can detect injury, scarring, pregnancy, and other important conditions. Although the procedure is simple, skill is required to interpret the findings and to examine the tract without injuring delicate tissues. Because the equine rectum is easily ruptured, rectal palpations should only be performed by trained personnel.

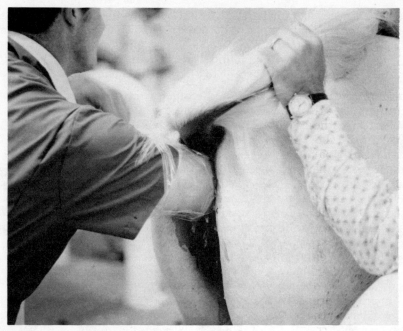

Rectal palpation to determine the condition of the reproductive tract is a routine part of the breeding soundness examination.

The palpator first passes a hand through the anus and along the rectum's length to locate the ovaries. Of primary concern are their size, texture, and follicular activity. Normal ovaries have a definite kidney-like shape and are about 2.4 inches long and 1.5 inches wide (6.1 centimeters long and 3.8 centimeters wide). During the natural breeding season, developing follicles may be felt as irregularities on the ovarian surface. After ovulation, the corpus hemorrhagicum can be palpated as a mushy depression on the surface of the ovary for about 24 to 36 hours. The corpus luteum, on the other hand, is more difficult to identify but can sometimes be felt as a small protrusion until about 72 hours following ovulation. (The corpus luteum is sometimes palpable for about 12 days on the ovaries of pony mares.) Small, hard, inactive ovaries are normal in young mares and in seasonally anestrus mares. However, during the natural breeding season, the ovaries of a mature mare should be active and of moderate size. The presence of unusually small ovaries in mature mares during the natural breeding season is associated with infantile reproductive tracts and other inherent reproductive abnormalities. Mares with enlarged ovaries should be reexamined 10 to 14 days later to ascertain the reason for the increase in size. Size increases can be caused by multiple follicles, ovarian hematoma, and granulosa cell tumor. Examination of the ovaries by rectal palpation cannot guarantee normal function but can insure that there are no gross ovarian abnormalities and, in the case of a prepurchase examination, that the mare has not been ovariectomized (i.e., that her ovaries have not been removed).

The fallopian tubes are difficult to locate by rectal palpation but can sometimes be identified by first determining the position of the ovaries and then locating the respective uterine horns. Normal fallopian tubes feel wiry with uniform consistency throughout. During the examination, they are rolled between the palpator's fingers to inspect for scar tissue or adhesions. As mentioned earlier, adhesions within the oviducts are unusual but can prevent the sperm from reaching the ovary or the fertilized ovum from entering the uterus.

The uterus should be palpated to determine its size, consistency, and location. After parturition (i.e., foaling) the uterus returns to near normal size through a healing process called involution. If involution does not occur or if it is incomplete, the stretched organ remains flabby, causing lowered fertility. In some cases only one horn of the uterus fails to involute, causing a condition known as uterine asymmetry. The size and consistency of the uterine body and horns may also indicate the presence of uterine infection. As described earlier, endometritis and acute metritis are characterized by an enlarged, doughy uterus. Pyometra is associated with the presence of an enlarged, pendulous uterus with thin, flaccid walls. Also, the walls of a pyometra uterus are often characterized by tough, fibrous, nonfunctional areas caused by extensive damage to the endometrium.

Occasionally, the palpator may detect hard spots, localized swelling, and other palpable irregularities. These unusual malformations may be caused by hematomas, cysts, tumors, scars, or other abnormal conditions. (The palpator must be familiar with the characteristics of a pregnant uterus in order to distinguish normal variation from abnormal changes.) In addition to size and consistency, the position of the uterus should be determined. Normally, the genital tract is suspended by the broad ligaments in such a way that adequate drainage and sperm movement are permitted. However, in older broodmares the ligaments lose some of their suspensory ability. For this reason, mares that have had numerous pregnancies should be checked carefully for stretched broad ligaments. (Refer to "Mare Infertility.")

The width and tone of the cervix are noted during rectal palpation to determine the current estrous state. For example, the cervix remains firm and tubular during diestrus, anestrus, and pregnancy, while proestrus and estrus are characterized by relaxation of the cervix and some loss of tubular definition. If cervical characteristics do not correspond with the stage of the estrous cycle that is indicated by the ovaries, the entire tract should be reexamined carefully for abnormalities and infections. The cervix is also checked for scars which might prevent proper closure and for adhesions which can prevent the entrance of sperm at breeding or prevent expulsion of the fetus at foaling. If cervical adhesions are detected during palpation, the veterinarian can assess their severity by visual examination through the vagina.

Finally, the roof of the vagina and vulva can be checked for tears and scars by rectal palpation, although this observation is more commonly made with the aid of a speculum. This area of the reproductive tract is more commonly damaged than any other because it serves as the birth canal.

With legs fully extended, the foal is forced through this restricted area during parturition (foaling). If the foal's limbs are not properly aligned, they may become battering rams powered by the mare's own contractions and may cause severe lacerations of the vagina and vulva.

SPECULUM EXAMINATION

During a detailed reproductive examination, the veterinarian views the condition of the mare's cervix and vagina through a speculum. Either a disposable speculum or a Caslick's speculum is used for visual examination. The disposable speculum is a long, hollow tube approximately 1.5 inches

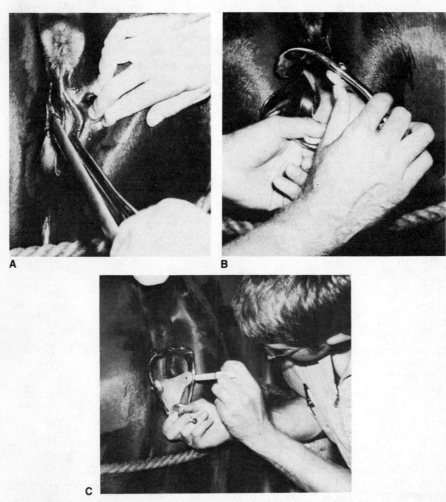

A sterile Caslick's speculum is (A) inserted between the labia and into the vulva and (B) opened. (C) In the open position the speculum permits visual examination of the vagina and cervix.

(3.8 centimeters) in diameter. The Caslick's speculum is adjustable so that it can be gently slipped through the labia and into the mare's vagina in its closed position and then opened to a wider position for easier viewing. The instrument must be sterile and should be used only by trained personnel, since the vagina and cervix are exposed to airborne debris and pathogens while the vulva is open. Because the mare's reproductive tract is exposed to contaminants during this examination, several precautionary measures should be followed:

1. The speculum examination should be carried out in a relatively dust-free area.
2. Prior to the examination, the mare's tail should be wrapped with sterile gauze.
3. The perineum and vulva should be thoroughly washed and rinsed before the speculum is inserted.

During the examination, the vagina and vulva are checked for evidence of bruising, irritation, scars, and tears. Normally, the vulval and vaginal membranes are a healthy pink color. Infected or irritated tissue is usually red and covered with cloudy mucus. White, grey, or yellow fluid on the vaginal floor indicates infection of the vagina, cervix, and/or uterus. The presence of urine on the vaginal floor or adhesions within the vulval or vaginal areas may also be encountered during a speculum examination. These conditions and others that threaten the mare's fertility are described in "Mare Infertility."

Occasionally, a persistent hymen that separates the vagina from the vulva is found during speculum examination of the maiden mare. This intact membrane should be opened in a strictly sterile manner by a veterinarian well in advance of the breeding season so that it is not torn painfully during the mare's first cover. Rarely, persistent hymen can be associated with other reproductive tract abnormalities. Because abnormalities may be complicated when the hymen is torn, the hymen should be opened only after a veterinarian has thoroughly palpated the reproductive tract.

The color and appearance of the cervix vary with the stage of the estrous cycle. For example, during diestrus the cervix is tight and pale. On the other hand, during estrus the cervix becomes relaxed, swollen, reddened, and covered with a clear mucous secretion. (Note: It is important that cervical changes associated with pregnancy be differentiated from normal cyclic changes. Refer to "Pregnancy.") Because the cervix plays an important role in preventing uterine infection and maintaining pregnancy, it should be checked carefully for signs of inflammation, tears, and adhesions. An inflamed cervical membrane indicates infection, especially when a cloudy discharge is also present. (During estrus, cervical secretions are normally clear and watery.) Cervical tears, caused by foaling difficulties or breeding accidents, may prevent proper closure of the cervix and predispose the uterus to serious infection. Both cervical infections and tears may cause adhesions. Depending on their severity, adhesions may completely block the cervix or may slow the passage of sperm. Minor adhesions are often corrected by a special surgical procedure, but normal cervical function is seldom regained if the adhesions are severe.

CULTURES

If a reproductive infection is suspected, cultures may be added to the examination procedures. Some breeding farms also require that outside mares be cultured before they are bred. Culturing is the process of collecting microorganisms and propagating those organisms on a special nutrient gel, or media. Specific organisms flourish on certain types of growth media and respond differently to certain environmental conditions (e.g., temperature, presence or absence of oxygen). Based on growth characteristics, organisms can be differentiated into basic types. These types are further differentiated by their size, shape, staining characteristics, and population growth patterns as seen through a microscope. Because successful identification of specific organisms requires skillful laboratory technique and a background in microbiology, cultures are often sent to special laboratories for analysis. If pathogens are identified and if an infection is indicated, the laboratory may also conduct antibiotic sensitivity tests to ascertain the most effective treatment.

Several methods can be used to collect a culture sample. Occasionally, the mare's uterus is flushed with a sterile fluid so that a representative sample from the entire uterus can be collected. This so-called irrigation culture is not employed as commonly as the swab technique, which involves collecting fluid on an absorptive, sterile material. Although methods do vary, the following description of one type of swab culture is provided as an illustrative example of the basic culturing procedure. (This description presents only an outline of the procedure and is not intended to serve as step-by-step instruction.)

1. Prior to culturing, the mare is placed in a dust-free, draft-free area. Her tail is wrapped in sterile gauze and tied or held to the side.

2. Because it is very important to avoid contamination of the specimen or the genital tract by external organisms, the mare's perineal area should be washed three times and rinsed thoroughly after each gentle scrubbing.

3. After donning a sterile, lubricated, obstetrical sleeve and glove, the practitioner inserts an arm into the vagina to introduce the sterile swab. When swabbing a specific area of the uterus, the practitioner maneuvers the swab through the cervix, and then removes the arm from the reproductive tract and inserts it into the rectum. In this position, the hand can be used to guide the swab to the desired location within the uterus. Usually, organisms from a specific area, such as the anterior cervix or the uterine body, are sought, and the swab must be protected from contamination until that area is reached. For this purpose, a special guarded swab contained within two sterile tubes may be employed. When culturing the uterus, for

example, the first tube is passed through the cervix and the second tube is telescoped into the uterus. The swab is then pushed out of the second tube far enough to touch the endometrium. By reversing the telescoping procedures, the swab is protected as the apparatus is removed from the reproductive tract. (When culturing the vagina or cervix, the swab may be inserted directly into the tract through a vaginal speculum, which protects the swab from contamination.)

4. When the swab contacts the reproductive tract's inner lining, it absorbs fluids and the organisms contained in those fluids.

5. After the sample is collected, the swab is applied to a special growth media, covered, and incubated for several days. If microbial growth occurs and if the media has not been swabbed too heavily, individual colonies may be apparent. These isolated areas of growth often have characteristic shapes and colors, especially when incubated under controlled conditions. (i.e., Changes in temperature, atmospheric pressure, humidity, and oxygen content change the appearance of these colonies.) Special laboratory techniques also include isolation of a pure culture, sensitivity tests, and microscopic analysis.

6. A swab from the uterus can also be smeared on a glass slide and stained for cytological examination. Infection causes an increase in the number of lymphocytes or neutrophils in the uterine mucus. Because cytological examination requires only a microscope, the appropriate stain, and slides, the veterinarian can base preliminary treatment on the results of the cytological examination until more complete information on the culture is available from the testing laboratory.

Most sources agree that culturing is an extremely valuable diagnostic tool when an infection is suspected or when clinical indications are present. Cervical culture through a sterile speculum is quick and simple to perform and is often used to screen a large number of mares for infection. However,

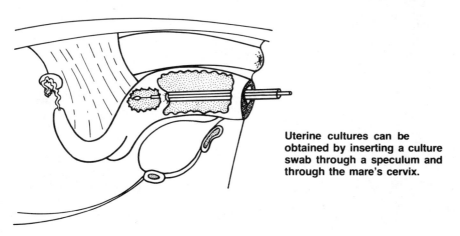

Uterine cultures can be obtained by inserting a culture swab through a speculum and through the mare's cervix.

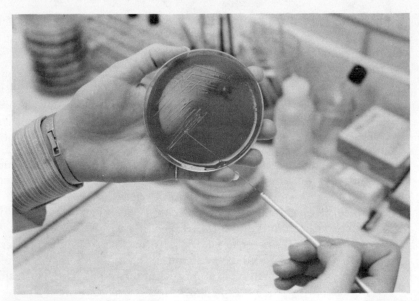

The uterine swab is applied to a plate containing growth media and then incubated.

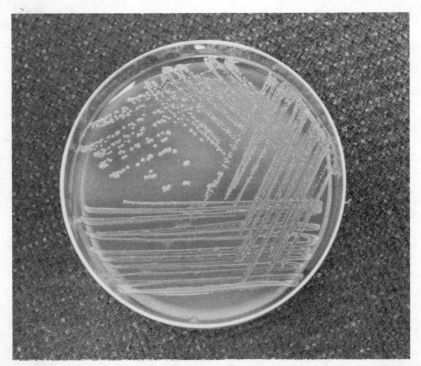

After a period of incubation under controlled conditions, colonies of bacteria may be apparent on the culture plate.

the use of routine cultures to check for infection is controversial, since pathogens are sometimes present even in healthy reproductive tracts. In addition, it is very important to remember that, although the cervix and vagina are often contaminated by pathogens, vaginal and cervical infections do not necessarily indicate that uterine infection is present. Routine cultures also increase the chances of contaminating the reproductive tract with airborne pathogens and debris. To counteract the effects of exposure, cultures are usually taken during estrus when a high level of estrogen in the blood increases uterine resistance to pathogens. Improper culturing of the mare during diestrus may cause or aggravate an infection if pathogens are introduced during this period of low resistance. (Also, miscellaneous non-pathogenic bacteria which flourish during diestrus may overshadow the primary pathogens on the growth media.)

Another advantage of culturing during estrus lies in the fact that increased mucous production during this period makes it easier to obtain a representative fluid sample from the uterus. In addition, the cervix is easier to penetrate when open and relaxed. Many veterinarians and technicians prefer to culture on the third day of estrus, but if a pathogen is present and if that pathogen is causing an infection, it is available for culturing during all stages of the mare's cycle.

UTERINE BIOPSY

For many years, detailed tissue studies (biopsies) have been used by researchers to study normal and abnormal cell physiology, but only recently has the uterine biopsy gained widespread recognition as a valuable tool for equine fertility evaluation. Consistent biopsy techniques and interpretations have been established so that individual mares can be tentatively evaluated for fertility on the basis of one or more tissue samples taken from the uterus. When combined with the results of a complete reproductive examination, endometrial biopsy provides a fairly accurate measure of fertility. Broodmares with questionable reproductive histories and those which represent a large investment should be biopsied to determine whether the uterus is capable of maintaining pregnancy. Endometrial degeneration and scarring are important causes of infertility and abortion but, unfortunately, such problems usually cannot be detected through rectal palpation, speculum examination, or cultures. Even when other clinical examination procedures fail to determine the cause of infertility or habitual abortion, samples of the uterine endometrium help to identify specific fertility problems and often provide vital information about the mare's ability to nurture fetal life. Barren mares, abortion-prone mares, mares with persistent uterine infections, and those that cycle irregularly during the natural breeding season are all candidates for uterine biopsy.

To collect a biopsy sample, a small tissue sample is taken from the inner lining of the uterus using a special tweezer-like instrument, referred to as forceps. The biopsy forceps are passed through the vulva, vagina, and cervix, and a pinch of tissue is taken from the uterine wall (usually at the horn-body junction) by closing the jaws of the instrument. It is extremely important to obtain a sample which is representative of the entire uterus. If uterine malformations are not palpated through the rectal wall, one tissue sample measuring approximately 1x2 centimeters is thought to be sufficient. However, palpable abnormalities should be biopsied separately.

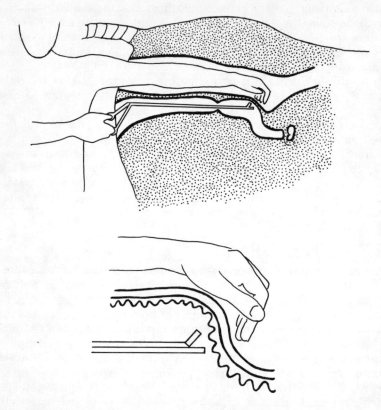

Uterine biopsies are taken with special forceps which are inserted into the uterus through the cervix. A small section of tissue is obtained by closing the jaws of the forceps while they are pressed against the uterine wall.

After the sample is retrieved, it is placed in a solution that preserves tissue structure until microscopic examination can be performed. During this examination, the pathologist looks for changes in cellular structure and function which may indicate specific disease processes such as inflammation, fibrosis, or necrosis. The mare's fertility is then evaluated according to the degree of tissue damage found in the sample. No long-term ill effects have been associated with biopsy, but in rare instances, the period of estrus following tissue collection may be prolonged and the following period of diestrus shortened. In most instances, however, biopsies can be performed during the same estrus that the mare is bred without any adverse effect on fertility.

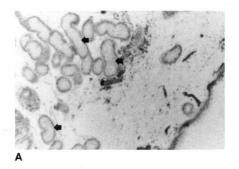

A

C   *Photos Courtesy of Kansas State University*
           Department of Surgery and Medicine

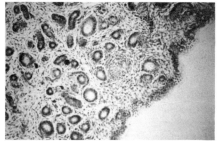

B

Uterine biopsies must be interpreted by a pathologist or a histologist. However, these three microscopic views of uterine biopsy samples show characteristic conditions found in the mare's uterus: (A) normal endometrial tissue containing well-defined uterine glands (marked by arrows), (B) metritis, (C) tissue severely scarred by infection with loss of uterine glands.

ENDOSCOPIC EXAMINATION

The endoscope is an instrument designed for viewing the inside of a canal or hollow organ. There is a wide variation among the currently used endoscopes. For example, hollow endoscopes are relatively inexpensive but are limited by their rigidity which makes them difficult to pass into the uterus and restricts the operator's viewing area. The presence of a light source on the terminal end of the instrument adds to this awkwardness.

The sophisticated fiberoptic endoscope, on the other hand, gives a clearer and more complete view of the uterus than does the hollow endoscope. The image is transmitted to a viewing window by flexible glass fibers which permit the examiner to bend the endoscope without losing the image. Because light can be transmitted along these fiber bundles, a light source does not have to be introduced into the mare's uterus. Although this expensive medical tool is used primarily as a teaching aid at universities and research centers, it is occasionally used by practitioners to visually examine the uterus for abnormalities which may not be detected by rectal palpation or by random biopsy. The complex optical system of the fiberoptic endoscope is easily damaged by misuse. Due to their cost and special handling requirements, these instruments are not yet used on a widespread basis.

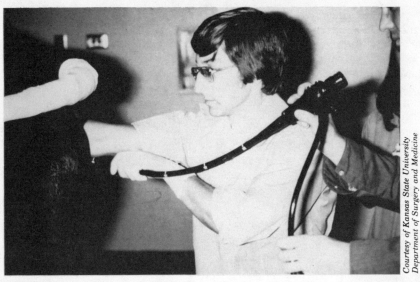

*Courtesy of Kansas State University
Department of Surgery and Medicine*

The flexible endoscope can be inserted through the vagina and cervix to view the interior of the uterus.

ESTIMATE OF FERTILITY POTENTIAL

After a complete reproductive examination is conducted, mares can be ranked on the basis of estimated fertility potential. One such system (Von Lepel) separates mares into five categories to simplify broodmare management.

VON LEPEL FERTILITY CHART

	REPRODUCTIVE HISTORY	ESTIMATED FERTILITY POTENTIAL
GROUP I	Pregnant or maiden mares without clinical abnormalities or infection	70-100 %
GROUP II	Mares without clinical abnormalities or infection but barren for one year	50-70 %
GROUP III	Mares without clinical abnormalities or infection but barren for more than one year	25-50 %
GROUP IV	Mares with clinical signs of genital disease	0-25 %
GROUP V	Mares with little hope for future breeding because of clinical manifestations of genital disease or old age	0 %

Because these classifications are only estimates of fertility, the mare may change categories from one breeding season to the next and a mare in any of the categories may conceive. Even if she has been designated a poor breeding prospect based on her reproductive history, the mare's barren state could be the result of factors such as stallion infertility, improper teasing techniques, or malnutrition. In addition to establishing categories for preconditioning purposes, these guidelines are helpful supplements to the mare's permanent file and might also be considered when culling individuals from a large broodmare band. Fertility guidelines provide the manager or veterinarian with only a general evaluation of the mare's reproductive potential. Thus, they should be considered with respect to her overall condition and environment.

Management for Increased Productivity

Although special management practices are often employed just before breeding to stimulate ovulation and prepare the mare's tract for conception, the most important preconditioning tools are utilized year-round. These include proper nutrition, preventive medicine, parasite control, exercise, safety, and many other common-sense husbandry techniques that encourage overall health and reproductive fitness. Much of the current equine reproduction research focuses on special techniques that may be used to improve the productivity of individual mares. These practices are exciting and innovative, but they are useless in the absence of sound husbandry practices.

On small breeding farms, special attention can be given to the management of each mare. However, when large numbers of mares are being handled, individual attention is more difficult. For convenience, broodmares are usually sorted into groups according to their reproductive status: maiden, barren, pregnant, or lactating. Not only are nutritional and management requirements similar within each group, but individuals within each reproductive category are also more likely to be socially compatible. If several new mares are to be added to a broodmare band, the introduction is usually smoother if they can be buddied up by pair bonding them in lots for several weeks prior to turning out to pasture. Two compatible mares that have been introduced in this fashion usually remain companions after being released within the broodmare band and are less likely to be singled out for bullying by more dominant members of the herd.

MANAGING THE MAIDEN MARE

Regardless of past training, showing or racing regimens, the maiden mare requires special consideration during the preconditioning period. She must be introduced to breeding farm routine in a manner that assures a healthy attitude toward her new career. If she is not stalled or if she cannot be pastured with a small group of other newcomers, the maiden mare should be introduced to an established herd very carefully since she may not mix well with settled broodmares. Ideally, maiden mares should be grouped together so that their estrous cycles can be observed and evaluated carefully. (Young maiden mares tend to show extremely erratic estrous patterns during their period of transitional receptivity early in the breeding season.)

New mares should be carefully introduced to the broodmare band, since they may be bullied by more dominant members of the established group.

The reproductive examination is important in any prepurchase or prebreeding situation but it is especially important with respect to the maiden mare. During this initial examination, rare but serious abnormalities (e.g., missing uterus, infantile tract, intersex, etc.) may be discovered. More importantly, the maturity of the young maiden's reproductive tract and the condition of the older maiden's reproductive tract can be evaluated carefully. Even though a young mare's tract is prepared to deliver ova and receive sperm, her growing body may not be able to handle the rigors of gestation and foaling. Individual and breed-related differences in rate of growth and development should be considered carefully before a young maiden mare is bred. Occasionally, an older maiden mare's genital tract may be suffering from, or damaged by, a long-term infection. For this reason, older maiden mares are sometimes a significant source of frustration to the breeder.

Ideally, preconditioning procedures for the maiden mare should begin during the summer or fall preceding her first cover. This period gives the mare time to adjust to new surroundings, different handlers, diet changes, and new companions. More importantly, the let-down period gives the mare time to unwind and to make a mental and physical transition from athlete to broodmare.

When a highly fit mare arrives at the breeding farm from racing or from showing she should be placed on a well-planned preconditioning schedule. Such a mare is usually very excitable and energetic; therefore, she should be kept calm with daily riding or hand walking, and her energy intake should gradually be reduced to maintenance levels. Initially, the mare should be confined to a stall, but as she becomes accustomed to a lighter exercise schedule and her new surroundings, she can be released for periods of free exercise in a pasture or paddock.

Although most mares have little difficulty adjusting to the slower pace of the breeding farm, there are several important problems that may be encountered during this transition:

1. Due to the mental and physical stresses associated with show and race careers, estrous cycles may be irregular until the mare's body adapts to diet and exercise changes.
2. During the adjustment period, the mare's body may be trying to eliminate drugs that have been administered in hopes of improving her performance. If the mare has been given hormones to suppress estrus, for example, her estrous cycles may be irregular until the causative agent is broken down and gradually eliminated from her body.
3. If corticosteroids have been used (e.g., to treat lameness), the mare's ability to overcome infection may be impaired. Corticosteroids quiet the natural inflammatory response to tissue injury and allow infections to persist unchallenged by the body's disease-fighting mechanisms. For this reason, mares treated with corticosteriods for extended periods are especially susceptible to long-term genital infections.

MANAGING THE BARREN MARE

Many breeders make the mistake of ignoring the barren mare until the breeding season begins. Because these mares may have reproductive problems that require frequent examination and a series of treatments, the time to concentrate on the cause of their barren status is when it is first determined that they are barren. If the barren mare is still cycling regularly, cultures, biopsies, and/or treatments can be performed during estrus when high estrogen levels increase the reproductive tract's natural disease resistance. If the mare is infected, treatment during several heat periods may be required to bring the infection under control. After treatment, several months of quiet anestrus give the reproductive tract time to recuperate and, if breeding is not attempted until regular estrous cycles become established, there is a reasonable chance of avoiding reinfection. It should also be noted that a concentrated attempt to correct reproductive problems well in advance of the breeding season allows the breeder to concentrate on foaling and breeding activities.

NUTRITION

If a mare is free of genital infection and reproductive abnormalities but remains barren, diet may be contributing to her fertility problem. Malnutrition can prolong winter anestrus and cause irregularities in the length of estrus. Both overweight and underweight conditions have been reported to alter normal ovulation patterns. Also, deficiencies of certain nutrients can reduce conception rates and increase the incidence of abortion. Deficiencies of vitamins A, D, and E are known to affect reproduction, but the therapeutic value of vitamin supplementation is often overrated. Vitamin supplements may increase fertility if the mare's ration is vitamin deficient, but research has not been able to show that dietary excess of any nutrient enhances reproduction. In fact, large doses of some nutrients can be detrimental, causing toxicity or making other nutrients unavailable by interfering with their absorption.

Although nutrient requirements vary with the mare's age, size, activity, and reproductive status, the following guidelines apply to balancing rations for any horse:

1. Determine the horse's nutrient requirements.
2. Analyze the ration and compare it to the horse's requirements.
3. Supplement the ration to meet deficiencies or adjust the ration to prevent wasteful and perhaps harmful dietary excesses of certain nutrients.

With these principles in mind, an appropriate ration can be formulated.

BARREN AND MAIDEN MARES

The effect of physical condition on fertility is very pronounced in mares. Studies show that when open mares are losing weight as the breeding season approaches, they begin cycling about 30 days later than mares that gain weight during this period. Reports also indicate that conception rates increase when open mares are kept trim during late fall and early winter and then given a high-energy ration several weeks prior to breeding. This unexplained phenomenon, referred to as flushing, is ineffective in obese mares. For this reason, it is necessary to keep open broodmares in trim condition through the fall and winter. (Note: Pregnant mares should not be flushed. Refer to "Pregnancy.")

In place of livestock scales, an inexpensive horse weight tape can be used to monitor the mare's weight while she is being conditioned for breeding.

Flushing can be accomplished in several ways, depending on the breeding farm facilities and the number of mares involved. The key to a flushing program is putting the mares in gaining condition several weeks prior to breeding. To accomplish this, energy intake is increased by reducing roughage intake slightly and by adding 2 to 3 pounds of grain to the ration about 4 to 6 weeks before breeding. Mares are usually kept in dry lots or stalls so that their intake can be monitored, but they can be maintained on good quality pasture if supplemental grain is provided. Unless the hay or pasture contains some legumes, it is also important to supplement the ration with protein. When pasture grasses become dry and mature in the winter, the mare's diet is very likely deficient in both protein and energy. Therefore, a supplement should be added to the ration to meet protein maintenance requirements and to increase energy levels. Whether the mare is kept in a stall, dry lot, or pasture, her diet should always contain a balanced calcium-phosphorus ratio as well as adequate levels of these minerals.

A maintenance diet is adequate for the open mature mare, since she is under little stress and tends to gain weight easily. The basic maintenance requirements are at least 8.5 percent protein, 0.3 percent calcium, 0.2 percent phosphorus, and 1.0 Mcal of digestible energy per pound of feed (2.2 Mcal/per kg). (Refer to the Appendix for a more detailed list of equine nutrient requirements.) When dealing with mature open mares, often the greatest challenge is to keep them trim but well-nourished. Since roughages are lower in energy than concentrates, good quality hay or pasture can be fed until about 60 days preceding the breeding season, when energy levels are increased to enhance conception rates. When pasture quality decreases in the winter, grain is usually required to fulfill energy demands. In extremely cold climates, grain may also be necessary to supply the mare with enough energy for body heat production during the winter. An ability to evaluate body condition is a great aid in estimating the amount of grain required during this period.

Cool season grasses, such as this wheat pasture, provide excellent winter grazing for barren mares.

Mares from two to three years of age require levels of protein, calcium, phosphorus and vitamin A that exceed the maintenance requirements for mature horses. If young mares are placed on a preconditioning program in anticipation of breeding at three years of age, the ration should contain at least 10 percent protein, 0.45 percent calcium, 0.35 percent phosphorus, and 800 international units of vitamin A. Nutritional deficiencies during this period of growth may result in prolonged anestrus or even stunted growth. Appropriate levels of calcium and phosphorus are particularly important since these minerals are stored in bone and released during periods of high demand (e.g., pregnancy and lactation). Depletion of these stores can cause brittleness of the skeletal structure and other metabolic problems.

PREGNANT MARES

The requirements of pregnant, non-lactating mares during the first eight months of gestation are identical to those of open mares. However, fetal growth during the last 90 days of gestation requires levels of energy, protein, calcium, phosphorus, and vitamin A well above maintenance requirements. In addition, the mare should be fed enough energy to store a moderate amount of fat before foaling, which can then be called upon to meet the tremendous energy drain of early lactation. Specific dietary needs and conditioning requirements of the mare during gestation are discussed in "Pregnancy."

LACTATING MARES

The nutritional requirements of mares with foals at-side should be considered independently of other mares on the farm. The lactating mare requires high levels of protein, energy, calcium, phosphorus and vitamin A for milk production. Three of these nutrients — calcium, phosphorus, and vitamin A — should receive special scrutiny when rations are planned since they are stored in the mare's body and are required in large amounts during late gestation and lactation. Although it is very important to supply adequate levels for normal milk production, the reader should note that oversupplementation can be harmful to the mare. (Refer to "Lactation," for a more detailed description of the nutrient requirements for mares in this category.) Some of the foal's nutritional demands on the lactating mare can be reduced by introducing creep feed, as discussed in "Suckling Foals."

VACCINATIONS AND DEWORMING

Most equine vaccination programs follow a continuous year-round schedule based on health problems that are common in a particular area. For example, influenza boosters are usually given in late winter, while Eastern, Western, and/or Venezuelan equine encephalomyelitis vaccinations may be given during the insect season. In some areas, preventive vaccinations for rhinopneumonitis (a respiratory disease which can cause abortion between the eighth and eleventh month of gestation) are given to each mare during the early stages of pregnancy. Tetanus toxoid is usually administered a month before foaling so that tetanus antibodies will be concentrated in the mare's colostrum. (Refer to the vaccination charts within the Appendix.)

Parasite infestations, which drain the mare of nutrients, cause tissue damage, and decrease her disease resistance, have a detrimental effect on fertility. Like the vaccination program, effective parasite control is a continuous process that involves proper hygiene, routine pasture management, and strict deworming schedules. Effective deworming programs rotate anthelmintics (dewormers) so that parasites do not build up a resistance to specific agents. Checking parasite levels between treatments helps the veterinarian and manager evaluate a dewormer's effectiveness and aids them in appraising the parasite control program's overall success. Specific methods of deworming the mare and appropriate dewormers should be chosen on the basis of the farm's location and the mare's reproductive status. Some deworming chemicals are harmful to the fetus and are not recommended for use during pregnancy. (For a helpful overview of breeding farm parasite control measures, refer to the Appendix.)

CASLICK'S OPERATION

If a mare's vulva is tipped so that it catches fecal matter or if the labia form a poor seal, it is essential that the vulval lips be sutured together to a point below the pelvic brim. This relatively simple procedure, referred to as Caslick's operation, involves trimming the labia to form raw edges and then stitching the edges together. After the area heals, a seal is formed, which prevents air and feces from entering the genital tract. A small opening at the lower V-shaped portion of the vulva allows urine to escape.

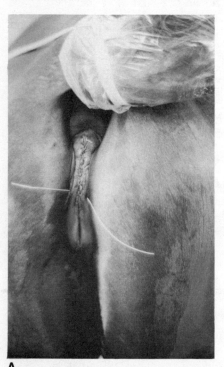 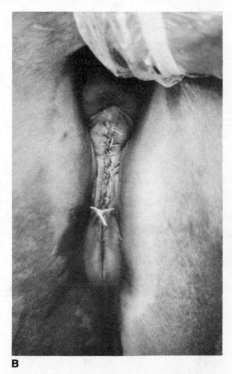

A   B

(A) A Caslick's operation was performed on this mare to prevent air from entering the vagina. A breeding stitch is inserted below the Caslick's suture line and (B) tied to reinforce the closure. The lower portion of the labia is not sutured, leaving an opening through which urine can be expelled.

(Note: If the mare tends to pool urine within her vagina, a small vulval opening may aggravate her problem.)

Caslick's operation is especially helpful to mares with perineal conformation that encourages pneumovagina, but the operation is performed routinely in many mares as added insurance against infection. Just prior to breeding, the vulva should be opened surgically or reinforced with a special breeding stitch, depending on the length of the sutured area and on whether the mare is to be bred by artificial insemination or live cover. If the mare is opened for breeding, her vulva should be sutured again after she conceives. To prevent serious damage to the vulva, it is extremely important that the sutured vulva be opened prior to her estimated foaling date.

CONTROLLING THE ESTROUS CYCLE

Using special management techniques, it is possible to control or interrupt the estrous cycle so that ovulation occurs at a desired time or in a predictable manner. One of the most common applications of estrous control is to induce an earlier breeding season using artificial light to lengthen the photoperiod. Postfoaling mares are also candidates for induced ovulation, as the discussion on prostaglandins points out. The many applications of estrous control also include heat synchronization to decrease the number of days that a large group of mares must be teased and bred (e.g., in an extensive artificial insemination program). Also, special techniques can be used to alternate several mares' periods of estrus throughout the breeding season, thereby protecting the stallion's fertility and potency by limiting the number of mares he must cover on a particular day. This aspect of estrous control is especially helpful when a large number of live cover bookings have been made to a single stallion. In addition, specific problems of the barren mare can sometimes be corrected or overcome by stimulating or interrupting the estrous cycle.

Manipulation of the estrous cycle requires a thorough understanding of the delicate endocrine balance that controls the normal cycle. In addition, accurate records, careful teasing practices, stringent health programs, and veterinary supervision are important prerequisites for any estrous control procedure.

LIGHT PROGRAMS

As mentioned in "The Mare Reproductive System," cyclic ovarian activity in the mare is affected by changes in daylight length. In response to a lengthening photoperiod, the hypothalamus secretes more gonadotropin releasing hormone (GnRH) which, in turn, stimulates the release of FSH and LH from the anterior pituitary. By mimicking the photoperiod of springtime, this chain of events is stimulated, hastening the onset of normal estrous cycles. Initiating early estrous cycles seems to have no long-term detrimental effects on fertility and does not cause early anestrus in the fall. In fact, the end of the breeding season can also be prolonged by employing artifical lights when daylight length begins to decrease. Delayed anestrus is not a common management tool but is sometimes used to gain several extra estrous cycles during which infected mares can be treated and cultured.

Although light programs vary between farms, there are several important guidelines to follow. First, the mare must be stalled or penned every night in a small pasture or paddock that is equipped with adequate light in order to regulate the photoperiod. If the mare is to be shipped to another farm for breeding, it is important that the light program be continued there. (An irregular light schedule will not induce estrus and may even throw mares back into anestrus.) Second, it takes a minimum of 45 to 60 days (usually longer) to induce ovulation with artificial light. If the breeding season starts on February 15, for example, lighting should begin no later than December 1. Many farms begin adding 30 minutes of light each evening starting sometime between November 15 and December 1. After a week, the artificial light period is extended an additional 30 minutes, and this period is gradually increased so that the total "daylight" length reaches 16 to 18 hours. At this point, the mare's reproductive system responds as though the natural breeding season had begun; irregular, transitional estrous cycles are gradually replaced by short, regular ovulatory cycles. This progressive increase in daylight length is both effective and popular, but recent research indicates that a gradual increase in artificial lighting is not necessary. Programs that begin around December 1 with 16 hours of light per day and that maintain the photoperiod until natural daylength is approximately 16 hours, or until the mare conceives, have successfully induced early estrous cycles in mares.

Courtesy of G. L. Morrow, DVM, Shelton Ranch

This specially designed barn is equipped for inducing early estrus in mares. Bulbs for supplemental lighting are placed well above the stalls for safety, and heating ducts regulate the barn's temperature.

The light source for these programs is also an important variable. A 200 watt incandescent bulb or four 40 watt fluorescent light bulbs are effective in stalls. Like all other electrical farm appliances, stall lights should be installed with the mare's safety in mind. Wire or clear plastic covers, proper wiring, and secure mounting are important safety precautions. Also, the manufacturer's guaranteed product lifetime should be noted upon installation, so that old bulbs can be replaced **before** they burn out.

Incandescent, mercury vapor, metal halide, sodium, or quartz fixtures on poles are suitable for lighting pens or paddocks. As a general rule, the intensity of outdoor lighting is adequate if a newspaper can be read at any point within the enclosure. Indoor and outdoor lighting systems may be controlled manually or with automatic timers. Since both methods are effective, the choice is usually a matter of economics versus convenience.

The lighting system can be equipped with an automatic timer to conveniently control the light period.

Another important aspect that should be considered is that lighting programs are sometimes ineffective when mares are in deep anestrus. However, response to an increased artificial photoperiod may be enhanced if mares are on a high-energy diet and in gaining condition. Careful teasing every other day may improve cyclic responses, but the teasing manager should not pressure unresponsive mares and risk their souring to the teaser. Heating the stall during this period may also enhance the mare's responses, but a definite relationship between environmental temperature and seasonal changes in ovarian activity has not been established.

PROSTAGLANDINS

In addition to its use for treatment of a persistent corpus luteum, which is discussed in "Mare Infertility," prostaglandin ($PGF_{2\alpha}$) is also used to regulate the estrous cycle. In a live cover operation, prostaglandin is sometimes used to shorten the interval between estrous periods. For example, if more mares came in heat on a given day than the stallion is able to breed, a few mares might be allowed to go out of heat without breeding. They would then be injected with $PGF_{2\alpha}$ on day 5 or 6 of diestrus. The majority of mares treated in this manner should return to estrus within two to four days.

Prostaglandin may also be used to aid in the detection of estrus and to synchronize estrus. If a group of mares are treated with two injections of $PGF_{2\alpha}$ 14 days apart, a high percentage will exhibit estrus within 6 days after the second injection. The first injection causes regression of any mature corpora lutea but does not affect mares in estrus or with immature corpora lutea (i.e., less than 4-5 days old). However, when the second injection is given 14 days later, both groups have mature corpora lutea and are sensitive to the luteolytic agent. Consequently, the corpora lutea regress, and the mares enter estrus.

Prostaglandin is also used to induce estrus in mares that have just foaled but cannot be safely bred on their foal heat (first estrus after foaling). By injecting $PGF_{2\alpha}$ on the 6th or 7th day postovulation (i.e., foal heat ovulation), the next ovulatory cycle is hastened, and the breeder does not lose valuable time waiting for the next period of estrus to occur. This delay can be especially frustrating if the mare experiences anestrus after foal heat.

Because the corpus luteum of pregnancy (the progesterone-producing body that maintains early pregnancy) is also sensitive to prostaglandin, the agent is sometimes used to induce abortion. Therefore, it is very important to determine whether the mare is pregnant before giving prostaglandin. Planned abortion might be instigated if the mare is misbred, if she is carrying twins, or if her health is somehow jeopardized by the pregnancy. (Refer to "Abortion.") Prostaglandin must be administered before the endometrial cups become established in order to induce abortion. On about day 36 of pregnancy, these endometrial cups form and produce pregnant mare serum gonadotropin (PMSG), a compound that supports early pregnancy.

The administration of prostaglandin and its effectiveness in initiating estrus in problem mares are examined within "Mare Infertility." These guidelines also apply to the use of prostaglandin as a routine management tool and should be studied carefully before such a program is initiated. It should be reemphasized that prostaglandin is often very helpful when used properly, but it is not 100 percent successful. Research reports indicate that prostaglandins are about 90 percent effective in causing regression of the corpus luteum. Occasionally, regression may be incomplete, resulting in a normal estrous interval or, in the case of a persistent corpus luteum, in failure to bring the mare out of diestrus. Reports also show that in some cases the corpus luteum fails to respond to prostaglandin or may regress and then recover suddenly so that estrus and ovulation are not promoted.

SALINE INFUSION

Warm saline (0.9 percent salt solution) can be used to infuse the uterus during diestrus to promote estrus and ovulation by stimulating the uterus to produce prostaglandins. If infused between the fifth and ninth day after ovulation, mares usually cycle three to eight days earlier than normal. The advantages of this treatment are obvious: there is little chance of saline solution upsetting delicate hormone balances and repeated treatments are not harmful or expensive. However, saline is ineffective if the uterus is physically incapable of producing or releasing prostaglandin.

HORMONES

The use of reproductive hormones to manipulate the mare's estrous cycle or to correct specific fertility problems is a new and promising concept. In fact, hormonal estrous control is probably one of the most researched and debated topics in the equine reproduction field. In spite of the active struggle to reveal the mechanisms that regulate the mare's endocrine system, there are still many unanswered questions, and routine use of most reproductive hormones is not yet recommended. Hormones that have received special recognition for their potential management and therapeutic capabilities and on which considerable research is being conducted include human chorionic gonadotropin (HCG), gonadotropin releasing hormone (GnRH), and progesterone.

HCG, which is found in the urine of women during the first 50 days of pregnancy, is very similar to PMSG (pregnant mare serum gonadotropin) and is sometimes used to trigger ovulation in mares. As described earlier, GnRH is secreted by the hypothalamus to stimulate the production and release of anterior pituitary hormones which, in turn, stimulate ovarian activity. In research studies where GnRH was given to mares once a day for three to four days, diestrus was shortened by about 1.5 days. More significantly, the agent shows promise for correcting ovulation failure (when the problem is caused by insufficient GnRH) and may also be helpful in stimulating follicular development on inactive ovaries.

The use of progesterone to hasten conception dates has also been documented, but the success of such treatment is variable. Theoretically, oral progesterone hastens the breeding season by shortening the duration of estrus and by decreasing the occurrence of split heat periods and anovulatory cycles. Workers have found that when progesterone treatment is followed by HCG treatment ovulation becomes more predictable. Progesterone/HCG therapy in association with an artificial light program effectively controls the long, erratic estrous periods that occur early in the breeding season. ♞

11

HEAT DETECTION

Two of the most important, but most frequently overlooked, aspects of equine production are the detection of estrus and the selection of breeding dates. As mentioned earlier, failure to detect heat and failure to breed the mare at a time conducive to conception are significant man-made causes of infertility in the equine. "The Mare Reproductive System" presented concepts in basic reproductive physiology to give the breeder a better idea of what estrus involves. "Mare Infertility" focused on specifics that interfere with the normal cycle, while "Broodmare Management" emphasized management techniques that enhance reproductive health and promote normal cyclic patterns. Each of these areas is of special importance, but without an ability to recognize and evaluate various stages of the estrous cycle, the aforementioned guidelines and suggestions are ineffectual.

Photo by Bill Wittkop

To achieve high conception rates, the breeder should carefully plan a heat detection program. Regardless of the operation's size or management conditions, personnel that are responsible for the detection of estrus should understand the essence of their task and the importance of performing those duties methodically. Individual circumstances dictate the type of heat detection program adopted on a particular farm. Many managers have special techniques that enhance their program but that would be impractical in another situation. This chapter is not presented to describe a single heat detection program, but to emphasize the importance of design and forethought when planning any operation's program. Honest evaluation of the farm's strong and weak areas, knowledge of each mare's estrous patterns and behavioral problems, and an understanding of the basic principles of estrous recognition are important to the development of an effective heat detection program.

Recognizing Estrus

As described earlier, hormones produced by the hypothalamus, the anterior pituitary, and the ovaries are responsible for both the release of ova and the overt signs of estrus in the mare. The characteristic changes that prepare the mare's genital tract for ovulation and conception are caused primarily by the hormone, estrogen. As estrus approaches, increasing estrogen levels stimulate the development of uterine endometrium and myometrium; the uterine tissue becomes swollen with fluids, and the entire tract becomes highly secretory. (The secretions are believed to facilitate entrance of the stallion's penis into the mare's vagina.) Estrogen also causes relaxation of the cervical folds, enabling a trained technician to determine the stage of the estrous cycle by rectal palpation and/or speculum examination.

In addition to influencing the tone and size of various reproductive organs, estrogen also stimulates the nervous system and causes the be-

Courtesy of Fairweather Farms

Both of these mares are receptive to the teasing stallion but the mare on the left exhibits the most obvious signs of estrus: squatting posture, urination, and raised tail.

This diestrus mare violently rejects the stallion's teasing efforts.

havioral patterns associated with estrus. In general, behavior during estrus is characterized by an obviously receptive attitude toward the stallion: squatting posture, raised tail, frequent urination, and spasmodic "winking" of the labia (i.e., eversion of the clitoris). ("Winking" also occurs after urination regardless of the mare's reproductive status.) Signs of diestrus range from neutral behavior to extreme violence, but the breeder should note that shy mares and mares with foals occasionally exhibit these passive or resistant patterns during estrus. In these instances, skillful observation, patience, and a variety of heat detection techniques are invaluable.

TEASING

While some mares require very little stimulation to show estrus, many mares exhibit these signs only in the presence of a stallion. Occasionally, mares "show" to geldings or to other mares, but most breeders rely on the teasing talents of a mannerly but aggressive stallion to elicit sexual responses from mares in estrus. If the mare is due to enter estrus according to her teasing records but is unresponsive to teasing, the handler should look carefully for any subtle reactions displayed by the mare as the teaser is approaching or leaving the area. Maiden mares, shy mares, and mares with foals at side may display signs of heat only when the stallion is at a safe distance. Although estrous patterns vary a great deal between mares, individual characteristics can often be identified, especially when each mare's behavior is observed and noted carefully throughout the season. If the manager is unaware of these individual tendencies, quiet mares may not be bred until late in the season, if at all. To insure that the teasing manager is familiar with each mare's estrous cycle, regular teasing should begin well in advance of the scheduled covers and should be continued during the breeding season. In addition, detailed teasing records should accompany the mare to any outside facility.

As mentioned earlier, abnormal estrous behavior and unpredictable estrus duration can be expected during the early breeding season. Unless special preconditioning programs have been employed, most mares do not establish regular cycles until late spring. Careful teasing during the early breeding season allows the teasing manager to follow the mare's transition from anovulatory receptivity to ovulatory receptivity and to determine when the mare is ready to be bred. Regular teasing after breeding allows the manager to catch mares that lose their pregnancies before the fortieth day of gestation. (Note: Because pregnant mares may show heat and because breeding may result in abortion, additional pregnancy tests should be carried out.) If the embryo is lost after 40 days, the mare will not enter estrus again for several months. (Postbreeding teasing has not been shown to increase abortion rates.) Continuation of irregular patterns throughout the summer, complete failure to show estrus during the natural breeding season, and other cyclic abnormalities can occur regardless of seasonal influence or breeding management techniques. It is important, however, that the handler differentiate between estrous abnormalities and failure to show to the teaser.

It is often stated that teasing stimulates normal estrous cycles in the mare, but this concept is a common misunderstanding. Teasing does not promote estrus or induce ovulation in the mare. The hormones that control the estrous cycle fluctuate with photoperiod, diet, stress, and perhaps with temperature. The importance of teasing lies in its ability to enhance the expression of estrous behavior in mares, making the mare's receptivity more apparent to an observer. The sole responsibility for identifying mares in heat should not be entrusted to the teaser, however. The teaser naturally concentrates on individuals that show strong positive responses, and he may ignore mares that are covert in their sexual behavior. For this reason, it is the responsibility of the handler to insure that each mare is teased and observed properly.

Variation in the teasing routine and special emphasis on those mares that are due to enter estrus help to insure that mares are teased properly. Routine teasing practices may not encourage the mare to show estrus, especially if she experiences the same stimulus day after day. The use of several different teasing methods, rotated throughout the season, may prove to be very helpful in some circumstances. It is also important to remember that maiden mares are easily frightened by aggressive stallions and should be teased by quiet, gentle stallions until they adjust to the heat detection regimen. Many older mares, on the other hand, seem to respond more quickly to stallions that squeal aggressively.

THE TEASER

On many farms, the teaser's performance has a significant effect on each season's conception rates. Before selecting a teaser, the breeder should evaluate the candidate's temperament, libido, and manners to insure that he is reasonably easy to handle, gentle with mares, and aggressive enough to encourage outward signs of heat in estrous mares. An ability to elicit estrous patterns from obstinate mares is an extremely valuable asset in the teaser. On the other hand, an ill-mannered horse requires constant restraint, is likely to injure mares and/or handlers, and may even intimidate mares that normally respond to teasing during the breeding season.

In some instances, the breeding stallion serves as a teaser. Naturally, the extent of his participation in the teasing regimen should not discourage normal sex drive. If a large number of mares must be teased regularly, the breeder may prefer to utilize a special teaser stallion. (Note: The mares must become accustomed to a particular teaser or, due to fear or curiosity, they may not respond.) For example, some breeders prefer pony stallions due to their size and relatively low maintenance costs, while others use intact or vasectomized horses. Vasectomy is a surgical procedure that involves cutting the vas deferens to prevent the passage of sperm from the testicles—an extra safety measure that protects against accidental breeding between mare and teaser. Because the testes are left intact, testosterone production is maintained and the vasectomized stallion has normal sex drive.

It is extremely important to keep the teaser in good mental and physical condition throughout the breeding season. Like other breeding farm stock, the teaser requires a balanced ration, adequate shelter, routine parasite control, and preventive medicine. Exercise is also a vital part of the teaser's health care program. Whether free or forced, exercise protects the teaser's physical well-being and encourages a healthy mental attitude. All but the most persistent teasing horses are subject to periods of low libido when used heavily. Many teasers become disinterested when asked to tease a large number of mares without rest or breeding. The most effective solution to this libido problem is to limit the number of mares teased daily, provide occasional periods of rest from teasing, and allow the teaser to cover mares periodically (or collect the teaser's semen with an artificial vagina 2-3 times a week). Each teaser is unique; the number of mares and length and frequency of his required rest periods must be determined on an individual basis. The ultimate goal, in any instance, is to utilize the teaser as efficiently as possible. Testosterone has been administered to both breeding stallions and teasers in hopes of improving libido, but research reports indicate that such treatment does not affect sex drive significantly and may even reduce fertility in the breeding stallion by inhibiting spermatogenesis. (Refer to "Stallion Infertility and Impotency.")

Even the number of teasers to which a mare is exposed may influence the efficiency of heat detection. A research study reported that when mares were teased with one stallion, sexual behavior was expressed in only 87 percent of the estrous mares. When two stallions were employed to tease the same group, the accuracy of heat detection rose to 96 percent. From these results, it seems that mares differ in their preference for stallions. Whether mares can discriminate against color or other physical characteristics of the teaser is unknown. The critical cues may even be sight, sound, touch, or smell, but little research has been done on this facet of equine behavior. However, to insure maximum efficiency in the heat detection program, normal cycling mares should be teased daily (or every other day) with one stallion, while irregular mares and those believed to be going in or out of estrus should be teased with two or more stallions.

TEASING PROCEDURES

Any teasing method that allows close association between the teaser and the mare for several minutes can be an effective means of detecting estrus. Factors that should be considered prior to designing a teasing facility include safety, efficiency, size of the broodmare band, availability of skilled farm labor, type of operation, and personal preference. Once these aspects have been evaluated carefully, teasing methods that best fit the operation's needs can be designed. The following methods are described to help the reader visualize how various teasing facilities differ.

Partitions

The use of a stall door, a wall, or a specially designed partition during the teasing process protects the stallion, the mare, and the handlers. The horses are led to the partition, and both handlers stand to one side of their charges, preferably behind separate partitions, to avoid injury should one of the horses strike or kick. The horses are introduced by a brief nose to nose encounter over the partition. At this point, the diestrus mare may lay back her ears, squeal, and strike at the stallion. The teaser can be allowed to gently nip and nuzzle the mare, but an overly rough and aggressive attitude may result in injury to the mare (and possibly to the stallion or the handlers) and should always be discouraged. Occasionally, a little gentle persistence on the part of the stallion may reveal that a resistant mare is actually in estrus. But if the mare's aggression continues, or if she totally ignores the teaser, she is probably not in heat. Hand teasing over a properly designed partition has two major advantages: it provides an added measure of safety to both horses and handlers, and it allows individual attention for every mare. Although this is the most commonly used teasing method, teasing each mare in this manner is time-consuming on a large operation, and some mares do not respond to teasers when forced into close association.

Walls designed and located for efficient teasing are seen on many breeding operations. Emphasis on safety and efficiency has led to the development of various types of teasing partitions, each with its own merits

Courtesy of Mark Gratny Quarter Horses

The handler stands well to the left side of the stallion when teasing mares along a stall aisle.

and handicaps, depending on the individual needs of the breeder and the demands related to that particular operation. For example, the teasing manager might lead mares past the stallion's stall, using the stall door as a partition or past a stall window that allows the stallion to present his head and neck. (It is very important that the door is high enough to prevent the stallion from jumping out of the stall during the teasing process.) Alternatively, a long aisle of sturdy stalls can be used to quickly tease a large number of mares. The teaser can be led down the aisle to tease mares in individual stalls. This method does not require handling every mare, but there must be an area where an observer can safely watch each mare.

The teasing wall found on many breeding farms is about 4 feet (1.25 meters) high and 10 feet (about 3 meters) long, and constructed of two heavy wooden layers. Boards at least 2 inches (5.1 centimeters) thick are preferable for teasing wall construction. A teasing wall's advantage over other teasing methods is that it allows the stallion to get his head and neck close to the mare, while protecting everyone involved in the procedures. When equipped with extensions that separate the handlers from their charges, a teasing wall is one of the safest ways to tease mares. For added safety, the teasing wall should be supported by heavy wooden or metal posts set in concrete. The wall is usually padded heavily to prevent injury to the mare (if she kicks) and to the stallion (if he strikes), and is sometimes equipped with a roller bar along the top to aid the stallion should he accidentally get a leg caught over the wall.

A heavily padded teasing wall helps prevent injuries to the stallion and mare during heat detection.

This teasing wall is designed to provide protection for the handlers, as well as for the mare and stallion.

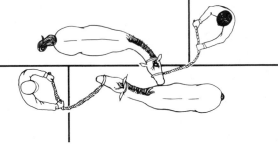

Another partition design surrounds a large teaser pen with smaller mare pens so that several mares can be teased at once. Similarly, a long continuous partition might be incorporated into a chute, designed to accommodate several mares at one time. The chute consists of two parallel fences 4.5-5 feet (1.4-1.5 meters) high and 3.5-4 feet (1 to 1.2 meters) apart and made of 2 x 6 inch (5 x 15 centimeter) planking. The chute should have gates at both ends so that a group of mares can be led in and tied at approximately 18 foot (5.5 meter) intervals. After teasing, these mares can be led through the exit gate and replaced with another group of mares. This teasing method requires two or three handlers and is suitable for a large group of mares.

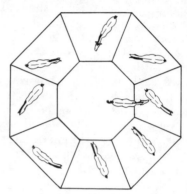

A central teasing pen surrounded by several enclosures allows the stallion to tease every mare within a group without interference from dominant individuals.

Courtesy of Texas A & M University
Department of Large Animal Medicine and Surgery

A teasing chute allows rapid, individual teasing of a large number of mares.

Pastures

When available labor and time are limited, any method that shortens the time invested in daily teasing without lessening its effectiveness is usually welcome. A pasture teasing program allows many mares to be teased in a short period of time with a minimum amount of labor. Some mares are more responsive to teasing in pasture, but the teasing manager should remember that this is an unsatisfactory arrangement for mares with foals at side due to the possibility of anxious or curious foals being injured by the teaser or by a distracted mare.

If the teaser is led along the fence line, the incorporation of several teasing partitions into the fence might help prevent serious accidents. Although some breeders allow their stallions to run free in alleyways located between pastures or paddocks with good results, the chances of injury to a good teaser, a good broodmare or to a breeding stallion should be considered carefully before a stallion is allowed to tease over a fence, with or without supervision.

On some farms, vasectomized stallions are allowed to run with broodmares, thus minimizing labor expenses and providing the mares with a fairly natural teasing situation. This method is not recommended, however, due to several important disadvantages: 1) it cannot be utilized safely when foals are pastured with their dams, 2) it may increase the spread of infection between mares via the teaser stallion, and 3) it increases the chances of injury to both the teaser and his mares.

Another procedural variation for teasing pastured mares involves the use of a special teasing cage. Instead of being pastured with the mares, the teaser is released in a sturdy, high-walled pen located inside the pasture or at the intersection of several pastures. The teasing cage is usually designed so that the teaser can put his head out to associate with the mares. Nuzzling, nipping, blowing, and other forms of physical contact are important elements of the teasing stimulus. Because the teaser need not be hand-held, one person can observe the mares and remove those found to be in heat. These estrous mares tend to seek out the teaser and display obvious signs of heat. When teasing cages are used, however, it is important that aggressive mares are not permitted to monopolize the teaser's attentions. When dominant mares are removed from the area, timid mares in estrus are more likely to approach the teaser.

Courtesy of Manx Farm

A teasing cage allows pasture teasing without requiring that a handler or the teaser enter the pasture.

A similar method, sometimes used to tease a large group of pastured mares, involves the use of several teasing cages located between larger mare pens (20-30 mares per pen). Mares can be driven in from pasture, teased by one or two stallions, and then driven back to the appropriate pasture. This technique allows many mares to be teased in a short period of time.

RECTAL PALPATION

Rectal palpation, which was described in "Broodmare Management," has been used for years to predict the time of ovulation in mares, cows, and other domestic livestock. As mentioned earlier, this procedure is an extremely helpful part of the mare's reproductive soundness examination. When used judiciously, it can also be an important tool for following the mare's estrous cycles, especially during the transitional period or when a mare fails to express estrus during the normal breeding season. For example, the palpation procedure described in "Broodmare Management" can be used to detect the changes that characterize normal cyclic activity:

1. During diestrus and pregnancy, the uterus is a firm tubular structure. In contrast, the uterus is flaccid and swollen during estrus. A knowledgeable and experienced palpator recognizes that a change in uterine texture is an indication of the mare's chances of conceiving. However, because these changes may or may not occur and because changes that do occur show great variability among mares, palpation of the uterus is not usually emphasized.

2. During diestrus and pregnancy, the cervix remains closed and, when palpated via the rectum, feels like a narrow, tense tube. Under the influence of estrogen, the cervical folds relax and the structure becomes easy to distinguish by palpation. Because cervical tone is a fairly accurate reflection of estrogen levels, rectal manipulation of the cervix is a reliable means of determining the stage of the estrous cycle.

3. Cyclic ovarian activity can also be detected by rectal palpation. At the beginning of estrus, for example, the maturing follicle is a tense structure approximately 1 cm in diameter. At this time, two or three follicles may be detected, but one is usually larger than the others. This larger (Graafian) follicle continues to increase in size until ovulation occurs. As a general rule, the detection of a follicle larger than 3 cm indicates that ovulation will probably occur within the next three days. Some sources report that the Graafian follicle softens within 24 hours prior to ovulation and conclude that this change in consistency is a better guide to ovulation than is follicular size. However, because follicular tone fluctuates considerably during estrus, softening of the follicle is not always an accurate indication of impending ovulation.

4. Because rectal palpation does not seem to affect the welfare of the fetus, it is also used during diestrus to determine if the mare is pregnant.

When regular cycles are established or when the cause of apparent anestrus is determined and resolved, the manager might prefer to focus on the teasing program so that palpation during estrus is not necessary. If palpation is required to overcome the effects of some fertility problem or to schedule a large number of covers to a single stallion, it should be conducted by, or under the supervision of, a veterinarian or trained technician.

VAGINAL SPECULUM

The use of a speculum by trained personnel to detect estrous-related changes in the mare's genital tract is also a helpful heat detection technique. (Refer to "Broodmare Management.") Through a sterile speculum, the observer notices whether the cervix is pale and tightly constricted

Photos by Pierre Lieux, DVM Courtesy of Pierre Lieux, DVM

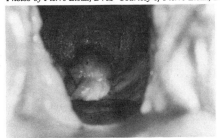

Cervix during diestrus

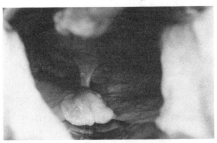

Cervix during proestrus

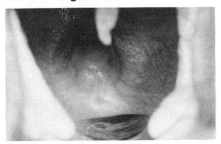

Cervix on first day of estrus

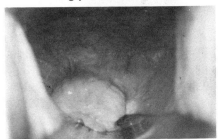

Cervix during estrus (close to ovulation)

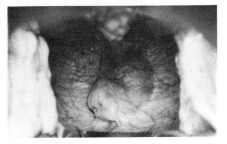

Cervix during pregnancy - 10 days after last breeding

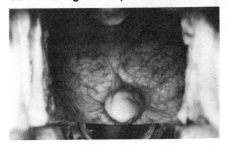

Cervix during pregnancy - 40 days after last breeding

(indicating diestrus) or relaxed and moist (indicating estrus). During diestrus, the cervical folds and the tissue around the cervical opening project slightly into the vagina, but during estrus these folds relax, and the cervix takes on a warm, pink flush. As ovulation draws near, the cervix becomes flaccid and increases to approximately three or four times its usual diestrus diameter. At this stage, the tissue surrounding the cervical opening sags to the floor of the vagina or onto the lower spoon of the Caslicks speculum. During estrus, the vagina also relaxes, becomes vascular, and secretes a thin, watery mucus. (Most of the fluid discharge comes from the cervix.)

HORMONE ASSAYS

The hormones that regulate the mare's estrous cycle are secreted into the circulatory system by endocrine glands. When blood samples are collected at various points during the normal cycle and analyzed for hormone content, typical cyclic hormone fluctuations become apparent. In a normal cycling mare, the outward signs of estrus coincide with high estrogen levels and low progesterone levels. High progesterone and low estrogen levels are characteristic of diestrus. Theoretically, a mare's reproductive status can be determined by analyzing a blood sample for progesterone and estrogen content. Analysis of reproductive hormone patterns has proven useful in research studies, but due to the expense and inconvenience involved in conducting such tests, hormone assays are not practical methods for routine heat detection.

Optimum Time To Breed

In the mare, ovulation occurs approximately 24 hours prior to the end of estrus but may take place anywhere from 48 hours before to 24 hours after this point. In 12 to 24 hours postovulation, the ovum begins to disintegrate within the oviduct. Sperm, on the other hand, normally retain their motility for at least 48 hours. To insure that a viable ovum and sperm are present for conception in the oviduct, the mare should be covered just prior to ovulation. Maximum conception rates are achieved when breeding takes place no later than two hours postovulation, and rates decrease rapidly following ovulation. It is suspected that fertilization of an aging ovum may cause embryonic abnormalities that result in early embryonic death, but research results to substantiate this claim are not available. Nevertheless, it is preferable to cover the mare just prior to ovulation. Since ovulation occurs without warning while the mare is still in heat (in most instances), many sources recommend covering the mare on her third day of estrus and then every other day thereafter until she is no longer receptive to the stallion. On this schedule, motile spermatozoa are present in the tract at all times, and the likelihood of fertilization is enhanced. However, unless artificial insemination is employed, such an extensive schedule is not always possible. A more judicious use of covers requires careful identification of each mare's reproductive status and well-timed matings. Ovum longevity, sperm longevity, and ovulation time are extremely important variables that determine the optimum time to breed a mare.

Records

The most important aspect of any heat detection program is the manager's ability to equate signs of heat with impending ovulation. An understanding of reproductive physiology gives the manager a good idea of what to expect but, because many mares deviate from the theoretical norm, special emphasis must be placed on analyzing the individual mare's estrous patterns and noting information about each cycle on a permanent record. Future reference to these individual files presents an invaluable overview of each mare's behavioral patterns, teasing habits and estrous irregularites. Individual teasing charts, for example, monitor the mare's estrual activity throughout her reproductive life. These charts can be simple and compact, grading the intensity of estrous behavior on a predetermined scale (e.g., 0=out, 5=hot). But any deviation from normal estrous behavior (e.g., unusual response to teasing or to a particular teaser) should be detailed so that these aspects do not surprise responsible personnel during subsequent breeding seasons. If palpation is also used in the heat detection regimen, palpation charts should be incorporated into the individual record. Follicular size and consistency, uterine tone, and cervical dilation should be observed and recorded during each examination. If at all possible, results should be recorded immediately to minimize errors. In addition to individual records, a wall chart for all mares booked to a particular stallion should be kept in the office, barn, or laboratory. If several stallions are standing, each should have a separate wall chart so that individual breeding schedules can be quickly evaluated. These charts allow personnel to tell at a glance when a mare is due to come into heat, how long she has been in heat, when she ovulated, when she was last bred and, therefore, identifies mares that should be observed very closely on a certain day. Unlike notebooks, loose sheets of paper, and file cards, it is unlikely that wall charts will be mislaid and estrous-related communications crossed. It is important, however, that each mare's permanent file be updated periodically with information from the heat detection wall charts. ♘

12

BREEDING METHODS AND PROCEDURES

There are three basic methods for breeding horses: pasture breeding, hand breeding, and artificial insemination. Each breeding farm has unique management characteristics, and similarly, each of the breeding methods has distinct advantages and disadvantages. After considering a number of factors, such as the number of mares bred yearly, the number of outside mares bred yearly, the number and value of stallions standing on the farm, and the facilities or space available for breeding activities, the most efficient breeding method for a particular operation can be determined.

Pasture Breeding

The economics of today's equine industry demand the most efficient management techniques in order to maximize production while minimizing costs. For this reason, very few breeders practice pasture breeding. Although pasture breeding programs can be very successful, they can also result in serious economic loss when used carelessly. In addition, pasture breeding may be ideal for one farm and yet impractical for another operation. The following discussion is presented to help the breeder understand what is involved in setting up and managing a pasture breeding program and to identify its limitations.

FACILITIES AND PROCEDURES

Stated simply, pasture breeding involves placing a stallion with a mare or group of mares in an area that is large enough to encourage natural breeding behavior. Ideally, this area should provide plenty of lush vegetation to minimize or eliminate the need for daily grain or roughage supplementation. In addition, the breeding pasture should be situated so that the stallion cannot contact other horses over the fence. Access to and through the pasture should be planned carefully. However, roads through the pasture should not be paved. If the mares will remain in the pasture all year, the area should also be designed and maintained for safe foaling. (Refer to "Foaling.") If foaling paddocks or stalls are used, the breeder should determine how and when the mares will be moved to these areas (e.g., through alleyways, in trailers, etc.).

Because horses can seldom be handled individually in pastures, management is often limited to preparing the stallion and mares for the breeding season, introducing new members to the breeding herd, and inspecting the herd periodically. Preparation for a pasture breeding season involves breeding shed training procedures (for an inexperienced stallion) and preconditioning techniques that the breeder or veterinarian feels are in order. The importance of using healthy horses in a pasture breeding program cannot be overemphasized. When a reproductive pathogen is introduced into a breeding herd, it may be transmitted throughout the herd via the stallion with devastating results on the group's fertility. To insure that the herd remains free from infection, the breeder may decide to restrict it to a group of mares that are known to be free of infectious organisms or that have no history of reproductive infection.

When the stallion is placed with a group of mares, the herd should be observed closely for several weeks, especially if any of the mares are unaccustomed to being pastured with other horses or if the stallion is experiencing his first pasture breeding season. An overanxious or inexperienced stallion should be introduced to the herd gradually, perhaps by leading him through the pasture and allowing him to learn that not all of the mares are going to be receptive and that the best approach is a cautious one. (Note: Walking an inexperienced or overanxious stallion through a group of mares is dangerous and should be attempted only by an experienced stallion handler who is familiar with the particular stallion.) During this learning period, the handler should permit the stallion to approach the left side of a gentle mare in deep estrus, and after a brief courting period, the stallion should be allowed to cover the mare.

Another method used to introduce the stallion to a herd is to pasture him with only one mare at first and then gradually increase the number of mares until the broodmare band is complete. Gradual introduction to pasture breeding cannot insure the stallion's safety, but it does lessen the chances of serious injury occurring during the initial adjustment period.

Although complex management procedures are seldom compatible with pasture breeding, the herd should be observed routinely to check each horse for injuries or illness and to make sure the stallion is servicing his mares. Additionally, periodic spot checks should be conducted to determine whether the stallion is settling his mares. It is important to remember that the stallion's natural instinct to herd and protect his mares can cause serious problems if he feels threatened by the observer's presence. Aggressive behavior is not exhibited by all stallions, but some individuals cause significant problems. Some stallions react viciously when another horse is ridden into their pasture and may even attack a human approaching on foot. For this reason, pasture breeding is not always a good choice for the novice stallion handler.

DISADVANTAGES OF PASTURE BREEDING

Although in many instances pasture breeding is a workable breeding method, the breeder should consider all aspects of a pasture breeding setup before establishing one. Allowing a herd to revert to natural breeding patterns is not in itself detrimental to fertility. Obviously, the species has survived for thousands of years with very few interruptions in evolutionary development and could probably persevere in a natural breeding environment with or without human assistance. Instead, many of the following disadvantages are actually limitations imposed by financial considerations and modern breeding practices.

The number of mares that a pastured stallion can settle is limited in comparison to the conception rates that a stallion can achieve in other programs that are carefully managed, especially if his semen is divided among several mares and inseminated just prior to ovulation. Although the size limit for any broodmare band depends on the individual stallion's fertility and sex drive, a range of 20 to 40 pastured mares per season is reasonable. (Young stallions should be started out with no more than 20 mares and allowed to gradually work up to their most efficient herd size.) When the prospect of 20 to 40 foals per reproductive year from pasture breeding is compared to the alternative of a valuable stallion producing 100 foals in an artificial insemination program, an important economic argument against pasture breeding is conceded. The difference between production rates achieved in pasture as compared to hand breeding programs is not as extreme. However, the ability to settle mares using only one or two covers per conception in a well-managed hand breeding program allows the breeder to significantly increase the size of a healthy stallion's book. When the stallion is used as efficiently as possible, the program's overall productivity is improved.

Close observation and care of pastured breeding stock is difficult and time-consuming. For example, removing an injured mare from a pasture can be troublesome and even dangerous if the stallion is overprotective and ill-mannered. Additionally, treating or palpating mares in pasture without the benefit of wash stalls or palpation stocks is impractical. Any management technique that requires physical restraint in pasture must be carried out with the aid of several competent assistants, careful forethought, and patience. In short, pasture breeding is not conducive to meticulous management procedures, and when the cost of close supervision is compared to the number of foals produced per season, it is obvious why pasture breeding has generally been abandoned for other breeding methods.

As mentioned earlier, the importance of infection-free stock in a pasture breeding situation cannot be overemphasized. If all members of a herd are healthy and if additions during the breeding season are either prohibited or carefully screened for reproductive infection, the spread of infection should not be a problem. On the other hand, one infected mare can contaminate the stallion who, in turn, exposes the other mares. Exposure does not necessarily result in an infection, but it does increase the likelihood of infection and subsequent infertility. If spot checks for pregnancy are not made throughout the season, an infection-related fertility problem could go undetected, and a large percentage of the mares could be left barren for that season. Also, long-term infection may cause infertility.

A very serious disadvantage, and often the first to be pointed out in an argument against pasture breeding, is the possibility of valuable breeding stock being injured. An overanxious or improperly trained stallion often develops impeccable manners when pastured with a group of mares, but unfortunately, these manners are usually acquired through negative reinforcement. That is, the stallion begins to associate pain with behavior patterns that the mares consider undesirable, and he avoids the pain by adopting the socially acceptable behavior patterns. Thus, many stallions used for pasture breeding acquire unsoundnesses (especially in the knees) and blemishes caused by annoyed mares, breeding or teasing accidents, attempts to service mares in a neighboring pasture, or over-the-fence fighting with neighboring horses. The most detrimental injuries, however, are those that affect spermatogenesis, erection, ejaculation, or the stallion's ability or willingness to mount a mare. (These injuries are often caused by breeding accidents.) Bruises on the penis, for example, are common in a pasture breeding situation. For these reasons, the possibility of a valuable stallion being injured is probably the most significant deterrent to pasture breeding. The situation should be evaluated on an individual basis (e.g., the mares' temperaments, the stallion's experience, availability of farm personnel to supervise the herd, etc.) before a decision to use the stallion in a pasture breeding capacity is made. In general, however, risking the fertility status or soundness of a valuable stallion by running him with a group of mares reflects disregard for that stallion's potential influence over the breeding operation's success or failure.

When studying the disadvantages of pasture breeding, the possibility of injury to a mare or foal should also be considered carefully. Stallion owners operating pasture breeding programs often discover that mare owners are very reluctant to place their mares in a pasture program, since pastured mares are sometimes injured during breeding (e.g., torn vulva, bruised cervix, injured vagina, etc.). Mares are also subject to social skirmishes which can affect the safety of their foals. In the wild, horses learn behavior patterns that enable them to survive and live harmoniously with other members of the herd, but without any previous exposure to the breeding herd hierarchy, a horse must learn important social patterns by trial and error. For example, the stallion's instinct to keep his mares together when their safety or his dominion are threatened may cause him to savage any mare that steps out of line. Most mares learn very quickly to avoid the stallion's disapproval without infringing on a domineering mare's territory, but inexperienced mares may be bruised and battered before these lessons of nature are learned. Some mares adapt well to a pasture breeding environment. These mares are usually dominant or are quick to avoid skirmishes and readily accept their positions within the hierarchy. Horses that have been raised on pasture are usually good survivors. Mares that have been stalled and cared for individually are usually slow to adapt and may even be rejected by the stallion. Although most mares with foals at side protect their youngsters from impending danger and avoid confrontations with other horses when possible, a mare that is not able to deal with several horses at once may be forced into a position that threatens the safety of her foal. For this reason, a mare's first pasture breeding experience is best carried out when she is without foal or after her foal has been weaned.

Courtesy of The King Ranch

Most mares are naturally protective of their foals. In a pasture breeding situation, however, the mare must also learn how to guard her foal from inquisitive mares without accidentally harming the foal.

It should also be mentioned that in a pasture breeding program it is difficult to determine exactly when a mare was settled. Although breed registries usually allow a period of exposure to the stallion to be reported as the breeding date, exact dates are helpful in determining when the mare should foal. The advantages and disadvantages of allowing mares to foal in pasture are presented in "Foaling."

ADVANTAGES OF PASTURE BREEDING

In spite of the aforementioned disadvantages, a well-managed pasture breeding program can be a very effective and convenient method for breeding a small number of healthy, farm-owned mares to a fertile, aggressive stallion. When conditions are favorable (e.g., plenty of space, nutritious pasture, and absence of reproductive infections), pasture breeding often results in relatively high conception rates. Although few studies have been performed in this area, breeders who use this method often report excellent results and suggest that close association between mare and stallion along with the absence of management-related stress enhances cyclic behavior in mares. Increased receptivity in shy mares and a relaxed attitude in mares that object strongly to restraint procedures used in a controlled breeding program are just two examples of the benefits that can result from pasture breeding.

The economic advantages of pasture breeding as compared with other breeding methods are limited. However, pasture breeding reduces the need for stalls, paddocks, and enclosed breeding areas; therefore the construction and maintenance costs associated with these facilities is minimized. If pasture is adequate, supplemental feed expenses and the cost of labor required to feed horses over a widespread area are also minimized. Although pastured horses must be observed routinely, the personnel requirements of a pasture operation are minimal when compared to the labor required to carry out routine teasing, palpations, prebreeding preparations, and breeding shed procedures in a closely supervised breeding program. When high-quality or abundant pasture is available but facilities and personnel are limited, pasture breeding may have a significant economic advantage over hand breeding or artificial insemination, especially if the stallion is normally bred to a small number of mares.

Hand Breeding

Today, most breeders supervise the equine breeding process using the so-called hand breeding method. Unlike pasture breeding, hand breeding programs allow fertility problems to be dealt with closely. Because each breeder must deal with a unique set of circumstances, hand breeding programs vary considerably from farm to farm. The equipment, the facilities, and even the step-by-step breeding procedures used reflect a farm's unique characteristics. Thus, the explanations and instructions provided in the following discussion are meant to give the breeder a good foundation for developing a successful hand breeding program. On the basis of this information, individual modifications can be adopted.

FACILITIES

Perhaps the most diversified aspect of hand breeding is facility design. However, the following guidelines should be considered when planning a breeding area:

1. The area should be reasonably dust-free to minimize airborne contamination of the male and female genitals.

2. The flooring should provide good footing to decrease the chances of horses slipping or falling and a soft landing to minimize the occurrence of injuries due to falls.

3. The breeding area should be located away from the mainstream of farm activities so that distractions are minimized. In some areas, laws may require that livestock breeding areas be screened from public view.

4. Ideally, the breeding area should be enclosed and sheltered from the elements. If this is not possible, the area should be well-drained and free from physical hazards (e.g., farm equipment, vehicles, and other horses).

5. If the breeding area is enclosed, it is important to have separate exits for the mare and stallion, preferably on opposite ends of the enclosure. This element of breeding shed design helps to insure the safety of the mare, the stallion, and their handlers.

6. To make the breeding process as efficient as possible, the breeding area's proximity to stallion and mare facilities should be considered. On some farms the breeding shed is an extension of the stallion barn, but the feasibility of such a set-up depends on many factors. In any instance, the breeding area's location should limit the time spent moving animals in and out while minimizing the distractions sometimes associated with a centralized location.

7. The need for wash areas and foal holding pens should also be considered when designing a breeding area. The wash area should have hot and cold running water, a nonslip floor surface that is easy to disinfect, adequate drainage, and breeding stocks for confinement and restraint. Proper foal pen design allows the mare to see her foal and, at the same time, protects the foal from possible injury.

8. Because differences between the mare's and the stallion's heights can cause difficulty during the breeding process, the breeding area should have a mound, ditch, or breeding platform to aid the stallion when he is asked to cover a mare that is much shorter or taller than he is.

A breeding platform permits a small stallion to safely mount a taller mare.

EQUIPMENT

Any equipment that is employed in breeding should be easy to locate and in good, usable condition. Because most of this equipment is designed to insure the safety and health of the mare and/or stallion, it should be used and maintained with care. For example, as discussed in "Stallion Management," stallion restraint equipment (e.g., halter, shank, chain, etc.) should be carefully inspected before each use. Similarly, any equipment used to control the mare during breeding (e.g., hobble, twitch, or leg strap) should be checked carefully prior to use. Equipment and supplies used to prepare the mare and stallion for breeding (e.g., wash buckets, bucket liners, disposable gloves, and tail wrap) are extremely important to the control of reproductive infection and should be handled accordingly.

Safety is a priority consideration in the breeding area. The experience and ability of those individuals handling the mare and stallion influence many aspects of the breeding procedure. However, even the most experienced handlers realize that special safety equipment, when properly used, is a valuable insurance measure particularly if either the mare or stallion is unpredictable. Regardless of the type of restraint used, it should always be applied with care since improper use of restraint can cause serious physical or psychological injury to the mare and/or stallion.

Although a mare in estrus usually stands quietly when mounted, bred, and dismounted, there is always a chance that she will suddenly refuse the stallion or show her disapproval with a sharp kick, especially during dismount. It is not unusual for her action to result in injury to the stallion's penis, sheath, or testes. In such cases, the stallion's fertility and/or potency may be affected. To protect the stallion the breeder may restrain the mare, using a method that provides for the safety of both animals.

The most commonly used mare restraint device is the breeding hobble, which consists of a leather strap that buckles around the mare's neck just in front of her shoulders. A rope or strap extends from the lower part of

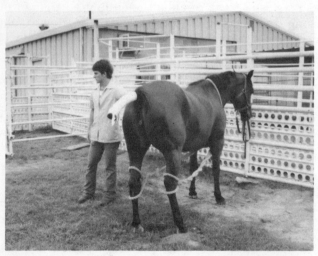

Hobbling prevents the mare from kicking the stallion during breeding.

this neck strap between the mare's front legs and attaches to the middle of a second rope or strap. The ends of this strap are connected to leather straps that buckle around the mare's hocks or pasterns, preferably the hocks since the mare is more likely to become entangled in the hobble when it is attached to her pasterns than when attached to her hocks. When restrained in this manner, the mare can walk but not kick. To prevent the stallion from becoming entangled in the hobble, the ropes should be adjusted to minimize slack and the hobble should be equipped with a quick release mechanism. Before a hobble is used for restraint during breeding, the mare should be accustomed to walking and standing with the apparatus in position.

Courtesy of Alduro   Photo by Jim Wright

As a safety precaution, the hobble should be secured with a quick-release knot.

Occasionally, an unruly mare must be restrained during breeding by forcing her to stand on three legs. One leg, usually the left foreleg, is held up by a leather strap that is equipped with a quick release mechanism. (Buckles are dangerous because they are difficult to release in an emergency.) This measure should be used with caution, since it may cause the mare to fall. The use of a leg strap may cause a nervous mare to feel even more insecure or an antagonistic mare to become more violent as the stallion approaches. If a leg strap is used, it is very important to release the leg strap after the stallion has mounted to help the mare maintain her balance, although the stallion cannot then be protected from a kick as he dismounts.

If the breeder feels that the risks involved in using hobbles or leg straps outweigh the benefits, a twitch or kicking boots may be considered. (Some breeders routinely use a twitch along with a hobble or leg strap.) When applied to a mare prior to breeding, the twitch usually turns her attention away from the stallion, allowing him to mount and breed her with minimal opposition. While kicking boots do not prevent kicking, they do lessen the severity of any blow the stallion might receive. These boots are made of heavy felt-like material and are designed to minimize the impact of a mare's kick. Regardless of whether kicking boots are used, the mare's hind shoes should always be removed prior to breeding.

Protecting the mare from injury is also important, especially when she is bred to an overly aggressive or poorly mannered stallion. The various types of tack used to control the stallion are discussed in "Stallion Management." Again, it should be emphasized that tack articles used during breeding, such as halters and lead shanks, should be reserved for that purpose. In addition, stallion restraint methods should not be so severe that the stallion is distracted from the mare or the job at hand. (Refer to "Stallion Management.")

In addition to stallion restraint equipment, there are several breeding aids that may be required to insure the mare's safety. If the stallion savages mares (i.e., bites them severely on the neck and withers), a leather neck shield should be used on the mares. The shield is held in place by three

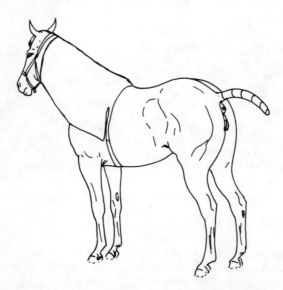

A leather neck shield protects the mare from a stallion that savages mares during breeding.

buckles and a girth strap and is designed to give the stallion something to grasp with his teeth without exposing the mare's neck to injury. If the stallion has a large penis and is bred to a small mare or a maiden mare, the mare may suffer an injury to the cervix. To prevent the cervix from being bruised or torn, a breeding roll should be used. The breeding roll is usually a padded cylindrical instrument about 5 inches in diameter and approximately 18 inches in length, although some breeding rolls are cone-shaped. A handle on one end of the roll allows the attendant to place the instrument between the mare and the stallion just above the stallion's penis. The roll should be covered with a sterile glove and placed in position as the stallion mounts to prevent the stallion from penetrating the mare too deeply.

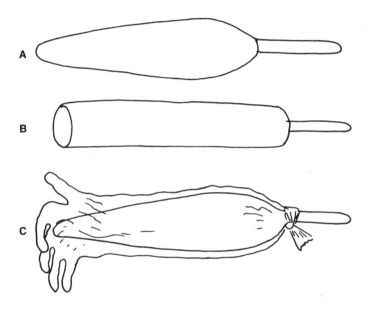

If the stallion's penis is very large, the mare's cervix can be protected from injury with (A) a cone-shaped or (B) a cylindrical breeding roll. (C) The breeding roll should be covered by a sterile obstetrical sleeve, which should be replaced before each service.

Hygiene is a very important consideration during breeding. Prior to breeding, for example, the mare's tail is wrapped to prevent her tail hairs from contaminating or lacerating the stallion's penis. A variety of products can be used to wrap the tail, including a clean tube sock, sterile gauze, a sterile plastic sleeve, or a track bandage. Wash procedures performed before and/or after the act of breeding also require special equipment and supplies. The table entitled 'Wash Equipment and Supplies' in this chapter lists the items needed to prepare the stallion for breeding. This list can also be applied to the mare's wash procedures.

Before breeding or reproductive examination, the mare's tail should be completely covered to protect the genital tract from contamination. A sterile, disposable obstetrical sleeve makes an excellent tail wrap that can be disposed of after use.

PROCEDURES

Although hand breeding is used to insure safety and efficiency, poor techniques are unproductive and extremely dangerous. The importance of adopting carefully planned breeding procedures cannot be overemphasized. Not all breeding techniques are applicable on all breeding farms; therefore, each breeder must consider the farm's circumstances and must deal with the unique behavior patterns and temperament characteristics of his horses. For this reason, the following step-by-step hand breeding instructions are provided as general guidelines for developing suitable breeding procedures. The methods can be adjusted to fit the breeder's individual requirements.

PREWASH TEASING

The steps involved in determining when a mare is ready to be bred were examined in the chapter "Heat Detection." Even though a mare's reproductive status has been established through teasing or palpation, she should be teased again just before breeding to encourage urination. After the mare's bladder has been emptied, wash procedures can be carried out with little threat of recontamination prior to breeding. If the stallion must also be used to tease the mare, preliminary teasing procedures should not be carried out in the breeding area. A well-trained stallion realizes that he is allowed to mount a mare only in a certain location and when special breeding tack is used.

WASHING THE MARE

Ideally, the mare and stallion should be prepared for breeding at the same time, thereby speeding up the prebreeding procedures and minimizing the possibility of the genitals becoming contaminated between washing and breeding. If it is impossible to prepare both at the same time, the mare should be prepared first. After her tail is **completely** covered with a track bandage, a disposable obstetrical sleeve, a tube sock, or some other tail covering, it should be held aside by one attendant while another washes the mare's hindquarters. Alternatively, the wrapped tail can be held to one side by a rope tied around the mare's neck just in front of her withers. In addition to washing the vulva, anus, and the skin between the thighs, the attendant should be sure that a large area surrounding the mare's perineal region is as clean as possible. The stallion's penis often contacts the mare's hindquarters before intromission, and if this area is contaminated with excrement or debris, the mare's reproductive tract may be exposed to a heavy dose of pathogenic organisms. To clean this critical area properly, the attendant should carry out the following wash procedures three times.

1. Wearing disposable gloves, dip several clean paper towels or pieces of sterile cotton into a bucket of warm, soapy water. (The wash and rinse buckets should be lined with clean, plastic liners.)

2. Scrub the mare's hindquarters thoroughly with the saturated towels. If more soap is required, dip only clean towels into the soap bucket. Used towels should always be discarded. Always begin washing at the vulva and continue in an outward (or expanding) circular pattern to minimize the chances of recontaminating a cleaned area.

3. Rinse the area thoroughly with lukewarm water. A clean cup can be used to dip water from a lined bucket and then to pour the water over the mare's hindquarters. It is important that the cup not be contaminated by contact with the mare during this process. A clean paper towel can be used to help rinse the area thoroughly.

4. After completing the preceding steps three times, dry the mare's hindquarters thoroughly with clean paper towels wiping in a circular pattern from the center outward.

(A) Assemble all of the equipment that is required for washing the mare prior to breeding: plastic obstetrical sleeve, adhesive tape, clean tube sock, clean dipper, mild soap, roll cotton, two buckets equipped with disposable bucket liners, and paper towels.

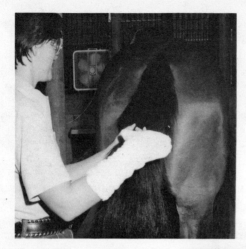

(B) With the tube sock inverted over one arm, grasp the end of the bony portion of the tail.

(C) Reverse the sock to cover the entire tail, and tuck any stray hair into the top of the sock. At the root of the tail, secure the tubular wrap with adhesive tape.

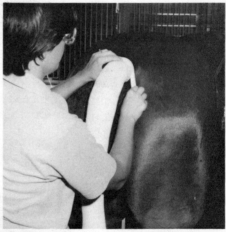

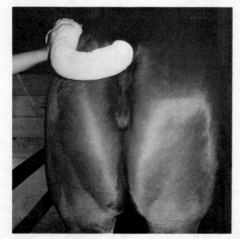

(D) This method of wrapping permits the tail to be drawn completely away from the perineal area during the washing procedure and prevents recontamination of the area by loose tail hair after washing. It also prevents tail hairs from being carried into the vagina by the stallion's penis.

(E) Saturate a generous piece of cotton or paper toweling with soapy water from the wash bucket. Once the material touches the mare's perineum, it should be discarded and never replaced in the wash bucket.

(F) Apply soapy lather to the inner portion of the perineum.

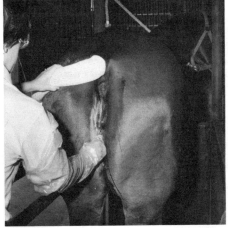

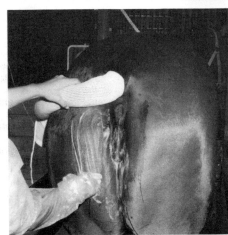

(G) Continue the washing process from the center of the perineal area, outwards, until the entire perineum and part of the mare's buttocks are thoroughly lathered.

(H) Carry water from the rinse bucket to the mare's perineum with a small, clean utensil. Disposable bucket liners reduce the chances of disease being passed from mare to mare via the washing procedure.

(I) Flush the soapy perineum and buttocks with water from the rinse bucket. Touching the mare with the dipper may contaminate the rinse water.

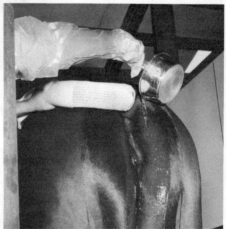

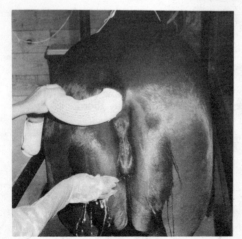

(J) While rinsing, use a clean piece of paper toweling or cotton to strip soapy water from the perineum and buttocks. Following the pattern established in the washing procedure, strip the inner portion first, followed by the outer portion of the perineum and buttocks. Never return to the inner portion with the same piece of toweling or cotton.

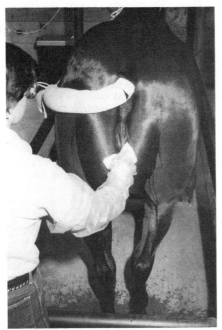

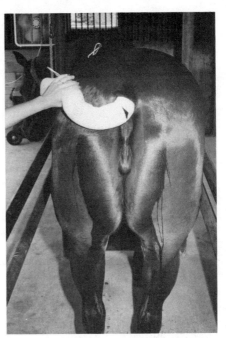

(K) After the perineum and buttocks are thoroughly rinsed, dry the entire area with clean paper towels.

(L) Once the area is washed, rinsed, and dried, the mare is ready for breeding.

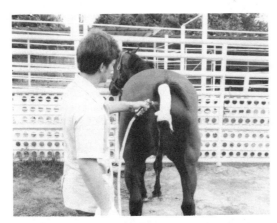

In place of the rinse bucket, a water hose and spray nozzle can be used for rinsing the mare, thus minimizing the chances of contaminating the mare during washing.

After the mare's tail has been wrapped and her perineal region washed, she should be led to the breeding area and restrained for breeding. The hobble or leg strap should be properly adjusted, kicking boots should be placed on the mare's hind feet, or the twitch should be ready for application.

WASHING THE STALLION

To encourage extension of the penis, the stallion should be allowed to tease the mare briefly and then moved to the wash area. The ideal location for a wash area is in or near the breeding area, since the mare's presence stimulates extension of the stallion's penis. After the penis is extended, it should be lathered gently with mild soap and clean, lukewarm water. The entire area should be rinsed carefully with clean water to remove all traces of soap and debris. Although some breeders prefer to simply rinse the stallion's penis prior to breeding due to the spermicidal effect of soap residues, the genitals should always be washed immediately after service to minimize the spread of disease.

A properly trained stallion is well-mannered during the wash process, but even the most tractable and well-trained stallion may prance or kick at this time. (Kicking is the stallion's natural response to handling of the external genitalia.) For this reason, it is important to acquaint a young stallion with the procedure. If the stallion seems to dislike the process, the handler should rinse the area and not attempt to wash it again that day. (If the stallion is washed more than once a day at first, his penis may become very sore.) The procedure is then repeated on the following day. If any area shows signs of irritation, the handler should stop washing the external genitals immediately.

By gradually removing the smegma, debris, and dead skin and by repeating the practice each day until the stallion accepts it readily, the handler can train most stallions in a few days. This gradual introduction helps to prevent physical irritation to the penis and limits the formation of any negative association (e.g., pain, discomfort, or fear) with the procedure.

At this point, the reader should note several problems associated with washing the stallion's penis. It has been reported that soaps and antibiotic solutions remove the naturally occurring bacteria that inhibit the growth of disease-causing organisms, although the significance of this concept is not yet clear. It is important to realize that vigorous or prolonged washing, the use of harsh soaps, or failure to remove all soap residues may irritate the penis. As a result, the stallion may become reluctant to breed. For this reason, close observation of penile skin condition is extremely important. (The use of iodine solutions in rinse or wash water should be avoided.) Finally, due to the well-known spermicidal effect of soap, rinsing is especially important around the glans to remove soap from the urethral diverticulum.

WASH EQUIPMENT AND SUPPLIES

clean lukewarm water (104-108°F)

mild soap solution: 1 part Ivory liquid soap to 1 part water in a clean
squeeze bottle

sterile cotton, preferably a roll of cotton

wash and rinse buckets (at least two, 2-3 gallon capacity buckets)

bucket liners (clean plastic trash bags are adequate)

disposable cups
sterile, disposable gloves
paper towels

WASH PROCEDURES

1. Prepare necessary equipment, observing strict hygiene:
   — one bucket (with clean plastic liner) containing warm soapy wash water (104-108°F),
   — one bucket (with clean plastic liner) containing warm rinse water (104-108°F),
   — open sterile cotton roll so that it is easily accessible but safe from contamination,
   — a clean disposable cup located near the rinse bucket — safe from contamination,
   — washer wearing sterile disposable gloves.
2. An attendant restrains the stallion while wash procedures are carried out. When trained properly, the stallion lets down soon after he is brought into the wash area. The presence of a mare in heat or the smell of urine from a mare in heat should stimulate extension of the stallion's penis.
3. After the stallion extends, he should be approached by the washer from the left side. Standing at the stallion's left shoulder and wearing sterile gloves, the washer applies soapy wash water to the stallion's penis and sheath using saturated cotton. Holding the penis gently but firmly in the left hand, the washer can use the cotton to wash the area, but once a piece of cotton is used it should be discarded. DO NOT RETURN USED COTTON TO EITHER BUCKET.
4. The penis and sheath should be washed thoroughly with special emphasis on the urethral diverticulum (recessed area around the urethral opening). It is important to remove all smegma from this area and from the preputial folds.
5. Rinse all soap and debris from the area by pouring clean, warm water from a cup onto the penis and sheath. DO NOT CONTAMI-NATE THE RINSE WATER. When rinsing is completed, fill the cup and submerge the glans penis to remove all soap from the urethral diverticulum.
6. Examine the prepuce and penis for signs of disease or irritation each time they are washed.
7. Water can have a detrimental effect on semen quality and should not be allowed to contact the semen. Clean paper towels or a sterile cloth can be used to dry the area. Some breeders prefer to remove excess water from the penis with a gloved hand and allow the area to air dry.

Courtesy of Manx Farm

The stallion's penis should be lathered gently using mild soap and sterile cotton.

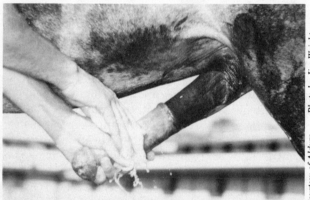

Photo by Jim Wright

Courtesy of Alduro

All soap should be rinsed from the penis with clean, lukewarm water and sterile cotton.

Photo by Jim Wright

Courtesy of Alduro

Special attention should be paid to the urethral diverticulum during rinsing to insure that no soap residue is left.

COVERING THE MARE

Before the stallion's mount is described, it is important to note that the location and alertness of every person involved in the process is essential to the completion of a safe, expedient cover. Each attendant should understand the cover procedure and should carry out each step attentively and calmly. If problems occur, carefully planned emergency procedures can prevent horses and attendants from being seriously injured. Appropriate actions for dealing with emergencies (e.g., releasing the hobble, backing the stallion, turning the mare quickly) should be planned and studied just as carefully as the routine breeding procedure.

After the stallion's genitals have been washed or rinsed and dried thoroughly, he must be allowed a period of sexual stimulation prior to mounting the mare. The length of time required to achieve an erection may vary from 20 minutes in young stallions to 15 seconds in more experienced stallions. As explained earlier, this period of stimulation is very important to the success of the stallion's mount. If the period is too short, the stallion may not be able to penetrate the mare; however if the period is too long, the stallion may lose interest in the mare. At this point, the experience of the stallion handler is extremely valuable. The handler should lead the stallion calmly but directly to the mare's left side. Inexperienced stallions are sometimes allowed to approach from the rear until they become more confident and skilled, but an approach at a 45° angle to the mare is more desirable in case the mare kicks.

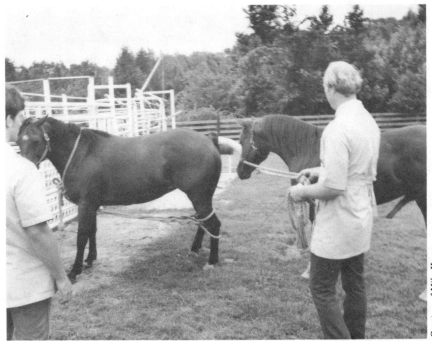

Courtesy of Mike Haney

The stallion should calmly approach on the mare's left side.

In any event, the stallion should not be allowed to charge the mare. If the mare is aware of the stallion's approach, she has time to brace herself and is not startled by sudden movement from the rear. The mare handler should stand just in front and to the side of her left shoulder. This individual is responsible for restraining the mare and must insure that the mare can see the stallion. The handler should be aware of the positions of both horses at all times and must be ready to stand back in case the mare or the stallion strikes.

Courtesy of Will A. Hadden III, DVM

After a short period of teasing, the stallion should position himself to mount.

Using a long lead shank, the stallion handler should stand to the stallion's left side and must maintain control over the stallion while standing at a distance away from the stallion. The handler should be ready to reprimand the stallion if he savages the mare and should stop him if he attempts to mount the mare before obtaining a full erection. The stallion can usually be discouraged from mounting the mare prematurely by tugging on the stallion's lead shank before he shifts his weight to his hindlegs. It is important to remember, however, that excessive correction during breeding can have serious psychological effects on the stallion, possibly resulting in reduced libido or refusal to mount the mare.

After the stallion has properly positioned himself on the mare, the handler may need to guide the stallion's penis (with a gloved hand) for proper intromission. (Note: Any manipulation or handling of the stallion's genitals, forelegs, etc., should be kept at a minimum since many stallions resent interference at this time and may become discouraged.) Preferably, an assistant should be available to hold the mare's tail to one side, direct the stallion's penis, and maneuver the breeding roll into position (if needed). If this third attendant is positioned on the stallion's left side, he or she must not interfere with the stallion handler's movement and must keep a close watch over both horses, ready to move out of the way quickly in case of an emergency. If the stallion handler holds the lead shank in one hand and braces the stallion's left foreleg with the other hand, the stallion's leg can be kept from slipping backward and hitting the assistant. When positioned on the stallion's right side, the assistant can also help push the mare's hindquarters into position but must be alert to avoid being caught in a dangerous position between the mare and the stallion.

A

B

Courtesy of Will A. Hadden III, DVM

(A) An experienced stallion will usually mount slightly to one side of the mare, but (B) he quickly positions himself behind the mare for intromission and ejaculation.

The number of thrusts that occur before the stallion ejaculates varies between stallions and throughout an individual stallion's season. Coitus may last anywhere from 30 seconds to 2 minutes. Ejaculation is signaled by flagging of the tail and, if the base of the penis is held gently, by pulsations along the urethra (i.e., along the underside of the penis). The stallion should be allowed to rest on the mare for several seconds and to dismount at his own pace. As the stallion dismounts, the mare should be turned to her left to protect the stallion from a possible kick, and the stallion should be backed and then turned away from the mare. Successful hand breeding is characterized by a good close cover followed by a relaxed dismount. A rapid dismount may cause injury to the mare's vagina or vulva and may allow large amounts of air to enter her reproductive tract.

Courtesy of Mike Haney

After dismounting, the stallion should be led away for postcover washing.

Before postcover procedures are discussed, it should be mentioned that the stallion may swoon, or faint, after ejaculation. This loss of consciousness is more common in stallions performing their first cover. Although the stallion may fall from the mare, his body is usually relaxed and unharmed by the landing. If the horse is injured by the fall or if he remains unconscious for an extended period, a veterinarian should be called.

POSTCOVER PROCEDURES

After breeding, the stallion should be returned to the wash area. His penis and sheath should be cleaned thoroughly with mild soap and then rinsed. Products other than mild soap, such as iodine or chlorhexadrene scrubs, are sometimes used for the postbreeding wash on farms that have had contagious equine metritis (CEM) outbreaks. However, these products are usually unnecessary and should be used conservatively.

Some breeders walk the mare quietly for 15 to 20 minutes after breeding to prevent straining and expulsion of semen. There is, however, no proof that this procedure improves conception rates, and authorities suggest that if a good close cover is executed, very little semen should be present in the vagina. The expulsion of semen from the vagina indicates that the ejaculate was not properly directed into the mare's cervix. Thus, walking the mare to counteract the effects of a poor cover is futile.

DISADVANTAGES OF HAND BREEDING

Although hand breeding is probably the most commonly used breeding method in the horse industry, there are a few disadvantages that should be noted. Without proper fertility studies and scheduling, hand breeding may result in overuse of the stallion, especially if he is bred extensively early in the season. In addition, scheduling problems may arise if several mares must be bred to the same stallion on the same day. However, both of these problems can usually be overcome with careful management. When compared with artificial insemination, hand breeding limits the number of offspring a stallion can sire within a breeding season. This situation is seldom a problem unless the stallion is very valuable and his services in high demand or unless his libido is very low, consequently limiting the number of mares he can cover naturally.

ADVANTAGES OF HAND BREEDING

There are a number of advantages to a hand breeding program; however, they can only be realized with good management.

1. Hand breeding allows more efficient use of the stallion's covers than does pasture breeding.
2. Routine wash procedures can be carried out in a controlled breeding situation; therefore, the spread of reproductive tract infection can be restricted. Attention to wash procedures is especially important when outside mares are bred on the farm, since the stallion can act as a potential carrier of pathogenic organisms from mare to mare.
3. In many instances, early conception dates for early foaling dates are important. Management techniques, such as light programs and hormone therapy, that enhance the chances of early conception dates are more practical in a controlled breeding situation.
4. When space is limited, far more mares can be bred in a hand breeding program than in a pasture breeding set-up, since space determines the number of mares that should be placed in a pasture breeding herd.
5. Use of the restraint techniques described earlier helps to insure the safety of everyone involved in the breeding process. This is especially important when the stallion is expected to cover unfamiliar or antagonistic mares.

6. Unlike pasture breeding, hand breeding allows the breeder to record the exact breeding dates and determine the probable foaling dates for each mare.
7. Close supervision of the stallion's covers allows the breeder to control the number of mares bred within a certain time period.

Artificial Insemination

When compared with other livestock industries, conception rates in horses are relatively low. In hopes of improving equine productivity, researchers and breeders have placed more and more emphasis on artificial insemination (the practice of injecting semen into the mare's reproductive tract using special insemination instruments and sterile techniques). In 1799, Italian physiologist Lazarro Spallazani performed the first scientific research on artificial insemination in animals. A century later, an American veterinarian used artificial insemination to settle mares that had failed to conceive by natural service, and a Russian physiologist developed methods for inseminating birds, horses, cattle, and sheep. These early methods lacked sophistication, but they formed the foundation for today's artificial insemination (AI) programs.

During the early 20th century, research emphasis shifted to AI in cattle and sheep due to increased demand for outstanding sires for breed improvement within the food industries. Until recently, however, information concerning equine AI techniques was limited. But as horse production gradually evolved into a multi-billion dollar industry, the need for modern breeding management tools became more and more important. As a result, research quickly filled the information gap, making AI a practical management tool on many large horse breeding operations.

Although artificial insemination has important advantages, it is prohibited by some breed registries and impractical on many smaller farms. Breeders should examine the feasibility of implementing an artificial insemination program based on the needs of the breeding operation. Because AI requires special facilities, laboratory equipment, and personnel with knowledge of AI techniques, such a program often requires a substantial investment. The individual breeder must decide if the cost is warranted. If, for example, the breeder owns a stallion whose services are in high demand, the stallion may be used more efficiently by extending his semen from one ejaculate to breed several mares.

The breeder should analyze all aspects of an efficient AI program, preparing a list of equipment requirements and a step-by-step description of procedures and safety precautions. AI can be used effectively to correct or overcome some fertility problems and to improve conception rates, but because the success of this breeding method depends on careful management, it can be equally disastrous if used carelessly. The following discussion is presented as a guide for establishing and operating an effective AI facility. After studying this material, the breeder should try to observe actual AI practices, either at a farm that has a well-established insemination program or at an AI course. Short courses are often presented by universities, agricultural extension agents, and breed associations.

FACILITIES

The layout of an artificial insemination facility affects the efficiency and safety of the entire breeding program. The design and cost of a facility can vary considerably, but the requirements of a functional facility are very basic. For example, the characteristics of a good breeding area also apply to the collection area. If the breeder plans to collect the stallion on a phantom mare, the collection area should contain a dummy that is about the same size as, or slightly smaller than, the stallion. The phantom should be padded and covered with a plasticized tarp or some other material that is durable and easy to clean. To help the stallion maintain his balance during the mount, the phantom mare's "neck" should be covered with a material that the stallion can grip with his teeth.

A sturdy, well-padded dummy (phantom mare) is usually used for the stallion to mount during semen collection. Soft footing should be provided in case the stallion falls from the dummy.

Due to the need for strict temperature control and careful semen handling procedures, a well-planned laboratory is essential to good AI technique. Before planning an AI laboratory, the breeder should review the discussion on laboratory design and safety within "Semen Evaluation." Since routine semen evaluation is essential to most AI programs, the laboratory should have an adequate power supply and good lighting. At the least, the laboratory should be designed so that evaluation equipment can be added in the future.

Depending on the size of the operation and the number of mares that may have to be inseminated at one time, the AI facility should have one or more insemination stocks. Ideally, these stocks, which may also be used to prepare the mares for breeding, should be near the collecting and semen handling facilities. The stocks should be designed for maximum efficiency and safety. For example, a row of stocks located side by side allows several mares to be inseminated quickly. Stocks should also be designed so that solid partitions separate individual mares, although simple pipe stocks without partitions can be separated by a buffer space of several feet.

Courtesy of Windward Stud

Well-designed examination stocks are an essential part of the artificial insemination facility.

One point should be remembered regardless of how the facility is organized: All areas should be constructed in such a way that they can be thoroughly cleaned. Hygiene is essential to the breeding operation. Therefore, to minimize the chances of infecting healthy horses, the breeder should make a conscientious effort to keep the AI facilities and equipment as clean as possible.

EQUIPMENT AND SUPPLIES

When setting up an AI facility, AI equipment and supplies form a significant part of the program's expenses. A transition from hand breeding to AI can usually be carried out with very few changes in breeding farm facilities, especially if the layout already includes a laboratory. However, much of the equipment used in an AI program must be replaced periodically. The cost of these supplies should be considered when the feasibility of artificial insemination is studied. Current prices can be ascertained by contacting a veterinary supply dealer.

EQUIPMENT CHECK LIST

Restraint Equipment

1. Halter and long lead shank, stallion chain, or snaffle bit;
2. Twitch, hobble, or other mare restraint equipment (if a jump mare is used when collecting the stallion);
3. Stocks or other equipment to restrain the mare during insemination.

Preparation and Collection Equipment

Refer to entries I and II under 'Semen Evaluation Equipment and Supplies' within "Semen Evaluation."

An incubator controls the temperature of artificial insemination equipment, as well as protecting the equipment from contamination during storage.

Laboratory Equipment

1. Filter assembly (gauze, milk line filter, etc.) if the artificial vagina does not include one;
2. Incubator, water bath, or hot plate;
3. Refrigerator for temporary semen storage.(Semen can be frozen for long-term storage, but this process is not yet used on a routine basis. Refer to the discussion on semen handling techniques within this chapter.);
4. Semen extender. (Refer to the extender charts for directions on preparing three types of AI extenders.);
5. Equipment for preparing extender: heat source, double boiler with smooth surface, pan that can be sterilized, candy thermometer, 20 ml glass tubes with plastic caps or stoppers, 100 ml beaker, stainless steel spoon, and a deep freeze;
6. Supplies for preparing extender: fresh skim milk and polymyxin B **or** half and half cream, Knox gelatin, and deionized water **or** non-fortified skim milk and Knox gelatin.

Insemination Equipment

1. Speculum (optional);
2. Roll obstetrical sleeves. (Disposable sleeves on rolls can be kept closer to sterile than boxed gloves.);
3. Sterile surgical gloves (optional);
4. Disposable plastic fusette with sterile plastic tubing **or** special AI catheter;
5. Flexible tip for catheter or fusette (optional);
6. Disposable syringes (one per insemination);
7. Sterile lubricant (not a chemically sterilized lubricant);
8. An opaque container to insulate and protect the semen-filled syringe.

EXTENDER FORMULAS*

FRESH SKIM MILK EXTENDER

1. Heat a measured volume (e.g., 100 ml) of skim milk in a double boiler for 10 minutes at 198-202°F (92.2-94.4°C). Approximately 10 ml of extender is used for each insemination, and a two to three monthly supply can be frozen for future use.
2. Allow the milk to cool.
3. Add 1,000 international units polymyxin B (an antibiotic) for every 1 ml of extender.
4. Divide the extender into 10 ml portions by transferring into 20 ml tubes. Stopper the tubes and store in a deep freeze.

5. Warm to approximately 100°F (38°C) before adding semen. Thawed extender can be stored in an incubator for about 24 hours.

CREAM GELATIN EXTENDER

1. Add 1.3 grams Knox gelatin to 10 ml deionized water.
2. Autoclave this mixture for 20 minutes.
3. Using a double boiler, heat about 100 ml half and half cream for 10 minutes at 198°F (92°C). Do not allow the mixture to boil or the temperature to exceed 204°F (95.5°C).
4. After heating, remove any scum from the milk surface.
5. Place 10 ml of the gelatin-water mixture into a 100 ml beaker and add half and half cream until a total volume of 100 ml is obtained.
6. Place 10 ml portions of the extender in 20 ml tubes and store in a deep freeze.
7. Warm to approximately 100°F (38°C) before adding semen.

SKIM MILK-GELATIN EXTENDER

1. Add 100 ml non-fortified skim milk to 1.3 grams Knox gelatin and shake gently for 1 minute.
2. Place the milk-gelatin mixture in a double boiler and heat at 197.6 to 203°F (92-95°C), swirling the mixture periodically. It is important that the temperature stay within the 197.6 to 203.0°F (92-95°C) range throughout the 10 minute heating period.
3. Place 10 ml portions of the extender in 20 ml tubes and freeze.
4. Warm the extender to approximately 100°F (38°C) before adding the semen.

* Colorado State University

PREPARATION

Careful preparation for artificial insemination enhances the program's efficiency and is just as important as the actual collection and insemination techniques. An important but frequently overlooked procedure is the routine evaluation of AI equipment. To make sure that all of the necessary supplies and tools for semen collection, semen evaluation (if performed on the farm), and insemination are ready for use, a thorough equipment check should be performed before the day's activities begin. AI equipment should also be checked carefully for soundness prior to each use. Even a small leak in the artificial vagina water jacket may ruin an entire ejaculate if water becomes mixed with the sample.

Any tool or agent that contacts the mare's reproductive tract or the stallion's genitals should be as clean as possible and preferably sterile. Anything that touches the semen sample must be sterile, free of spermicidal residues, and prewarmed to 100°F. Sterile disposable insemination units (containing a syringe, a fusette or catheter, and gloves) can be purchased in sealed packages. Reusable equipment must be sterilized prior to each use and should be stored carefully to prevent contamination. Sterile wraps, autoclaves, incubators, and dust-free cabinets are commonly used for equipment storage. Special attention should be paid to sterilization techniques, since semen samples may be contaminated by nonsterile equipment or the equipment may act as a reservoir for infection. The temperature within the incubator or water bath can influence semen quality (e.g., motility and longevity) and should be monitored closely. A precollection check list may also include the following:

1. Check the supply and age of frozen extender (if a semen extender is used). Although frozen extender is probably safe to use within a year, it is best to make a fresh batch every three months.
2. Prepare reagents and stains if the semen sample is to be evaluated.
3. Calibrate the spectrophotometer and standardize the pH meter (if these instruments are used).

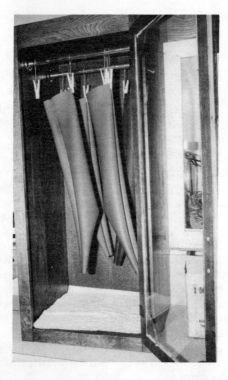

Artificial vagina liners are usually stored in a well-sealed cabinet to protect the liners from dust.

Another important preliminary procedure is the scheduling of collections and inseminations. First, mares must be selected on the basis of teasing and palpation findings. Ideally, heat detection procedures are carried out in the morning, leaving a full day for preparation, semen collection, semen evaluation, and insemination. The correct stallion or stallions for the mares to be bred that day should be scheduled for collection and a tube of frozen extender should be removed from the freezer (unless raw semen or fresh extender is used). The use of a carefully planned schedule can save time and prevent mistakes. Factors that influence scheduling include the number of mares to be bred, the number of stallions to be collected, and the number of times that one stallion must be collected on a single day.

Although there are several methods of collection, the only semen collection device recommended for AI is the artificial vagina (AV). There are many types of artificial vaginas, and each has unique merits and disadvantages, but because each would require a lengthy discussion, a complete description of the assembly of only one AV model, the Colorado AV, is presented.

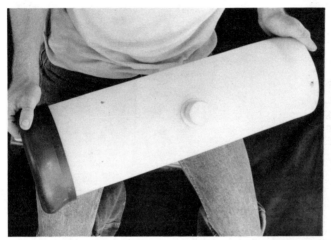

(A) The tubular cylinder is designed to give the assembled AV structural rigidity. Notice the capped opening that allows the AV to be filled with water.

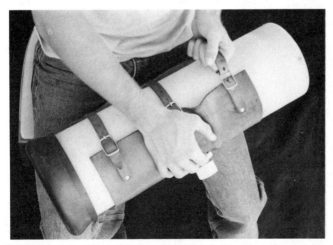

(B) The leather casing provides added insulation and gives the attendant something with which to grip the apparatus during collection.

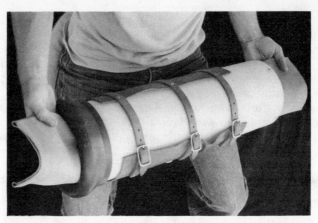

(C) Insert the outer liner inside the AV.

(D) Turn one end of the outer liner over the rubber-covered end of the AV.

(E) Turn the other end of the outer liner over the bottom of the AV and secure the liner with a heavy rubber band.

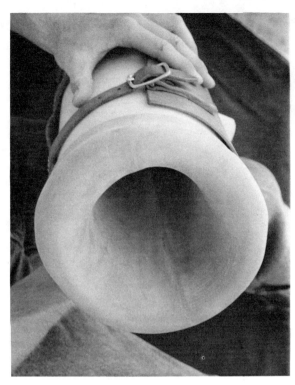

(F) When properly applied, the outer liner should form a smooth, unwrinkled passageway through the AV.

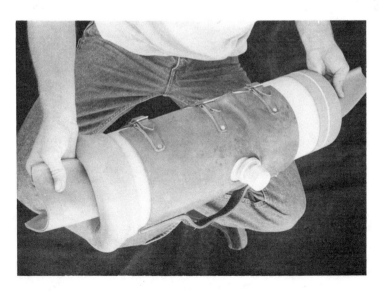

(G) Insert the inner liner inside the outer liner.

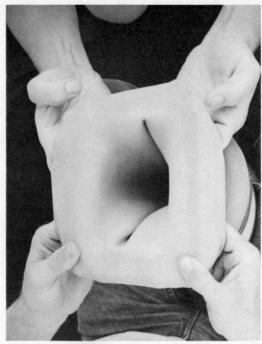

(H) Turn the upper end of the inner liner over the rubber-collared end of the AV being careful not to rip the liner.

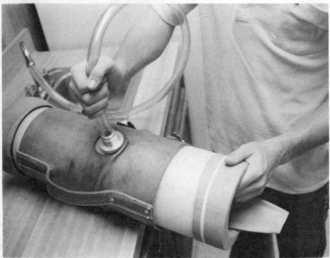

(I) Fill the AV with warm water at 120 -140° F (49-60° C). The initial water temperature must be warm enough to compensate for heat loss to the cooler AV and any cooling that occurs before collection. The amount of water placed in the jacket controls the AV pressure and should be adjusted to fit the individual stallion's preferences. After determining the amount of water that causes the most favorable response from a particular stallion, the breeder should note the weight of the AV so that the same pressure can be obtained for future collections from that stallion. When filled with water, the Colorado AV usually weighs between 20 and 25 pounds (9.07-11.34 kg).

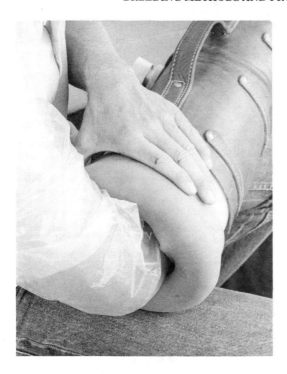

(J) Insert a sterile, lubricated obstetrical sleeve.

(K) Turn the open end of the sleeve back over the AV's rubber collar to protect the inner liner from contamination.

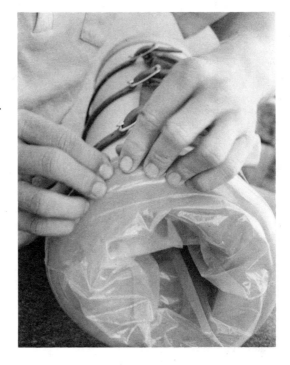

(L) Place a thermometer in the sleeve to monitor changes in the AV's temperature. The final temperature should be between 111°F and 118°F (44°C and 48°C).

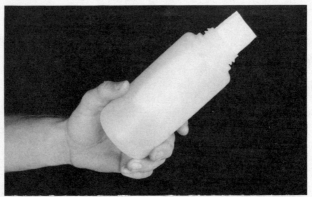

(M) A prewarmed collection bottle is used to collect the semen sample, and a filter is usually inserted inside the bottle to separate the gel fraction from the remainder of the semen sample.

(N) When the AV is completely assembled, a closed pathway is provided for the sperm to run through the vagina and into the receptacle for collection.

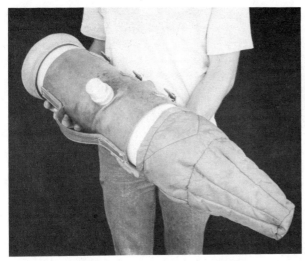

(O) If the AV must be used in a cool area or if the collection bottle must be exposed to sunlight, a prewarmed and insulated cone can be used to protect the collected sample.

Efforts to keep the AV equipment as clean as possible should not be undermined by failure to prepare the breeding animals properly. After the AV is assembled and while its temperature is stabilizing, the stallion should be prepared in the same manner described earlier in this chapter. The stallion is stimulated and washed in a routine fashion, with special emphasis on rinsing and drying the penis and sheath completely. If a jump mare is used during the collection process, she should be washed, and her tail should be wrapped as if she were being prepared for live cover. Preparation of the mare for insemination is identical to the procedure presented within the discussion on hand breeding, with one exception. After the mare is washed and her tail is wrapped, she is usually left in (or placed in) a stock. She can be restrained by other methods, but stocks are preferable in most instances.

SEMEN COLLECTION

The need for a calm, well-directed collection procedure is important not only to the stallion's attitude but also to the safety of all concerned. When a jump mare is used, the sequence consisting of approach, stimulus, and mount is similar to that performed during hand breeding. When a phantom mare is used, a mare in heat may be required to stimulate erection, but the breeding sequence follows the usual pattern. Each attendant's position and function are very important to the efficiency of the procedure and can minimize the chances of accidents or injuries. Regardless of how the collection procedure is carried out, each participant should understand the exact routine and should be alert for problems. Because the stallion is expected to cooperate in an unnatural breeding situation, it is imperative that the collection routine be well planned.

When a jump mare is used, restraint procedures should be selected according to the mare's temperament. For example, a docile, trustworthy mare that is used routinely for collection (e.g., gentle nymphomaniac or mare on special hormone treatment) may not require hobbling, but the breeder must make this decision with careful consideration of the stallion's safety. Even when the jump mare is in strong heat, some breeders prefer to hobble and twitch her. Obviously, the use of a dummy mare minimizes the chances of a stallion's being injured, but special safety precautions should always be observed.

1. The phantom should be generously padded for the stallion's comfort.
2. The body of the phantom should be covered with a material that can be cleaned and that minimizes the chances of the stallion's being pinched or "burned" during collection.
3. The phantom should have a special cover on the neck area that allows the stallion to grip the dummy with his teeth, thereby positioning and balancing himself during the mount and collection procedures.
4. The phantom should have a sturdy base that can be adjusted for the stallion's height. It must also be able to safely carry the stallion's weight.
5. The stallion should not be allowed to charge the dummy. The approach should be characterized by the same behavior required during a live cover.

If the mare handler, the stallion handler, and the collector are positioned along the animals' left sides, in an emergency the mare can be pulled to

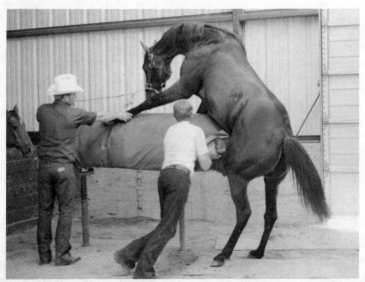

The safety precautions for collecting semen using a jump mare and a dummy are similar. Safe footing should be provided for the stallion. In this case, a portion of the floor is covered with mats to provide traction. In addition, the stallion handler should support the stallion's forefoot to protect the collector.

When a jump mare is used, the stallion should approach to the left and rear of the mare. The mare should be hobbled to protect the stallion. Her tail should be wrapped, and her perineal area should be washed as thoroughly as if she were to be bred by the stallion.

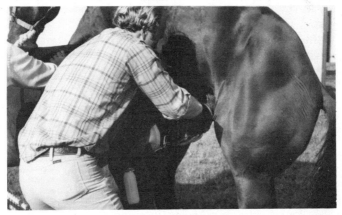

As the stallion mounts, his penis should be deflected into the artificial vagina.

Photos courtesy of Mike Haney

After the stallion ejaculates, the artificial vagina should be tilted to allow the semen to run into the collection bottle attached to the inner liner.

her left so that her hindquarters swing away from the stallion and handlers. This action prevents the stallion and his handler from being kicked by the mare. Additionally, the stallion handler can steady the stallion's left foreleg so that a flailing hoof does not strike the collector. As the stallion mounts the phantom or jump mare, the collector should move quietly along the handler's right side to the stallion's left side. The collector must be ready to deflect the stallion's penis with a gloved hand, guiding it gently to the AV. (Note: The presperm semen fraction, which drips from the stallion's penis when he becomes excited, contains a high concentration of bacteria and debris and should not be collected.) The AV must not be forced on the stallion, since the stallion usually recognizes this type of movement as abnormal and becomes discouraged or distracted. Once the stallion enters the AV, the instrument should be steadied against the mare's thigh at the point of her buttock with the collection bottle end pointed down about 30° from a horizontal plane that passes through the point of the buttock. If the stallion attempts to dismount, he should not be followed with the AV as this would only reinforce any doubts he might have about the procedure.

The collector may have to hold the AV in both hands, but if a thumb and index finger can be wrapped around the base of the stallion's penis, pulsations along the lower side of the penis can be detected when ejaculation occurs. Ejaculation is approximately a 10 to 12 second process, but about 75 percent of the sperm is present in the first three spurts. The postsperm fraction is a grey viscous material. Because most of the sperm cells are present in the first few spurts of ejaculate and because it is important to separate the gel and sperm-rich fractions, the end of the AV to which the collection bottle is attached should be slowly lowered as the stallion ejaculates. This causes the majority of the sperm fraction to run through the inner liner and into the collection bottle.

After the stallion ejaculates, he should be allowed to rest on the dummy or jump mare and dismount when he is ready. As the stallion dismounts, the AV is removed and 50 to 75 percent of the water is released from the pressure jacket so that the semen moves quickly into the collection bottle. The AV should be taken to the laboratory as quickly as possible and, unless the semen is to be used immediately, the collection bottle should be placed in an incubator or water bath.

LABORATORY TECHNIQUES

Immediately after dismount, the semen sample should be covered with a protective cone or wrap to protect it from light and temperature change and then taken to the laboratory. The filter apparatus must be removed as quickly as possible to prevent the gel from seeping through to the sperm-rich fraction. If a filter device is not included in the collection apparatus, the sample should be filtered through sterile gauze or a milk line filter immediately after collection. (Filtering the sample as it is collected is preferable, however.) Techniques such as washing the sperm cells and separating semen portions by spinning the sample in a centrifuge at a designated speed, have been developed for research purposes but are impractical on the breeding farm. For most practical purposes, sterile gauze placed over the mouth of the collection bottle is effective in preventing gel from moving into the collection bottle.

After removing the filter, or filtering the sample, the total gel-free volume should be measured in a prewarmed graduated cylinder. The cylinder must be tilted and the sample poured slowly to avoid damaging the sperm cells. The sample should then be studied to determine how much semen must be included in each insemination dose. Two values, motility and concentration, must be determined before the insemination dosage can be calculated. (Refer to "Semen Evaluation" for instructions on motility and concentration evaluation.) If any semen is left over after insemination is complete, morphology studies, white blood cell counts, and other semen evaluation procedures can be performed.

Authorities recommend that each mare be inseminated with 100 to 500 million progressively motile sperm cells. If the semen is refrigerated for over two hours or if the fertility status of either animal is significantly low, 500 million progressively motile sperm are preferable. Studies have shown, however, that insemination dosages exceeding 500 million sperm do not enchance conception rates. To determine the number of progressively motile sperm, multiply the sample's percent motility by the number of sperm cells per milliliter:

$$\begin{matrix} \textbf{\% progressively} \\ \textbf{motile sperm} \end{matrix} \quad \textbf{x} \quad \begin{matrix} \textbf{sperm cells} \\ \textbf{per ml} \end{matrix} \quad = \quad \begin{matrix} \textbf{progressively} \\ \textbf{motile sperm per ml} \end{matrix}$$

To calculate the proper insemination volume, divide the number of motile sperm required by the number of motile sperm in each ml of semen:

$$\begin{matrix} \textbf{progressively} \\ \textbf{motile} \\ \textbf{sperm required} \end{matrix} \quad \div \quad \begin{matrix} \textbf{progressively} \\ \textbf{motile} \\ \textbf{sperm per ml} \end{matrix} \quad = \quad \begin{matrix} \textbf{insemination} \\ \textbf{volume} \\ \textbf{in ml} \end{matrix}$$

If the sample is to be used immediately or within two hours, its temperature should be stabilized at about 100°F (38°C) in an incubator or water bath. The semen must be gradually cooled and then refrigerated if the sample must be used later in the day (e.g., 2 to 24 hours later). Some reduction in the number of live cells and in sperm motility is to be expected when the sample is rewarmed prior to insemination, but the use of refrigerated semen can be successful when the handling guidelines described in "Semen Evaluation" are followed carefully. Freezing techniques designed to allow long-term semen storage have become an important part of the cattle breeding industry, but only a few equine breed registries allow the use of frozen semen. Perhaps in the near future, long-term semen storage will be a realistic and effective management tool for horse breeders.

In many instances, raw semen is divided into individual dosages and inseminated soon after collection. Although this is an effective method and certainly one that should not be discouraged, there are several advantages to adding an extender to the sample. For example, extenders that contain antibiotics or antibacterial agents allow the control of pathogenic organisms within the semen and, consequently, aid in the control of reproductive infections. In addition, the extender's nutrient base (e.g., skim milk or gelatin) seems to prolong sperm survival and is especially helpful when handling samples from subfertile stallions. Occasionally, the sample is highly concentrated in a small volume, and because small samples are difficult to handle, a significant number of sperm cells may be lost. Semen extenders ease these handling problems and reduce the chances that a large percentage of the sample's sperm cells will be lost as the sample is transferred from one container to another or inseminated in the mare.

Many extender formulas have been described by researchers, and as more is learned about the effects of various agents on equine sperm, their recommendations change. Today, milk-gelatin extenders described earlier in this chapter seem to provide good results and are relatively easy to prepare. Other extenders can be used, but the reader should keep in mind that glycerol, an agent that was once a common additive in equine semen extenders, has been shown to be spermicidal. It is also important to note that concentration studies must be performed on raw semen before it is diluted. Although motility studies are often performed on extended semen, the cream and whole milk extenders contain fat globules which interfere with microscopic examination of sperm motility.

INSEMINATION

The introduction of semen into the mare's reproductive tract cannot be a completely sterile process, but to be an effective breeding tool, the techniques used should minimize contamination. All equipment used in this process, especially the instruments that enter the mare's tract, must be sterilized. Also, a sterile path from the exterior through the mare's

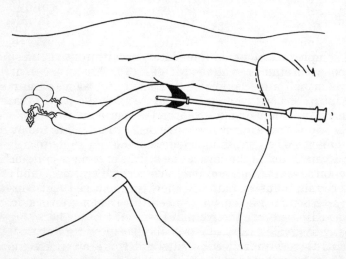

Four inches of the semen-filled catheter are inserted through the cervix so that the catheter extends two inches into the uterus.

vagina into the cervix and uterus must be established through which to introduce semen. This path usually consists of sterile plastic tubing and a 22 inch (55.9 centimeter) catheter, insemination pipette, or fusette which enters the cervix and deposits semen directly into the uterus. A rubber elastrator band located 4 inches (10.2 centimeters) from the tip of the catheter helps the inseminator determine when the catheter extends 2 inches (5.1 centimeters) into the uterus. Because these instruments contact the mare's tract and/or the semen sample, their sterilization is extremely important.

Insemination also requires a method of measuring the necessary volume of semen and injecting that volume through the sterile path to the uterus. A syringe can be used for this purpose, but it is important to use the sterile, disposable type without attempting to wash and reuse the syringe. Special insemination kits that include the syringe, tubing, and catheter can be purchased in special sealed packages. These kits are sterile, easy to store in an incubator, and very popular on many AI operations.

Although the insemination procedure can be carried out by one person if the mare is secured by stocks, an assistant can help the process flow more smoothly by preparing and presenting the equipment to the inseminator. Efficiency is especially important when several mares must be inseminated at the same time.

If a small volume of semen is used (e.g., without an extender), 2 additional ml of semen drawn into the syringe will account for the percentage of the sample that is left within the insemination tubing. Once the syringe is prepared, it should be protected from sunlight and maintained at about 100°F (38°C).

The arm that enters the mare's reproductive tract must be covered with a sterile, lubricated sleeve. To lubricate the sleeve, apply sterile, nonspermicidal lubricant to the area of the knuckles and wrist. As the inseminator's arm enters the tract the lubricant will spread evenly over the glove.

Holding the catheter or flexible fusette tip in the palm of the hand, the technician introduces the insemination unit into the mare's reproductive tract. If the arm cannot be introduced because of Caslick's sutures, a speculum with a light can be used to guide the insemination unit into the cervix visually. Because a flexible tip cannot be used with this method, the inseminator must use caution to avoid injuring the mare's tract with the catheter. If the mare urinates before the catheter tip enters the cervix, cover the tip with a finger. After urination has ceased, remove the arm, change sleeves, and continue the procedure.

As the hand reaches the cervix, the degree of cervical dilation should be checked. If the cervix is closed, the mare may be pregnant, and no attempt should be made to inseminate until her reproductive status can be checked closely. The cervix may be open only slightly or may be so loose that it actually falls over onto the vaginal floor.

Some individuals prefer to check the status of the cervix first with a finger and then withdraw the arm slightly to pick up the insemination unit so that the catheter or fusette can be passed through the vagina with the finger over the tip. In any event, the inseminator must locate the cervix, insert an index finger into the cervical opening, and guide the catheter into the uterus. If a flexible fusette tip is used, the inseminator must curve the tip so that it enters the appropriate horn (i.e., depending on where the Graafian follicle has developed). When the fusette-flexible tip junction or the rubber elastrator band reaches the posterior end of the cervix, the finger should be removed from the cervix and the catheter steadied from within the vagina.

The syringe is then attached to the tubing or the catheter, and the semen is injected with slow, even pressure over about a 15-second period. If the semen is expelled too rapidly, a significant percentage of the sperm may be damaged. If resistance is felt when the sample is injected, the end of the catheter tip may be pressed against the uterine wall. If this is the case, the catheter should be retracted slightly. If the mare urinates while the catheter is in the cervix or uterus, the inseminator should hold the cervical opening closed and continue expelling the semen in a slow, steady manner.

After the sample is expelled, the insemination unit should be retracted until the tip reaches the palm of the hand. Then the arm should be removed in a slow downward motion to prevent a sudden rush of air into the reproductive tract. The reduced pressure within the tract causes a vacuum, but by reducing the pressure within the tract gradually, the inseminator minimizes the back-flow of air. Holding the vulva closed above the arm also limits the influx of air.

DISADVANTAGES OF ARTIFICIAL INSEMINATION

As illustrated throughout this discussion, artificial insemination is a progressive management tool designed to maximize efficiency on the breeding farm. It should be noted, however, that a successful AI program is the result of careful planning, proper facilities, and sterile collection and insemination techniques. When performed carelessly, artificial insemination can have a disastrous effect on the farm's overall productivity. Even when the procedures involve forethought and conscientious semen handling techniques, an AI program's first year is often characterized by reduced conception rates. After a season of trial and error procedural adjustments, the program usually flows more smoothly.

To minimize these early problems, the breeder should study other AI programs and consult experts in the field before establishing this type of program. When studying the feasibility of an AI program, the breeder should also consider the following disadvantages of artificial insemination as an equine breeding tool.

1. Perhaps the most important disadvantage of AI is its detrimental effect on fertility when performed improperly. Human intervention in the equine breeding process was described earlier as one of the most significant causes of infertility in both mares and stallions. In an AI program, human error and management-related infertility are considerable problems.

2. Another important drawback of artificial insemination is that it is prohibited by some breed registries.

3. The expense often associated with establishing and maintaining an AI program can also be a significant disadvantage. If the farm already has modern facilities for semen evaluation, teasing, palpation, and live cover, the initial expense should be minimal. Unlike pasture breeding and live cover, however, equipment plays an important role in maintaining high conception rates and preventing infection. The labor involved in maintenance procedures must also be considered along with that required for preparation, collection, and insemination procedures.

4. If AI is used to overcome specific fertility problems, the breeder should consider the possibility of propagating inherent reproductive or endocrine abnormalities by encouraging problem animals to produce offspring. Although selection for fertility is minimal (and usually impractical) in today's equine industry, low fertility is in itself a means of culling certain abnormalities from a population.

5. A popular argument against AI states that the breeding technique invites fraud and error with respect to parentage. Because the semen may be handled several times before it is used to impregnate mares, there is a significant opportunity for error, either intentional or accidental. This possibility can be minimized, however, by the use of blood typing and meticulous record keeping.

6. Some breeders and registries are also concerned that if the more popular stallions are bred to a greater number of mares, there may be a marked decrease in the demand for less popular stallions.

7. Similarly, some breeders fear that if a stallion is allowed to sire a significantly larger number of foals through the use of AI, the average value for each foal may decrease.

8. Another argument against AI involves the possibility of concentrating a stallion's genetic faults within the breed by using him extensively in an AI program. These faults would then be difficult to cull from the breed. Also, if certain bloodlines are concentrated in a population, the genetic variation within that population may decrease. (Genetic variation is necessary for improving a breed through selection.)

9. Another point that is sometimes mentioned as a disadvantage of artificial insemination is that AI does not provide the stimulus necessary for the production of oxytocin in the mare. Research indicates that natural service causes the release of oxytocin which, in turn, induces uterine muscle movement.

ADVANTAGES OF ARTIFICIAL INSEMINATION

In spite of the arguments against AI, this breeding technique is rapidly becoming an important part of the horse breeding industry. As a business, horse breeding demands efficient management techniques and maximum productivity. When used properly, AI is efficient and allows maximum control over breeding hygiene and safety. Although the following arguments for AI outnumber those against, it should not be assumed that AI is the answer to all breeding management or fertility problems, or that it is the best technique for all breeders.

1. One of the most important advantages of AI is that it minimizes the spread of reproductive infections by eliminating physical contact between the mare and the stallion. Also, the concentration of bacteria and other pathogens within the ejaculate is reduced by dividing the sample into insemination dosages and diluting each dose with an extender, particularly if the extender contains antibiotics.

2. Another important advantage of AI is that it greatly reduces the chances of injuring the mare or the stallion. If a phantom mare is used to collect the semen sample, contact between the breeding animals is eliminated, and the chances of injury to either animal during collection or insemination are minimal. AI is especially useful for stallions that tend to savage mares or reject mares with certain characteristics and for mares that are particularly ill-mannered.

3. Artificial insemination can also be used as a tool for enhancing conception rates. This breeding technique not only improves the percentage of mares in foal at the season's end but also increases the number of mares that settle on the first heat cycle. Again, it should be emphasized that AI is beneficial only when the guidelines developed through research and experience are followed carefully. It is also important to remember that the farm's conception rates may not improve during the first few breeding seasons.

4. By eliminating the physical stress associated with live cover, artificial insemination allows certain types of problem mares to be bred quickly without excitement or strain. For example, if due to lameness, a mare is unable to support the weight of the stallion during live cover, AI may be extremely helpful.

5. AI is an important breeding aid for stallions that have arthritis, injury, and age-related weakness. Some breeders have actually trained lame stallions to ejaculate into an artificial vagina without mounting a dummy or jump mare. However, before making this

type of allowance for a handicapped stallion, the breeder should consider the ethical responsibility of all breeders to eliminate inherent weaknesses within the breed.

6. If insemination dosages are calculated for each ejaculate, an artificial insemination program encourages close observation of the stallion's fertility status throughout the breeding season. Changes in the number of sperm cells per ejaculate and in sperm motility or maturity can be detected immediately, and corrective measures can be implemented so that any period of decreased productivity is minimized.

7. Because semen quality can be closely monitored and because an average semen sample can be divided to impregnate as many as 10 mares, AI helps to prevent overuse of the stallion. This is especially significant when several mares must be bred on the same day.

8. Although the benefit of this point is somewhat controversial, AI can be used to increase the number of foals sired by a particular stallion. It is important, however, that the market breeder control the number of foals produced so that the value of each foal is not reduced. The basic business principle of supply and demand should be considered before the stallion's book is increased.

9. AI is a safe method for overcoming height differences between the mare and the stallion.

10. Another argument for artificial insemination is that AI programs encourage careful record keeping, especially when the breed registry enforces strict requirements with respect to the registration of AI offspring.

11. Although frozen semen is seldom used at this time, advocates of AI point out that with frozen semen stallions can be utilized on a worldwide basis without the risk and expense of transporting breeding animals. However, frozen semen is seldom used at this time.

Embryo Transfer

Embryo transfer is the process of removing a fertilized ovum from a donor mare and introducing it into a recipient mare's uterus, a procedure that can be carried out with or without surgery. The first attempt to transfer an equine ovum was made by two Japanese researchers in 1972. This early attempt did not result in any recipient pregnancies, but in 1974 the same researchers recovered 18 ova from 29 mares, transferred 15 embryos, and confirmed 6 pregnancies. In 1978, researchers at Texas A&M University performed the first successful embryo transplant by nonsurgical methods in the United States. A request to register the resulting foal, a Quarter Horse filly named Miss T Eyes, prompted an important ruling by the American Quarter Horse Association. The AQHA ruled that embryo transfer foals could be registered with certain restrictions. The ruling restricts the number of foals produced per mare, identifies requirements for donor mares, and requires certain tests and applications. This decision has encouraged other registries to consider the feasibility of embryo transfer and has encouraged further research efforts. Although this process offers tremendous possibilities as a research and breeding management tool, techniques are still in the formative stages, and the method has not yet gained widespread acceptance within the industry.

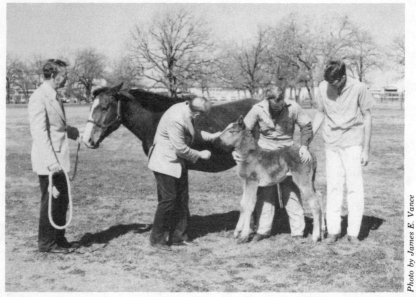

Photo by James E. Vance

Courtesy of D. C. Kraemer, Texas A & M University

The result of the first successful nonsurgical embryo transfer in the United States was the American Quarter Horse filly, Miss T Eyes, shown here with the recipient mare.

PROCEDURES

To be successful, embryo transfer must be performed on mares with synchronized estrous cycles. In other words, the donor and recipient mares must be within 24 hours of the same stage of estrus at the time of transfer. Estrus synchronization is usually achieved through the use of prostaglandin but can also be accomplished by maintaining a large number of recipient mares to increase the chances that at least one mare will be ready when needed. (Refer to the discussion on estrus synchronization within "Broodmare Management.")

If breeding is successful, the ovum is fertilized and begins to develop in the donor mare's oviduct. The fertilized ovum reaches the mare's uterus by the sixth day after ovulation but does not become attached to the uterine wall for about five to six weeks. Between the seventh and ninth days after ovulation, the fertilized ovum (at that point, called a blastocyst) is removed from the uterus using a special flushing technique. (Surgical techniques for removing the fertilized ovum have also been described, but the flushing technique is less traumatic and far more practical.) To insure that most of the fluid used to flush the mare's uterus is recovered, a special catheter (Foley catheter) equipped with a balloon-like cuff is inserted through the cervix. The cuff is inflated with air or water and is pulled back against the

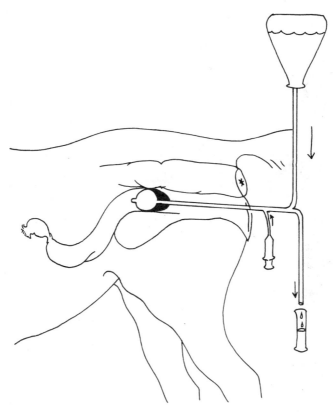

The embryo collection apparatus consists of a special catheter equipped with an inflatable cuff, which prevents fluid from escaping the uterus around the catheter, and tubing through which fluid and the flushed embryo pass into a collection receptacle.

cervix, permitting fluid to escape the uterus only through the catheter. A quart of fluid (e.g., physiological saline) is allowed to flow through the catheter and into the uterus by force of gravity. After a few minutes, the fluid is drained from the uterus, collected in small glass containers, and carefully checked for the presence of a blastocyst, which can be detected visually as a shiny speck at seven to nine days after ovulation.

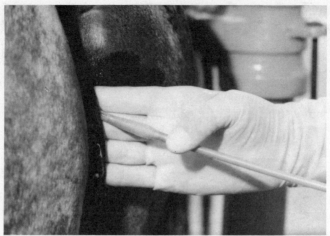

Courtesy of D. C. Kraemer, Texas A & M University

The instrument used for flushing is a Foley catheter. The cuff, which is not inflated until after the catheter is inserted, is shown to the left of the thumb.

Courtesy of D. C. Kraemer, Texas A & M University

The fluid from the uterus is recovered through flexible rubber tubing and collected in small dishes for examination.

Although the flushing procedure may have to be repeated three times, the fertilized ovum is usually found in the first infusion. (Uterine palpation via the rectum may be required to help remove fluid from the uterus during the second and third infusions.) A microscopic check is used to positively identify the blastocyst. For example, the technician may locate the cellular mass by studying the sample through a stereoscope, an instrument that permits the viewer to see a magnified three-dimensional field. When the blastocyst is located, it is usually placed in a small bowl containing a special nutrient medium.

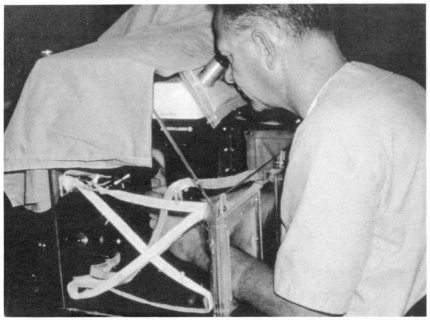

Courtesy of D. C. Kraemer, Texas A & M University

Although a seven-day-old embryo is sometimes visible to the unaided eye, it appears only as a tiny speck. Therefore, the fluid recovered from the uterus is carefully examined under magnification to locate the embryo.

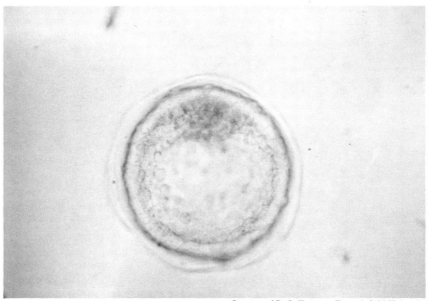

Courtesy of D. C. Kraemer, Texas A & M University

Seven-day-old equine embryo.

The collected blastocyst is implanted in the recipient mare's uterus within 2 hours, regardless of whether a surgical or nonsurgical technique is used. Surgical transfer requires general anesthesia, an incision in the recipient mare's flank or ventral midline, and the introduction of the blastocyst through a small pipette. Although this technique seems to result

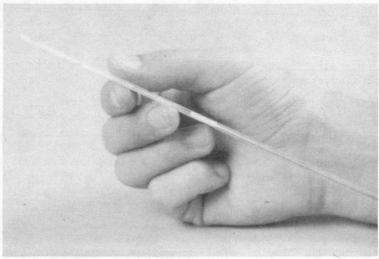

Courtesy of D. C. Kraemer, Texas A & M University

The recovered embryo is inserted into the recipient mare through her cervix by means of a hollow embryo transfer straw.

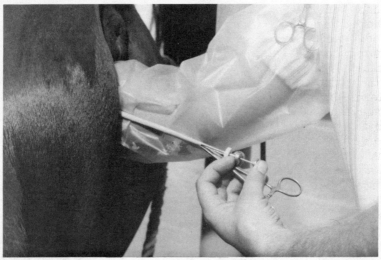

Courtesy of D. C. Kraemer, Texas A & M University

Transfer is completed by expelling the fluid and the embryo within the straw into the mare's uterus.

in higher pregnancy rates than the nonsurgical method, it is more stressful for the mare, more time-consuming for the veterinarian, and more expensive for the mare owner. Preparation for the nonsurgical implant method is very similar to that employed for routine artificial insemination. The mare is washed, her tail is wrapped, and she is placed in a palpation stock. The hygiene and safety guidelines described in the discussion on artificial insemination are observed to improve the chances of a successful transfer. After preparation procedures are completed, the blastocyst and several cubic centimeters of the nutrient medium are drawn into a special insemination catheter very carefully. Wearing sterile sleeves, the technician passes the catheter through the mare's cervix and deposits the fertilized ovum directly into the mare's uterus.

APPLICATIONS OF EMBRYO TRANSFER

Perhaps the most important application of embryo transfer is to obtain foals from mares that are physically incapable of carrying a foal to term. For example, a mare may have suffered a chronic uterine infection that damaged cellular structure to such an extent that much of her uterus is unable to produce prostaglandins or transport nourishment to a developing fetus. If this mare can be stimulated to produce an ovum and can be bred successfully, she may continue to produce offspring in spite of her pathological infertility. Similarly, a mare may have scar tissue within her cervical opening as a result of a past injury or infection. If the presence of this scar tissue prevents proper closure of the cervix during pregnancy, the uterus and the fetus may be subjected to contamination by urine, airborne debris, bacteria, etc. If she is valuable, this type of mare is a possible candidate for embryo transfer. (There are a number of possible causes of pregnancy failure. Refer to "Abortion.")

Embryo transfer also allows older mares to produce foals without the risk of age-related pregnancy and foaling problems, problems that could affect the mare's fertility or, in rare cases, result in her death. Reports indicate that 85 percent of the deaths that occur in pregnant and postpartum (i.e., after foaling) mares are caused by gastrointestinal tract torsion or uterine arterial hemorrhage. The occurrence of uterine arterial hemorrhage increases five-fold in older mares. For this reason, embryo transfer could be used as a measure of added insurance for owners of older broodmares.

The transfer of a fertilized ovum to a recipient mare could also be performed to provide a genetically promising foal with the best possible maternal environment. It is known that some mares are better milk producers than others and that the size of a mare's uterus can limit her foal's mature size regardless of its genetic potential. In fact, even the dam's attitude has an important effect on the foal's tractability since the youngster often mimics its dam's behavior patterns. If a mare's inherent attributes (e.g., conformation, performance, etc.) are desirable but her mothering ability somewhat limited, embryo transfer could be a means of giving her foal every possible advantage.

Embryo transfer could also allow mares to produce offspring without ending or interrupting their race or show careers. This breeding technique could play an important role in shortening the equine generation interval by eliminating the delay caused by athletic and show careers and by allowing two-year-old fillies to produce offspring. Even though most fillies begin cycling by two years of age, pregnancy is usually detrimental to both offspring and dam due to the mare's physical and psychological immaturity.

Another application of embryo transfer is the production of several full siblings within a single reproductive year. This application could be especially helpful to breeders who are trying to develop or preserve a specific breed or type. On the other hand, many breeders are concerned that this type of increase in production may have an undesirable economic impact on the operation. It should be noted, however, that the number of ova produced by one mare during a single breeding season is limited and that for economic reasons mass production through embryo transfer would not, at least in the foreseeable future, reach the magnitude of production rates obtained in some artificial insemination programs.

As embryo transfer techniques are developed, other applications may become feasible. For example, embryos could perhaps be collected, frozen, and shipped to another farm. (Embryos have been successfully transported in the oviducts of rabbits.) Another possibility is the use of embryo recovery techniques to assess a stallion's fertility status. The stallion's fertility could be evaluated on the basis of the number of fertilized eggs recovered from a group of test mares. Scientists are also enthusiastic about the use of embryo transfer as a research tool, especially with respect to the study of inherited disease processes, such as CID. If a number of full sibs are produced, genetic studies, which are very time-consuming, could be hastened. Other research areas that could be enhanced by embryo transfer include the effect of uterine environment on the foal's physical characteristics, genetic improvement within the breed, and sex determination and selection at the embryonic stage.

Breeding Policies of United States Breed Registries (1981)

ASSOCIATION	ARTIFICIAL INSEMINATION									EMBRYO TRANSFER					Comments*
	Accepts A.I. Foals	Requires Bloodtyping of Stallion	Requires Bloodtyping of Mare	Requires Bloodtyping of Foal	Veterinarian must collect & inseminate	Breeder must submit special forms to association	Allows use of frozen semen	Allows transport of semen off farm	Allows storage of semen	Accepts embryo transfer foals	Breeder must submit special forms to association	Number of registerable foals per mare per year	Age restrictions on donor mare	Requires bloodtyping of stallion, recipient, donor, foal	
American Association of Owners and Breeders of Peruvian Paso Horses	yes*	yes	yes			yes	no			no policy					A. I. - only 30 mares per stallion each year; additional mares may be bred by natural service.
American Bashkir Curly Registry	no policy									no policy					
American Buckskin Registry Association, Inc.	yes	only if parentage is questioned				yes	no	no	no	no policy					
American Connemara Pony Society	yes				yes	yes	yes*	no		no policy					Use of frozen semen allowed only with individual approval from association.
American Council of Spotted Asses	no policy									no policy					
American Donkey and Mule Society	yes					yes				no policy					
American Fox Trotting Horse Breed Association, Inc.	yes									yes					
American Gotland Horse Association	no policy							no		no policy					
American Hackney Horse Society	yes*									no policy					A. I. must take place in the presence of person authorized to sign breeder's certificate.

ASSOCIATION	ARTIFICIAL INSEMINATION									EMBRYO TRANSFER					Comments*
	Accepts A. I. Foals	Requires Bloodtyping of Stallion	Requires Bloodtyping of Mare	Requires Bloodtyping of Foal	Veterinarian must collect & inseminate	Breeder must submit special forms to association	Allows use of frozen semen	Allows transport of semen off farm	Allows storage of semen	Accepts embryo transfer foals	Breeder must submit special forms to association	Number of registerable foals per mare per year	Age restrictions on donor mare	Requires bloodtyping of stallion, recipient, donor, foal	
American Morgan Horse Association, Inc.	yes	yes						no		yes		1		yes	
American Mule Association	no policy														
American Mustang Association, Inc.	yes				yes			yes		no policy					
American Paint Horse Association	yes			only if parentage is questioned			no	no	no	yes*	yes	1	6 yrs.	yes	Donor mare must have failed to conceive for 3 consecutive years.
American Quarter Horse Association	yes						no	no	no	yes*	yes	1	6 yrs.	yes	Donor mare must have failed to conceive for 3 consecutive years.
American Quarter Pony Association	yes									yes					
American Saddle Horse Breeders Association	yes*				yes	yes	no	no		no					Collection & insemination must take place in presence of person authorized to sign breeder's certificate.
American Shetland Pony Club	no							no		no policy					
American Walking Pony	yes				yes	yes		no		no policy					

ASSOCIATION	ARTIFICIAL INSEMINATION									EMBRYO TRANSFER					Comments*
	Accepts A.I. Foals	Requires Bloodtyping of Stallion	Requires Bloodtyping of Mare	Requires Bloodtyping of Foal	Veterinarian must collect & inseminate	Breeder must submit special forms to association	Allows use of frozen semen	Allows transport of semen off farm	Allows storage of semen	Accepts embryo transfer foals	Breeder must submit special forms to association	Number of registrable foals per mare per year	Age restrictions on donor mare	Requires bloodtyping of stallion, recipient, donor, foal	
Andalusian Horse Registry of the Americas	no policy									yes*		1	yes		Donor mare must have aborted 2 years consecutively.
Appaloosa Horse Club, Inc.	yes	yes						no		no policy					
Arabian Horse Registry of America, Inc.	yes	yes	no	no	no	yes	no	no		yes	yes	1	no	yes	
Belgian Draft Horse Corporation of America	no									no					
Chickasaw Horse Association, Inc.	no policy								no	no policy					
Colorado Ranger Horse Association	no policy								no	no policy					
Galiceno Horse Breeders Association	no policy									no policy					
International Arabian Horse Association										no					Artificial insemination rules follow those of the purebred registry.
International Buckskin Horse Association, Inc.	yes									no policy					

ASSOCIATION	ARTIFICIAL INSEMINATION									EMBRYO TRANSFER					Comments*
	Accepts A.I. Foals	Requires Bloodtyping of Stallion	Requires Bloodtyping of Mare	Requires Bloodtyping of Foal	Veterinarian must collect & inseminate	Breeder must submit special forms to association	Allows use of frozen semen	Allows transport of semen off farm	Allows storage of semen	Accepts embryo transfer foals	Breeder must submit special forms to association	Number of registerable foals per mare per year	Age restrictions on donor mare	Requires bloodtyping of stallion, recipient, donor, foal	
International Miniature Horse Registry	no policy									no policy					
International Trotting and Pacing Association	yes						no	no	no	no policy					
National Quarter Horse Registry	yes					yes				yes	yes				
Paso Fino Owners and Breeders Association, Inc.	yes	yes	yes			yes*	yes*	yes		yes*					
Percheron Horse Association of America	yes							no		no policy					A. I. - signed statements must be received from veterinarians at collection & insemination. Embryo Transfer - considered on an individual basis.
Peruvian Paso Horse Registry of North America	yes*	only when frozen semen is used			yes	yes	yes*	no	yes*	yes*	yes			yes	A. I. & Embryo Transfer are currently test programs.
Pony of the Americas Club, Inc.	yes							no		no policy					
Standard Jack and Jennet Registry of America	no policy									no policy					

ASSOCIATION	ARTIFICIAL INSEMINATION									EMBRYO TRANSFER					Comments*
	Accepts A. I. Foals	Requires Bloodtyping of Stallion	Requires Bloodtyping of Mare	Requires Bloodtyping of Foal	Veterinarian must collect & inseminate	Breeder must submit special forms to association	Allows use of frozen semen	Allows transport of semen off farm	Allows storage of semen	Accepts embryo transfer foals	Breeder must submit special forms to association	Number of registerable foals per mare per year	Age restrictions on donor mare	Requires bloodtyping of stallion, recipient, donor, foal	
Tennessee Walking Horse Breeders and Exhibitors Association	yes							no		no					
The Pinto Horse Association of America, Inc.	yes				yes	yes				no policy					
United States Trotting Association	yes	yes	only if parentage is questioned				no			no policy					
Welsh Pony Society of America, Inc.	yes*									no					A. I. only accepted for reinforcement of live cover.

13

FETAL DEVELOPMENT

The period from conception to birth is a critical but frequently overlooked aspect of equine production. The complex patterns of growth and development that change a two-celled organism into a self-sufficient individual present an excellent example of how beautifully complex biological processes can be. Although most broodmare managers are very conscientious about their management programs and strive to give each unborn foal the best possible advantage, few are familiar with the intricate and delicate changes that occur inside the mare's uterus during pregnancy. An understanding of the needs and sensitivities of the foal during its prenatal (before birth) stage is just as important as knowing how to care for the foal after it is born. A knowledge of the changes that take place during gestation leads to an understanding of why the mare's environment and condition are extremely important to the health of her fetus. In this respect, the following discussion on equine gestation is presented as an introduction to the next chapter, one which emphasizes important management practices that help to insure the health and safety of both the mare and her offspring during pregnancy.

Conception

Throughout the first part of this text, conception has been an important focal point. It is the object of much planning and preparation during most of the reproductive year. Conception rates are commonly used to evaluate certain breeding management practices, and conception failure is a very important obstacle to economic efficiency on the breeding farm. If conception occurs, the first breeding management phase ends, and attention turns to the care of both mare and foal throughout the foal's various stages of development. Before the foal's first stage of life is studied, however, its conception should be examined more closely.

After the ovum leaves the ruptured follicle, it is caught by the infundibulum and guided by the whip-like action of minute cilia (i.e., finger-like projections) to the ampulla (i.e., the enlarged tubular portion of the oviduct which serves as the site of fertilization). When a sperm cell contacts the ovum, it releases a special enzyme that breaks down the protective layer (zona pellucida) which surrounds the ovum. As this layer is penetrated by the sperm cell, it reacts immediately with the cellular membrane of the ovum (vitelline membrane) to prevent penetration by other sperm cells. After the penetrating sperm cell passes through the zona pellucida, the genetic material within the sperm and that within the ovum unite. This fertilization process triggers cellular division of the ovum and marks the creation of a new individual.

The combination of genetic material from the sire and dam is a systematic process that initiates embryonic development and determines the resulting foal's inherent characteristics and capabilities. In order to understand the significance of what occurs during conception, a few basic genetic concepts should be reviewed:

1. The genetic messages that dictate how and when a cell divides or functions are called genes.

2. Thousands of genes are involved in the development and maintenance of a living organism. Each cell contains a full set of these genes but responds to only a few selected messages.

3. The genes are located on long protein strands, or chromosomes. Each chromosome is paired with another of similar size and genetic content. Corresponding genes on these chromosome pairs control the same characteristics but may influence those characteristics differently. For example, one gene may cause brown eye color, while the corresponding gene causes blue eye color. Because the brown eye color gene is dominant over the blue, the individual will have brown eyes.

4. Each cell within the horse's body contains 32 pairs of chromosomes.

5. When the sperm and ovum are formed within the testes and ovaries, a special cell division process reduces the number of chromosomes to one from each pair. In other words, the sperm and ovum contain 32 chromosomes each.

When fertilization occurs, the stallion's 32 chromosomes pair with 32 chromosomes from the mare. The moment that the normal chromosome number (32 pairs) is established, the offspring's inherited characteristics and genetic potential are determined. For example, one pair of chromosomes (designated as XX in the female and XY in the male) carries the genes that determine the individual's sex. The female carries two X chromosomes in every body cell, while the male carries an X and a Y chromosome. When the ovum or sperm is formed, only one chromosome from the sex chromosome pair is received. In other words, each ovum

produced by the mare normally carries one X chromosome, while 50 percent
of the stallion's sperm cells carry X chromosomes and 50 percent carry Y
chromosomes. When the sperm and ovum unite, the sex of the resulting
zygote is determined by whether an "X" sperm or a "Y" sperm fertilizes the
ovum.

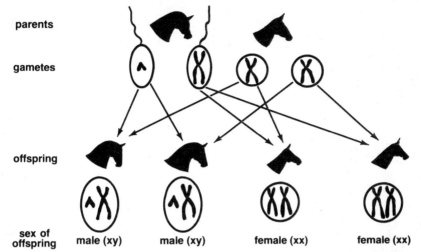

At conception, the sex chromosomes that are carried by the sperm and the ovum
combine to determine the sex of the new individual.

Thousands of gene relationships are established at conception,
but the complexity of the basic gene functions and the presence of many
complicating factors (e.g., dominance, recessiveness) place this topic beyond
the scope of this text. It should be remembered, however, that many
management practices are carried out to insure that an individual's genetic
potential, which is established during conception, is realized. No amount
of care or concern can change undesirable inherited characteristics. For a
more detailed analysis of these important genetic concepts, refer to the
text, EQUINE GENETICS & SELECTION PROCEDURES.

Embryology

The study of an individual's development from conception to birth is
referred to as embryology. In order to discuss basic embryology, it is
necessary to examine general systems (e.g., the placenta, the circulatory
system) as if they were independent of one another. This is not a totally
accurate presentation, however, because fetal growth is actually a complex
series of events that involves many interrelationships between systems.
For example, normal organ growth can occur only when the placenta

functions properly, and the placenta's role is fulfilled only when there is adequate attachment with the uterus. Each phase of fetal growth is also influenced by how well a previous stage was completed, which is determined by genetic messages and by environmental factors, such as nutrition or the presence of toxic agents. Because fetal development is characterized by increasing levels of complexity and because this chapter only attempts to present a basic review of embryology, the discussion is organized into very general categories, and systems within each category are examined independently.

EARLY GROWTH

Within about 24 hours after conception, the fertilized ovum (or zygote) undergoes its first division to form a two-celled structure. Division continues, and in about 98 hours postconception a tight group of cells, referred to as the morula, is formed. During this very early stage of embryonic development, the cytoplasm of the ovum is apportioned equally between each new daughter cell so that progressively smaller cells are formed by each cell division. A very small amount of the ovum's cytoplasm is expended as fuel for the cell division process, but the overall size of this primitive body remains almost unchanged. Although the cellular size decreases during this period, each cell has a complete set of chromosomes. Thus, this early growth period is characterized by a considerable increase in the amount of genetic material without a significant increase in overall size.

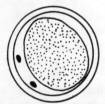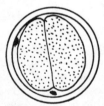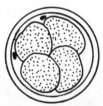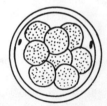

During the first three days of development the fertilized ovum progresses from one cell to the multi-celled morula shown at the far right. The size of the fertilized ovum remains nearly constant during this period of early growth, although the number of cells increases.

As these early changes are taking place, the cellular mass migrates through the oviduct and reaches the uterus about 6 days postconception. Smooth muscle contractions, movement of cilia, and the flow of fluids within the oviduct encourage this migration, but the fertilized ovum seems to have an important influence over its own movement to the uterus. Unfertilized ova remain in the fallopian tube (oviduct) until they degenerate, but recently fertilized ova are capable of passing these degenerating cells on their way to the uterus. The exact mechanism for this phenomenon is not clear, but several theories on ova transport have been proposed. One suggests that ova must reach a certain stage of development before they can complete their journey along the fallopian tube. (Ova smaller than the 16-cell stage have never been found in the uterus.) Another theory proposes that the movement is stimulated or enhanced by a semi-fluid substance produced by the fertilized ovum.

The zona pellucida restricts growth of the fertilized ovum until the morula stage is reached. Then it begins to thin and expand as it absorbs fluids from the uterus. These uterine fluids are the earliest outside source of embryonic nutrition. Some of the cells inside the zona pellucida remain attached to the expanding perimeter, forming an inner cellular lining for the enlarging fluid-filled cavity. The membranes that comprise the placenta develop from this inner lining/zona pellucida boundary. Most of the growing cell mass congregates at one pole of the morula, forming the so-called inner cell mass. This mass is the earliest recognizable evidence of the developing embryo. These primitive embryonic structures (the inner cell mass, the inner cell lining, the zona pellucida, and the fluid-filled cavity) are referred to collectively as the *blastocyst*.

By the eighth day following fertilization, a group of cells called the inner cell mass are grouped at one pole of the fertilized ovum. At this stage of development the ovum is referred to as a blastocyst.

CELLULAR DIFFERENTIATION

Cell structure changes very little during the early division processes, but as the blastocyst stage is reached, certain cells or cell groups begin to specialize and pursue different routes of development. During this critical stage of early growth, the groundwork is established for the development of important body structures such as tissues and organs. Initially, the developing cells are very adaptable, and the death of a few cells does not interrupt normal embryonic development. As growth continues, however, the genetic material within each cell adopts certain developmental roles. In other words, each cell reaches a point where it must "determine" which genetic messages it will obey. As a result, the identity of a cell or group of cells becomes fixed, tissues become specialized, and the growing cellular mass begins to develop into a self-sufficient individual.

The identities of all cells do not become established at the same point in embryonic development, and even when determination does occur, the cells do not necessarily resemble those of the tissues they will eventually form. Researchers suspect that determination begins as early as the morula stage, but it is very difficult to identify exactly when this "declaration" of developmental roles occurs. Observable changes in cell structure and in cellular growth patterns are referred to as *differentiation*. As cells differentiate, they begin to assume the individual characteristics and anatomical positions of the organs and tissues that evolve during later stages of embryonic development. Both determination and differentiation are governed by genetic messages, but during this period the cells are highly sensitive to physical insult. The various drugs, chemicals, diseases, and other factors that can alter the course of normal development are examined later in this chapter.

GERM CELL LAYERS

The earliest evidence of cell differentiation is the development of three primary cell layers within the blastocyst. These so-called germ cell layers are the predecessors of all fetal tissues and also form the membranes which surround the embryo and fetus during gestation. The outermost cell layer is the zona pellucida, or ectoderm. The second layer to appear is the inner cellular lining described earlier. This layer of cells, referred to as the endoderm, is closely associated with the ectoderm. Together, these two cell layers form the yolk sac, the earliest structure to tap external nutrient sources (uterine fluids) for the benefit of the embryo. A little later in gestation, a cell layer called the mesoderm forms between the endoderm and the ectoderm. The embryo's connective tissue (e.g., bone, cartilage, and blood) originates from this middle germ cell layer.

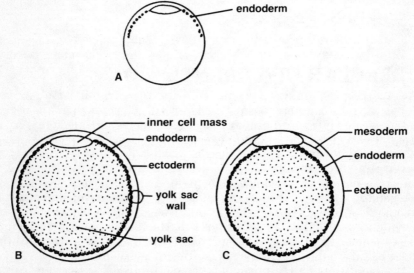

(A) The endoderm forms inside the ectoderm between days 9 and 12. (B) The association of these two layers forms the yolk sac wall. (C) Around day 14, the mesoderm develops within the yolk sac wall. Blood vessels originate from the mesoderm and provide a nutrient transport system for the embryo.

By the 16th day postconception, the ectoderm begins to fold over the inner cell mass, which at this stage can be identified as the developing embryo. Two projections of ectoderm meet over the cellular mass and fuse to form a small fluid-filled vessel, referred to as the amnion. As a result of this structural change, the embryo, the yolk sac, and the amnion are surrounded by ectoderm. Other structural changes transform this simple 3-layer capsule into a complex supportive membrane system.

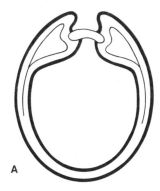 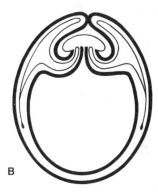

The amnion results when (A) the ectoderm folds over the embryo and (B) fuses to form a fluid-filled cavity surrounding the embryo.

EXTRAEMBRYONIC MEMBRANES

The specialized organ that regulates the embryonic environment (e.g., protects the embryo from concussion and provides the developing individual with nutrients) is referred to as the *placenta*. The fetal portion of the placenta (i.e., the extraembryonic membranes) consists of the yolk sac, the amnion, and the allantois.

The yolk sac is the earliest extraembryonic membrane to appear, developing by about day 12 postconception. In most mammals, including the horse, the yolk sac is believed to be the remnant of a nutrient storage vessel that may have been functional during early evolutionary development. This form of nutrient storage is still seen in birds and reptiles (i.e., as the egg yolk). In most mammalian embryos, however, this vessel functions as a temporary nutrient absorption and transport organ. Nutrients (i.e., uterine fluids) are absorbed as the yolk sac develops during the morula-blastocyst stages. At about 16 days postconception, blood vessels, originating from the mesoderm, proliferate within the yolk sac wall, forming a complex, vascular system that transports nutrients from the yolk sac to the developing embryo. The yolk sac functions until at least day 20 of gestation but regresses somewhere between day 21 and day 35 as other extraembryonic membranes begin to carry out this important nutrient transport process.

As mentioned earlier, the continuous layer of ectodermal tissue that surrounds the embryo is the amnion. The fluid-filled cavity that is formed when the ectoderm folds over and fuses around the embryo serves as a liquid cushion that protects the embryo from concussion and prevents adhesions between the embryo and surrounding tissues. The normal point of attachment between the amnionic sac and the embryo gradually becomes restricted to form the umbilical ring. The amnion is perhaps more commonly known as the *embryonic sac* that is observed during foaling.

The allantois is a pouch that emerges from the embryonic hindgut and surrounds the amnion on about day 23 of gestation, forming a second fluid-filled cavity around the embryo. One side of the allantois fuses with the outside of the amnion. The resulting amniotic-allantoic membrane shields the embryo from fetal waste products that accumulate in the allantoic cavity. The outside of the allantois fuses with the ectoderm to form the allantochorion, a highly specialized membrane that assumes the nutritional transport functions of the now regressing yolk sac. As the allantoic pouch fills with fluid, the outer walls are invaded by a network of blood vessels that transport nutrients from the dam to the embryo's umbilical cord. The allantochorion also plays an important role in waste and gas

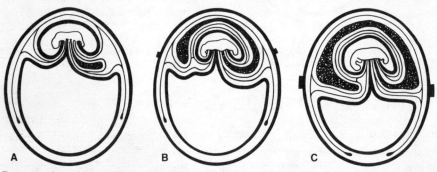

Between days 20 and 30 the allantois emerges from the embryo and gradually surrounds the amnion. (A) The allantois originates from the embryonic hindgut. (B) The inner layer of the allantois and the outer layer of the amnion fuse to form the amniotic-allantoic membrane, while the outer layer of the allantois joins the ectoderm to form the allantochorion. This outer wall is highly vascular and participates in nutrient transport for the embryo. The band of folds known as the chorionic girdle begins to form about this time at the junction of the allantois and the yolk sac. (C) By day 30 the yolk sac is obviously regressing as the allantois is expanding. This characteristic marks the period during which the embryo's nutrient transport functions shift from the yolk sac to the allantochorion.

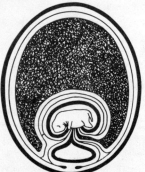

The allantois is the dominant structure within the conceptus by day 40, and the yolk sac is significantly regressed. The temporary bond between the chorionic girdle cells and the mare's uterus is lost and the allantochorion is only tenuously attached to the uterus.

transport. It is commonly known as the *water sac* that breaks just prior to delivery of the foal. The release of allantoic fluids (i.e., when the mare "breaks water") lubricates the reproductive tract prior to foaling.

PLACENTATION

As the embryo develops, a more complex method of nutrient-waste transport is required. The shift from simple absorption of yolk sac fluids to a more intimate relationship with the dam's uterus is called placentation. During this period, the allantochorion of the fetus and the epithelium (cellular lining) of the mare's uterus join and establish an exchange mechanism that carries out all of the duties of the non-functional fetal lungs and digestive tract.

After leaving the oviduct, the conceptus (developing embryo and surrounding membranes) enters the uterine horn, where the earliest placentation efforts occur at about day 24 or 25 postconception. (Because this site is the lowest point in the mare's reproductive tract, gravity probably accounts for its selection.) Although the extraembryonic membranes are well-developed by day 80 of gestation, they do not become firmly attached to the maternal tissues until about the fifth month of gestation. This pre-implantation period is relatively long in the horse when compared with other mammalian species.

EARLY ATTACHMENT

The earliest sign of placentation occurs on about day 25 of gestation when a band of shallow folds develops on the allantochorion. This so-called *chorionic girdle* encircles the fetal chorion at a point near the uterine horn-body junction. Specialized cells within this girdle elongate and engulf uterine epithelial cells, forming a temporary localized attachment between the mare's uterus and the conceptus. This early attachment ends on about day 38 post-conception as the fetal cells migrate into the maternal endometrium and detach from the allantochorion. Although the conceptus can migrate from the uterine horn after the chorionic girdle is released, it usually returns to this site for firm implantation.

After invading the uterine endometrium, the fetal cells differentiate and form hormone-producing structures on the uterus. These cup-like bodies (endometrial cups) form a band around the inside of the uterus near the horn-body junction, corresponding to the previous position of the chorionic girdle. The hormone produced by the endometrial cups (PMSG) plays a very important role in early pregnancy maintenance, which is examined later in this chapter. The endometrial cups gradually degenerate and begin to separate from the endometrium around day 90 of gestation. The reason for this separation (or sloughing) is not clear, but it may be that the dam's body recognizes the fetal cells as foreign bodies and rejects them. The sloughed material may remain in the uterus or may form an allantochorionic pouch, a pocket-like structure in the allantochorion. Researchers believe that these pouches can be absorbed by the fetus, but it is not uncommon for these bodies to persist throughout gestation.

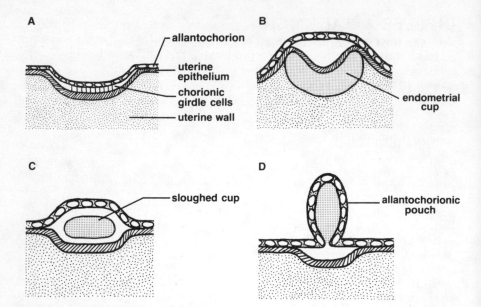

(A) Temporary attachment of the developing placenta to the uterus occurs when chorionic girdle cells positioned on the allantochorion begin to invade the uterine endometrium. (B) Around day 39, the chorionic girdle cells detach from the allantochorion to form endometrial cups, and by day 40, the endometrial cups begin to produce PMSG. (C) The endometrial cup sloughs from the endometrium around day 90 and (D) may form an invagination in the allantochorion called an allantochorionic pouch.

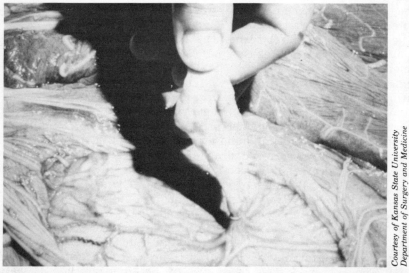

Courtesy of Kansas State University Department of Surgery and Medicine

The invagination that is formed in the wall of the allantoic cavity by a sloughed endometrial cup is sometimes found in the placental membranes at birth. This structure is called an allantochorionic pouch.

DIFFUSE ATTACHMENT

By the time temporary attachment between the chorionic girdle and the endometrium is lost (day 36-38), the fetal allantochorion begins to cling tenuously to the uterine epithelium. This attraction is made possible because the surface of the fetal membrane is covered with tiny projections (microvilli) that become embedded within the uterine epithelium. The tendency for adjacent tissues to stick together may also contribute to this transitional attraction between fetal and maternal tissues.

Firm attachment of the allantochorion to the uterine epithelium is a gradual process that occurs between day 45 and day 150 of gestation. At day 45, tissue folds (early macrovilli) can be observed on the surface of the chorion. As these macrovilli develop, they fuse into tuft-like arrangements and gradually become established inside endometrial "pockets." Each endometrial pocket contains small crypts that receive the tiny projections of the fetal tufts. When this "pocket-tuft" relationship is fully established, the resulting structure (i.e., the macrovilli tuft and the opposing pocket) is referred to as a microcotyledon. Firm placental attachment is accomplished when microcotyledonary development is completed, an event that occurs at about day 150 postconception.

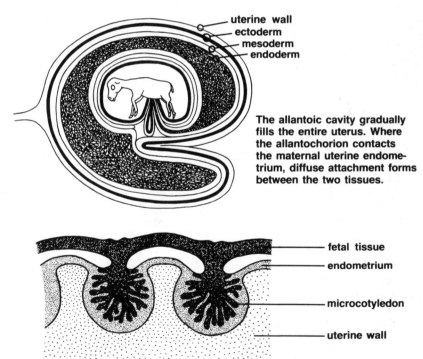

uterine wall
ectoderm
mesoderm
endoderm

The allantoic cavity gradually fills the entire uterus. Where the allantochorion contacts the maternal uterine endometrium, diffuse attachment forms between the two tissues.

fetal tissue

endometrium

microcotyledon

uterine wall

The diffuse attachment between the allantochorion and the maternal uterine endometrium is formed by microcotyledons. These structures are composed of tiny tufts on the allantochorion, which become established within corresponding pockets in the endometrium. The microcotyledons are well established by day 150.

In addition to providing a means of secure attachment between fetal and maternal tissues, the microcotyledons greatly increase the surface area of the placenta. This is an extremely valuable arrangement since it also increases the area for nutrient, waste, and gas exchange. Microcotyledonary attachments form over the entire surface of the placenta, with the exception of areas where the fetal allantochorion does not touch the uterine epithelium (e.g., over the cervical entrance, over the oviduct entrances, and within placental folds). At each of these attachments, the maternal and fetal blood vessels come into very close association. Some materials (e.g., amino acids, glucose, and some waste products) cross the placental barrier, but the transfer of most blood-borne solids and bacteria is blocked, primarily because of their large size. This selective transfer process protects the developing fetus from many substances that could be harmful, but it may also prevent the fetus from forming immunities prior to birth.

ORGAN DEVELOPMENT

When the germ cell layers differentiate during early gestation, they leave the compact inner cell mass at one pole of the blastocyst. The transition of this primitive embryo into a fully developed foal is a remarkable process that involves two important phases: embryonic development and fetal growth. Most of the cell differentiation and organ development take place during the embryonic phase. By about day 50 of gestation, almost all of the organs are present (at least in a rudimentary state), and the individual has a recognizable equine form. The second phase involves continued growth and maturation of the organ systems. Because these two phases are gradual and interrelated, the identification of a specific date for the transition from embryo to fetus is difficult to make. Researcher's estimates for the end of the embryonic stage vary from day 30 to day 60 of gestation. For the purposes of the following discussion, the end of embryonic development is designated as day 40.

Because information on equine embryology is limited and because a discussion on embryonic development and fetal growth in the horse would require extensive interspecies comparisons and assumptions, a detailed analysis of these growth processes is not attempted. Rather, the reader is provided with a review of the basic changes that occur when two important systems are established. The circulatory and reproductive systems have been chosen to illustrate early growth processes because their development is not completed until after birth. For this reason, the following discussion provides important background information for a later discussion on physiological adjustments at birth. (Refer to "The Neonatal Foal.")

CIRCULATORY SYSTEM

During the embryonic stages, adequate nutrients are received by simple absorption of uterine fluids. As the embryo enlarges, however, its organs become more complex, and a system for transporting nutrients to the inner parts of the body and for shuttling waste products away from the embryonic tissues is needed. This nutrient/waste transport requirement is met by the development of a well-defined circulatory system, an event that corresponds closely with placentation.

Blood Vessels

As mentioned earlier, blood vessels arise from the middle germ cell layer (mesoderm) of the early conceptus. During very early embryonic development, a complex network of small vessels in the yolk sac wall is connected to the developing heart by two major vessels. Additional circulatory pathways are diverted to areas of rapid growth so that, by the time the yolk sac has regressed, the extraembryonic membranes contain a well-developed circulatory system. This fine network of blood vessels transports fetal blood to and from the villi of the microcotyledons and plays a very important role in the nutrient/waste transport mechanism. Although several capillaries supply each microcotyledon, each of these areas is drained by a single vein. This arrangement slows the blood flow through the microcotyledon and allows more complete exchange to take place.

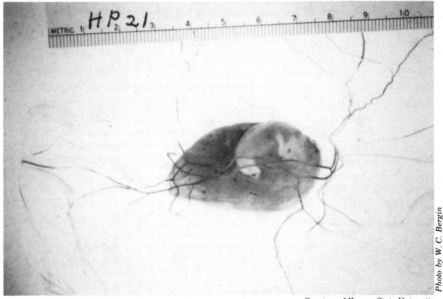

Photo by W. C. Bergin

Courtesy of Kansas State University
Department of Surgery and Medicine

Blood vessels are clearly seen radiating from this 21-day-old embryo. A circulatory system is established in the yolk sac wall very early in gestation. A complex network of blood vessels gradually develops in the allantochorion and in the embryo to meet the growing embryo's increasing nutritional needs.

Blood transport between the placenta and the fetus is provided by special umbilical vessels. For example, the umbilical arteries pass through the umbilical cord and branch into many small capillaries which carry blood to the microcotyledons. Confusion often arises when the terms *vein* and *artery* are used in reference to fetal circulation. Arteries are commonly thought of as the transport system for oxygen-enriched blood, while veins are associated with oxygen-depleted blood. By strict definition, however, arteries carry blood from the heart, while veins carry blood to the heart. Because the fetal lungs are non-functional and because gas transport is carried out in the placenta, oxygenated fetal blood is carried by veins (i.e., to the heart), while deoxygenated blood approaches the placenta via the umbilical arteries.

Blood

Blood is the transport medium for nutrients, waste products, and gases. This important fluid begins to develop on about day 14 post-conception, originating from special blood manufacturing colonies within the yolk sac. Both blood cells and blood manufacturing colonies are carried to the embryo by the developing circulatory system. The embryo's first blood cell-producing colonies are established in areas of rapid growth and development. When the nutrient needs decrease or level off in a certain area, blood production centers move to more active (i.e., tissue-building) sites. Eventually, the blood production sites become established in the bone marrow, where they remain active throughout the individual's life.

Heart

The heart originates from a pair of tubes located in the embryo's chest cavity. During embryonic development, these tubes fuse together and form a primitive heart. Surrounding structures force the heart to bend into a loop, giving it the characteristic heart shape. Gradually, this muscular structure is partitioned into four chambers: the right and left atriums and the right and left ventricles. This separation process divides the heart into two pumps; the right side of the heart pumps deoxygenated blood to the lungs, while the other side sends oxygenated blood back into the body. It is important to note, however, that because the fetal lungs are not functional during gestation, this dual pump system is not important until after birth. To insure that all of the heart chambers receive an adequate blood supply and that each chamber is conditioned throughout gestation to pump a full load of blood, two by-pass mechanisms are present in the fetal heart.

REPRODUCTIVE SYSTEM

Although the tissues that give rise to the foal's reproductive structures are present in the very early embryo, the individual's sex cannot be distinguished visually until about day 40-45 of gestation. By about day 55, prominent vulval lips are visible on the female fetus, and two symmetrical structures develop between the male's hind legs at a point where the scrotum will later develop. By day 77, the male prepuce is visible, and by day 100-200, the female clitoris has moved to its permanent position between the labia (i.e., vulval lips). These developments are just a few examples of the gradual changes in tissue structure and organ position that occur as the external portion of the reproductive tract becomes established.

The embryological development of the tubular portion of the male and female reproductive tracts is also a gradual process, one that transforms indifferent primitive reproductive tissues into either the female oviducts, uterus, cervix, vagina, and vulva or into the male's epididymis, vas deferens, seminal vesicles, and urethra, depending on the individual's sex. During early embryonic development, both the male and female are characterized by the presence of two duct systems (Müllerian ducts and Wolffian ducts) and a urogenital sinus. In the male, the Wolffian ducts give rise to the epididymis, the vas deferens, the ampulla, and the seminal vesicles. As the Wolffian ducts develop into these tubular structures, the Müllerian ducts regress until they become a nonfunctional vestige. This arrangement is reversed in the female. The Müllerian ducts develop into the oviducts, the uterine horns, the uterine body, the cervix, and the vagina, while the Wolffian ducts degenerate. At one end of the tract, the Müllerian ducts remain small and separate, forming the oviducts. The diameter of the ducts increases to form the uterine horns, and then the enlarged ducts fuse to form one continuous tube that develops into the uterine body, the cervix, and the vagina. The terminal end of the male and female tracts originates from the urogenital sinus. In the male, this sinus gives rise to the prostate and bulbourethral glands. In the female, the urogenital sinus develops into the vulva. In both the male and female, this embryonic structure forms the bladder and the urethra (which serves as the ejaculatory duct in the stallion).

The gonads (i.e., testes or ovaries) originate from an area referred to as the genital ridge. Growth of the fetal gonads (by an increase in the number of interstitial cells) is very slow until about day 100 postconception. At this point, both the male and female gonads grow very rapidly, becoming unusually large when compared with gonadal size at birth. At one point in their development, the fetal gonads are 10 times as heavy as they are in the newborn foal. In fact, the weight of the fetal gonads actually exceeds that of the maternal ovaries after about day 150 of gestation. Neither the cause nor the effect of this size increase is clear, but researchers suspect that it is associated with the condition of the maternal ovaries and hormonal concentrations in the mare. After about day 250 of gestation, the gonads decrease in size as the interstitial cell mass begins to break down. Most of the interstitial cells shrink and degenerate, but some persist and can be found in the mature gonads.

Although the seminiferous tubules are the areas of sperm production in the mature testes, they are found in both male and female gonads at about day 60. These tubules continue to develop in the fetal testes, and by day 150 a complex network of tubules is distributed throughout the interstitial cell mass. In the developing ovaries, however, the seminiferous tubules regress and are very difficult to find by day 100 of gestation. At this time, the female gonads begin to differentiate, forming a unique arrangement of germinal (ova-producing) and nongerminal tissue. Unlike the seminifer-

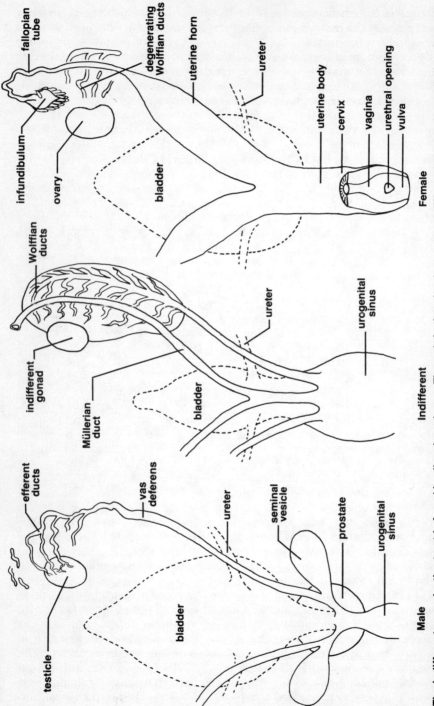

The indifferent reproductive tract is found in all early embryos. Through further development and differentiation during gestation, features that are characteristic of male and female tracts arise from the indifferent tract.

ous tubules, which begin to produce sperm cells when the male reaches puberty, the germinal tissue of the ovaries completes the first steps of oogenesis (ova production) by about day 150-160 of gestation. In other words, the female is born with a defined number of oocytes (primitive egg cells), while the male is capable of producing an indefinite number of sperm. (Note: This oocyte supply is far greater than that which the mare could possibly use throughout her reproductive life.)

Controlling Factors

With the onset of pregnancy, the mare's cyclic activity normally ceases, and her reproductive tract is quieted to insure that the conceptus is provided with a stable environment. Several regulatory agents seem to be involved in this quieting process, and their presence or absence may also cause other changes associated with pregnancy. The endocrinology of gestation is the study of how and why the production of these controlling factors fluctuates and how these fluctuating patterns influence the biological mechanisms that support pregnancy. An understanding of how hormone levels and relationships affect gestation can be extremely helpful in determining the causes of early embryonic death, some cases of repeated abortion, and certain developmental abnormalities. For example, there are several periods during gestation when the hormone balance is very delicate and pregnancy maintenance is at a point of minimal security. These periods seem to occur when the source for a particular hormone is changing and when hormone production is maintained at a very low level. Because researchers still have many unanswered questions regarding the endocrinology of gestation, a detailed study on this topic would involve the presentation of many theories and the comparison of numerous studies and somewhat contradictory data. For this reason, the following discussion presents only the basic functions of four important regulatory factors: progesterone, prostaglandin, pregnant mare serum gonadotropin, and estrogen.

PROGESTERONE

Progesterone levels are directly related to the suppression of estrus and the maintenance of the quiet reproductive state associated with pregnancy. In addition to calming uterine muscle activity, progesterone stimulates changes in the endometrium which encourage implantation of the embryo and cause the uterine glands to produce the fluids that support early embryonic life. As explained in "The Mare Reproductive System," progesterone is produced by the corpus luteum (i.e., the yellow body that develops within the ruptured follicular cavity). Because prostaglandin is not released by the uterus in the pregnant mare, the corpus luteum does not regress as it does during a normal estrous cycle. Between day 30 and 40, progesterone levels begin to rise in response to the activation of another progesterone source (secondary corpora lutea). The primary and secondary corpora lutea supply all of the progesterone that is needed for pregnancy maintenance during the first 100-160 days of gestation.

It is important to note that the fetus also contributes to pregnancy maintenance. On about day 80, the placenta begins to produce significant levels of progesterone, and when the corpora lutea regress (on about day 160), the placenta prevents abortion by producing enough progesterone to maintain pregnancy. Progesterone levels are highest between day 56 and 160. After reaching peak levels on about day 120, however, progesterone begins a gradual decline. When the placenta is the only source of progesterone, levels are maintained at their lowest point throughout gestation. On about day 300 of pregnancy, progesterone levels surge and do not fall again until a few days before foaling.

PROSTAGLANDIN

As explained in "The Mare Reproductive System," prostaglandin is an agent that causes the corpus luteum to regress on about day 14 post-ovulation if conception does not occur. If the ovum is fertilized and if the embryo is present in the uterus between day 10 and day 15 postconception, prostaglandin does not achieve a significant level in the blood, and the corpus luteum persists longer than its normal lifespan. Although the exact mechanism for CL maintenance is not known, researchers have suggested that an unidentified agent may inhibit the production or release of prostaglandin by the uterus. Estrogen is capable of prolonging the life of luteal tissue in some species and acts, along with progesterone, to maintain the firm uterine tone and tight dry cervix that are associated with pregnancy. Researchers anticipate that future studies will show estrogen to be at least partly responsible for inhibiting prostaglandin's production or blocking its action. If this is true, further studies may find that the conceptus plays an important role in pregnancy maintenance, since it is capable of producing estrogen quite early in gestation.

PREGNANT MARE SERUM GONADOTROPIN

Although research is not conclusive at this time, an agent has been identified as the probable cause of secondary follicular development. Pregnant mare serum gonadotropin (PMSG), a hormone that is produced by the fetal cells that invade the uterus when endometrial cups are formed, is closely associated with follicular growth in the mare. PMSG blood levels rise rapidly after day 40 of pregnancy but begin to decline gradually after day 65. By day 180, a point that marks the end of corpora lutea activity in the pregnant mare, PMSG is almost undetectable in the mare's blood.

PMSG levels vary considerably between individual mares and seem to be influenced by several factors. For example, PMSG levels are nearly twice as high in a mare that is carrying twins as compared to a mare that carries only a single fetus. Because each fetus is enclosed in a separate placenta, two chorionic girdles (each with complete and separate endometrial cup development) are present when PMSG production begins. Differences in PMSG levels have also been noted between pregnant mares and pregnant jennies. When horses and donkeys are interbred, the resulting hybrid fetus influences the levels of PMSG in its dam. For example, when a mare carries a mule (mare x jack) fetus, her PMSG levels are lower than when she carries a horse fetus. When a jenny carries a hinny (jenny x stallion) fetus, her PMSG levels are higher than when she carries a donkey. Other factors such as the mare's age, size, breed, and possibly the season or climate may also account for some individual differences in PMSG production, but more research is needed to substantiate these as contributing factors.

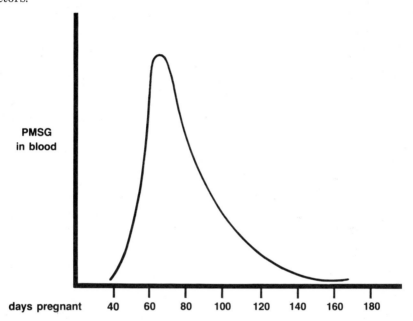

Changes in the level of PMSG in the mare's blood reflect the progress of endometrial cup development. The cups are fully developed by day 40 and within 20 to 30 days are producing high levels of PMSG. After a brief peak, PMSG levels decline as the endometrial cups begin to separate from the endometrium.

ESTROGEN

As mentioned earlier, estrogen has been suspected as a possible inhibitory agent for prostaglandin. It is produced by the conceptus very early in gestation (as early as day 12) and is capable of prolonging the life of luteal tissue in other species. Estrogen is also associated with the growth and development of the fetal reproductive tract. Although the mare produces substantial amounts of estrogen during the first half of gestation, research indicates that the fetal gonads are the prime sources of estrogen during the second half of gestation. Fetal estrogen is carried through the placental barrier to the dam's blood and is present in high concentrations in the mare's urine, especially during the second half of gestation. Both the size of the fetal gonads and the concentration of estrogen in the dam's blood and urine increase during the second half of pregnancy. Regression of the fetal gonads (at about day 280) is accompanied by a drop in estrogen concentration in the maternal bloodstream.

Maternal Influence

Although the dam and the sire of a foal contribute equally to its genetic makeup, the dam provides the foal's environment throughout gestation. Thus, the dam significantly affects the foal's early development. A number of maternal characteristics and their impact on the foal are summarized here:

MATERNAL CHARACTERISTIC	EFFECT ON THE FOAL
OVUM	The nucleus carries the dam's genetic contribution to her offspring, and the cytoplasm provides structural material for early embryonic growth.*
UTERUS	Fetal growth is restricted by the surface area available for placentation and by the physical capacity of the uterus.
AGE	The immature mare is unable to support her own growth and the growth of her fetus without making some sacrifice (e.g., loss of condition, loss of the fetus). The aged mare may be unable to support fetal development if her tract has suffered extensive damage due to recurring or chronic infection, foaling problems, or numerous pregnancies.
LACTATION	The ability of the mare to provide adequate nutrients (especially colostrum) to her foal has a very important impact on the foal's health and development. Milking ability is an inherited trait.
TEMPERAMENT	The foal mimics the dam and may pick up desirable and/or undesirable behavioral patterns during the suckling period.
MOTHERING ABILITY	The mare's instinct to protect her foal affects the youngster's safety, especially when the foal must learn to deal with a social hierarchy in a pasture situation.

*Although some sources support the idea that maternal cytoplasm increases the dam's influence over her offspring (as compared to the sire's genetic influence), this concept is still somewhat controversial. It has been noted, however, that the X chromosome is much larger and carries more chromosomal material than the Y chromosome.

Fetal Nutrition

Studies on the effect of poor nutrition during early pregnancy indicate that fetal nutrition is most critical from days 25 to 31 of gestation. It is not clear, however, whether the embryonic losses that occur during this period result from nutrient imbalances or specific nutritional deficiencies or whether the changing relationship between the endometrium and the embryonic membranes contributes to the embryo's apparent sensitivity to nutritional deficiencies during this period. Later, the growing fetus responds to a low plane of nutrition by decreasing its growth rate, thereby reducing its nutrient requirements. As a result, the foal's birth weight is reduced significantly. Because the fetus depends on nutrient transport from the mare to build its own energy reserves, poor nutrition during late gestation can result in the birth of a very weak foal. (The fetus normally stores glycogen to use as an immediate source of energy after birth.) Although the fetus is capable of adjusting to a reduced plane of nutrition, extreme nutrient deficiencies may result in abortion. When the mare's diet is insufficient, she will sacrifice her own nutrient supplies for her fetus only until her life is threatened. At that point, fetal death and abortion usually occur.

Although the mare's diet during pregnancy plays an essential role in fetal nutrition, the ability of available nutrients to reach the fetus is also very important. Initially, the conceptus relies on nutrients present within its own cellular fluid. This gel-like fluid, also called cytoplasm, is primarily maternal. (i.e., It originates from the ovum.) After the developing embryo reaches the uterus, the absorption of uterine fluids begins, and from this point embryonic survival depends not only on the ability of the uterine glands to produce uterine milk but also on the ability of the yolk sac to absorb this essential fluid. Later, the establishment of special nutrient/waste transfer attachments (microcotyledons) helps the developing fetus secure the nutrients needed to satisfy its increasing requirements.

There seems to be a direct correlation between the degree of placental attachment and the rate of fetal growth. As the fetus becomes larger and its nutrient demands increase, placental attachment normally becomes more and more extensive. If the placental attachment is hindered by the presence of scar tissue within the uterus or if a significant percentage of the placental attachment is nonfunctional, the fetus must compensate by one of two methods. First, the fetus may redirect placental circulation to undamaged areas of the mare's uterus. If redirection of the placental circulation fails to provide enough nutrition resource, the fetus then resorts to its second option and reduces its growth rate. One study showed that when 50 percent of the placental attachment is nonfunctional, the average birth weight drops by as much as 40 percent. Although the nature of placental attachment in the equine (i.e., diffuse attachment) enables the fetus to survive when large areas of the placenta are unattached or nonfunctional, extensive endometrial scarring can result in repeated failure of the mare to carry a foal to term.

Teratology

Teratology is the study of abnormal growth and development during gestation. Normally, growth occurs in a complex but orderly manner; cells respond to selected genetic messages, change structures, and perform special functions. For example, a cell may produce chemicals that participate in an important chain reaction (e.g., as an energy source) or may change structures and reproduce rapidly to form specialized tissues. During periods of rapid growth and development, these cells are very susceptible to the effects of certain deleterious agents, or teratogens. Teratogens may cause developmental abnormalities (congenital defects) by damaging cells, by arresting cell growth, or by changing the normal growth patterns for a particular area. Each organ or tissue area is particularly sensitive to teratogens during a specific period of development. As a case in point, an excess of vitamins A and D in the dam's diet before fetal limb differentiation is completed may cause serious limb abnormalities. If high levels of these vitamins are not present in the fetal circulation until after the critical period for limb development, the limbs may only be slightly affected or may escape the detrimental effects entirely.

Because growth and differentiation is a continuous process, the periods of teratogen susceptibility of different organs vary and are difficult to pinpoint. In addition, the exact actions of many teratogens are not clear. Many complicating factors may be involved in teratogen activity, and the effects of several teratogens may combine to produce a totally different defect. Although various types of congenital defects are described in the following discussion, the reader should note that descriptions of all the congenital defects known to occur in horses are not included. An in-depth review of congenital defects in horses would require a lengthy discussion even though the causes of most of these defects are unknown.

INCIDENCE AND SEVERITY OF CONGENITAL DEFECTS

When a developmental abnormality occurs, the fate of the affected individual lies in the type of abnormality (or abnormalities) present and the degree of its severity. In a study of 200 births, 20 percent of the newborn foals were identified as congenitally abnormal. (This survey was conducted with a limited population in a restricted geographical area. The incidence of congenital defects in other groups or areas depends on a number of factors, including exposure to teratogens and the intensity of inbreeding.) Half of the affected foals had only mild defects, many of which were difficult to detect without a thorough examination. Examples of less severe abnormalities include minor limb deviations, parrot mouth, and umbilical hernias. In about 4 percent of the births studied, the congenital defects were severe enough to cause death at or immediately after birth. For example, the absence of part of the digestive tract or a serious heart defect usually results in the newborn foal's death. If vital organs are affected during gestation, death may occur before birth. (Dead embryos are usually absorbed by the uterus and may go undetected, while the dead fetus is usually aborted or carried to term in a degenerated and softened or mummified form.)

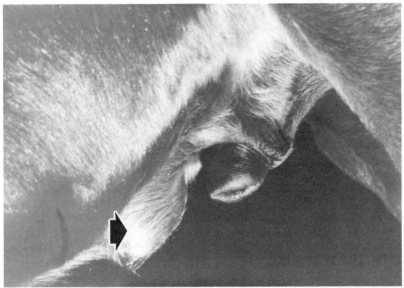

Many of the congenital defects that are detected in foals are not severe or life-threatening. For instance, umbilical hernias are often seen in newborn foals, but they are usually corrected naturally as the foal matures.

CAUSES AND CONTROL OF CONGENITAL DEFECTS

The causes of congenital defects can be categorized as genetic or environmental, but it is important to remember that many of these abnormalities are caused by the interaction of both genetic and environmental teratogens. To add to the confusion, a number of teratogens can cause similar defects, and many of these causative agents are very difficult to identify. The causes and control of congenital defects are not emphasized on most farms since harmful or fatal defects are rare and since many of the milder defects go unnoticed. When an unusually high incidence of birth defects are noticed in a population, however, a serious effort should be made to determine why these abnormalities are occurring. If the teratogen, or teratogens, can be identified (or even categorized), it may be possible to control or eliminate the defect by changing certain environmental characteristics or by reevaluating the selection/culling program. Obviously, the underlying purpose behind this type of effort is to maximize the usefulness and marketability of each foal crop. Allowing congenital defects to appear without making an honest investigation into their origin is an economically unsound practice that can have a significant effect on the breed as a whole. Although the following review of genetic and environmental teratogens is very basic, it provides the breeder with important background information that can be used when a specific investigation is initiated.

GENETIC TERATOGENS

When the frequency of a congenital defect is higher in the offspring of a particular mare and/or stallion, the presence of a genetic teratogen is very likely. In these instances, the offspring inherit genes that alter normal protein synthesis so that a change in cell structure or function occurs. Depending on the extent and location of these changes, the structure and function of important organs or tissues may be altered so that the individual develops an undesirable trait.

The first step in controlling or eliminating an inherited defect is to determine that the trait is, in fact, inherited and to categorize the trait's controlling gene (or genes) according to its relationship with other genes (e.g., dominant, recessive). Unfortunately, the identification of an inherited trait may not always be possible, especially within only a few generations. A parent may pass a genetic teratogen to only one out of several offspring, and since the generation interval (i.e., birth to reproductive age) is 2-3 years or longer in horses, it may be very difficult (if not impossible) to determine whether there is a pattern of inheritance involved. There may also be a number of complicating factors that influence the appearance of inherited traits. For example, the presence of a gene can be masked by the action of another gene. As a result, the hidden gene may be passed from parent to offspring for several generations and then appear suddenly as a congenital defect without a readily noticeable pattern in the horse's ancestry. In addition, some inherited characteristics may be the result of a sudden change in the parent's genetic material. These changes, or mutations, can occur spontaneously but may also occur in response to environmental factors such as radiation, exposure to certain chemicals, and viral infections. To complicate matters further, some genetic teratogens cause early embryonic death, which is frequently mistaken for low fertility. (For a complete description of defects that have been identified as inherited or that are suspected of being inherited, refer to the genetics text, EQUINE GENETICS & SELECTION PROCEDURES.)

Depending on whether the defect is recurrent and whether it has a detrimental economic or practical impact on the breeding program, the breeder may decide to cull breeding stock that are known to carry the causative gene. The feasibility of this procedure varies with the breeding program's size and with the breeder's goals and financial circumstances. If, for example, the breeder is trying to instill a desirable trait into a large breeding herd but finds that the appearance of an inherited defect has been encouraged by the selection program, the program's goals may need to be reevaluated. If only a small percentage of the breeding herd is affected, the problem may be traced back and the carrier animals culled with little financial impact on the program as a whole. Because the elimination of an inherited defect is not always a simple process and because the culling of breeding stock on the basis of one defect can have a serious impact on a previously established breeding goal, a brief discussion on "how to remove a trait from a herd" is dangerously simplistic. For this reason, the breeder that is confronted with this type of problem is advised to refer to a more complete analysis of possible equine selection/culling programs in EQUINE GENETICS & SELECTION PROCEDURES.

ENVIRONMENTAL TERATOGENS

Although embryonic growth and differentiation are initiated and monitored by genetic messages, these important developmental processes can be influenced by a number of environmental factors. Numerous environmental teratogens have been recognized, and researchers are challenged by the complexities of tracing and matching cause/effect relationships between the environment and embryonic development. Because teratology is a relatively new science, research evidence on the effects of specific environmental factors on the equine embryo is scarce. For this reason, the following discussion on environmental teratogens includes general information drawn from data on other species.

Nutrition

The impact of a nutritional deficiency, imbalance, or toxicity on normal embryonic development depends on two important factors. First, the type of nutrient involved determines which area of the body or which developmental processes will be involved. If several nutrients are involved, their relationship (e.g., calcium-phosphorus ratio) may also have an important effect on the type of abnormality that results. The second factor that influences the impact of a nutritional problem during gestation is timing. The stage of development is critical in determining the severity of an environmental teratogen's actions. Most developmental problems occur during early gestation when cells are differentiating to form specialized body structures. Because the nutrient demands of the early embryo are relatively low, however, serious developmental abnormalities, caused by poor nutrition, are rare unless the dam is subjected to a severe nutrient deficiency or imbalance.

Vitamins are the most critical nutrients during the period of cellular differentiation. The B vitamins are required for normal organ development and red blood cell formation, and vitamin K is essential to the synthesis of blood-clotting factors. Vitamin E is believed to be necessary for development of normal ova and could be indirectly related to congenital defects if the ova's ability to migrate or divide is affected. Oversupplementation of certain vitamins can also cause abnormalities. Water soluble vitamins seldom present problems, but because fat soluble vitamins (A, D, E, and K) are stored in the dam's body, they can reach levels that may damage the embryo or fetus. Of particular concern are the vitamins A and D. An excess of these two vitamins interferes with calcium-phosphorus utilization and increases the occurrence of congenital birth defects.

Adequate levels of iodine are required for normal utilization of energy sources, and either an excess or a deficiency of this important mineral can cause congenital goiter. Except in severe cases, goitrous foals are seldom characterized by enlarged thyroid glands, but they are usually weak (or stillborn) and may have long hair, contracted tendons, and poor muscle development. Goiter due to an iodine deficiency is not a common problem in horses since most salt supplements contain adequate iodine. It is extremely important that iodized salt be added to the rations of horses living in areas where the soil is notably deficient in iodine. Goiter caused by iodine toxicity is usually the result of oversupplementing the mare's diet with seaweed or kelp products. A total intake of greater than 83 mg of iodine per day seems to increase the incidence of fetal goiter significantly.

During late gestation, energy and protein become more critical than vitamins and minerals due to the rapid rate of overall body growth characteristic of this period. (This is not to imply that vitamins and minerals are unimportant during this period, however.) Severe deprivations of these two important nutrients have been reported to increase the incidence of abortion between days 25 and 31 of gestation, but because the need for both protein and energy increases sharply during the third trimester of pregnancy, deficiencies are most likely to occur during this last developmental stage. There is some evidence that the mare is able to offset the effects of protein and energy deficiencies by sacrificing her own body protein and energy reserves. The mare donates these nutrients to her fetus only until her own life is threatened. At that point, the fetus is usually aborted.

Drugs, Chemicals, and Toxic Plants

Many agents that are potentially harmful to the fetus are found in anesthetics, therapeutic drugs, salves, pesticides, and occasionally as residues on grain and in water. Teratogens can also be found in some plants at certain stages of maturity. Unfortunately, very little is known about the actions of most drugs, chemicals, and toxic plants on the equine fetus. Most of the information on these types of teratogens is extrapolated from research on laboratory animals and man. It is known, however, that the age of the embryo and the level of exposure determine how severely the fetus is affected. The most sensitive period seems to be during the first six weeks of gestation.

Several mechanisms seem to protect the fetus from some harmful substances. For example, the mare is capable of breaking down some teratogens so that the level of exposure to the fetus is reduced. In addition, the placenta is capable of screening many substances on the basis of size. The mare/fetus attachment operates like a sieve, sorting out large particles and permitting the smaller ones to pass through. Many drugs, including some antibiotics, can pass through the natural placental barrier, however. Because the placental waste/nutrient exchange mechanism is unable to remove blood-borne teratogens from the fetal circulation efficiently and because the fetal liver is unable to detoxify blood until after birth, teratogens that do pass through the placental barrier may become more and more concentrated in fetal tissues as gestation progresses. It is important to remember that the teratogen levels required to cause signs of toxicity in the mare (if the agent is toxic to the mare) is far greater than the levels required to affect the fetus. Any agent that must be administered to the mare during pregnancy should be given cautiously, especially during the first six weeks of gestation. Examples of products that are normally restricted during pregnancy include corticosteroids, ACTH (adrenocorticotropic hormone), phenylbutazone, anabolic steroids, and tranquilizers. The best policy for administering drugs and dewormers during pregnancy is to restrict the use of medications as much as possible unless the product has been proven safe for use in pregnant mares. The control of other toxic agents, such as pesticides and toxic plants, should be incorporated into the basic management program.

Disease-Causing Organisms

Although the placental barrier protects the fetus from most disease-causing organisms, some viruses can penetrate the barrier due to their exceptionally small size. More commonly, infectious organisms reach the conceptus via the cervix as a result of improper breeding or examination techniques. After invading the embryo or fetus, these organisms may persist throughout gestation, causing abnormal development or possibly fetal death and abortion. Infections that occur during the later stages of gestation are more likely to cause fetal death and abortion, while those that occur before critical cell differentiation is complete are more likely to cause congenital deformities. Some infections can also damage the placenta, resulting in a reduction in or a complete loss of the organ's ability to screen out harmful substances.

Diseases that impair the dam's liver function may have a serious effect on the fetus if drugs, toxic agents, and waste products build to high levels in the maternal circulation, increasing fetal exposure to these harmful agents. However, even infectious agents that are unable to cross the placenta can also influence the fetus indirectly, if the mare suffers a high fever or reduces her feed intake as a result of infection.

Physical Factors

A discussion on teratology would not be complete without mentioning several physical aspects that influence the success or failure of the foal's first developmental stage. An adequate oxygen supply is vital to the well-being of the fetus, and any condition that lowers the oxygen level in the mare's blood or that decreases the rate of transfer across the placenta can halt the growth processes. The fetus also requires enough space within the uterus to move the head, neck, and limbs. Paddling of the limbs is a normal movement that is very important to proper limb development. If mobility is restricted, the fetus may develop limb stiffness, "fixed" limbs, or limb deformities. These conditions are apparent at birth and are usually irreversible. Temperature extremes and reduced atmospheric pressure have caused birth defects in laboratory animals, but their effects on the equine fetus have not been established.

Gestation Length

The average gestation length in the mare is 340 days, but healthy foals are sometimes born as early as 300 days. A gestation of 360 days is considered extremely long, but normal foals are sometimes presented well after a year of gestation. The mechanism that causes pregnancy to terminate is not clear, but several factors are known to influence the length of gestation.

Environment affects gestational length by altering the ideal conditions for fetal growth. For example, season changes the length of gestation significantly. Foals born during early spring are carried an average of 10 days longer than those born in late summer. The most common environmental causes of abnormally shortened or lengthened gestation periods are unthriftiness or malnutrition in the dam and diseases of the uterine endometrium or placenta. Dead or defective fetuses can also delay parturition. This delay is suspected to be caused by the fetus's inability to produce some critical stimulus that signals the end of gestation. These fetuses can also shorten gestation, probably because the uterus rejects the decomposing material.

The age of the sire and dam seem to have no effect on the foal's gestation length, but the genotype of the sire, the dam, and the fetus are important factors. Some mares consistently carry the fetus for longer or shorter durations (i.e., when compared with an average for that breed). Because gestation length is known to be a heritable trait, the average gestation length could (theoretically) be changed by selecting stallions on the basis of their daughters' gestation lengths and by selecting mares on the basis of their own gestation lengths. The sex of the fetus also influences gestation length; colts are carried from two to seven days longer than fillies. In addition, the fetus is believed to have some endocrine control over gestation through its pituitary and adrenal glands. This effect may explain why a hybrid fetus (e.g., resulting from a horse-donkey cross) changes the normal gestation length of its dam. For example, a mule fetus lengthens the normal gestation period of the mare, while a hinny fetus shortens the normal gestation period of the jenny. ♞

14

PREGNANCY

The mare's reproductive role is emphasized throughout this text, illustrating that she is in the mainstream of breeding farm activities throughout the year. After the mare is declared "safe in foal," she is usually placed on a basic management program, one that includes an adequate diet, required vaccinations, strict parasite control, and routine observation. At this point, the breeder's attentions are usually diverted to other farm procedures, such as evaluating barren mares, weaning foals, and conditioning yearlings. Although the mare's pregnant state indicates that her reproductive status is generally acceptable, her health and safety should not be ignored during this period. This "quiet" phase of the mare's reproductive year is of utmost importance to the developing foal's future health and productivity and can have a great impact on the mare's future reproductive ability. The complex physiological events that occur during pregnancy have been examined to give the reader a basic understanding of the demands placed on the mare during this period. With this essential background information in hand, pregnancy can be explored from a management viewpoint.

Pregnancy Tests

On many breeding farms, the mare's reproductive status is evaluated several times throughout the reproductive year. Prior to the breeding season, for example, the mare's fertility status is studied carefully to determine whether special preconditioning techniques will enhance her chances of settling. As the mare enters the breeding season, her cycles are scrutinized to determine when she has the best possible chance of conceiving. To detect the end of the first reproductive phase (preparation for conception) and the beginning of the second reproductive phase (production of a healthy foal), the mare's reproductive status must be evaluated once again.

The primary reason for performing a pregnancy test is obviously to determine whether the mare is pregnant. From a management standpoint, there are also several secondary reasons why pregnancy tests are helpful:

1. Pregnancy tests help prevent breeding during gestation. (i.e., If the pregnant mare comes into heat, she will not be mistaken for an open mare.)
2. They increase the chances of getting a mare settled before the season is over, since early tests indicate whether the mare is pregnant at about 18-21 days postovulation (i.e., when she is due to cycle once again).
3. They help the breeder evaluate the operation's efficiency as the season draws to a close by providing data for the calculation of conception rates.
4. Pregnancy tests also minimize the time that an outside mare must spend at the breeding farm, thereby minimizing the mare owner's mare care fee.
5. If the breeding contract has a "safe in foal" restriction, pregnancy tests help determine if and when the stallion fee is payable.

There are several methods that can be used to evaluate the mare for pregnancy. Most of these methods have a defined period of effectiveness, and each varies with respect to cost, convenience, and accuracy. Although pregnancy tests are usually performed by the farm veterinarian, the farm manager often has a significant input with respect to when and how these tests are performed. For example, some breeders insist that pregnancy tests be carried out as soon as possible (e.g., 18-21 days postovulation) so that any open mares can be rebred during the following cycle. In order to make responsible recommendations, the breeder should be familiar with the limitations and advantages of the different types of tests and should understand when each can be utilized.

EXTERNAL CHARACTERISTICS

The outward indications of pregnancy can be helpful, but should not be used as the basis for a positive or negative pregnancy diagnosis. The small contracted vulva and the pale, dry appearance of the vulva, vagina, and cervix are characteristic of, but not restricted to, pregnancy. Similarly, failure to show heat at 17-18 days after the last estrus is only suggestive of pregnancy, since some pregnant mares show signs of estrus throughout gestation, while some nonpregnant mares fail to show estrus due to a number of possible physiological and psychological problems. As a pregnancy testing method, teasing is only about 60 percent accurate. Tests that detect the embryo, the fetus, or a pregnancy-related hormone are far more dependable methods of diagnosing pregnancy.

RECTAL PALPATION

Rectal palpation is a very effective (and the most common) method of diagnosing pregnancy, since it allows the detection of physical changes that occur within the mare's reproductive tract during gestation. Experienced palpators can diagnose pregnancy at about 21 days postconception with about 90 percent accuracy, but many breeders and veterinarians have experienced the frustration of confirming a mare's pregnancy by rectal palpation at 21 days only to find that she is open several months later. Although human error should not be ruled out completely, these contradictory evaluations are usually due to embryonic resorption or fetal loss rather than misdiagnosis. (There is no evidence that normal palpation increases the incidence of fetal death.) For this reason, early palpation should be used to screen mares for pregnancy, and any positive findings should be verified several weeks later to insure that the embryo has not been resorbed during early gestation. Used in this manner, rectal palpation helps to identify mares that have not conceived so that breeding procedures can be repeated as soon as possible.

Before a mare is palpated, she should be positively identified. Identification is especially important when a large number of mares must be palpated or when the palpator is not familiar with the horses. Each mare's teasing and breeding records should be reviewed before she is palpated, and the examination findings should be noted immediately on the daily record chart which, in turn, can be used to update the mare's individual record. The basic palpation procedures described in "Broodmare Management" are used when evaluating the mare's tract for pregnancy. Although rectal palpation is a relatively simple procedure, it should only be performed by a veterinarian or trained technician. The untrained palpator usually has difficulty distinguishing between pregnancy-related and abnormal conditions of the reproductive tract and may be unable to differentiate between the stages of pregnancy or between the various organs within the abdominal cavity. In addition, the mare's rectum is very thin and is easily irritated or ruptured during palpation. Rectal rupture is often followed by death due to shock and/or peritonitis (i.e., infection of the abdominal cavity). Injury to the embryo is also possible if the reproductive tract is handled roughly.

When evaluating a mare for pregnancy, the palpator usually concentrates on physical characteristics of the uterus and cervix. Throughout pregnancy, the cervix should remain firm and tubular so that it maintains a tight seal and protects the reproductive tract from contamination. The degree of cervical closure may vary somewhat without adverse effect, but complete loss of cervical tone signals an endangered pregnancy. Uterine tone and thickness increases slightly following ovulation. If conception does not occur, this tone is lost in about 17 to 21 days. By 17 to 21 days postovulation, the uterus of the nonpregnant mare is flaccid and edematous, and she begins to express behavioral patterns associated with estrus. On the other hand, the uterus of the pregnant mare remains firm and tubular due to continued secretion of progesterone. At around 21 days postconception, the gravid uterus undergoes a marked increase in tone. Although this change is sometimes difficult to detect, it is a reliable indication of pregnancy in most mares and is especially well-defined during a mare's first pregnancy. At 21 days, the palpator may also feel the conceptus, which at that point is a soft bubble of fluid located at the junction of the gravid horn and body of the uterus. The conceptus is difficult to grasp, but as the palpator's hand passes over the horn-body junction, the small bulge can sometimes be detected.

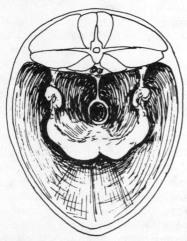

The healthy, nonpregnant uterus remains firm and tubular during diestrus.

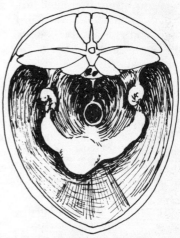

Around day 24, the conceptus (i.e., the embryo and the fluids and tissues which surround it) can be detected as a small bulge in the gravid uterine horn.

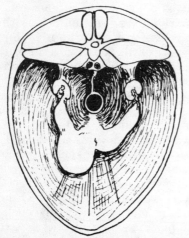

At 40 days of pregnancy, the bulge caused by the conceptus is enlarged, and the gravid horn has begun to lose its tone.

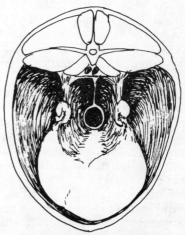

The conceptus begins to soften by about day 85, and the palpator can begin to distinguish the fetal head and body.

During early gestation, the conceptus may migrate from one horn to the other, but between days 50 and 70 postconception, this free-floating state is prevented by the initiation of firm implantation. By about day 50, the gravid horn is completely filled, and the conceptus begins to occupy the uterine body. At this time, the conceptus becomes elongated, measuring about 3 inches (8 centimeters) in length and 2.4 inches (6 centimeters) in diameter. Although the uterine bulge can be felt as early as day 21 of gestation, the fetal body cannot be grasped until about 85 days postconception. When the operator attempts to palpate the fetus before this time, the tiny body slips through the allantoic fluid away from hand pressure. However, the palpator can tap the bottom wall of the rectum as early as day 60 and feel the fetus as it gently bumps against the uterus (i.e., detection of the fetus by ballottement). At about day 85 of gestation (range: 70 to 120 days), allantoic tension decreases, and the conceptus becomes softer and easier to compress. When the uterine bulge is palpated at this time, the fetal head and body can be distinguished. Later, the limbs and neck can be differentiated, and fetal movement can be detected.

The fetus can be detected by palpation until its own weight begins to pull the uterus down (at about day 150). As the uterus moves to a lower position in the abdominal cavity, the fetus moves from a crosswise to a lengthwise position and begins to protrude into the opposite uterine horn. By about 6 months postconception, the uterus and ovaries are usually out of the palpator's reach. This situation, combined with the presence of tense broad ligaments, is indicative of pregnancy. (Note: These characteristics could also indicate severe uterine infection.) At 7 months, the uterus is large enough that it can be palpated via the rectum once again. Palpating the mare at 6 or 7 months to determine whether she is still pregnant helps to prevent the loss of an entire reproductive year. Even if it is too late to rebreed the mare, the manager has advance notice of the mare's preconditioning needs for the following season.

LABORATORY TESTS

In addition to the physical methods of detecting pregnancy, there are several laboratory tests which detect hormonal or tissue changes associated with pregnancy. Analytical methods are generally more expensive and more time-consuming than the diagnostic methods described earlier, but these laboratory techniques may be very helpful when the mare cannot be palpated. Because these methods require a background in biochemistry and an ability to collect blood, urine, or tissue samples, they are usually performed by a veterinarian and/or a lab technician. For this reason, the following discussion presents only a description of each test and does not attempt to present step-by-step laboratory instructions.

CERVICAL SMEARS

Once pregnancy is established, the cells of the vagina and cervix undergo structural changes and begin to secrete a sticky gum-like mucus. A pregnancy test based on these changes has been developed by Japanese workers but is not well known in the United States. This so-called mucin test (or Kurosawa method) involves 1) collecting mucus from the cervical os, 2) placing the sample on a slide, and 3) preparing the smear with a special stain. When viewed through a microscope, smears taken from the cervix of pregnant mares are thick and dark and contain globules of mucus and epithelial cells. (Smears from anestrus mares are also characterized by these globules of mucus but do not contain the epithelial cells.) Because this test has not been used extensively, estimates on its reliability are not available at this time. It is known, however, that these microscopic evaluations may give false positives if embryonic resorption or fetal death has occurred.

MIP-TEST

There are several pregnancy tests that are based on patterns of PMSG production during pregnancy. The most commonly used and the easiest test to perform is the Mare Immunological Pregnancy Test (MIP-Test, Diamond Laboratories, Inc.). The MIP-Test is a commercial kit that contains all of the required equipment and solutions: disposable test tubes, test tube rack, special neutralizing solutions, antibodies (anti-equine gonadotropin serum), cell suspension (sheep red blood cells coated with PMSG), eye droppers, etc.

This test is based on the reaction that occurs between an antibody to PMSG and PMSG-coated sheep red blood cells. Normally, the antibody and the coated cells agglutinate (clump together) when exposed to one another. In a round-bottomed test tube, the clumped cells then settle and cover the bottom with an even mat. However, when the antibody is first exposed to PMSG, agglutination does not occur between the antibody and the cells. In this case, the cells fall to the bottom of the test tube and form a doughnut-shaped pattern.

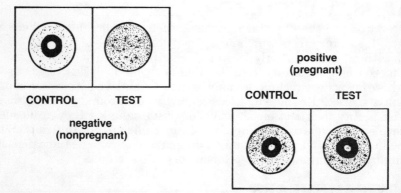

The Mare Immunological Pregnancy Test involves the comparison of a control sample and a test sample. The control sample should always form the doughnut-like sedimentation pattern that is associated with pregnancy. However, if the tested mare is not pregnant, the test sample shows a negative pattern.

For pregnancy diagnosis, a blood sample is drawn from the mare. The blood serum from the sample is added to a test tube already containing an antibody to PMSG. PMSG-coated sheep red blood cells are then added to the solution. If the mare is not between 40 and 100 days pregnant, her blood serum does not contain PMSG. Thus, the antibody and cells agglutinate and settle in an even mat. If the mare is between 40 and 100 days pregnant, however, her blood serum contains PMSG. The PMSG reacts with the antibody and prevents agglutination. As a result, the characteristic doughnut-shaped pattern of a positive test forms as the sheep red blood cells settle in the test tube.

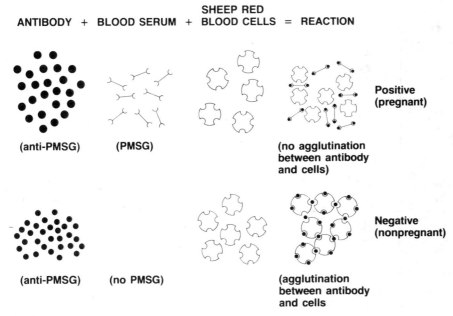

SHEEP RED

ANTIBODY + BLOOD SERUM + BLOOD CELLS = REACTION

(anti-PMSG)   (PMSG)   (no agglutination   Positive
                                                            between antibody   (pregnant)
                                                            and cells)

(anti-PMSG)   (no PMSG)   (agglutination   Negative
                                                            between antibody   (nonpregnant)
                                                            and cells

As shown in this representation of the agglutination reaction, PMSG in the pregnant mare's blood serum prevents agglutination from occurring between the antibody and the sheep red blood cells. Thus, lack of agglutination is a positive reaction for pregnancy.

The MIP-Test is about 97 percent accurate from day 40 to day 100 of pregnancy when PMSG levels are elevated significantly. When these hormone concentrations get too low (between days 100 and 140), the test can no longer be utilized. Because the endometrial cups continue to produce PMSG until day 120-140 even when pregnancy has been terminated, the test may give a false positive.

ASCHHIEM-ZONDEK TEST

Another method of detecting the presence of PMSG in the mare's blood is based on the response that this hormone elicits when injected into the abdominal cavity of an immature female rat. When fresh whole blood containing high levels of PMSG is introduced into the rat's abdominal cavity, her uterus and ovaries become enlarged. If a postmortem examination (performed 72 hours after the injection) reveals these changes, a positive pregnancy diagnosis is indicated. The test is fairly accurate between days 40 and 120 of gestation, but the best results (95 percent accuracy) are obtained during peak PMSG production between days 50 and 80 of pregnancy. This test is very useful in certain situations but is more expensive and more time-consuming than rectal palpation and the MIP-Test.

FRIEDMAN TEST

An adaptation of the Aschhiem-Zondek test is the Friedman test. The basic procedures for these two tests are similar, but the Friedman test utilizes an immature female rabbit instead of a rat. Both the Aschhiem-Zondek and the Friedman tests are highly accurate during a 30 day period (i.e., day 50 to 80 of gestation) but are used infrequently due to the added expense of laboratory animals and supplies and because rectal palpation is also an accurate pregnancy detection method during the 50 to 80 day period.

FROG TEST

A third test used to detect PMSG in the mare's blood is based on the response that the hormone elicits when injected into a male frog. When male frogs are stimulated by courtship or by a gonadotropic hormone (e.g., PMSG), sperm cells are produced and released. After the mare's blood serum is injected into the frog's dorsal lymph sac, his cloaca (genital tract opening) is examined carefully. The presence of spermatozoa is indicative of elevated PMSG levels in the mare's blood.

CUBONI TEST

Between days 120 and 290 of gestation, the fetal gonads produce a large amount of estrogen, which enters the mare's bloodstream and is excreted in her urine. Reasons for enlargement and increased activity of the fetal gonads during this period are not known, but the two characteristics seem to be directly correlated.

There are several analytical methods of detecting estrogen in the mare's urine, but the most commonly used technique is the Cuboni test. This test involves collecting a urine sample and carrying out a special laboratory procedure that creates a green fluorescence (evident under ultraviolet light) in samples collected from mares that are 120 to 290 days pregnant. The Cuboni test is relatively simple to perform when the proper laboratory equipment, supplies, and experience are available, and it is about 98 percent accurate between days 120 and 150. The test is most accurate between days 150 and 250, but its reliability drops rapidly as the fetal gonads regress and as estrogen levels decrease after about day 250 of gestation.

The disadvantages of this test are 1) pregnancy can be detected much earlier by rectal palpation; 2) it is not a practical on-the-farm testing method; and 3) the test does not differentiate between increased estrogen levels caused by pregnancy and increased levels caused by abnormal conditions such as ovarian cysts, nymphomania, and chronic genital infections.

ULTRASONIC PREGNANCY DETECTION

One of the most significant problems associated with pregnancy detection is that many of the tests do not provide proof that the fetus is alive. For example, fetal loss prior to day 120 of pregnancy cannot be detected by the MIP-Test, since PMSG production continues regardless of events that may occur after the establishment of the endometrial cups. Movement of the fetus can be detected by rectal palpation, but not until about day 70-85 of gestation. Ultrasonic instruments that detect fetal pulse as early as 42 days postconception are not only useful for pregnancy diagnosis but also provide an excellent means of monitoring the viability of the fetus throughout gestation. Some of these specialized instruments are also capable of differentiating between fetal heartbeat, fetal movement, blood flow through the umbilical vessels, and blood flow through the uterine arteries. From about day 90 to day 270 of pregnancy, fetal pulse is readily detectable, and because the fetal heartbeat is much faster than the mare's heartbeat, fetal pulse is easily distinguished from maternal pulse. After day 270, however, fetal pulse cannot be detected.

Ultrasonic pregnancy detection is based on the Doppler principle: when ultrasound strikes a moving object, it is reflected at a different frequency. For instance, one ultrasonic instrument uses a transducer (sending and receiving instrument) to direct ultrasonic soundwaves toward the mare's uterus. Soundwaves striking the uterus are reflected back to the transducer at a frequency that is determined by the organ's tone and density. Because the uterine fluid build-up that occurs during pregnancy affects tone and density, pregnancy can be detected by analyzing the ultrasonic soundwaves that are reflected by the fluid-filled organ.

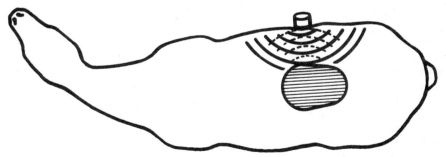

Ultrasonic pregnancy detection instruments operate on the principle that ultrasound directed toward the fetus is reflected back to the source at a different frequency. Some detection transmitter/receivers, such as the one illustrated, project ultrasonic sound through the flank, although some models can be placed in the mare's rectum.

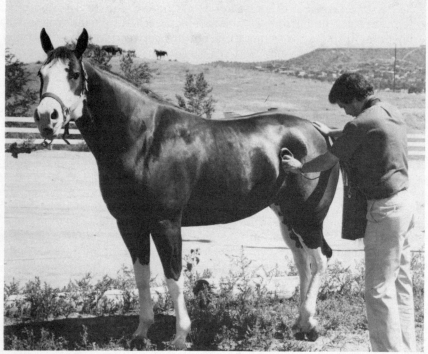

Courtesy of Animark

The Boveq, a commercial ultrasonic pregnancy detection instrument, diagnoses pregnancy by detecting the density and tone changes that occur in the pregnant uterus.

Care of the Pregnant Mare

Throughout gestation, the broodmare manager's primary responsibilities are to protect the mare's future reproductive capacity and to give the developing fetus every possible advantage for a healthy, productive life. In order to meet these challenges, the manager must recognize the changes that are occurring within the mare's reproductive tract and must plan a health care program that enables the mare to protect and nourish her fetus without sacrificing her own well-being. As will be pointed out later in this chapter, the chances of foaling problems are also minimized by keeping the mare in good condition throughout pregnancy. In addition, the mare's ability to recover soon after foaling and be bred successfully during that breeding season depends not only on the occurrence or absence of foaling problems but also on the mare's physical condition and reproductive health prior to foaling. Good husbandry techniques accompanied by special pregnancy management procedures help to insure the mare's health throughout pregnancy. Specifically, the manager must concentrate on exercise, nutrition, parasite control, and vaccination programs for the pregnant mare.

Courtesy of Earl and Barbara Lang Quarter Horses

This mare was kept in good condition throughout pregnancy, as evidenced by a healthy haircoat and adequate body fat.

EXERCISE

Both absence of physical activity and over-exertion can be harmful to the pregnant mare. A moderate exercise program (e.g., self-exercise in a large pasture) enhances the waste/nutrient transport mechanism between the dam and her fetus by improving circulation to the uterus. As explained in the previous chapter, the fetus is totally dependent on the close association that develops between its blood system and that of its dam. Exercise increases the flow of blood to the uterus through the utero-ovarian artery and thereby increases the supply of oxygen and nutrients to the fetus. Light exercise also helps to minimize pregnancy edema and stiffness. (Refer to 'Conditions Associated With Pregnancy' within this chapter.) In addition, a physically fit mare is more likely to foal without complications, since her efforts are not hampered by large amounts of internal fat or by weak muscles. Because physical fitness plays such an important role

throughout the mare's reproductive career, the manager should make sure
that each mare remains active throughout pregnancy.

Although pregnant mares frequently become sedentary, they will stay fairly fit if given access to free exercise in a safe pasture.

Courtesy of Paradise Farm

The ideal level of physical activity for a particular mare depends on her
overall condition and on how extensive her exercise program was before
she was bred. Mares that are maintained on pasture receive adequate
exercise while grazing, but if a mare is accustomed to regular forced
exercise (e.g., riding or driving), her level of physical activity should be
reduced gradually. This is not to imply, however, that an exercise program
cannot be continued throughout gestation. Obviously, if the mare was not
accustomed to two hours of daily riding before pregnancy, this type of
exercise program should not be pursued without a gradual conditioning
program. Total stall confinement is not recommended for broodmares, but
if it is unavoidable, the stalled mare should be exercised daily until the
last few days prior to foaling.

Because abortion has been attributed to exhaustion due to both nervousness and over-exertion, pregnant mares should not be subjected to
extreme psychological or physical stress. As a case in point, the manager
should avoid transporting mares long distances during late gestation.

Shipping may result in fatigue and loss of appetite. During the last 60 days of gestation, when the mare's nutrient requirements are increased, shipping stress can tax the mare's maternal resources.

NUTRITION

When planning a pregnant mare's diet, it is important to remember that nutrient requirements vary with the mare's age, size, physical activity, and stage of pregnancy. For this reason, dietary needs during pregnancy must be evaluated on an individual basis. (Refer to the nutritional requirement tables within the Appendix.) Ideally, pregnant mares should be fed separately. During the first six to eight months of gestation, mares fall into two major categories: those that do not have foals at side (i.e., previously maiden or barren) and those that do have foals at side. The nutrient requirements of the second group are much higher than the first due to the demands of lactation. In fact, the lactating mare's energy requirements are only exceeded by those of the suckling foal and the horse that is worked heavily. Therefore, lactating mares should be fed separately from nonlactating mares during early gestation when their diet must be supplemented. Because the pregnant mare's requirements change significantly during the last three months of gestation, it is also helpful to group pastured mares according to stages of pregnancy.

Anyone who designs or manages a broodmare feeding program must 1) understand each animal's nutrient requirements, 2) understand important nutrient characteristics of various feeds, 3) be able to recognize quality feed, and 4) know how to balance a ration in such a way that the needs of both the mare and fetus are met. This type of education requires a dedicated and continuous effort, along with a willingness to consider new feeding practices. The following discussion is an overview of important points to consider when planning a diet for the pregnant mare. (For a review of the lactating mare's nutrient needs during pregnancy, refer to "Lactation.") As with other diets, the nutrients of principle concern are protein, energy, vitamins, minerals, and water.

PROTEIN

During the first eight months of gestation, the protein requirement of the nonlactating mare is the same as that of a mature horse. Good quality legume hay or well-managed grass pasture provides adequate protein during early pregnancy, but legumes or mixtures of legumes and grasses are preferable, since they are reliable sources of several other important nutrients (e.g., calcium, vitamin A). During the last three months of gestation, the mare's protein requirements increase significantly due to rapid growth of the fetus. A **good quality** legume hay provides sufficient dietary protein to meet her protein requirement, but when grasses are stored as hay, the protein content is reduced. For this reason, broodmares that are fed grass hay should also receive legume hay or some other high-quality protein supplement, especially during the last three months of pregnancy. Stored hay should be subjected to laboratory analysis to determine its protein content. Current research indicates that 8.5 percent protein is required during the first 8 months of pregnancy and that 11 percent protein is required during the 3 months prior to foaling. If there is any question as to whether the ration is supplying adequate protein, a supplement should be added for a margin of safety against protein deficiency.

There are two main sources of supplemental protein: 1) animal products, which include fish meal and other meat by-products and 2) plant products, such as soybean meal, cottonseed meal, and linseed meal. Animal proteins are very good supplements, but they are also expensive feedstuffs. Fortunately, horses can utilize the more economical plant proteins quite efficiently. The criteria for evaluating the quality of a protein source is its complement of amino acids. The absence of one or more essential amino acids causes a protein deficiency even though the percentage of digestible protein is adequate. Corn, milo, barley, oats, cottonseed meal, and linseed meal are all lacking at least one important amino acid, but soybean meal contains all of the essential amino acids for horses, a characteristic that makes it an ideal protein supplement.

Photo by Bill Witthop

During the last three months of pregnancy, broodmares require increasing amounts of energy and protein. The concentrate portion of the ration must be properly mixed to supply the correct balance of these nutrients.

ENERGY

During early gestation, the energy requirements of the nonlactating mare can also be obtained from good quality hay, if the mare is idle (e.g., self-exercised on pasture or exercised lightly for less than one hour per day). If the mare is a poor doer or if she is very active during this early period, an energy supplement (e.g., grain) should be added to her diet at a level that keeps her in good condition. The skilled eye of an experienced manager can be very helpful when evaluating body condition, but if livestock scales are available, regular weighing is the best method to determine whether the mare's diet is supplying the proper amount of energy. (Weight gain should be minimal during early gestation.)

During the last three months of gestation, an average sized mare (1100 lbs) can be expected to gain a little over 0.5 pound per day. This weight gain represents fetal growth only and not a significant increase in maternal fat. Because fetal size is determined by uterine capacity and because the uterus of an obese mare is compressed by layers of fat, extremely overweight mares tend to produce small foals. If excessive internal fat places pressure on other tissues or organs, the powerful muscle contractions that occur during labor may damage these areas, resulting in internal hemorrhage. On the other hand, insufficient dietary energy forces the mare to draw from her own body reserves. As a general guideline, pregnant mares should maintain a moderate fat cover over the ribs so that the ribs can be felt but not seen.

In order to insure that the mare's energy needs are met during this period of rapid fetal growth, the manager must provide an energy supplement. Increasing the hay ration to supply the required energy is ineffective since the amount of hay required to supply the mare's energy needs may exceed her maximum possible roughage intake. In other words, even if hay were available free-choice, the mare might not be able to consume enough to meet her daily energy requirements. A practical solution is to decrease the amount of hay in the ration and eliminate the energy deficiency by adding grain. Unless high-quality alfalfa is fed, a protein supplement may have to be added to eliminate any protein deficiency.

VITAMINS AND MINERALS

In many species, vitamins and minerals play essential roles in a number of important physiological processes. Unfortunately, the functions of many of these nutrients as they relate to equine health and development have not been established, and the effects that these micronutrients may have on the mare during pregnancy are not known. In spite of this lack of information on physiological mechanisms, minimal daily requirements for most of these nutrients have been established for the horse. (Refer to the Appendix.) To insure that the pregnant mare's micronutrient needs are met, it is important not only to understand what these needs are but also to evaluate available feeds and supplements to determine whether the mare's vitamin and mineral needs are met once protein and energy have been balanced. Mineral and vitamin content of hay or grain, for example, varies with plant species, the soil's mineral content, plant maturity when harvested, the methods used to harvest, and the conditions present when the crop was harvested.

When studying the mare's micronutrient needs, special emphasis must be placed on vitamin A, calcium, and phosphorus. Vitamin A is important to healthy epithelium (cell layers that form tissues such as the skin and the endometrium) and is, therefore, essential for normal reproductive function. It is important to note that vitamin A is stored in the liver. This process insures that the animal has an adequate vitamin A supply when the diet is deficient in this vitamin. Feeding excess vitamin A over a prolonged period may result in abnormalities such as bone fragility and excess bone growth. The most practical and the safest way to supply vitamin A to the mare throughout pregnancy is to feed fresh, green forage or a legume hay of good quality.

Although calcium and phosphorus are essential to the development and maintenance of healthy bone, they also perform many other important functions. As components of protein and fat compounds, calcium and phosphorus are necessary for the development and maintenance of healthy soft tissues (e.g., muscles, organs, blood cells, etc.). These minerals are also vital to normal relaxation and contraction of muscles and to the characteristic irritability of nerves. Normally, calcium and phosphorus are stored in the bone and released as required for these life-sustaining functions.

Severe deficiencies of these important minerals during pregnancy can cause birth defects (e.g., weak foals, abnormal bone development, abnormal tissue growth), but the mare is usually able to sacrifice her own calcium and phosphorus stores to meet the needs of her fetus. Excessive demands make the mare's skeleton brittle and highly susceptible to injury. To insure that the mare's calcium and phosphorus intake is sufficient to meet both her needs and the needs of her fetus, the manager must study the mineral contents of available feeds and balance the mare's ration so that she receives approximately 23 grams of calcium and 14 grams of phosphorus per day during the first 8 months of gestation. During the last 90 days of gestation when fetal growth accelerates, these requirements increase to approximately 34 grams of calcium and 23 grams of phosphorus per day.

The ratio of calcium to phosphorus is very important to proper utilization of these minerals. Excess phosphorus, for example, interferes with calcium absorption. In adult horses, a diet that contains a calcium to phosphorus ratio between 1:1 to 6:1 is acceptable. However, a ratio of 2:1 for these minerals is considered ideal. It is important to note that because grain is relatively high in phosphorus and because grain must be increased in the ration to meet the escalating energy demands of late gestation, calcium supplements (e.g., bone meal or ground limestone) may need to be added to the mare's diet during the last three months of pregnancy. (To provide other important minerals such as magnesium, iodine, copper, manganese, iodine, salt, and potassium, trace-mineralized salt should be offered free choice.)

WATER

Because of its vital role in many body processes, water is the most important nutrient. Pregnant mares should be expected to drink 10 gallons or more per day (depending on roughage intake and level of activity), and it is best if this volume is consumed in frequent, small quantities. For this reason, mares should have constant access to clean, cool water at all times. The use of ponds or other sources of ground water is discouraged unless they have been tested and declared suitable for livestock consumption. During the winter, water sources should be free-running or heated to prevent freezing. It is important to note, however, that horses actually prefer slightly chilly water (40-45°F).

PARASITE CONTROL

As pregnant mares graze, nibble spilled grain, or pick at their bedding, they ingest parasite larvae and are continuously reinfected with a variety of parasitic organisms. Good husbandry techniques and proper management of facilities help minimize this continuous process of reinfection. A discussion of parasite control techniques can be found in the Appendix.

Parasite populations in pastures, lots, and stalls can be reduced with good husbandry practices, but when animals are concentrated in small areas, it becomes very difficult to control these populations through facilities management alone. Because many horse farms are space limited, the well-planned deworming schedule has evolved as an important weapon against parasites. Each farm's facilities, management circumstances, and horse concentration dictate the type of deworming program that should be

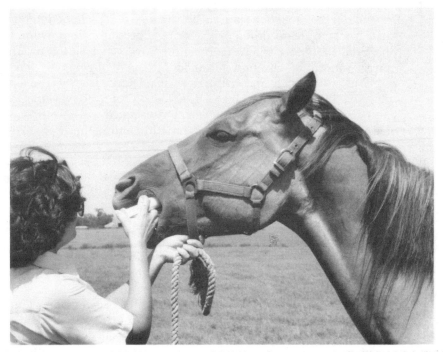

Pregnant mares should be kept on a regular deworming program until one month before foaling.

adopted. In any event, it is very important to select products that are effective against internal parasites but that have a wide margin of safety (i.e., the difference between the effective dose and the toxic dose). Manufacturer precautions pertaining to pregnant mares should be carefully noted before administration. Cambendazole, for example, is a very popular horse dewormer, but it cannot be administered to mares during the first three months of gestation. On the other hand, products containing pyrantel pamoate or fenbendazole are safe for use in pregnant mares. Unless the mare is in her last three months of pregnancy, organophosphates (e.g., dichlorvos and trichlorfon) can be incorporated into the botfly deworming program. Organophosphates are very effective against strongles, pinworms, ascarids, and bots but are not recommended for use during late pregnancy since they interfere with nervous control of smooth muscle, a condition that may stimulate uterine contractions and trigger abortion. Most deworming products should **never** be administered during the last month of pregnancy. Information on specific deworming agents is included in the Appendix.

VACCINATIONS

Although vaccination programs are usually planned and carried out by the farm veterinarian, there are a few points that should be noted with respect to the mare during pregnancy:

1. In some areas, abortion caused by the equine rhinopneumonitis virus is a significant problem. (Refer to "Abortion.") To prevent an abortion epidemic, the veterinarian may either build up the mare's immunity through planned infection during the early stages of gestation or may utilize a rhinopneumonitis vaccine. The "planned infection" method is the least preferred vaccination method since there are many risks associated with its use. A modified live virus vaccine (Norden - Rhinomune) has been used extensively, but proof that the vaccine prevents abortion has not been established. A new vaccine that uses a killed (or inactivated) virus (Fort Dodge - Pneumabort-k) is currently available, and research suggests that this type of vaccine will prevent abortion. In pregnant mares, the recommended vaccination schedule is the third, fifth, and seventh months of gestation. This vaccine offers a distinct advantage in that it can be given at any stage of gestation without risk to the fetus. The modified live virus or live virus (planned infection) vaccines cannot be given during very early or late gestation without endangering the pregnancy.

2. The mare is usually given a tetanus toxoid booster one month before her expected foaling date to insure that plenty of tetanus antibodies are present in her colostrum (i.e., to insure that the foal receives passive immunity to the tetanus organism immediately after birth).

Conditions Associated with Pregnancy

Condition	Comments
Pregnancy Edema	Pregnancy edema is the retention of fluid that causes swelling in the limbs and lower abdomen. This condition is usually caused by poor circulation (e.g., lack of exercise) during pregnancy, but severe abdominal edema during the last few months of pregnancy may be the result of ruptured prepubic tendons. Mild edema can be relieved by increasing the mare's physical activity.
Limb Stiffness	Limb stiffness is another condition associated with inactivity during pregnancy. Stiffness is relieved by increasing physical activity.
Ruptured Prepubic Tendons	The prepubic tendons attach abdominal muscles to the pubic bone and help to keep the pelvis in position. In rare instances, pressure from edema and increased weight of the conceptus causes these tissues to rupture. The resulting condition is characterized by tense, painful edema on the abdominal floor. (Ruptured prepubic tendons are seen most frequently in idle, overweight draft mares and may also be caused by injury to the prepubic area.) This condition requires immediate veterinary attention.
Signs of Heat	Signs of estrus are not unusual during the first few months of pregnancy. Occasionally, pregnant mares cycle throughout gestation or, more commonly, between days 90 and 150 of pregnancy when estrogen levels are high.
Displacement of the Large Colon Beneath the Uterus	This is an unusual condition (characterized by signs of colic) that involves the trapping of the large intestine beneath the enlarged, distended, gravid uterus. This condition must be diagnosed and treated by a veterinarian.

Torsion of the Uterus

Occasionally, the gravid uterus rotates along its long axis, resulting in severe colic. This condition is seen most often after the fifth month of pregnancy. Torsion of the uterus is usually discovered by the veterinarian during diagnosis of colic.

Pseudopregnancy

Occasionally, an open mare's reproductive tract imitates a state of pregnancy. The reason for false pregnancy is not clear, however. ♞

15

ABORTION

The term *abortion* refers to the expulsion of the embryo or fetus between day 30 and 300 of pregnancy and is sometimes used synonymously with the terms *fetal death* and *absorption*. Prior to day 30, loss of the embryo is termed early embryonic death. Because the embryo is small and boneless during this period, it is usually absorbed by the mare's uterus rather than expelled through her vulva. In most cases, absorption occurs very early, and the mare may return to heat without any indication of ever having been pregnant. When embryonic loss occurs before day 37 of pregnancy, the mare usually returns to heat and may be rebred if her reproductive tract passes a soundness examination. A mare that aborts after day 37, however, may not return to estrus until day 120 or 140 post-conception due to continued production of PMSG and the resulting persistent corpora lutea. (If fetal loss is detected during this period of PMSG production, some mares may be induced to return to estrus through hormone therapy.)

Expulsion of the fetus between days 150 and 300 is considered a late abortion. If the foal is delivered after a gestation of more than 300 days, it is designated as stillborn, premature, or full term, depending on the circumstances of the delivery. Stillbirth, for example, is the expulsion of a dead fetus that has reached what would normally be considered a viable stage of development. Foals that are born after 301 days but before 325 days of gestation are considered premature, while those that are born after a gestation of more than 325 days are considered full term. A late abortion is generally more complicated and harder on the mare than an early abortion or a full term delivery because the mare's pelvic muscles and cervix lack the relaxation needed to deliver a large fetus. For this reason, the chances of damage to the mare's reproductive tract are considerable, and it may be several months before the mare's breeding soundness is restored after a late abortion.

Depending on the cause of an abortion, the mare may or may not give the breeder any warning signs. Usually, the mare exhibits no premonitory signs before an early abortion. In fact, the breeder may not even be aware that the mare has aborted. The mare may exhibit premature udder development and may drip milk prior to a late abortion, especially when twins are aborted. Abortions caused by uterine infections are usually preceded by a vaginal discharge. Many mares in late pregnancy exhibit signs similar to "prefoaling colic," but this behavior should not be confused with signs of impending abortion.

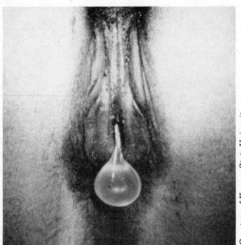

Courtesy of Kansas State University Department of Surgery and Medicine

Expulsion of the fluid-filled amnionic sac may be noted in the initial stages of abortion.

Causes of Abortion

The diagnosis of an abortion is based on an examination of the aborted fetus and its surrounding membranes, cultures from the mare, and the mare's reproductive history. In addition, bacteriological and microscopic examinations of tissues taken from the abortus (aborted fetus and fetal membranes) are performed to rule out possible causes. To aid in these diagnostic procedures, the breeder should note or try to remember any indications the mare may have given prior to the abortion. Until proven otherwise, all abortions should be considered infectious and handled accordingly. The direct causes of most abortions are 1) a fetus that is dead, 2) a fetus that is hopelessly diseased, and 3) a uterus that is incapable of supporting fetal life. These causes, in turn, may be due to a variety of factors, including injury, nutritional imbalances, infections, or fetal abnormalities. In some instances, the cause of an abortion cannot be determined due to the condition of the fetus or to the absence of any detectable fetal or maternal abnormalities.

VIRAL INFECTIONS

The term *virus* is used to describe a parasitic organism that is less than 300 millimicrons (mμ) in size and contains only one type of genetic building block, or nucleic acid. Viruses are described as parasitic because they must penetrate another living cell in order to reproduce. Hundreds of viral particles are produced by the invasion of a single cell. Viral infections cause many different types of changes within the host cells, and quite frequently these structural changes result in cellular death. Viruses can also cause rapid cell proliferation as in the formation of a tumor, but of concern at this point is the ability of two viral strains, equine herpesvirus I and equine arteritis virus, to induce abortion in mares.

EQUINE VIRAL RHINOPNEUMONITIS (EVR)

Rhinopneumonitis, sometimes referred to as equine viral abortion, is a highly contagious disease caused by equine herpesvirus type I (EHV I). The virus causes a mild upper respiratory infection in young horses and may also cause abortion in pregnant mares. "Abortion storms" involving up to 90 percent of the pregnant mares on a single farm have occurred following outbreaks of EVR. Affected mares usually show no signs of infection but do exhibit positive serological reactions (blood tests). Although equine viral abortion usually occurs after the eighth month of pregnancy, exposed mares have been known to abort as early as the fifth month of pregnancy. Many sources suggest that the lesions found on an EHV I-infected fetus are part of the fetal response to the presence of the virus and not merely a direct effect of the virus.

Rhinopneumonitis is spread by the ingestion or inhalation of virus-contaminated material, and outbreaks of this disease are likely to occur in areas where horses congregate. For example, foals often contract EVR in the fall when they are weaned and grouped in close quarters. At this time, susceptible broodmares may catch this highly contagious disease from the young horses. Although the mare usually shows no outward signs of infection, the virus may invade the placental membranes and attack her fetus. The fetus becomes heavily infected and is usually aborted in about 3-12 weeks when it is no longer able to function physiologically. Examination of the aborted fetus reveals the presence of excess fluid in the pleural (lung) cavity, cellular death within the liver, and evidence of hemorrhage on some of the internal organs. A slight yellowing of the mucous membranes of the eyes, nose, and mouth may also be present.

EVR-related abortions may be controlled through the immunization of horses against the equine herpesvirus (type I). Following recovery from an EHV I infection, the horse develops an immunity that lasts 3-6 months. Following an EHV I abortion, mares may remain immune to future infections for as long as two years. EVR is a common disease, and frequent exposure promotes the maintenance of a satisfactory level of immunity in most horses. However, the best way to promote immunity and prevent "abortion storms," is to utilize a rhinopneumonitis vaccination program designed by the farm veterinarian.

EQUINE VIRAL ARTERITIS (EVA)

Equine viral arteritis is an infectious disease that affects horses of all ages. This disease is usually introduced through the arrival of an ill or convalescent horse and is spread by direct contact, the inhalation of virus laden droplets, or by the ingestion of contaminated material. An EVA infected horse is usually characterized by inflammation of the mucous membranes of the eyes, teary eyes, a nasal discharge, and a fever of between 103 and 106°F. (Refer to the discussion on EVA within the chapter "Suckling Foals.") Infected mares may or may not abort, but when abortion does occur, it usually takes place during the febrile or early convalescent period of the disease (1-14 days after the onset of the first signs). Mares that are exposed to the virus between the fifth and tenth months of pregnancy are more susceptible to abortion. It is important to remember, however, that mares may abort without showing any premonitory signs. Examination of the aborted fetus reveals small, round hemorrhagic areas on the heart and on the surfaces of the pleural and abdominal cavities. A vaccine against the EVA organism is available, but outbreaks of this disease are rare in most countries, including the United States.

BACTERIAL INFECTIONS

Bacteria are one-celled plants that represent one of the most primitive forms of life known to man. These organisms can be classified within the plant kingdom according to size, shape, and physiological characteristics. The scientific name for each organism is composed of two Latin words, not necessarily descriptive terms. The first name identifies the organism's genus and always begins with a capital letter. (The genus of the horse, for example, is Equus.) The second name designates the organism's species and is not capitalized. Several disease-causing bacterial species have been described in the chapters on infertility. Many of these same species affect the viability of the fetus and are capable of inducing abortion in mares.

STREPTOCOCCUS ZOOEPIDEMICUS

Streptococcus zooepidemicus is a common cause of infertility and abortion in mares. As explained in "Mare Infertility," this particular streptococcus organism is often found on the genitals of healthy mares. Normally, the mare is resistant to streptococcal infections, but if her defense mechanisms are rendered less effective by factors such as stress, pneumovagina, or urine pooling, a streptococcal infection may become established within her reproductive tract. If the mare conceives when a streptococcal infection is present in the uterus, the placenta also becomes infected, a condition referred to as placentitis. Once the placenta is infected, a chronic or acute infection of the fetus usually follows.

Chronic infections of the fetus occur when *Streptococcus zooepidemicus* enters the uterus via the vagina and cervix. The infected portion of the placenta is either thickened, leathery, and grey/yellow in appearance or is covered with a reddish brown exudate. The difference between healthy and diseased placental tissue is easily noted upon examination, but it may be difficult to determine whether the abortion was caused by a streptococcal infection or by a mycotic (fungal) infection without special laboratory tests. In both instances, the fetus is starved gradually as a result of impaired uterine function and appears small, emaciated, and very pale. Abortion due to a chronic *Streptococcus zooepidemicus* infection may occur at any stage of pregnancy but is most common between the third and sixth months.

A rare form of *Streptococcus zooepidemicus* infection causes acute fetal disease and subsequent abortion. In these cases, the organism's path of entry to the placenta is through the mare's bloodstream. The chorion of the aborted fetus is usually dotted with small hemorrhagic areas, and the fetus may be characterized by slight tissue degeneration or by extensive softening and degeneration of tissue. Abortion caused by an acute *Streptococcus zooepidemicus* infection is most common between the third and sixth months of pregnancy.

Although many factors are involved in determining a mare's level of susceptibility to infection, unsanitary breeding and foaling practices increase the mare's chances of acquiring a streptococcal infection. The utilization of sound techniques for breeding, foaling, and reproductive examinations is important to both the prevention and eradication of streptococcal infections. Mares seldom recover from a *Streptococcus zooepidemicus* infection without assistance. Treatment usually involves uterine flushing and antibiotic therapy.

ESCHERICHIA COLI

The organism *Escherichia coli* is normally found in the intestinal tract and feces of all livestock, including the horse. If this organism is introduced into the mare's reproductive tract (either before or after conception), placentitis and acute fetal disease may result in abortion. Although abortions due to the presence of *E. coli* may occur at any stage, they are most common during the second half of pregnancy. The chorion of the aborted fetus has a reddish brown discolored appearance, and the fetus shows various signs of autolysis (i.e., self destruction of cells by their own enzymes). Fatal septic metritis may follow an *E. coli* abortion in the mare.

LEPTOSPIRA

The spiral-shaped bacteria of the genus Leptospira cause infectious disease in both man and animals. Over fifty different serotypes have been identified, but each host species tends to be affected by different serotypes. The horse, for example, is susceptible to *L. pomona, L. canicola, L. icterohemorrhagia, L. grippotyphosa,* and *L. sejroe. Leptospira pomona* is the most common cause of leptospirosis in horses and is believed to be an occasional cause of abortion in mares. Affected mares may be asymptomatic (without clinical signs) or may display a slightly elevated temperature (102-103°F), depressed appetite, a slight yellowing of the mucous membranes, and 3-4 days of general depression. The organism may be spread to susceptible horses through contact with the nasal secretions or urine of infected or carrier animals. (Cattle are often responsible for spreading the disease to horses.) Under the proper conditions (e.g., wet, marshy soil), leptospira organisms may survive in the soil for up to 100 days. If exposed to the disease between the seventh and eleventh months of pregnancy, the mare may abort one to three weeks later. The aborted fetus is often characterized by jaundice and tissue degeneration.

Treatment of *L. pomona* infections involves the administration of tetracycline or penicillin and streptomycin for 4-5 days to prevent abortion and recurring attacks of temporary but eye-damaging blindness (periodic ophthalmia) that may result in permanent loss of sight. The infected horse should be isolated to prevent the spread of disease to healthy animals. Following recovery, mares can shed the organism in their urine for 70-120 days postinfection. Horses may be vaccinated against *L. pomona*, but the vaccine does not protect the horse against leptospirosis caused by other serotypes.

BRUCELLA ABORTUS

Brucella abortus causes poll evil and fistulous withers in horses and infectious abortion (Bang's disease) in cattle and is sometimes responsible for abortion in mares. Affected horses develop pus-forming inflammation in the bursae at the poll or withers. Mares most often contract this condition when pastured in a contaminated area (e.g., with infected cattle). Brucella abortus may be ingested in feed and water contaminated by the secretions or excretions of infected animals. Because the organism lives for approximately 70 days in the soil and for about 35 days in water, horses should not be pastured in areas that have been used by infected cattle until at least three months after the cattle have been removed.

A vaccine is not available for protection against *Brucella abortus*, and treatment is often unsatisfactory. Drainage of the bursa accompanied by the administration of antibiotics is the treatment of choice in most cases.

SALMONELLA ABORTUS-EQUI

Salmonella abortus-equi, introduced to the United States in 1886, was once a dreaded cause of abortion in mares. Today, the disease is controlled through the use of vaccines and other protective measures. As with many other bacterial infections, salmonellosis is contracted through the ingestion of contaminated material. Affected mares may be asymptomatic or may have a fever as high as 104°F accompanied by colic, diarrhea, loss of appetite, and a purulent vaginal discharge.

Abortion is most common between the sixth and ninth months of pregnancy and is often preceded by mild colic. Injuries and lacerations during abortion are common due to the size of the fetus and the lack of relaxation of the pelvic muscles. The abdominal and chest cavities of the aborted fetus contain excessive amounts of cloudy fluid, and the fetal digestive tract is often characterized by hemorrhagic inflammation. The chorion is swollen with fluids and covered with areas of dead, dirty-grey tissue. *S. abortus-equi* infected mares may deliver live foals, but these foals are usually weak and often die within a few minutes or weeks after birth. Following abortion, the mare usually sheds the Salmonella organism for about 30 days. In most cases, the uterus returns to normal, and the mare may be rebred on her second postabortion estrus. Prior to breeding, the mare should be cultured to minimize the chances that she will contaminate the stallion.

A vaccine is available for the control of *S. abortus-equi*, but it is seldom used due to the scarcity of this organism in most areas. Vaccination against *S. abortus-equi* involves the administration of a three-dose series in the fall of each year.

KLEBSIELLA PNEUMONIAE

Klebsiella pneumoniae is one of the most dangerous organisms that cause infection and abortion in the mare. Infected mares are characterized by inflammation of the cervix and uterus accompanied by a purulent discharge from the vagina. Klebsiella infections interfere with the mare's estrous cycle and may result in sterility. This bacteria is readily transmitted through the act of breeding and the use of contaminated examination and insemination instruments. The conception rates for infected mares is very low, but if an infected mare does conceive, her uterus is usually unable to maintain the pregnancy to term.

Klebsiella infections are very persistent and very difficult to eradicate. If an infection is discovered and treated early, neomycin may be an effective therapeutic agent. If the infection is well-established, however, the eradication of Klebsiella is often followed by other infections due to reduced uterine resistance to other organisms.

PSEUDOMONAS AERUGINOSA

Pseudomonas aeruginosa, normally a harmless skin bacteria, is the cause of deep tissue infections in both man and animals. First isolated from an aborted equine fetus in 1949, this bacteria is sometimes found in the genital tract of both horses and cattle and may be spread through the act of breeding. The presence of the organism in the uterus causes placentitis, fetal malnutrition, and subsequent fetal death and abortion. Examination of the abortus reveals a deep red discoloration of the placenta's cervical star, discoloration of the allantois, and a grey necrotic chorion. Abortion due to a *Pseudomonas aeruginosa* infection is sometimes followed by fatal septic metritis.

Although experimental vaccines have proven successful, *Pseudomonas aeruginosa* infections do not occur often enough to warrant immunizing procedures. Treatment of a *Ps. aeruginosa* infection involves the administration of gentamicin or chloramphenicol.

HAEMOPHILUS EQUIGENITALIS (CONTAGIOUS EQUINE METRITIS)

Haemophilus equigenitalis is responsible for acute inflammation of reproductive tissues and subsequent infertility in mares with contagious equine metritis (CEM). Although CEM-infected mares seldom conceive, some contaminated mares that show few clinical signs conceive and then abort within 60 days of conception. (Refer to the discussion on CEM within "Mare Infertility.")

MYCOTIC (FUNGAL) INFECTIONS

Allescheria boydii, Aspergillus, and *Mucor sp.* are the organisms most commonly responsible for mycotic placentitis and abortion in the mare. Because the mare's uterus is very resistant to fungal infection (except after long-term intrauterine antibiotic therapy), mycotic abortions are believed to be the result of fungal entry through the mare's gut or lungs. Once the organism enters the dam's bloodstream, it can move to the placenta. The resulting infections do not seem to interfere with conception, and the fetus is rarely attacked. However, fungal infections do cause inflammation of the placental villi, resulting in the separation of the placenta from the uterine wall. In most cases, the fetus eventually dies from malnutrition and a build-up of toxic metabolites and is aborted during the later stages of pregnancy. Examination of the aborted fetal membranes reveals large, thickened areas in the placenta. The outermost surface of the placenta is covered with a brownish exudate, and the amnion is covered with irregular areas of necrosis and may adhere to the fetal skin. The aborted fetus is small and emaciated and, in rare cases, small greyish white nodules are present in the fetal lungs. Following abortion, a heavily fungus-laden discharge is expelled from the vagina for several days. Most mares completely shed the fungi within a few months and may be rebred (i.e., following precautionary cultures). A persistent infection may be treated by irrigating the uterus with Betadine or nystatin. Following recovery, the mare's breeding soundness is usually unimpaired.

TWINNING

Twinning is probably the single most common cause of abortion in mares. Although 18 percent of all conceptions are estimated to be twin conceptions, only 0.5 to 1.5 percent of all births involve twin foals. Usually, one or both of the twin embryos die, and even if the twin fetuses survive through the fourth month without being aborted or absorbed, approximately 90 percent of them are aborted during the later stages of pregnancy. Because available space and nourishment are limited in the equine uterus, the twin fetuses outgrow their blood and nutrient supply towards the last trimester of pregnancy. In the struggle for nourishment, one fetus generally develops more rapidly and acquires most of the available space. The smaller fetus usually dies as a result of malnutrition, and its abortion often brings about the expulsion of the other fetus from the uterus. In some cases, however, the dead fetus undergoes mummification and is not expelled until the birth of its twin.

Twin abortion is often preceded by premature udder development. The mare may drip or run milk for several days or weeks before the abortion. In some instances, these signs regress and then recur in a few weeks accompanied by abortion. The first incidence of lactation signals the death of the first fetus, while the second incidence accompanies the death and abortion of the second fetus.

The diagnosis of twin abortions is based on the presence of two aborted fetuses. One of the fetuses is usually smaller and less developed than the other, reflecting the lack of space and nourishment that caused its death. This smaller fetus is usually emaciated and edematous and may be in an advanced stage of decomposition or mummification. If one or both fetuses are missing, twin abortion may be detected by the presence of two placentas. The areas of contact between these placentas is devoid of placental villi.

BREEDING THE PREGNANT MARE

Most mares stop cycling once they have conceived, but some individuals cycle once or twice during pregnancy, and others may even cycle regularly throughout gestation. Even though a mare expresses signs of estrus when she is pregnant, she usually refuses the stallion's services. If the pregnant mare is hobbled and force-bred during apparent estrus, the cervical seal that maintains uterine stability may be disrupted. As a result, abortion may occur in a few days. To guard against this possibility, any mare that returns to heat after she is bred or declared "safe in foal," should be palpated by a veterinarian or an experienced technician to determine her reproductive status prior to rebreeding.

PREGNANCY IN THE UTERINE BODY

Because the available space within the uterine body is inadequate for normal fetal growth and development, implantation of the fetus in the body of the uterus rather than in one of the uterine horns usually results in abortion. The aborted fetus displays the classical signs of placental deficiency, including small size and emaciation. The body of the placenta appears abnormally large, while the placental horns are small and atrophied.

NUTRITIONAL FACTORS

MALNUTRITION

Although little research has been done on the relationship between nutritional factors and equine abortions, studies have shown that poor nutrition during the early stages of pregnancy may cause early embryonic death. (i.e., A high incidence of early embryonic death in association with malnutrition has been reported between days 25 and 31 of pregnancy.) It is doubtful that nutritional deficiencies cause abortions in mares after day 35 of pregnancy except under the most extreme conditions. Up to a point, the mare gives her fetus first priority nutritionally. If nutritional deficiencies are extreme and if starvation reaches a dangerous level, this situation is reversed, and the fetus is deprived of nutrients in order that the mare may save her own life. If this point is reached, the fetus dies and is aborted. Although abortions due to malnutrition are very rare except in cases of extreme neglect, abortions due to toxic compounds within a particular feedstuff are notable.

FESCUE TOXICITY

Fescue is a hardy, fairly nutritious grass that is popular as pasture forage in many parts of the United States. However, fescue has been associated with equine abortion, abnormal foaling, and failure to produce milk. Fescue toxicity is caused by an alkaloid present in the grass during the lush stages of growth. For this reason, pregnant mares should not be allowed to graze on fescue pastures during the spring and early summer. Although fescue toxicity has been reported in horses, it is seldom a problem on pastures that contain only a small percentage of fescue. Because this grass is less palatable than most pasture grasses, horses usually refuse to eat it unless nothing else is available.

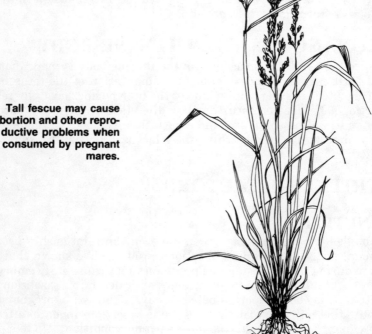

Tall fescue may cause abortion and other reproductive problems when consumed by pregnant mares.

SUDANGRASS POISONING

Sudangrass is a tall, hardy grass that is common as both a pasture grass and a hay crop in some areas. However, caution should be exercised when pasturing horses on sudangrass due to the danger of prussic acid (HCN) toxicity. Sudangrass may develop toxic levels of HCN in the new growth that develops during a period of warmth or heavy rainfall following a frost, a period of drought, or a period of heavy trampling or other physical damage. Horses grazing on toxic sudangrass may develop chronic inflammation of the bladder and a loss of muscular coordination. These effects are sometimes accompanied by abortion. Because toxicity declines in more mature plant growth, the danger of prussic acid toxicity is reduced by grazing sudangrass after it reaches the height of 18 inches (46 centimeters). The potential toxicity of sudangrass hay is as yet unclear but it seems to vary with the maturity of the cutting as well as the curing method. However, sudangrass is not a popular hay for horses and is normally only utilized for grazing.

Prussic acid toxicity may occur in broodmares grazing lush sudangrass.

ERGOT

Ergot is a parasitic fungus found on grasses and cereal grains. This toxic fungus appears as a black, banana-shaped mass (1/4 to 3/4 of an inch long) that replaces some of the grass or grain seeds. Rye, wild rye, bromegrass, and dallisgrass are most commonly affected by ergot infestation. The ingestion of large quantities of ergot at one time causes acute ergot poisoning, a condition that is characterized by paralysis of the limbs and tongue, gastrointestinal disturbances, and abortion. The ingestion of small quantities of ergot over a long period of time causes the animal's tail, ears, and hooves to slough off. This effect of cumulative ergot poisoning is often followed by delirium, spasms, paralysis, and death.

The black masses of ergot that are found in diseased plants are clearly visible in the enlarged inset view of this perennial ryegrass plant.

Ergot poisoning can be avoided by keeping the seed heads mowed off in the late summer and by checking hay and grain carefully for the presence of the seed-like fungus. It should be noted that only a small amount of ergot is necessary to produce toxic effects. Grain containing more than 0.3 percent ergot by weight is considered toxic and should not be fed. The treatment of ergot poisoning varies with the condition's severity. Mildly affected animals may recover completely when their toxic feed is replaced with good feed. Although severely affected animals are usually beyond medical assistance, tannin drenches and vasodilator drugs may be beneficial.

LOCOISM

Several varieties of locoweed are known to be poisonous to all domestic livestock species, including horses. Due to the presence of a toxic substance (miserotoxin), these plants cause a condition known as locoism. The most common form of the disease causes stiffness, nervousness, loss of directional sense, incoordination, abortion, and congenital defects. In severe cases, the animal may also be characterized by extreme loss of muscle tissue, a long hair coat, and blindness. Locoism is seen most often in the spring on desert range in the western United States, where locoweed is abundant. In some areas (i.e., areas with seleniferous soils), locoweeds may also accumulate dangerously high concentrations of selenium.

In addition to abortion, locoweed can cause disorientation and muscle degeneration.

TOXIC CHEMICALS

The relationship between chemicals and abortion has not yet been researched extensively, but certain agents have been identified as possible causes of abortion in mares. These chemicals include phenothiazine, thiabendazole, lead, and organic phosphate insecticides and dewormers. (For information on chemical dewormers that may be harmful to the fetus, refer to the Appendix.)

HORMONE IMBALANCES

Progesterone deficiencies are believed to be an important cause of habitual abortion in mares. This form of abortion usually occurs between the fourth and fifth months of gestation but may take place anywhere between 75 days and 8 months. During the first five months of pregnancy, progesterone is produced by the corpus luteum (and by accessory corpora lutea after about day 40). Progesterone production is then taken over by the placenta. If this transition from one progesterone source to another is not smoothly synchronized, progesterone levels may drop temporarily below that required for pregnancy maintenance, and abortion may result. Early abortion may be due to failure of the mare to establish an effective corpus luteum or failure to form accessory corpora lutea. At this early stage, the embryo is very small, and abortion may occur without detection or may be observed as failure to conceive.

Many veterinarians recommend progesterone injections to help maintain pregnancy in problem mares (i.e., habitual aborters). Recommended dosages vary from 100 mg given intramuscularly at 42 days of gestation to 500-1000 mg injected intramuscularly on day 42 and given at 7-10 day intervals until the tenth month of pregnancy. Studies indicate, however, that progesterone therapy (in the dosages usually recommended) is of little or no value in the prevention of abortion. In the early stages of pregnancy, injections of follicle stimulating hormone followed by injections of luteinizing hormone may stimulate the development of additional corpora lutea. Injections of 2-10 mg of stilbestrol, an estrogenic compound, at 2-4 day intervals during the fourth and fifth months of pregnancy have also proven successful in preventing abortion due to hormone imbalances.

CERVICAL INCOMPETENCE

The tubular body of muscle that separates the vagina and uterus forms a tight seal, which protects the uterus and the fetus from external contamination throughout pregnancy. Normally, the cervix remains in a contracted state (except during estrus and foaling), but an injury, the presence of scar tissue, or a hormone imbalance may prevent the establishment or maintenance of this seal. Abortions due to an injured or weakened cervix are believed to be more common than once thought. Because cervical injuries are difficult to detect without a thorough reproductive examination, they are often overlooked and left untreated. Palpation of the cervix accompanied by visual evaluation through a speculum may be necessary to locate the defect. Once the problem has been identified, surgery may be required to repair any damage. If an injury occurred during foaling because the cervix lacked sufficient relaxation to permit the passage of the foal, the cervix is very likely to retear during subsequent foalings.

UTERINE INCOMPETENCE

Inability of the uterus to protect and nourish the developing fetus is an important cause of habitual abortion in problem mares. Although uterine incompetence primarily affects older mares, it may occur in any mare that suffers chronic or repeated uterine infections. Cumulative damage to the uterus (through years of foaling, several infections, etc.) causes the formation of scar tissue. If scarring and fibrosis are extensive, the exchange of nutrients across the placenta may be impeded, and the fetus may die of malnutrition. The size of the aborted fetus and the point in gestation when abortion occurs depend on the degree of uterine damage. There is no known treatment for uterine incompetence, and little can be done to prevent this condition other than promptly treating and eradicating any reproductive infection that the mare may acquire during her reproductive career. (Mares that develop this condition might be considered as candidates for embryo transfer.)

HYPOPLASIA OF THE CHORIONIC VILLI

Hypoplasia, or underdevelopment, of the chorionic villi is another condition that impedes the exchange of nutrients between the dam and her fetus. In the normal placenta, the chorionic villi attach the placenta to the uterus and provide a mechanism for nutrient, gas, and waste exchange between the dam and the fetus. Examination of the aborted placenta affected by hypoplasia reveals a chorion with a thin, sparsely villated membrane and wider than normal separation between the villar stalks. Many of the villi are stunted and show little or no branching. (Refer to the discussion on placentation within the chapter "Fetal Development.") The point at which the placental exchange mechanism can no longer support the fetus depends on the condition's severity. (As fetal size increases, the nutritional shortage becomes more critical.) Like uterine incompetence, hypoplasia of the chorionic villi is seen primarily in older mares but may affect mares at any age.

TWISTED UMBILICAL CORD

Twisted or tangled umbilical cords are very unusual causes of abortion in the equine. Two or three twists in the umbilical cord are common and

The umbilical cord of this aborted fetus contains a normal number of revolutions.

Courtesy of Kansas State University Department of Surgery and Medicine

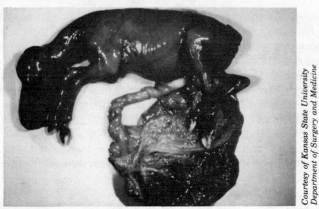

Courtesy of Kansas State University
Department of Surgery and Medicine

This aborted fetus died as a result of a twisted umbilical cord. Note the dark, congested appearance of the fetus and the umbilicus.

seem to present no threat to fetal life. However, when the cord is very edematous and has an excessive number of twists (up to twenty in some cases), fetal blood supply may be impaired or cut off, and the fetus may be aborted. The cord may also become wrapped around the fetal neck, head, or body, causing death and subsequent abortion. The cause of these unusual conditions is unknown, but fetal movement is believed to be a contributing factor.

TRAUMA

Because the fetus is surrounded and protected by a cushion of shock-absorbing fluid, trauma or injury is rarely a cause of abortion in mares. However, a severe kick, blow, or jolt that hits the fetus directly may cause abortion. The aborted fetus usually displays a distinctive bruise at the site of injury. Although any life-threatening injury to the mare, whether it concerns the fetus or not, may cause abortion at any point during pregnancy, mares have been known to sustain severe injuries during pregnancy without showing any inclination to abort. Severe stress due to prolonged shipping, overwork, fright, or a severe illness is sometimes responsible for abortions, especially during the later stages of pregnancy. Studies indicate that nervous and high-strung mares are more likely to abort under stress than are more placid mares.

INHERITED LETHALS

When a new individual is created, chromosomes from the sire and the dam pair to form new gene combinations in the offspring. These gene pairs control fetal development and determine the physical characteristics of that individual. Occasionally, an embryo receives a gene combination that is detrimental to intrauterine or extrauterine survival. The presence of these lethal genes causes early embryonic absorption, abortion, stillbirth, or death of the newborn foal, but unless the lethal trait occurs with enough frequency that a pattern of inheritance can be detected, the reason for embryonic or fetal death is very difficult to determine. Although very little is known about the inheritance patterns and physical actions of lethal genes, the following example is included to give the reader a general idea of the effect a lethal gene can have on a breeding program.

Dominant white is the term used to describe the completely white (not cream) coat color that occurs with pink skin and colored eyes (e.g., blue, brown, hazel, or amber). The gene combination that causes this color has been designated as Ww. Unlike many other species, complete absence of pigment in the skin, hair, and eyes (true albinism) does not exist in the horse. Authorities believe that the dominant white gene is capable of producing albinism when the gene combination WW is inherited, but because this gene combination also has a lethal effect on the fetus, albinism is never seen in the horse. Resorption of the lethal white fetus occurs very early in gestation, and the loss appears as a failure of the mare to conceive. Theoretically, this conception loss can affect one-fourth of all conceptions when dominant white (Ww) horses are bred:

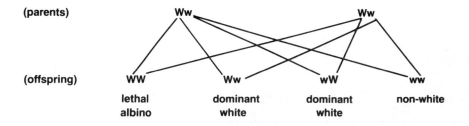

Embryonic death and abortion in humans have been associated with a variety of chromosome defects. Chromosome abnormalities may also cause severe fetal anomalies that result in early embryonic death or abortion in the mare. For more information on chromosomes, chromosome aberrations, and mutations, refer to the text EQUINE GENETICS & SELECTION PROCEDURES.

Induced Abortion

If an unplanned breeding results in an unwanted pregnancy or if palpation reveals that the mare is carrying twins, the breeder may decide to have a veterinarian induce abortion. Before this procedure is carried out, the mare's stage of pregnancy must be considered. An early abortion can be performed with less threat of damage to the mare's reproductive tract and is usually followed by a more rapid recovery than a late abortion. As a general rule, abortions should not be induced after the sixth month of pregnancy. After this point, the large size of the fetus greatly increases the chances of tearing or otherwise damaging the mare's reproductive tract.

Reports indicate that it may be difficult to get the mare in foal for 6-8 months following an induced abortion, but after this period, fertility does not seem to be affected.

There are two basic methods of inducing abortion in mares: hormone stimulus and dilation of the cervix. Large, repeated doses of estrogen can be given to induce abortion, but this treatment may also cause the formation of cysts on the ovaries. Studies have shown that intramuscular injections of prostaglandin $F_{2\alpha}$ administered at 12 hour intervals will induce abortion at any stage of pregnancy. Researchers speculate that this agent causes abortion by stimulating uterine contractions. The second abortion method, manual dilation of the cervix, should always be followed by a sterile saline/antibiotic flush to help prevent infection. Due to the risks involved in forcing the mare to abort, these procedures should be carried out by a veterinarian. An improperly induced abortion may impair the mare's fertility permanently. After the mare has aborted, the attendant should follow the veterinarian's mare care instructions very carefully. Proper care coupled with several weeks or months of sexual rest may be necessary to restore the mare's reproductive health.

Postabortion Care

Although the mare may recover spontaneously from some forms of abortion, special postabortion treatment is often necessary to restore the mare's reproductive soundness. Depending on the cause of abortion, postabortion care may range from a short period of sexual rest and observation to several months of sexual rest and intensive antibiotic therapy. Postabortion care is especially important and complex if the mare suffered a late abortion or an abortion due to uterine infection. For example, if the abortion occurred during the later stages of pregnancy, surgery may be required to repair tears and lacerations within the reproductive tract. In addition, late abortions are often associated with retention of placental tissues. If left untreated, retained afterbirth constitutes a serious and often fatal condition. (Refer to "Foaling.") If the abortion was caused by uterine infection or if the cause of abortion is unknown, the following precautionary steps should be taken immediately after the abortion:

1. The fetus and all fetal membranes should be sent to a diagnostic laboratory.
2. Any surface that contacted the aborted material should be disinfected and limed.
3. The mare should be isolated, and her stall should be thoroughly disinfected and kept vacant for at least three weeks.
4. If the abortion occurred in pasture, all mares in that pasture should be isolated and observed closely until they foal.
5. All isolated horses should be examined carefully before being allowed to rejoin a breeding herd.

Prevention of Abortion

It has been estimated that approximately 50 percent of all abortions could have been prevented through sound breeding management practices and the use of sanitary examination, breeding, and foaling techniques. A conscious effort to prevent abortion should include the following measures:

1. Thorough prebreeding examinations should be carried out to insure that only healthy mares with healthy reproductive tracts are bred.
2. Only "clean" (i.e., contamination free) stallions should be used in a live cover situation.
3. Strict sanitation should be observed during examination, breeding, and foaling procedures.
4. If an abortion-causing disease (e.g., rhinopneumonitis or viral arteritis) presents a problem in the area, a special vaccination program should be utilized.
5. Pregnant mares should be fed a well-balanced diet, one that corresponds with the needs of both the fetus and the mare during various stages of gestation.
6. Care should be taken to avoid the possibility of mares consuming toxic or abortion-causing substances. For example, the use of insecticides and herbicides on pastures should be carefully controlled, and the use of abortion-causing dewormers should be avoided.
7. Pregnant mares should receive regular exercise but should not be fatigued.
8. Any action that causes stress or excites the mare should be avoided when possible. This is especially important in nervous mares and during the last few months of gestation.
9. If a mare has a history of twin ovulations, her ovaries should be palpated prior to breeding to determine whether two mature Graafian follicles are present. If there is a chance of twin conceptions the breeder may decide to postpone breeding until the next estrous cycle. ♞

16

FOALING

Foaling is the visible culmination of approximately a year of effort invested in breeding and caring for the mare. Fortunately, the birth of a foal is not a complicated process, and foaling difficulties are considered unusual. In fact, greater than 90 percent of all equine births proceed

normally without assistance from an attendant. However, when problems do occur, they can threaten the life of the mare and her foal very quickly.

Because there is considerable variation in the circumstances in which mares may foal, perhaps the most important aspect of prefoaling preparation is education. Knowledge of the normal birth process is essential in detecting abnormalities and for assisting the mare, if necessary. Therefore, this chapter should serve as a guide for preparing for foaling and for charting the mare's progress during foaling. (Even when a mare foals unattended, there are certain procedures that must be carried out after foaling of which the breeder should be aware.) Although this chapter can also be used as a quick reference at foaling time, it should be reviewed well in advance of the first anticipated foaling date in order to prepare the breeding farm's facilities, equipment, and personnel for the foaling season.

Although the attendant must always be prepared, foaling usually proceeds successfully without assistance. This mare foaled in a clean, grassy paddock during the early morning hours and presented a normal, healthy foal without the aid of an attendant.

Facilities

The selection and preparation of a foaling facility deserve special consideration. The farm's size and the number of mares expected to foal on the farm are only two of many factors that determine the type of foaling area that is suitable for a particular operation. Each of the three basic foaling facilities (pasture, paddocks, and stalls) examined in this discussion has distinct advantages, but regardless of where the mare foals, three important points should be noted.

1. Hygiene is of utmost importance, since both the mare and foal are very susceptible to infection after foaling.

2. The area should be free of physical hazards that might endanger the mare's safety or injure her newborn foal.

3. Ideally, the mare should reside in the same area (i.e., on the same farm) for at least two months prior to foaling. This period of exposure allows her to build antibodies against organisms in that area and concentrate those antibodies in her colostrum. As a result, colostrum supplies her foal with protection against many of the disease-causing organisms that may be encountered in that area.

PASTURES

Although most breeding farms have special indoor foaling areas that allow close supervision of the birth process, many breeders allow their mares to foal in pastures. A clean, grassy pasture is an excellent place for most mares to foal, since over 90 percent of all foalings occur without problems. When managed properly, a pasture minimizes the possibility of infection and provides the mare with privacy that she might not have in a stall or paddock. A large pasture is also the most natural place for a mare to foal, but the breeder should note several important characteristics that may affect the success of a pasture foaling program:

1. A clean, grassy pasture provides maximum protection against infection during and immediately after foaling. However, a pasture that has not been managed properly (e.g., one that is overgrazed, unharrowed, or swampy) may harbor significant sources of contamination for both the mare and foal.

2. The pasture should be checked carefully for safety hazards. Water-filled ditches or small puddles, for example, may seem relatively unimportant, but it only takes a few inches of water to drown a newborn foal.

3. Inadequate fencing may permit the mare to deliver her foal into a neighboring pasture.

4. Other horses in the pasture may disturb or injure the foaling mare or her newborn foal.

5. The main drawback to pasture foaling is that birth often occurs unobserved. Morning tours of the pasture allow workers to check on postfoaling mares and to treat newborn foals, but if a mare has suffered foaling difficulties during the night, even a short delay in assisting her may have serious consequences.

PADDOCKS

As an alternative to pasture foaling, foaling paddocks provide some of the advantages of stalls. The use of paddocks allows each mare to be isolated during foaling, thereby protecting the foaling mare from inquisitive horses and preventing the accidental exchange of foals. By far, the most significant advantage of individual foaling areas is that the term mare can be checked periodically by an attendant in case she requires assistance. Even when foaling stalls are available, paddocks can be used for observing pregnant mares during the last two weeks before foaling. Each mare can then be placed in a stall immediately before she begins to foal. In this way, a few well-equipped foaling stalls can be used efficiently for a large group of mares, and the foaling stalls can be kept clean until foaling.

As with all foaling facilities, paddocks must be kept clean and free of safety hazards. Ideally, the paddock should be a clean, grassy lot surrounded by a small-weave mesh fence and located away from heavy farm traffic or unusually loud activity. Flood lights may prove to be very helpful in the event that foaling difficulties do occur and the mare must be assisted. Water and feed containers should be designed and located with the safety of both the mare and her newborn foal in mind. The mare's compatibility with neighboring horses is also an important consideration.

STALLS

Because many of the foaling management procedures presented throughout this chapter are practical only when the mare is confined, references to foaling stalls are made frequently. Foaling stalls provide maximum control over the mare's environment and allow close supervision during foaling. Although most mares foal without difficulty, many breeders prefer to keep a close watch over the term mare to insure that help is available, if needed. For example, the mare may be removed from the pasture and allowed to spend her nights in a roomy foaling stall two weeks prior to her projected foaling date. In some areas, foaling stalls also provide early foals (e.g., those born in January, February, or March) with protection from extreme cold. These stalls are frequently equipped with special monitoring devices that alert the attendant to an impending foaling or enable the attendant to check the mare periodically throughout the night. (Refer to 'Foaling Detection.') Even if foaling is completed normally, the attendant can perform routine foal and mare care procedures soon after birth, thereby insuring that both mare and foal have every possible advantage during the critical period immediately following foaling.

As with pasture and paddock facilities, the usefulness of foaling stalls is directly related to their design and management. The following guidelines should be considered carefully with respect to the use of foaling stalls.

1. The size of the foaling stall should be at least 16 x 16 feet to provide adequate room for delivery and for assisting the mare, if necessary. In addition, the mare may be very nervous prior to foaling and should have plenty of room to move about.

2. The lower portion of the foaling stall walls should be solid. If partition boards are spaced more than a few inches apart, the mare's leg or the foal's head or body may become caught in the space.

3. The stall should be checked carefully for splinters, loose nails, and other safety hazards.

4. Feed and water containers should be positioned in such a way that they cannot interfere with foaling or injure the newborn foal as it struggles to its feet and learns to walk.

5. Tartan and rubber floors are popular on some farms because they can be thoroughly disinfected. However, these surfaces become slippery when damp and some mares are reluctant to lie down on them. Clay floors are used by many breeders and are acceptable in foaling stalls. This type of flooring is difficult to disinfect, however, and requires frequent maintenance.

Photo by John Noye

Courtesy of The Thoroughbred Record

A well-designed foaling stall should be roomy. It should also have adequate lighting and flooring which can be thoroughly cleaned, such as this Tartan surface.

6. Since the mare is confined in a relatively small area, hygiene is of utmost importance. Before a new mare is introduced to a foaling stall, the floor and walls should be cleaned thoroughly. After the bedding is removed, the walls and floor (unless earthen) should be scrubbed with hot, soapy water. A wire brush can be used to scrub walls and mangers. After the stall has dried, it should be disinfected with povidone-iodine (a disinfectant iodine compound), pine oil, or some other product recommended by the farm veterinarian. Floors that cannot be scrubbed (e.g., clay or dirt) should be sprinkled with lime, or if heavily soiled, they should first be replaced and then sprinkled with lime. After the stall has been thoroughly disinfected, it should be left empty for several days before clean bedding is spread.

7. Opinions on the type of bedding that is most suitable for a foaling stall are varied. Generally, wheat straw is considered the best bedding material. Rye straw is usually avoided since it is susceptible to ergot (a fungus that may cause abortion when eaten). Wood shavings tend to adhere to the wet, newborn foal and may irritate the foal's eyes and mucous membranes or interfere with its respiration. Regardless of the type of bedding used, the material should be spread generously but not so deep that it would hinder the foal's attempts to stand or walk. Some researchers feel that bedding is a source of irritation and contamination of the mare's reproductive tract and recommend rubber or Tartan floors without bedding for the foaling stall. It is important to note, however, that when bedding is not used, the floor must be cleaned everytime the mare urinates or defecates to prevent her from lying in soiled areas.

Foaling Detection

To make sure that an attendant is present during foaling, the breeder should design a foaling detection program that fits the needs of that particular operation. If, for example, the breeder has only one or two pregnant mares each season, it might be practical to simply use an alarm clock to check on the term mare periodically throughout the night. On the other hand, if the breeder has several broodmares, this night vigil could become very exhausting. To reduce the need for periodic checks, the term mare can be fitted with a special electronic monitoring device that alerts the attendant when the mare lies flat on her side with legs fully extended — the normal position during the second stage of labor. One such device, the Baby Buzzer™ (Locust Farm, Inc., Kirkland, Ohio) consists of a small radio transmitter attached to a girth-like strap. The transmitter sends a signal to a receiver that emits an audible signal to alert the attendant. The radio signal is transmitted 100 to 150 feet from the mare but can be transmitted even farther by a remote station or audio speaker system.

Some breeders keep watch over their foaling mares and observe parturition on closed circuit television sets. The presence of a video camera in the foaling stall allows the mare to be watched closely without sacrificing her privacy. The camera should be mounted on the stall ceiling and can be equipped with a special housing to protect it from heat, cold, dust, and moisture. This video camera runs off a 115 volt outlet or power transformer and requires that the stall be well lit. Because the camera transmits only over short distances, the monitor must be situated in a nearby area (e.g., tackroom or house).

As an alternative to remote detection devices, stalled mares can be observed directly by a foaling attendant. Since many mares foal during the night or early morning, some breeders keep an experienced attendant on duty all night to observe mares and assist foaling, if necessary. On farms where large numbers of mares foal each year, foaling barns are often equipped with special viewing areas. The area should be designed so that an attendant can monitor one or more expectant mares in stalls surrounding an observation room. The arrangement usually consists of an informal living area or lounge with a window into each foaling stall. The window may be equipped with one-way glass so that the attendant can view the mare without her being distracted by activities in the observation room.

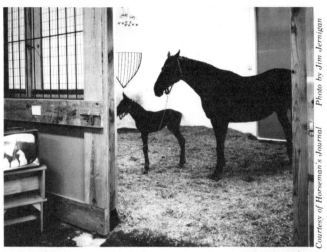

A television monitor permits the foaling mare to be observed from a distant location. Several screens placed in a central area such as a tackroom or office enable one attendant to monitor several expectant mares at the same time.

An observation area positioned next to the foaling stall allows an attendant to keep constant watch over the foaling mare.

Equipment

In addition to planning and preparing a foaling area, the breeder should study possible foaling equipment needs. A foaling kit should be prepared well in advance of the foaling season and stored near the foaling area. The contents of the kit may vary from farm to farm but the kit should be well planned and replenished after each foaling. The following list includes items that are commonly employed before, during, and immediately after foaling.

FOALING KIT

Item	Comments
tail wrap	
electric lantern	if adequate light is not available
3 buckets	one for warm soapy water, one for warm rinse water, and one for the expelled afterbirth
sterile plastic bags	to line buckets
mild, nondetergent liquid soap in a squeeze bottle	to wash mare's hindquarters
paper towels	
roll cotton	
sterile, disposable sleeves	to minimize contamination during examination or while assisting the mare
nonirritating, antiseptic solution	for attendant's hands and arms if sterile sleeves are not used
sharp, straight scissors	to open a sutured mare or to open fetal membranes in an emergency
mare's halter and lead	in case the mare must be encouraged to stand
clean, dry towels	for rubbing or drying the foal
Lugol's solution (5% iodine) or tincture iodine (7% iodine solution)	in a small wide-mouth container for saturating the foal's umbilical cord
foal enema	to prevent or correct retained meconium
oxygen	should be used only if attendant knows how to use this highly flammable gas
obstetrical chains and handles	if attendant knows how to use these tools to correct an abnormal presentation

colic or sedative mixture	to relieve colicky pains in the mare after foaling
antibiotics	if recommended by the farm veterinarian
tetanus antitoxin	for administration to the foal

Depending on the geographic area, supplemental heat might be considered for the foaling area. Although most foals can withstand cold weather, heat lamps are sometimes used to help take the chill off a foaling stall and to allow the foal to rest comfortably. (Note: Do **not** use ultraviolet lamps.) Heat lamps get very hot and should be mounted securely on the stall ceiling and covered with a special wire cage to prevent accidents. The lamp should be directed toward one side of the stall so that the foal can move out of the light if the heat becomes uncomfortable. Ventilation is far more important than temperature, and regardless of how cold the weather is, the mare and foal should receive plenty of fresh air even if fans are required to force air in and out of the barn. (A well-ventilated stall is not the same as a drafty stall, however.)

Preparation Check List

1. Estimate the mare's foaling date according to the gestation table within the Appendix.
2. Check the mare's vaccination record. If a tetanus booster is given during the month prior to foaling, the mare's colostrum provides protection to the newborn foal. (The foal is usually given tetanus antitoxin during the first few days of life to supplement the passive immunity received in colostrum.)
3. Notify the veterinarian of the estimated foaling date. Select and notify a second veterinarian in the event that the farm veterinarian cannot be reached when needed.
4. The foaling area should be ready to receive the mare at least two weeks before her estimated foaling date.
5. All equipment should be cleaned and stored in a dust-free cabinet at least two weeks before the mare's estimated foaling date.
6. If the mare's vulva has been sutured, it should be opened with a pair of sharp, sterile surgical scissors along the line of scar tissue. This procedure can be performed a few hours or minutes before foaling (up to the second stage), but it is usually done about one month before the expected foaling date in case the mare foals unattended. The sutures may be opened by a veterinarian, but the attendant should be familiar with the procedure.

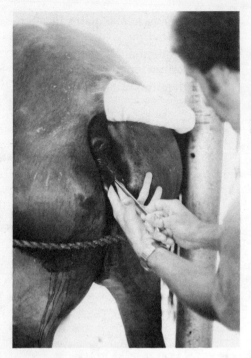

If the mare's vulva has been sutured, it should be opened with a sharp pair of sterile surgical scissors prior to foaling.

7. Any mare that has produced a foal with neonatal isoerythrolysis should be tested before foaling by a veterinarian to determine whether hemolytic antibodies are present in her colostrum.

8. The foaling attendant should be familiar with the mare's foaling characteristics (e.g., signs of foaling, time between onset of signs and breaking of water, presence of small pelvic cavity, etc.) and any history of foaling problems.

9. If possible, the mare should be introduced to her foaling paddock or stall about one or two weeks before her projected foaling date so that she is familiar with her surroundings and relaxed during foaling. During this period, a regular free-exercise program should be continued to prevent pregnancy edema.

10. Some sources recommend that the mare's feed be cut back about a week before foaling to reduce her milk flow and prevent scours in the newborn foal. Others recommend that the mare be given a bran mash to help reduce intestinal contents prior to foaling. However, the farm veterinarian should be consulted before any significant changes are made in the term mare's diet.

11. When signs of impending foaling become evident, the mare's tail should be covered with a clean sock, a sterile gauze roll, or a roll bandage to prevent contamination of the reproductive tract and to keep tail hairs away from the foal during delivery.

12. If time allows, the mare's genitals should be washed in the manner described within "Breeding Methods and Procedures." The mare's udder should also be cleaned before the foal nurses to reduce its bacterial population, and to remove built-up wax and debris. Washing also helps accustom a young mare to having her udder handled before the foal nurses. Because unusual odors may confuse the foal, unscented soap should be used.

Factors That Control Foaling

The mechanisms that initiate the birth process have not been studied thoroughly in the mare. Although the hormones involved in parturition have been identified in some other mammalian species, it is difficult to transpose this type of information and apply it to horses. Certain differences and similarities have been noted, however. Points of special interest with respect to the initiation of parturition in the mare include the release of cortisol by the fetus, changes in progesterone and estrogen levels prior to parturition, changes in prostaglandin levels during parturition, and the release of oxytocin during the second and third stages of labor.

CORTISOL

In other species, the release of the steroid hormone cortisol by the fetal adrenals initiates parturition. Although this mechanism has not been identified in the equine, there is some evidence of increased corticosteroid concentration in the newborn foal, suggesting that a similar process is responsible for the initiation of parturition in the mare. In other words, it seems likely that the fetus controls the gestational length, possibly through a response of the fetal hypothalamus and anterior pituitary. However, the mare is capable of delaying parturition to a certain extent. (Most mares foal between 7:00 PM and 7:00 AM.) The mare may be able to delay parturition by suppressing voluntary abdominal contractions that aid in the expulsion efforts.

ESTROGEN AND PROGESTERONE

There is evidence that the fetus is responsible for the changes that occur in progesterone and estrogen concentrations prior to parturition. After about the seventh month of gestation, progesterone levels in the mare's bloodstream rise gradually and then drop sharply just after parturition. The placenta supplies most of the fetal and maternal progesterone during this period. High levels of placental progesterone are believed to suppress uterine contractions during late gestation. The estrogen level declines during late gestation and is closely associated with a decrease in the size of the fetal gonads. (The fetal gonads are the primary sources of both fetal and maternal estrogen during the second half of gestation.) Unlike other species, estrogen levels in the mare drop sharply just prior to parturition. Researchers are not yet certain whether the precipitous decrease in estrogen levels before parturition stimulates labor or, instead, results when the placenta separates from the mare's endometrium.

PROSTAGLANDIN

Research studies indicate that prostaglandin levels increase gradually during the mare's last three to four months of pregnancy, but these levels remain very low and sometimes undetectable until just minutes before parturition. A significant increase in the prostaglandin level occurs in response to myometrial (smooth muscle layer of the uterus) stretching and cervical dilation as labor begins. Levels are highest as the foal enters the birth canal. In fact, the levels measured when the foal is in the birth canal may be over four times as high as those measured 30 minutes prior to parturition. The exact role that prostaglandin plays during foaling is unknown, but the agent has been used to induce abortion in mares.

OXYTOCIN

Oxytocin is known to stimulate milk let-down and myometrial contractions in many species, and the few studies that have been performed on horses indicate that the hormone plays a similar role in the mare. Very little oxytocin is released during the first stage of labor, but significant quantities of the hormone are released during the second and third stages when myometrial contractions are most violent. The use of oxytocin to induce foaling is examined under 'Induced Labor' in this chapter.

Signs of Impending Foaling

As parturition approaches, changes in the mare's appearance and behavior are usually observed. The signs of impending foaling are stimulated by changes in certain hormone levels, which were discussed previously. Although the hormone changes may be initiated as early as three weeks prior to foaling, the onset of observable prefoaling characteristics is extremely variable between mares. Most mares exhibit at least a few of the signs described in this discussion, but because of inconsistency between mares, the appearance of these characteristics can only be used to roughly predict the onset of labor. A review of the mare's foaling history may reveal a consistent prefoaling pattern, but the attendant should keep a close watch over any mare that shows even one of the signs of impending foaling that are discussed in the following sections.

DISTENDED UDDER

During the last month of pregnancy, the mare's udder usually enlarges and undergoes changes in shape, temperature, and texture. During this period, the mare's udder may fill up at night while she is resting and shrink during the day while she exercises. When the udder remains full throughout the day (i.e., a day of free exercise), parturition is probably imminent, and the mare should be watched closely. Although the hormones that initiate and maintain lactation have been identified, their mechanisms are not completely understood. (Refer to "Lactation.")

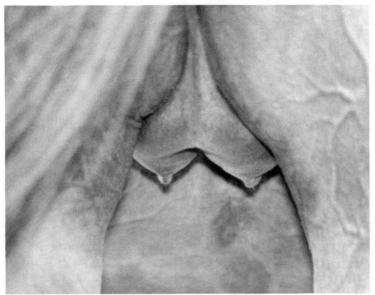

A few weeks before foaling, the udder gradually fills and enlarges. This is one of the early signs that parturition is approaching.

FILLING OF THE TEATS

As the udder enlarges, the upper portion of the teat is stretched in such a way that it is difficult to distinguish from the rest of the udder. For a while, the lower portion of the teat remains small, but as foaling draws nearer, the warmth and tenseness in the area increases. Gradually, the teat enlarges and is deflected outward by pressure from within the udder.

RELAXATION OF THE PELVIC AREA

The muscles in the mare's pelvic area begin to relax about three weeks before parturition. This change permits the fetus to pass smoothly through the birth canal (i.e., cervix, vagina, vulva). Although relaxation is gradual and is not even noticeable in some mares, it usually causes a distinctive change in the mare's appearance. A hollow develops on either side of the root of the tail as muscles between the point of the hip and the buttock relax. If this area is felt each day as the mare is fed or handled, the change in muscle tone can usually be detected.

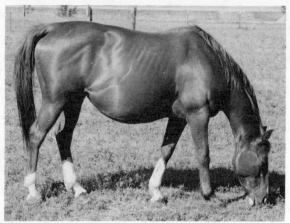

The muscles of the pelvic area gradually relax in preparation for foaling, causing the area surrounding the tail head to appear hollow.

CHANGES IN THE ABDOMEN

During the last few months of gestation, the mare's abdomen becomes increasingly large and pendulous. About a week prior to foaling, however, it may appear to shrink as the foal begins to shift into the birth position. This change in abdominal size and shape is difficult to detect in some mares.

WAXING

Many breeders believe that the appearance of a honey-colored, wax-like bead of dried colostrum at the end of each teat indicates that the mare will

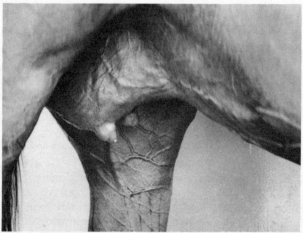

Small droplets of colostrum often appear at the teat openings before foaling. Although it is not a reliable indicator, many attendants watch for waxing to roughly determine the time of foaling.

foal within 12 to 36 hours. However, waxing may actually occur a week or two before foaling or may fail to occur at all. Some mares wax more than once. Others wax normally, but the wax falls off before it is detected. Because waxing shows so much variation between mares, it cannot be used as a reliable indicator of a specific foaling time. Thus, mares that do show wax should also be watched closely.

RELAXATION OF THE VULVA

Within 24 to 48 hours before foaling, the mare's vulva swells and relaxes so that it is capable of stretching to several times its normal size. Both relaxation of the vulva and loss of muscle tone in the pelvic area ease the passage of the foal to its extrauterine environment.

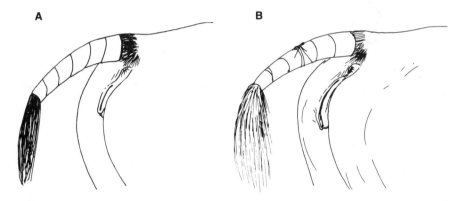

(A) In the properly conformed mare, the vulval lips seal to form an effective barrier to contamination. (B) However, in preparation for parturition, the vulva lengthens and relaxes to permit expulsion of the fetus.

DILATION OF THE CERVIX

As the cervical opening enlarges in preparation for foaling, a slight discharge may be seen on the mare's vulva. This mucous cervical secretion may be streamed with small amounts of blood, but there is no cause for alarm unless blood loss becomes significant.

MILK FLOW

After the wax falls from the ends of the teats, droplets of milk may appear. Although waxing and milk secretion usually indicate that delivery will occur very soon, many mares foal without the appearance of wax or milk, while some mares drip or stream milk for several days prior to foaling. Unfortunately, mares that stream milk before foaling may lose large amounts of colostrum, the vital first milk that contains antibodies and a laxative for the newborn foal. Mares showing spontaneous milk flow should be kept under close surveillance, not only to watch for the onset of parturition but also to determine how much colostrum is lost during this period. If a mare is losing a significant quantity of colostrum, it should be collected and frozen. The colostrum can then be thawed and fed to the newborn at birth.

RESTLESSNESS

The mare's behavior may also change during the last few weeks of gestation. She may become cranky and restless, and as she enters the first stage of labor, the mare usually wants to be left alone. She may appear uneasy, walk continually in the pasture or stall, switch her tail, look at her sides, and kick at her abdomen. These signs are also indicative of colic, but if the mare eats, drinks, defecates, and urinates frequently, the first stage of labor is probably in progress.

SWEATING

As parturition nears, the mare often breaks into a sweat. Like restlessness, sweating may indicate that the mare is in the first stage of labor. The mare's neck or flanks may merely feel warm and damp, or hot sweat may cover her entire body.

Parturition

The normal progression of events that occurs during foaling is easy to recognize once the basic physical processes are understood. These processes can be divided into three distinct stages: positioning of the foal, delivery of the foal, and expulsion of the placenta. An ability to recognize each stage and to follow the normal chain of events that occurs during each phase allows the attendant to determine whether the mare needs assistance. If the second or third stage of labor is delayed or if the changes that normally occur during a particular stage are in some way altered, the attendant should carry out the procedures recommended in "Abnormal Foaling." Fortunately, these stages occur without deviation in over 90 percent of all equine births.

POSITIONING OF THE FOAL

The first stage of labor involves the following internal preparations for delivery of the foal:

1. Special hormones cause cervical softening, vulval relaxation, and increased secretion within the vagina.
2. Hormonal changes also cause increased activity of the uterine muscles. These wave-like uterine contractions pass down the uterine horns toward the cervix.

3. Uterine muscle contractions place pressure on the foal and its surrounding fluids.
4. Uterine pressure and the mare's movements help move the foal into a position for smooth delivery.
5. Pressure on the fetus and its surrounding membranes causes the placenta to protrude through the cervix and gradually dilate the opening of the already softened structure. Normally, the portion of the allantochorion that corresponds with the cervix, called the cervical star, is relatively thin. It is this portion of the allantochorion that ruptures so that the allantoic fluid is released.

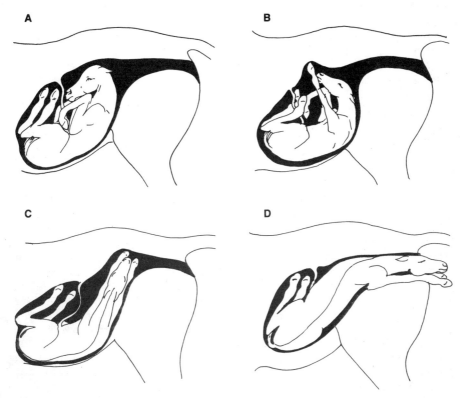

A B

C D

During the first stage of labor, the fetus gradually shifts from a position on its back, rotating until its head and forelimbs are extended in the birth canal.

The outward signs that indicate the onset of the first stage of labor include restlessness and sweating at the flanks and behind the shoulders. As uterine contractions become more severe, the mare may become very nervous. Usual signs of restlessness include pacing in the stall, looking back at her flanks, and kicking at her abdomen. She may also paw the ground, lower her hindquarters, or get up and down several times to help position the foal. A pastured mare usually moves away from other mares and may seek an isolated corner or hidden spot in the pasture, but she also shows this characteristic restlessness and discomfort during the first stage of labor. Unlike the mare with colic, the mare in the first stage of labor usually eats, drinks, defecates, and urinates normally. Some mares show few signs during this period, while others show marked distress for several hours. Transitory contractions that occur without cervical dilation cause the mare to show signs of distress and then "cool off" several times before the foal actually enters the birth canal.

Once the first stage is recognized, the attendant should cover the mare's tail with sterile gauze, a roll bandage, or a sock and clean her hindquarters and udder. When these tasks have been completed, the attendant should leave the stall and watch the mare quietly or view her progress over a closed circuit television. The end of the first stage is marked by rupture of the allantoic membrane and the sudden release of allantoic fluid, a process that lubricates the birth canal. This is commonly called breaking water and usually occurs within one to four hours after the onset of this preparatory phase.

DELIVERY OF THE FOAL

The second stage of labor is characterized by very strong contractions of the abdominal and uterine muscles. The mechanisms that initiate these contractions are not known, but increased muscular activity is apparent soon after the mare breaks water. During this period, the mare usually positions herself on her side with her legs fully extended to facilitate voluntary straining that aids in the natural expulsion efforts. She may get up and down several times to help position the foal and may even stand and move around when the foal's head and legs are protruding. (If labor continues while the mare is standing, someone should catch the foal and lower it gently to the ground.) If the mare lies down next to a wall or fence, the attendant should make sure that there is plenty of room for the foal's delivery. If the mare's perineal region is too close to an obstacle, the attendant should make the mare rise and then allow her to find a new position.

The foal is normally presented in an upright position with its head tucked between extended forelegs. (Many foaling attendants routinely check the foal's position by inserting an arm in the mare's vagina after she breaks water. This should only be done when wearing a sterile sleeve or after scrubbing the hands and arms thoroughly with an antiseptic solution.) As the head and neck appear, enclosed in the bluish-white amnion, the foal's shoulders pass through the pelvic opening. One foot is usually positioned slightly in front of the other to help reduce the circumference of the foal's shoulders and thereby ease its passage through the pelvic opening. After this critical period, the mare usually rests for a short time and then delivers the rest of the foal with relative ease. The fetal membranes (amniotic membranes) are usually broken as the foal emerges or as it first attempts to lift its head. If the membrane is not broken immediately after the foal's delivery, the attendant should tear the membrane and peel it away from the foal's nose to prevent suffocation.

After the foal's hips have passed through the mare's pelvis, the mare usually rests once again. The foal's hindlegs may remain in the mare's vagina for several minutes. The rest period allows the foal to receive essential blood from the placenta via the umbilical cord and should not be interrupted. In most instances, stage two (from the point that the allantochorion breaks to the postdelivery rest period) is completed in about 15 minutes, but a range of 10 to 60 minutes is considered normal.

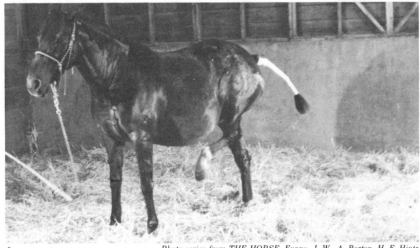

A

Photo series from THE HORSE. Evans, J. W., A. Borton, H. F. Hintz, L. D. Van Vleck. San Francisco: W. H. Freeman and Company, 1977.

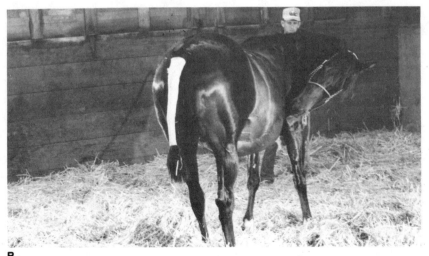

B

Early in the first stage of labor, the mare is usually very restless, and often (A) kicks at her belly and (B) looks at her sides.

C

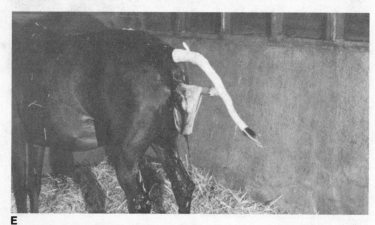

D

E

(C) During the first stage of labor, the mare often shows colic-like signs and may even roll. (D) As parturition progresses the amnionic membrane appears, (E) and it is soon followed by the foal's forelegs.

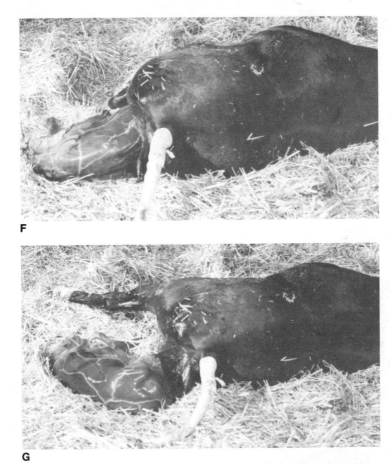

F

G

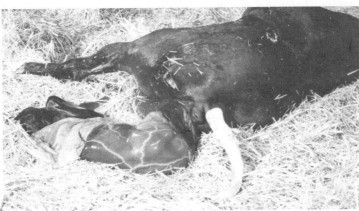

H

(F) The foal's shoulders and then (G) its hips are expelled by the mare's contractions. (H) If the amnion is not torn by the foal's forelegs, it should be drawn away from the foal's head to uncover the nostrils.

(I) The foal usually rests with its hind legs still in the mare for a short period. **(J)** When the mare rises, the navel cord usually breaks.

EXPULSION OF THE AFTERBIRTH

The final stage of labor involves expulsion of the fetal membranes, a process that normally occurs within three hours. (The normal range is from 10 minutes to 8 hours.) During this period, uterine contractions continue, and the mare may show signs of discomfort. The contractions are believed to originate at the tip of the uterine horns, causing inversion of the placenta. In other words, the placenta is expelled inside-out. Uterine contractions continue after foaling and help cleanse the tract of fluid and debris and return the expanded uterus to its normal size.

Since this phase of parturition may last for several hours, the attendant should tie the afterbirth in a knot that hangs above the mare's hocks. This not only prevents the mare from stepping on the membranes and tearing them out prematurely but also adds gentle pressure that aids in the expulsion efforts. The attendant should **not** attempt to pull the placenta from the mare's reproductive tract. Manual removal of the placenta can cause irreparable damage to the mare's reproductive tract and may cause part of the placenta to be retained. Retention of even a small piece of the afterbirth is a potentially serious condition.

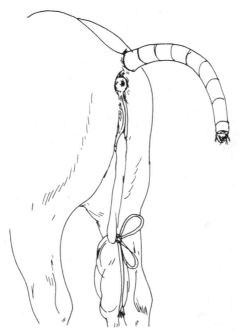

The afterbirth should never be pulled from the mare's vulva. Instead, the membrane should be tied up in a knot above her hocks to prevent it from being stepped on and torn.

After the placenta is expelled, it should be weighed and examined as soon as possible. Some equine specialists believe that placental weight correlates with the condition of the mare's reproductive tract. The normal placental weight in light horses is 10 to 13 pounds. If the placenta exceeds 13 pounds, foal heat breeding is not usually recommended. The texture and condition of the expelled placenta are probably of even more importance than its weight. If the membrane is thick and tough or if hemorrhagic spots are evident, placental infection might be suspected. When the placenta has been infected, the fetus will often display some abnormality at birth.

Close examination of the afterbirth also helps to determine whether part of the placenta is missing. Tears in the afterbirth can sometimes be detected by pouring water into the placenta after it is completely expelled. Any leak should be carefully inspected to determine whether a piece of the placenta is missing. If it is torn or abnormal, the placenta can be kept moist in a plastic-lined, covered bucket with a small amount of water until a veterinarian can examine it. The amnion should have a translucent white appearance, while the allantochorion is normally red and velvety on one side and smooth and light-colored on the other side. Small areas on the villous allantochorion (areas of nonattachment) may have a smooth appearance, but this characteristic should be prominent only at the cervical star (the torn area through which the foal exited).

Intricate vascular systems can be seen within both the allantochorion and the amnion. The umbilical cord contains larger vessels (e.g., the umbilical vein, arteries, and urachus) and is attached to the portion of the placenta that corresponds with the gravid (pregnant) horn of the uterus near the horn-body junction. The gravid horn is usually longer than the nonpregnant horn, but variations have been observed.

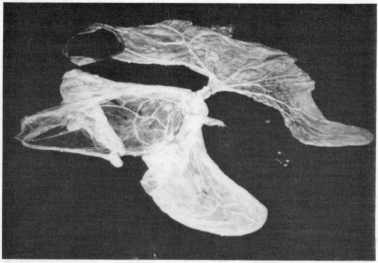

Courtesy of Kansas State University Department of Surgery and Medicine

The placental membranes consist of the allantochorion (upper right) and the amnion (lower left), which is the translucent membrane that encloses the emerging foal.

Close observation of the allantochorion's smooth surface (i.e., the surface that forms the outermost wall of the allantoic cavity during gestation) reveals the presence of several pouches. These invaginations form over large blood vessels, remnant endometrial cups, and over areas of incomplete fusion between the mesoderm layers of the chorion and the allantois. The allantoic cavity may also contain beige, yellow, or brown bodies, referred to as hippomanes. These bodies usually have a putty-like consistency and are believed to originate from minerals and proteins that are deposited in the allantoic cavity during gestation. The hippomanes are sometimes expelled with the allantoic fluid (i.e., when the mare "breaks water"). The attendant may also discover other harmless curiosities when examining the placenta. For example, the remnant of a calcified yolk sac is sometimes mistaken for a twin. Soft pads are sometimes found in the remaining amniotic fluid. These so-called golden slippers cover the fetal hooves and protect the mare's reproductive tract during late gestation.

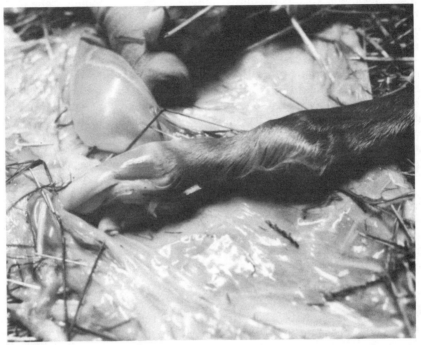

Soft pads known as golden slippers are sometimes found covering the hooves of the newborn foal.

Induced Labor

The induction of parturition in the mare has been an important research and teaching tool for many years. As more information is obtained with respect to the endocrinology of parturition and the pros and cons of various induction techniques, this process could also become a routine management tool. Although inducing labor cannot prevent foaling problems or guarantee the birth of a healthy foal, it can reduce the severity of problems such as dystocia, laceration of the mare's reproductive tract, retained placenta, and suffocation of the foal, by insuring that someone is present during foaling to assist the mare. Induction has proven to be useful when gestation is significantly prolonged, or when the pregnant mare experiences uterine atony (absence of uterine tone), impending rupture of the prepubic tendon, or other problems that might interfere with normal birth. Induced foaling can be a very helpful technique, but it is a procedure that requires a carefully controlled environment, veterinary supervision, and trained personnel.

When parturition is induced several hours before the mare would have foaled naturally, it is a relatively safe procedure that allows the veterinarian and/or the foaling attendant to be present. In most instances, the mare's induced labor pains are not any more severe than those experienced by that particular mare during a natural birth. (Some researchers have suggested that the severity of labor is related to the type and amount of induction agent administered, but further studies are required to substantiate this theory.) Most of the problems that have occurred as a result of induction are associated with experimental induction techniques, lack of supervision, or failure to follow special precautionary measures described later in this discussion. Induction has been associated with absence of the first stage of labor and failure of the allantochorion to rupture. However, most of these problems can be dealt with by an attendant or veterinarian that has equine obstetrical experience. Uterine rupture can occur any time that the mare's violent labor contractions force a malpositioned foal against the maternal pelvis, but the chances of uterine rupture occurring during an induced foaling are minimal if an experienced supervisor is present to promptly correct the foal's position. Reports also suggest that foal birth weights are lower when parturition is induced, but induction has not been shown to reduce disease resistance or to increase the incidence of postfoaling problems in the mare or foal.

PREREQUISITES

In order to reduce the risks involved in forcing a mare to foal, every effort should be made to insure that both the mare and foal are prepared to meet the physical challenges involved in the birth process. For this reason, three important prerequisites should be met before routine induction is performed:

1. Routine induction should not be carried out before 330 days of gestation. A foal born earlier in gestation may be insufficiently developed to survive in its new environment. Many veterinarians prefer to delay induction until the mare is very close to foaling as indicated by mammary development and cervical relaxation.
2. Before induction is carried out, the mare should also have adequate amounts of colostrum and milk in her udder. If induced foaling is planned, the attendant should monitor changes in the mare's udder daily. The udder usually appears full several days before foaling and its secretions gradually change from a cloudy, straw-like color to yellow or yellow-white. One to four days before foaling a bead of wax-like, dried colostrum often forms at the teat opening. At this point, the veterinarian should be notified so that the mare's progress can be evaluated.

3. The third prerequisite for inducing parturition is the presence of a softened cervix. Once the first two prerequisites have been met, the veterinarian checks to make sure that the mare's cervix is relaxed in preparation for foaling. If the cervix is still too constricted, the veterinarian may decide to postpone induction. Occasionally, estrogen compounds are administered to relax the cervix and the vulva in preparation for induction, but when possible, it is preferable to allow these physical changes to occur naturally.

INDUCTION PROCEDURES

If the mare meets the three induction criteria, she is ready to foal and can be induced safely and quickly. Although there are several agents that stimulate uterine contractions in the mare, the hormone oxytocin is usually administered for routine induction. This agent has been used successfully in many species, including the horse, and the response is usually very rapid. Most mares foal within 30 to 40 minutes (the normal range is 15 to 60 minutes) after an intramuscular injection of 40 to 60 international units of oxytocin. The synthetic corticosteriod dexamethasone has also been used to induce early parturition in the mare, but this agent must be injected daily for four or more days (i.e., until parturition occurs). Injections of pituitrin and estrogen (to soften the cervix) may also be used to induce early parturition. Prostaglandin is a known abortifacient (an agent that causes abortion), but its usefulness for inducing labor in mares has not been determined. Under normal circumstances, oxytocin is currently the recommended induction agent.

The mare usually becomes nervous and begins to sweat at the flanks and shoulders about 10 minutes after the oxytocin injection. She may also lie down at this time but should be forced to stand for vaginal examinations. The mare does not usually resent precautionary examinations but she should be made to stand with a handler at her head any time an examination is required during the first 10 to 20 minutes after injection. If the cervix is adequately dilated, the veterinarian may be able to position the foal's head and limbs without breaking the fetal membranes. (Occasionally, the fetal membranes rupture when the foal's position is manipulated, but this does not seem to have any adverse effect on the foal.) The foal's forelimbs should be guided carefully through the cervix to prevent cervical bruising. If the allantochorion has not ruptured when the foal enters the birth canal, it should be opened with a pair of small forceps. (For more information on correcting abnormal presentations, refer to the chapter "Abnormal Foaling.")

The hormone oxytocin also stimulates expulsion of the placenta. For this reason, the third stage of induced labor usually occurs without problem. (Oxytocin is frequently administered to encourage expulsion of retained placentas.) Because this hormone also helps the uterus return to its normal size, uterine prolapse (i.e., expulsion of the uterus) is rarely seen in induced mares.

Care of the Mare After Foaling

Unless disturbed, the mare usually remains recumbent and tranquil for as long as 30 minutes after delivery. During this period, the mare frequently lifts her head to check on her foal, often nickering to it or licking it as much as she can without rising. (Moving the foal around to the mare's head is unnecessary.) This is the most critical period for the development of a strong bond between the mare and her foal. The smell of the fetal membranes seems to play some role in the mare's identification and acceptance of her foal. Although the following discussion recommends that certain procedures be carried out immediately after foaling, human intervention should be minimal during the first two hours after foaling.

If an episiotomy (i.e., incision of the sutured vulva) was performed immediately before foaling, the veterinarian may decide to resuture the mare soon after the foal's delivery. Because the incision is still raw, the vulva readily reseals. In addition, if this procedure can be performed within 30 minutes after birth, local anesthesia is unnecessary. Alternatively, the

After a short rest, the mare usually begins to show interest in her new foal. Normally, during the hours following foaling, a strong bond forms between the mare and her foal, and the mare begins to recognize and accept the foal.

upper portion of the vulva may be temporarily clipped after the placenta is expelled. The clips should be removed within 24 to 48 hours and replaced with permanent sutures. Either procedure minimizes aspiration of air into the reproductive tract when the mare rises after foaling and thereby helps to minimize contamination of the tract.

As the mare rises or as the foal struggles to stand, the umbilical cord usually breaks. (Because the foal receives a significant amount of blood from the placenta via the umbilical cord, the cord should not be ruptured prematurely. Procedures that should be carried out immediately after the cord ruptures or in the event that the cord does not rupture naturally are described in "The Neonatal Foal.") Once the physical bond between the mare and foal is broken, the cord and the amnion should be tied in a knot above the mare's hocks to prevent her from tearing the placenta by stepping on or kicking at the dragging afterbirth. The weight of this knot also adds gentle pressure that aids the natural removal of the placenta from the endometrium. Unlike some other mammalian species, mares do not normally eat their afterbirth.

Once the routine postfoaling procedures have been completed, the attendant should quietly rebed soiled areas of the foaling stall. Although the mare and foal should not be disturbed for several hours, they should be checked periodically or observed quietly from outside the stall to make sure 1) that the foal nurses within several hours, 2) that the foal defecates normally, 3) that the mare and foal are healthy, 4) that the mare and foal follow normal postfoaling behavioral patterns, and 5) that the placenta is retrieved for careful examination.

After the placenta is expelled, it is very important that the membranes be examined carefully. The following points should be noted and recorded for future reference:

1. time required to expel the placenta after the foal's birth,
2. absence of any pieces (e.g., fill the allantochorion with water to check for tears or holes),
3. condition of the membranes (e.g., weight, color, thickness, and presence of hemorrhage spots).

During the third stage of labor, the mare may experience some discomfort caused by contractions of the uterus as it expels the placenta. Occasionally, the mare also shows signs of colic after the third stage is complete. If the pains caused by cramping of the empty uterus are severe, the mare may be given a mild sedative normally used to relieve intestinal discomfort. In some cases, the veterinarian may give intravenous injections of an antispasmodic agent to help the mare through this adjustment period.

The mare should not be given any grain or water after foaling until the placenta is expelled. (Note: If the membranes are retained for more than three hours, the mare should be fed and watered.) Because the mare is usually hot and feverish after foaling, she should be given small quantities of lukewarm water. Laxative feeds (e.g., oats and wheat bran mash) may be fed following foaling since sore abdominal muscles may make defecation painful to the mare. To encourage bowel movement, the veterinarian may administer 1 gallon (3.78 liters) of light mineral oil by stomach tube to help prevent impaction. For this same reason, many breeders cut the mare's feed in half before foaling and then gradually increase the amount fed until the mare is back on full feed by about 10 days after foaling. (For a description of the mare's postfoaling nutritional needs, refer to "Lactation.") Some breeders also believe that decreasing the mare's feed intake reduces the amount of rich milk produced during early lactation and, as a result, helps to prevent foal scours.

Small tears in the allantochorion that might otherwise be overlooked, can frequently be detected by filling the sac with water.

The mare should be watched carefully for four to five days after foaling. It is normal for the mare to have a dark red discharge for six to seven days, but a yellow discharge may indicate infection. Because a discharge is not necessarily present in cases of reproductive infection, the mare's temperature should be checked periodically throughout the postfoaling adjustment period. (The normal temperature is 100.5°F / 37.5°C.)

Adequate exercise improves the newborn foal's muscle tone and encourages uterine involution in the mare.

Exercise helps to empty the mare's uterus, stimulates involution (return to normal size) of the uterus, reduces pregnancy edema, prepares the mare for foal heat breeding, and improves the newborn foal's muscle tone. The mare and foal should be turned out in an empty paddock and allowed limited periods of self exercise until both animals appear ready to be turned out to pasture or placed on a regular exercise program. The first few outings should be limited to help prevent the foal from tiring the mare or vice versa. For example, the mare and foal may be given a 15 minute period of self-exercise on the morning after foaling (assuming that the mare foals at night) and another 15 minutes of freedom that afternoon. Periods of exercise should be increased gradually to suit the condition of the mare and foal. Some breeders prefer to keep the mare and foal stalled for 24 hours after foaling to insure that the mare receives a sufficient rest period and that the foal receives adequate colostrum. In addition, the newborn foal can be protected from inclement weather. Other breeders feel that both the mare and the foal should be turned out as soon as possible regardless of weather condition. This decision should be made only after carefully evaluating the condition of both mare and foal and the circumstances surrounding the birth (e.g., prolonged foaling, retained placenta, or a weak foal).

17

ABNORMAL FOALING

Before abnormal foaling is described, an important point must be re-emphasized: more than 90 percent of all foalings occur without significant deviation from the normal birth process. During normal foaling, the educated foaling attendant can relax and observe parturition from a distance. It is not necessary to assist the mare or her foal. If a minor problem is encountered, minimal but immediate assistance should be given. Although most foalings proceed normally, the possibility that the mare or foal will encounter some problem during this critical period should not be taken lightly. In addition to the normal foaling preparations described in the previous chapter, the breeder should devise a special "plan of action" to be carried out in the event that a problem is recognized or suspected. These emergency procedures may vary between farms but should be designed to meet even the most serious abnormalities.

Even breeders that hire expert foaling attendants should have a veterinarian on call during the foaling season. Despite the fact that a qualified attendant has years of foaling experience and is capable of handling all but the most serious foaling problems, it is wise to call the farm veterinarian the minute abnormal parturition is suspected. Because the second stage of labor occurs rapidly and with great force, complications may become serious in a very short period. If the attendant corrects the problem before the veterinarian arrives, the mare may still require a thorough examination to determine whether any postfoaling complications have resulted from the difficult labor. If the attendant cannot correct the problem, an early call to the veterinarian may save the life of the mare and her foal.

It is also extremely important that the attendant be able to distinguish abnormal conditions that can be corrected easily from those that require immediate veterinary assistance. The following description of foaling abnormalities is presented to help the attendant make this distinction. The individual attendant must carefully scrutinize his or her ability to help the mare through a difficult labor but, more importantly, should be familiar with emergency techniques that may save the mare and/or her foal from serious injury or death while waiting for the veterinarian's arrival. The attendant should also be able to recognize problems that may occur or become apparent after foaling. Although postfoaling complications in the mare are described in this chapter, problems that may be observed in the newborn foal are examined in "The Neonatal Foal."

Dystocia

Dystocia is a term used to refer to any abnormality that prevents delivery of the foal by the mare's efforts alone. These abnormalities can be categorized as maternal or fetal dystocias, depending on their causes. An unresponsive (atonic) uterus, for example, causes maternal dystocia, while failure of the fetus to become positioned properly prior to delivery results in fetal dystocia. Unfortunately, a large percentage of dystocias result in dead, seriously weakened, or oxygen-deficient foals. The time element is critical during abnormal foaling, not only because of the tremendous physical stress placed on the foal during prolonged foaling but also because of the foal's need for an external oxygen supply if the placenta begins to detach from the uterus or if the umbilical cord is pinched inside the birth canal. Thus, to improve the chances of obtaining a healthy foal, corrective measures must be carried out quickly and carefully. In order to assist the mare as soon as possible, the attendant should watch for the following warning signs of dystocia:

1. failure of the white amnion to appear soon after the mare "breaks water,"
2. appearance of the red chorioallantois instead of the white amnion at the beginning of the second stage of labor,
3. appearance of the white amnion without both limbs and/or without the foal's head,
4. absence of straining for prolonged periods during the second stage of labor,
5. presence of straining without any progress with respect to the foal's delivery,
6. repeated shifting and rolling of the mare during any stage, but especially during stage two.

The problems that can cause these warning signs are complex and numerous. The most significant abnormalities, however, include malpresentations and malpositions of the foal, placenta previa, a small pelvic opening, uterine torsion, and uterine inertia.

Mutation (manipulation of the foal's position) and traction (pulling the foal as contractions occur) can be used to correct some forms of dystocia. In

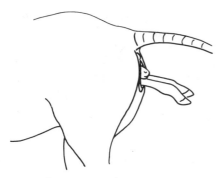

In a normal delivery, an examination of the vulva should reveal both of the foal's forelegs (soles down) with the nose lying on top of the legs.

many instances, the attendant can perform these techniques successfully, especially when dystocia involves the simple misplacement of a forelimb or a slightly rotated position. Specific methods of correction are discussed later in this chapter.

It should be noted that mutation and traction become less effective when assistance is delayed. Even if an experienced attendant is present, veterinary assistance is usually required when dystocia is discovered after the mare has been in labor for an extended period. As parturition continues without progress, the uterus tends to contract down around the fetus, making mutation more difficult than in early stages. If necessary, the veterinarian may administer a drug to relax the uterus and to relieve the mare's discomfort.

Another problem associated with prolonged parturition (described in the discussion on uterine inertia) is a dry birth canal. Normally, the allantoic fluid that is released at the end of the first stage of labor lubricates the foal's passageway, but if expulsion of the foal is delayed, the birth canal may become very dry, making parturition even more difficult. In these cases, mineral oil is sometimes used for lubrication so that the foal can slip through the birth canal.

If mutation and/or traction are unsuccessful in correcting dystocia or if it is apparent that these techniques will not be successful, the veterinarian may elect to utilize one of two surgical methods of delivery. Fetotomy, which is usually reserved for dead or extremely deformed fetuses, involves dividing the body into sections that can be passed easily through the cervix. This procedure requires special equipment and proper training, and the sharp wires and difficult maneuvering required to dispart the fetus present a definite risk of damaging the mare's cervix and uterus. When performed skillfully, however, fetotomy is less stressful to the mare than is Caesarian section.

If dystocia cannot be corrected by manipulation of the foal or if a dead fetus cannot be removed by fetotomy without endangering the mare, the veterinarian may elect to deliver the foal by Caesarian section. This surgical procedure requires that an incision be made in the mare's flank or ventral midline, through her uterine wall, and through the placenta. The foal is then removed, and the incision is closed by a special procedure that helps to reduce hemorrhage of the uterine wall. If the operation is carried out under ideal surgical conditions prior to the onset of labor, the chances that the foal will survive are high. When Caesarian section is used as a last alternative in cases of prolonged dystocia, however, the success rate drops considerably. Although mare mortality is less than 10 percent (unless complications, such as severe infection or uterine torsion, are present), Caesarian section predisposes the mare to metritis and to the development of scar tissue. For this reason, Caesarian section should not be used in place of more simple corrective measures, such as mutation or fetotomy, unless it is the only means of saving the mare or her foal.

Before the various abnormal deliveries are studied, it is important to note that the mare's reproductive tract is very susceptible to infection during parturition due to the stress of placental detachment and because the dilated tract may be exposed to airborne organisms. When fetal posture must be manipulated, it is very likely that the mare's tract will be stressed and contaminated during the procedure. For this reason, sanitary conditions should be observed anytime a foaling mare is assisted. Asepsis (i.e., freedom from infective organisms) is most desirable, but problem deliveries are often conducted on an emergency basis in less than ideal surroundings. The preparatory techniques described earlier (e.g., wrapping the mare's tail, disinfecting the foaling stall, etc.) help to minimize the chances of infection in both the mare and her newborn foal but are especially important when the attendant must enter the mare's genital tract to check or correct the foal's position. When assistance is required, the attendant's arms and hands should be scrubbed thoroughly with a nonirritating antiseptic solution and/or placed into sterile obstetrical sleeves.

When an abnormal presentation of the fetus is suspected, the attendant should not hesitate to enter the mare's tract (after scrubbing and/or donning sterile sleeves) to check the foal's posture. The reader should note that the mare's contractions become extremely forceful during the second stage of labor and that these contractions are actually capable of breaking a hand or arm. Forcing the mare to stand helps to reduce the force of these

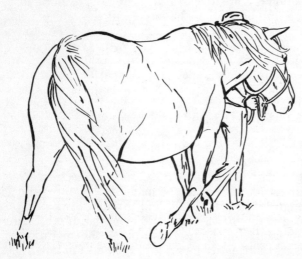

If the handler discovers that the mare's dystocia will require veterinary correction, she should be kept walking until assistance arrives. Walking the mare is especially important if correction requires manipulation because it discourages further uterine contraction.

contractions and, in many instances, allows the foal to drop back into the mare's uterus, giving the attendant more room to manipulate the foal. If an arm must be inserted through the cervix alongside the foal's body, special care should be taken to prevent wedging the arm against another bony structure, such as the foal's leg or shoulder, during a forceful contraction. The careful attendant can feel an oncoming contraction in plenty of time to reposition the hand or arm. If the dystocia requires veterinary assistance, one of the most effective actions that can be taken until the veterinarian arrives is to keep the mare on her feet and walking. This activity can delay parturition for 30 to 60 minutes.

MALPRESENTATIONS AND MALPOSITIONS OF THE FOAL

The most common causes of dystocia in the mare involve malpresentation (i.e., part of the body that is presented first during delivery) and malposition (i.e., overall posture) of the foal. A few hours or days before birth the fetus normally moves from its position in one uterine horn and the uterine body to a position predominantly in the uterine body. Just prior to delivery, the foal should be in an upright position with its front feet and nose extended toward the cervix so that it can slip easily through the dilated birth canal when labor contractions become severe. If the fetus (or fetuses in the case of twins) is arranged in some variation of the normal parturient position, its movement into or through the birth canal is usually impeded, and violent contractions against the immovable body may cause serious injury to the foal or to the mare's reproductive tract. For this reason, many foaling attendants routinely check the foal's position soon after the mare "breaks water." If this preliminary examination is not performed, a problem delivery should be suspected if the foal's forefeet fail to appear within 10 to 15 minutes after the allantochorion ruptures or if any presentation other than head betwen the forelegs (with pads of the hooves pointed down) is observed. If the first or second stages of labor are delayed beyond the normal time ranges given in "Foaling," the foal's position should be checked as quickly as possible. The longer a malpresentation or abnormal position is neglected, the harder it is to correct.

ABNORMAL FORWARD POSITIONS

To manipulate the foal into a normal parturient position, the location of the head and forelimbs should be evaluated carefully. As the fetus leaves the uterus, it is very easy for its nose or forelegs to be deflected downward before entering the cervix. As a result, the foal becomes wedged against the mare's pelvis. If the head is not presented, it may be found tucked beneath the foal or turned back beside the foal's body. (If the head cannot be located, the legs should be checked for the presence of hocks to determine whether the foal is in a backward position.) Similarly, if one or both forelimbs are not visible within the amnion, they may be found flexed or extended beneath the foal. If the appearance of one limb is delayed significantly or if both feet are presented alongside the head, a partially flexed elbow (or elbows) may be caught at the mare's pelvis. If delivery is impeded as the foal's chest emerges from the vagina, its hind feet may be pointing forward underneath the foal in such a way that they become caught at the mare's pelvis. Occasionally, extended forelegs become crossed over the foal's head, and if the foal enters the birth canal in this position, violent labor contractions may force the forelimbs through the vaginal wall and into the rectum, causing severe rectovaginal lacerations. (Refer to 'Abnormal Conditions Associated with Parturition' within this chapter.)

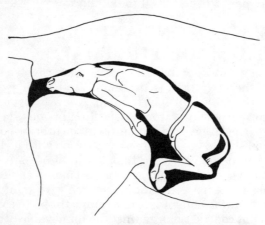

One or both forelegs may fold under the foal during delivery. In order to correct this position, the foal should be pushed back into the uterus, and each leg should be extended so that the foal is aligned correctly for delivery. While manipulating the leg, the foal's foot should be covered with one hand so that the uterine wall is not damaged.

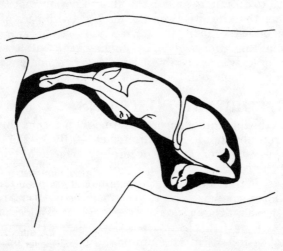

When the foal's nose is not found in the birth canal, it may be found tucked downward. The foal must be pushed back so that its head can be lifted.

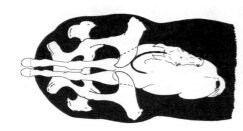

If the nose is not readily found in the birth canal or beneath the foal, the entire head and neck may be turned back alongside the foal's body. If only the head is turned back, it may be brought forward by minor manipulation. However, veterinary assistance is usually required in order to correct the position when both the head and neck are turned back.

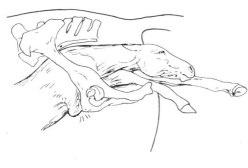

The foal can become lodged at the pelvic brim when its forelegs are not extended properly. The flexed elbow increases the foal's diameter and impedes delivery. The foal should be pushed back until its forelegs can be stretched forward.

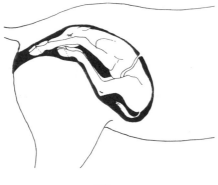

The foal's hind legs prevent delivery if they become lodged against the mare's pelvic brim in a tucked position. This type of dystocia is very difficult to correct if labor has progressed very far. If the foal is found in this position upon initial examination of the mare, a veterinarian should be called immediately.

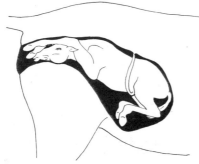

Damage to the birth canal sometimes occurs when the foal's forelegs are positioned over its head. This malposition also halts the foal's progress through the pelvic outlet. The foal may be pushed back and its legs uncrossed and repositioned. While correcting this position, the attendant's hands should cover the fetal hooves to protect the reproductive tract wall.

When discovered before violent contractions begin, many of the abnormal postures that occur in the forward position can be corrected by pushing the foal back toward the mare's head, grasping the misplaced limb or head, and maneuvering the body into the normal parturient position. As mentioned earlier, the mare should be forced to stand for examination or manipulation of the foal. A standing position allows the foal to drop back into the mare's uterus (unless it is lodged at the pelvis) and reduces the force of the contractions against the foal.

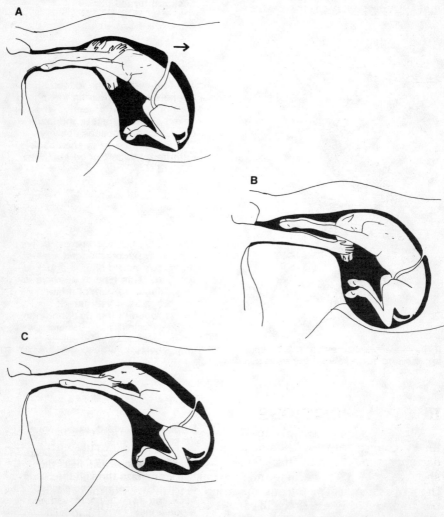

In order to correct many abnormal positions, (A) the foal is gently pressed back into the uterus to obtain working room for manipulation. (B) The misplaced portion of the foal (in this instance, the head) is placed in the normal position for delivery, and (C) the foal is guided into the birth canal.

After the appropriate position is achieved, the mare usually expels the foal with a few contractions. In some cases, however, gentle traction on the foal's forelegs is required. Traction should only be applied when the mare is contracting and should follow a downward arc toward the mare's hocks as soon as the foal's head emerges from the vulva. The purpose of traction is to aid the mare's contractions, not to forcibly remove the foal from the uterus. Therefore, the ropes and winches that are sometimes used on calves should not be used on the foal. The force of traction should never exceed that which can be exerted by one person. Excessive force may injure the mare's reproductive tract and also risks injury to the foal.

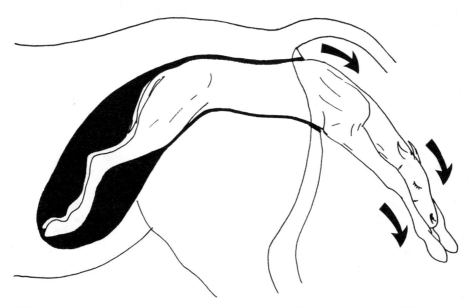

If the mare must be aided in her expulsion efforts, traction on the foal should follow an arc out and down toward the mare's hocks.

ROTATED POSITIONS

Although a true ventral (upside-down) position rarely occurs, the foal may become rotated to one side prior to delivery. Any deviation from the dorsal position (i.e., foal's back facing the mare's spine) may cause the foal's shoulders or hips to become impacted at the mare's pelvis. If the soles of the foal's front feet point any direction but toward the mare's hocks, its position should be checked. To correct this type of abnormal position, the attendant must first determine which direction to turn the foal. For example, if the foal's back is 30° to the right of the mare's spine, the foal should be turned counterclockwise. To rotate the foal, cross its forelegs, grasp its withers, and apply firm force in the appropriate direction.

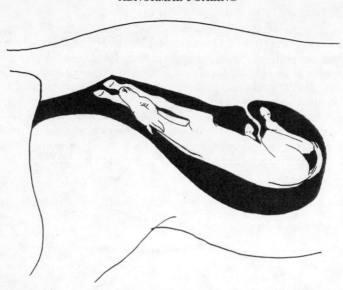

When presented in a ventral position, the arch of the foal's back does not complement the upward curve of the birth canal. Before delivery can proceed, the foal must be repositioned by rotation.

BACKWARD PRESENTATIONS

The foal may also be presented backwards with the hindlegs extended back through the birth canal ("true" backwards presentation) or located underneath the foal's body (breech presentation). A "true" backwards presentation is characterized by the "soles up" position of the fetal hooves and, upon examination, by the presence of hocks rather than knees on the presented limbs. The foal can be delivered successfully in this position if parturition proceeds rapidly. When the foal's hind feet leave the uterus first, the umbilical cord is compressed against the mare's pelvis before the shoulders and head leave the uterus. This pressure limits the foal's maternal oxygen supply and stimulates respiration. If the foal's first few breaths occur while its head is still in the fluid-filled amnion, it may suffocate. For this reason, a backward presentation must proceed rapidly with the attendant contributing to the mare's efforts by exerting strong traction (i.e., one person pulling on the foal's hindlimbs in conjunction with the expulsive efforts). As with forward traction, the foal should be directed in such a way that the arch of the foal's back follows the arch of the birth canal.

Backward delivery may be difficult since the internal organs are pushed toward the rib cage, causing this area to expand before it passes through the birth canal. (In a normal presentation, this situation is reversed so that the rib cage folds to facilitate smooth passage through the birth canal.) If the foal is turned toward either side in the backward position, it must be rotated so that its back faces the mare's spine. In any other position, the arch of the foal's body is not paired with the arch of the birth canal, and the birth process is impeded. Rectovaginal lacerations and injury to the

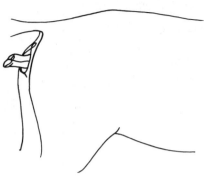

When the foal's feet emerge from the mare's vulva with the soles up, it is important to determine whether the foal is emerging backwards or forwards in order to select the appropriate correction method.

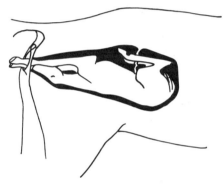

If the foal is in a ventral forward position, the next joint on the foal's leg will be its knee. This can be distinguished by slipping one hand into the mare's vulva and along the foal's leg until the joint is reached. The first two joints (knee and fetlock) bend in the same direction on the forelegs. Before the delivery can proceed the foal must be rotated into a normal, upright position.

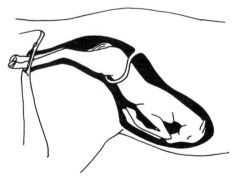

The first two joints of the hindleg (hock and fetlock) bend in opposite directions. In this manner the true backward position can be identified.

foal's back or neck may result if the mare attempts to deliver the foal in a rotated backward position. (Note: The foal that is delivered in the backward position often requires resuscitation. Refer to "The Neonatal Foal.")

When the foal is presented backwards with its legs folded underneath its body, its tail may be observed or felt in the vagina. Unless the foal is very small with respect to the mare's pelvic opening, it cannot be delivered in this so-called breech presentation. Manual correction of this position should be handled by a veterinarian, since surgical removal of the foal may be required. To minimize impaction of the foal against the maternal pelvis, the mare should be forced to stand and walk until the veterinarian arrives.

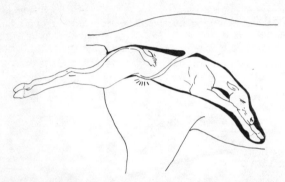

In a true backward position the foal's umbilicus is compressed against the pelvic brim before the foal's head is free of the amnion. This can cause suffocation if delivery does not proceed quickly. Therefore, the attendant should assist the mare's expulsive efforts once the foal's hips emerge. If the delivery is difficult, the birth canal can be lubricated with mineral oil, and the foal can be rotated slightly from side to side until the rib cage, shoulders, and head slip through the birth canal and out of the mare's vagina. The amnion should be torn and pulled away from the foal's nostrils as quickly as possible.

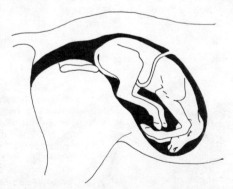

Because extensive manipulation or even surgery may be required to correct a breech position (foal presented backward with legs tucked beneath its body), veterinary assistance should be obtained as quickly as possible. This position may be corrected by placing one hand over the hock and the other hand over the hoof to protect the uterus and then by straightening the tucked legs.

OTHER ABNORMAL DELIVERIES

Live, healthy fetuses usually become arranged properly for parturition during the final weeks of gestation, but weak or dead fetuses often fail to become oriented in the normal parturient position. For this reason, mares carrying these fetuses are very likely to experience dystocia. Although extended gestation periods do not necessarily indicate the presence of weak or dead fetuses, the breeder should be alerted to the possibility of a difficult birth if the mare experiences an unusually long pregnancy.

In rare instances, dystocia may be caused by a congenital abnormality of the foal (e.g., hyperflexion of the forelegs, wry neck, hydrocephalus) that prevents a normal parturient posture. If these abnormalities are severe, Caesarian section may be required to save the foal and to prevent damage to the mare's reproductive tract. Once again, the importance of calling a veterinarian the moment abnormal foaling is suspected must be emphasized. In the event that the foal's presentation or posture cannot be corrected by the attendant, an early call to the veterinarian increases the chances of the foal's survival and lessens the likelihood of injury to the mare.

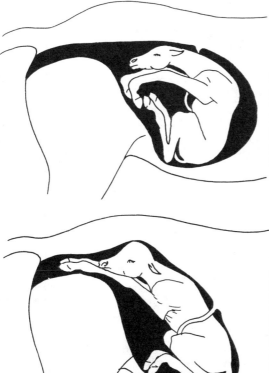

Congenital contracted flexor tendons prevent the foal's forelegs from extending properly for expulsion. If the contraction is severe, the veterinarian usually removes the foal by Caesarian section or by fetotomy to avoid damaging the mare's reproductive tract.

The enlarged forehead associated with hydrocephalus may prevent the foal's head from passing through the birth canal. This condition is usually discovered when the attendant examines the foal's position. If the forehead impedes the progress of delivery, the veterinarian usually removes the foal by fetotomy or Caesarian section.

Twin fetuses are often presented in positions that preclude normal delivery. With two fetuses in the uterus, neither has room to maneuver into the birth canal, and crowding in the uterine body makes it very difficult for the attendant to rearrange them. If one or both foals are facing the cervix, it may be possible to push one foal back into the uterus in such a way that the other foal can be guided through the birth canal. Dystocia due to twins is very difficult to correct if the situation is not recognized promptly. However, when one twin is delivered, the other usually follows fairly easily unless it is in an abnormal position.

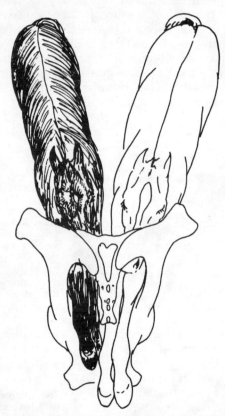

The presentation of twins often prevents normal delivery of either foal. It is often necessary to press one foal back into the uterus in order to permit the other foal to enter the birth canal.

Other variations of the normal delivery position involve transverse presentations in which the foal's body is wedged crosswise in the uterus. One to four feet or a combination of one foreleg and one hind leg may be present in the cervix, or the foal may be positioned in such a way that neither limbs nor head can pass through the cervix. Because the limbs or head must enter the vagina to stimulate abdominal contractions, transverse presentations often result in a prolonged first stage of labor (i.e., absence of abdominal straining). Parturition cannot proceed when the foal is in a transverse position. As with breech presentations, transverse positions are difficult to correct, and it is recommended that the attendant walk the mare until a veterinarian arrives. If the fetus lies in a transverse position for a lengthy period during gestation, its neck gradually bends to conform to the curvature of the uterine walls. This position can cause atrophy and contraction of the fetal neck muscles. The resulting malformation, referred to as wry neck, prevents the foal from assuming a normal parturient position.

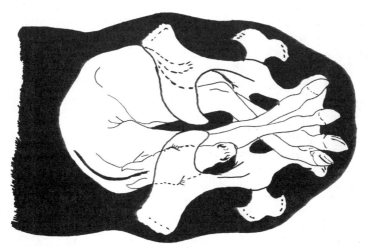

A crosswise position of the foal in the uterus distinguishes the transverse presentation. The foal's hindlegs must be pushed back into the uterus so that the forelegs and head can be brought into the birth canal.

Disproportion, or fetal oversize, is uncommon in horses. Normally, the size of the dam's uterus limits fetal size regardless of the offspring's inherited capacity for mature size. In rare instances, however, the mare may carry a fetus that is affected by hydrocephalus (enlarged brain), shistosoma reflexum (fetus turned partially inside out), or some other abnormal enlargement or deformity that prevents its passage through the pelvic opening.

PLACENTA PREVIA

As described earlier, vigorous uterine contractions that occur toward the end of stage one cause an area of the allantochorionic membrane, referred to as the cervical star, to enter the cervix. Normally, pressure on the fetal membranes causes the cervical star to rupture and, consequently, release several gallons of lubricating fluid. If the allantochorion is unusually thick or if it releases from the uterus prematurely, labor contractions may force the intact membrane through the cervix and into the vagina. This condition, which is referred to as placenta previa, is characterized by presentation of the thick red allantochorion at the vulva.

The appearance of a red membrane at the vulva should alert the attendant to the following possible complications:

1. Premature separation of the placenta from the endometrium terminates the vital exchange of oxygen between the maternal and fetal circulatory systems, a process that stimulates the onset of respiration in the foal. Unless the membrane is opened and the foal's head exposed promptly, the foal will suffocate in the amniotic fluids.

2. If separation of the placenta from the uterus began during the latter stages of gestation, the fetus may have suffered the effects of long-term oxygen deficiency (e.g., brain damage).

3. Degeneration of the allantochorionic attachment may be the result of a disease process within the mare's reproductive tract. In such cases, the fetus may be affected by the disease and, as a result, may be malpositioned during parturition and weak at birth.

When placenta previa is recognized, the attendant should rupture the allantochorion immediately with a hand or with a pair of sharp scissors so that the foal can move into the birth canal. Because delivery must proceed rapidly to minimize the chances of suffocation, the attendant should check the foal's presentation and make any necessary corrections gently but quickly. The membranes surrounding the foal's head should be peeled back away from its nostrils as soon as possible. It is also important that a veterinarian be summoned to examine both the mare and foal for complications associated with placenta previa.

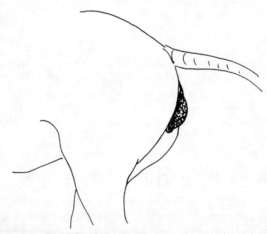

Normally, uterine contractions rupture the allantochorion during stage one of labor and the amnion enveloping the foal protrudes from the vulva. Occasionally, however, the thick, red allantochorion fails to rupture and appears in the birth canal. This membrane must be torn before the fetus can be expelled.

SMALL PELVIC OPENING

Occasionally, the mare's pelvic opening is unusually small or malformed in a manner that prevents normal parturition. Failure of the pelvis to reach a size that corresponds with uterine capacity may be caused by nutritional deficiencies or imbalances during early growth and development. In addition, failure of the pelvic bones to realign properly after a fracture may also obstruct the pelvic opening. A mare that is known to have an unusually small pelvic opening or a pelvic malformation should not be bred, but if she is bred inadvertently, dystocia can be avoided by removing the foal by Caesarian section before the onset of labor. If the condition is not detected until parturition, fetotomy or Caesarian section will probably be required.

UTERINE TORSION

Twisting of the uterus, or uterine torsion, is rare in the pregnant mare but should be mentioned as a possible cause of maternal dystocia. Reports indicate that the chances of lengthwise rotation of the gravid uterus are greater during the last trimester of pregnancy when the size and weight of the fetus increase rapidly. It has been suggested that uterine torsion may be caused by an accident (e.g., sudden fall) or by excessive rolling (e.g., during severe colic) and that the degree of rotation may reach 360° or more. In addition to blocking the foal's entrance to the birth canal, uterine torsion may cause circulatory disturbances that result in dehydration and shock in the mare. Uterine degeneration, uterine congestion, and a decrease in uterine circulation may threaten fetal survival and delay the onset of parturition.

Because uterine torsion is often characterized by intermittent signs of colic, the possibility of this disorder should be considered anytime the mare shows signs of gastrointestinal disturbance during the last few months of gestation. Diagnosis of this condition is made by rectal palpation of the uterine ligaments or, in the foaling mare, by direct observation of the twisted uterus through the dilated cervix. If the condition is detected prior to the onset of labor, the veterinarian may first try to correct the twist by sedating the mare and rolling her in the direction of the torsion. In most instances, however, uterine torsion must be corrected by rotating the organ to its proper position through an incision in the mare's flank. If, during this procedure, the veterinarian discovers that the fetus is dead or that the uterus is congested, removal of the fetus is usually recommended.

If uterine torsion is not discovered until after labor has begun, the chances of the foal's survival decrease significantly. In some instances, uterine position can be corrected by inserting an arm through the cervix and rotating the foal. Most cases of dystocia caused by uterine torsion require Caesarian section, however. It has been estimated that approximately 60 percent of the mares and only 30 percent of the foals survive when uterine torsion is not detected until after the onset of labor. Mares that do survive usually show no significant decrease in fertility as a result of the torsion, and there is little evidence to suggest that uterine torsion will recur during subsequent pregnancies.

UTERINE INERTIA

If the second stage of labor is prolonged or delayed with an absence of normal contractions, the attendant should consider the possibility of uterine inertia. Uterine inertia, or failure of the uterus to contract with enough force to expel the foal, is often associated with other foaling problems. In most instances, this absence of or decrease in uterine movement is caused by exhausted uterine muscles. For example, if the second stage of labor is extended due to malposition of the foal, the uterine muscles may become fatigued and fail to expel the foal even though its position has been corrected. This condition is often associated with an absence of voluntary straining due to a painful or prolonged second stage of labor. Uterine inertia may also be caused by physical changes in the uterine muscles that occur with age or as a result of prolonged uterine infection.

If the mare is not assisted quickly (e.g., by correcting the foal's position and applying traction), the uterine muscles may form rings that contract tightly around the foal, thereby complicating the relief of dystocia. In addition, allantoic fluids are lost, and the birth canal becomes very dry, making it difficult to remove the foal without damaging the mare's reproductive tract. If the uterus and cervix are contracted around the foal or if the birth canal cannot be lubricated adequately, Caesarian section may be recommended by the veterinarian. Complications that often occur in association with uterine inertia include infection, retained placenta, and failure of the uterus to return to its normal size after foaling. It is also important to note that forced extraction of the foal (e.g., by more than one person) may cause uterine rupture.

Abnormal Conditions Associated with Parturition

The primary concern of the foaling attendant is to protect the well-being of both the mare and foal during and immediately after parturition. To carry out this duty, the attendant must be able to recognize problems quickly and provide immediate assistance or to summon a veterinarian as soon as possible when medical supervision is needed. It is equally important that the attendant be able to recognize complications associated with parturition that may threaten the mare's well-being. These conditions vary in severity and may occur before, during, or shortly after parturition, but each may have serious long-term consequences with respect to the mare's health and fertility. In most instances, immediate veterinary attention is essential. By carefully observing the mare during and after parturition, the educated attendant can identify these conditions early, thereby improving the chances of successful treatment. (Note: Abnormal conditions associated with parturition that may affect the foal's well-being are examined in "Neonatal Foals.")

HEMORRHAGE

One of the most serious conditions associated with parturition is hemorrhage within the reproductive tract or from one of its associated blood vessels. The severity of this condition varies with the location of the bleeding and the amount of blood that is lost. Acute hemorrhage, which involves the sudden loss of large amounts of blood, is often fatal, while a slower, chronic hemorrhage can often be treated successfully when the condition is recognized promptly. For the purposes of this discussion, birth-related hemorrhage in the mare is divided into two categories: external and internal.

External hemorrhage is any bleeding that occurs within the reproductive tract in such a way that it either collects in the tract or is expelled through the vulva. This type of hemorrhage is often caused by lacerations within the birth canal which, in turn, may be caused by the foal's feet or by the attendant when manipulation of the foal's position is required. If hemorrhage causes a swelling, or hematoma, that blocks the pelvic cavity, defecation may be very difficult for the mare. If a major vessel is ruptured, an attempt to reduce the bleeding should be made either with a pressure pack (e.g., sterile cotton) or by a special surgical procedure if a veterinarian can be summoned quickly. If a large amount of blood is lost, the mare may require treatment for shock and loss of body fluids.

External hemorrhage may also be caused by failure of the blood within the allantochorion (fetal side of the placenta) to be transferred to the foal due to uterine inertia or to premature rupture of the umbilical cord during forced extraction. This type of bleeding does not affect the dam but may cause a serious oxygen deficiency in the newborn foal. Because of the manner in which the placenta is attached to the mare's uterine wall, seeping from the endometrium is unusual, unless the placenta is torn out prematurely with marked force. This type of bleeding can be diagnosed and treated with some success. In rare instances, rupture of a major uterine vessel, such as the posterior uterine artery, causes external hemorrhage from the mare's uterus. This type of bleeding usually requires surgical correction, restoration of the circulating blood volume, and treatment for shock.

Internal birth-related hemorrhage usually involves rupture of the utero-ovarian artery, middle uterine artery, external iliac arteries, or the uterine wall. As the vessels lose their elasticity, stretching during pregnancy predisposes them to rupture during parturition. Exercise or stress immediately after parturition also increases the chances of rupturing a weakened blood vessel. If vessels within the broad ligament rupture in such a way that the blood is contained within the tissue layers of the ligament, the mare may survive the resulting shock and anemia. However, if blood is released into the peritoneal cavity or a major vessel is involved, the condition is almost always fatal. Fatal hemorrhage is seen most frequently in older mares and is usually associated with aneurysms (dilated artery) or degenerative changes in the arteries. When uterine rupture is the source of internal hemorrhage, bleeding may or may not be profuse, depending on the extent of damage to the uterus. (Refer to 'Uterine Rupture' within this chapter.)

Initially, the hemorrhaging mare usually appears uneasy and is reluctant to rise. Internal hemorrhage is also characterized by violent rolling and profuse sweating and may be difficult to distinguish from the discomfort associated with the second and third stages of labor. However, these signs persist and intensify, and the mare's mucous membranes become pale and then yellow. If blood is escaping into the peritoneal cavity, the mare usually staggers and falls and may have muscle spasms and a retracted upper lip. When the signs of internal hemorrhage are recognized, a veterinarian should be summoned, and the newborn foal should be removed from the immediate area for its own protection. Although the mare's chances of survival are slim, prompt surgery is sometimes successful. If the mare dies during labor, the foal should be removed by traction or through an incision in the mare's flank.

When hemorrhage is suspected, the mare should be kept as calm as possible, since increased blood pressure may aggravate the condition. The mare should be confined, and the veterinarian may administer tranquilizers or drugs to reduce her blood pressure. Occasionally, mares can be saved through proper care, but they remain susceptible to rupture.

RUPTURED PREPUBIC TENDON

The prepubic tendon is an elastic structure that holds the pelvis in position and serves as a point of attachment for the abdominal muscles which, in turn, help to support the uterus during pregnancy. Although a ruptured prepubic tendon is unusual, it is sometimes seen in obese and

Courtesy of Kansas State University Department of Surgery and Medicine

This mare shows the characteristics of a ruptured prepubic tendon: tilted pelvis, stretched stance, and a pendulous abdomen.

aged mares as the result of either trauma or increasing weight of the fetus during the last few months of pregnancy. Rupture of this important tendon may occur suddenly as the result of an accident or of abdominal straining but is more often a gradual process that is preceded by partial edema of the abdomen. Once rupture is complete, the mare is characterized by a tilted pelvis, a stretched stance, increased respiration, a weak pulse, and sweating. Eventually, the mare collapses and dies. If she is near term when this occurs, the foal may be salvaged by Caesarian section, but in most cases of complete rupture, both the mare and her foal are lost. If the impending rupture is recognized before it is complete, it can sometimes be delayed by supporting the abdomen with a sling until the fetus can be delivered by Caesarian section. This is seldom successful, however, unless the mare is already very close to the end of gestation. Even if preventive measures and Caesarian section are successful, the damaged tendon cannot be repaired, and the risk of rupture during subsequent pregnancies eliminates the feasibility of rebreeding the mare.

UTERINE RUPTURE

Rupture of the uterine wall is another unusual complication of parturition that occurs as the result of difficult labor or faulty obstetrical techniques. Although uterine rupture may also be caused by an accident, it is more frequently associated with severe dystocia when expulsion efforts are extremely vigorous or when correction is delayed. Sources have suggested that spontaneous uterine rupture may be caused by abnormalities such as uterine torsion, gross uterine distention (e.g., caused by the presence of twins), excessive fetal size, or breech presentations. The severity of uterine rupture varies. A small tear, for example, may not even be noticed until an adhesion is palpated during a subsequent reproductive examination. If uterine rupture results in a large tear, the foal and large amounts of blood may enter the abdominal cavity. In such instances, uterine contractions cease, and the attendant may suspect uterine inertia. The size and site of the lesion dictate whether the veterinarian will allow delivery to proceed via the birth canal or elect to perform a Caesarian section. If uterine rupture is not recognized and corrected after parturition, a portion of the intestines may protrude into the uterus and through the vulva. If the uterus is not extremely contaminated during labor, most lesions can be corrected surgically.

INTESTINAL RUPTURE

Ruptures of the large intestine (cecum and colon) have also been observed in mares during parturition. The affected mare is characterized by sweating, trembling, increased heart rate, fever, and depression. Straining efforts are diminished, and the progress of parturition is impeded. Because the digesta within the intestinal tract is a rich source of contaminants, intestinal ruptures allow infections to become established in the abdominal cavity. Infections in this area are often fatal and, therefore, require immediate veterinary attention. Sources have suggested that the chances of intestinal rupture increase when the mare is given large amounts of feed prior to parturition. For this reason, some broodmare managers limit the mare's feed intake for a short period prior to her estimated foaling date. Although it seems likely that this measure helps to reduce intestinal pressure, the effectiveness of this precaution has not yet been established.

PERINEAL LACERATIONS

Damage to the vagina and/or vulva is one of the birth-related complications commonly seen in the mare. As with other postparturient problems, perineal lacerations vary in severity. First degree lacerations involve only the mucous lining of the vagina or vulva. These injuries are not serious, but swelling may persist for several days after parturition. Injuries that affect deeper tissues within the vagina and vulva are referred to as second degree lacerations.

First and second degree lacerations may be caused by malposition or malpresentation of the foal, failure to open the sutured mare prior to foaling, an unusually large or deformed foal, manipulation or forceful extraction of the foal, or by the instruments used during fetotomy. Occasionally, one or more of the foal's hooves may completely penetrate the vaginal wall, entering the rectum. These so-called rectovaginal fistulas are sometimes referred to as third degree lacerations. If uterine contractions continue to propel the foal, the shelf that separates the birth canal from the rectum may be completely torn as the foal's feet exit through the mare's rectum. These extensive fourth degree lacerations may look hopeless upon initial examination, but they are rarely life-threatening. Although severe lacerations may cause conformational changes that predispose the mare to urine pooling and pneumovagina, most mares can be rebred after reconstructive surgery. It is important to note, however, that perineal lacerations require immediate veterinary attention to assess the damage and to clean

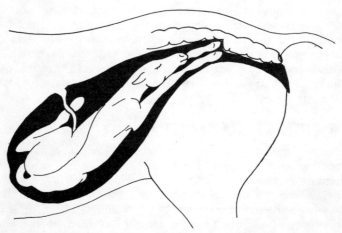

Rectovaginal fistulas often start when the foal's feet are deflected upward toward the rectum.

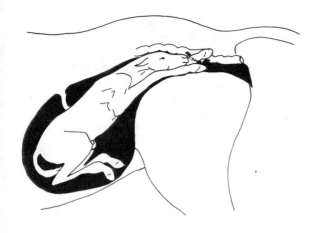

Uterine contractions force the foal's feet out of the vagina and into the rectum.

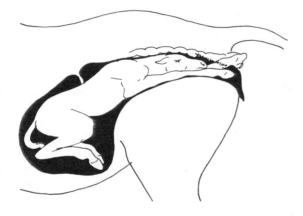

Unless the feet are pressed back into the vagina and guided through the birth canal, they continue to tear through the shelf which separates the rectum and the vagina.

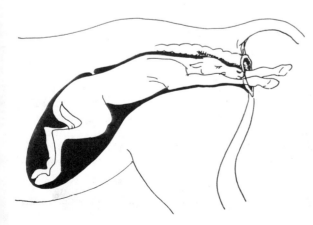

Uterine contractions continue to force the foal's feet through the shelf and form a rectovaginal fistula.

the torn surfaces, thereby minimizing the chances of infection. Occasionally, lacerations can be corrected immediately. However, after a few hours the damaged tissue swells considerably so that reconstruction becomes impossible. If the lacerated area is already swollen, surgical repair must usually be postponed for several weeks.

UTERINE PROLAPSE

Uterine prolapse is an unusual birth-related abnormality that involves the inversion of the uterus in such a way that it protrudes through the vagina and the vulva. This condition may be the result of forcing the placental membranes from the uterus or excessive straining during prolonged dystocia. Once uterine prolapse is recognized, it is extremely important to wrap the uterus in a clean towel or sheet that has been moistened with warm water. This procedure prevents drying of the uterine tissues, helps to protect the uterus from injury and contamination, and provides some support for the displaced organ. It is also important to keep the mare standing quietly until a veterinarian arrives. Forcing the mare to stand helps to reduce the force of any contractions that may aggravate the prolapse and also eliminates the possibility of uterine lacerations caused by rubbing or catching the organ on objects such as stall walls, fence posts, or feed containers.

Replacement of the prolapsed uterus must be performed as soon as possible, not only because of the risk of damage to the uterus but also because other complications (e.g., prolapse of the intestines into the normal uterine location, hemorrhage, edema) may follow. These secondary conditions complicate replacement of the uterus and place additional stress on the mare. The veterinarian may administer special drugs to combat shock, reduce straining, and minimize the chances of infection. The uterus is then cleansed gently with warm saline and worked back through the pelvic opening. This procedure is usually very difficult and care must be taken to prevent uterine rupture (i.e., by a finger penetrating the uterine wall). If the uterus is replaced successfully without recurring prolapse, the mare's chances of conceiving and carrying a foal the following year are very good.

RETAINED PLACENTA

During the final stage of labor, blood drains from the placental villi, which are embedded in the endometrium. As a result, the projections shrink and slough away from the endometrial crypts. As the intimate attachment between placenta and the endometrium is lost, uterine contractions and the weight of the fetal membranes hanging from the mare's vulva expel the afterbirth from the uterus. Normally, the placental membranes separate from the endometrium and are expelled within 30 minutes to 3 hours, but occasionally, the membranes are retained longer. Retention almost always involves a localized area of attachment in the nongravid uterine horn and has been closely associated with abortion, Caesarian section, and dystocia. Authorities suggest that the condition is caused by abnormal uterine muscle activity or hormone imbalances.

If the placental membranes are not expelled within 6 hours or if examination of the afterbirth reveals that pieces of the placenta may have been left behind, a veterinarian should be summoned immediately. The presence of degenerating placental tissue within the uterus predisposes the mare to acute metritis, laminitis, and blood poisoning. The severity of these conditions varies with the amount of tissue present and the length of time the membranes are retained. For this reason, placental retention should be dealt with quickly by a veterinarian.

Gentle traction may help to loosen retained tissues, but there is always a danger of leaving behind a small tag that is tightly adhered to the endometrium. In addition, traction increases the chances of uterine prolapse. Since the retained placenta is usually attached at the tip of one uterine horn, pulling on the placenta may turn the uterus inside out. Infusion of the uterus helps to stimulate the release of these areas of attachment so that the membranes can be retrieved within 24 to 48 hours. The hormone oxytocin has also proven to be helpful in stimulating the expulsion of retained membranes. If the placenta has been retained for less than 18 hours, the veterinarian may infuse several gallons of liquid into the placenta and close off the membranes in such a way that the fluid-filled tissues balloon inside the uterus. This pressure stimulates uterine contractions within 5 to 10 minutes, and in most instances, the entire placenta is released within 15 to 30 minutes.

DELAY OR ABSENCE OF UTERINE INVOLUTION

After the foal is delivered, the placental membranes are expelled, leaving the uterus stretched and edematous. Uterine contractions continue after stage three of labor is complete, aiding in the expulsion of pregnancy-related secretions, blood, and debris. The appearance of a discharge from the vulva is normal for several days after foaling unless pus or large amounts of blood are present. As these postparturient contractions continue, the uterus becomes less vascular and begins to shrink. Immediately after foaling, the uterus begins returning to its normal, healthy, nonpregnant state. This healing process is referred to as involution. During this period, the mare may experience postparturient colic (e.g., rolling, kicking at her belly, sweating) caused by the physiological changes that occur during involution. (Note: Severe abdominal pain may be caused by other birth-related abnormalities, such as hemorrhage or constipation.)

In general, uterine involution is influenced by nutrition, general health, history of uterine infections, ease of foaling, the mare's age, and the number of times she has been pregnant. Although the rate of uterine involution varies greatly between mares, uterine size on the seventh day after foaling is usually about two to three times that of a barren mare. Some mares are ready to conceive and maintain the conceptus during their foal heat, while other mares may show only limited involution by their second postpartum heat. Nursing promotes uterine involution by stimulating the release of oxytocin which, in turn, stimulates uterine contractions. Exercise also advances involution by improving muscle tone and promoting the expulsion of pregnancy-related secretions.

Delayed involution has been associated with abortion, dystocia, placentitis (i.e., infection within the placenta), and retained placentas. During the first few days postpartum, the mare may be characterized by signs of mild pain, loss of appetite, and depression. Because the cervix may close tightly before involution is complete, rectal palpation of uterine tone is a more reliable means of determining the degree of uterine involution than the degree of cervical closure. Uterine biopsy is also becoming a valuable aid in diagnosing uterine involution, since the regeneration process involves identifiable changes on a cellular level. If the veterinarian feels that involution is not proceeding normally, oxytocin may be administered and exercise recommended to stimulate uterine regeneration. With advanced age and numerous pregnancies, the uterus may gradually lose its ability to regenerate. Once the uterus reaches this state, attempts to stimulate involution are often futile. In such cases, the ability of the mare's uterus to support fetal growth decreases significantly.

Maternal Behavior Problems

Throughout the previous chapter, the importance of allowing the mare and foal to establish a strong bond has been emphasized. Although a specific equine nurturing pattern is imbedded in each mare's genetic make-up, there are both physiological and environmental factors that can impede or completely inhibit maternal instinct. In addition to abnormal behavior patterns, problems may occur when normal maternal responses are repressed or challenged by a domestic environment. In order to differentiate between true and imposed behavioral problems, the breeder must understand the mare's preservation instinct, which is perhaps at its highest point during and immediately after foaling.

In the wild, survival of the horse depended on its ability to flee from predators. Because both the mare and her foal were vulnerable to attack during and immediately after foaling, inborn behavior patterns that maximized their chances of survival developed throughout the horse's evolution. For example, the equine birth process is normally brief and intense, and its onset can be determined to some extent by the mare who seeks an isolated hidden area and then waits until protective darkness falls. Soon after her foal is born, the mare encourages the establishment of a close bond through a series of stimuli and responses. During this process, the foal becomes aware of its surroundings, attains marked agility, receives nourishment, and learns to recognize its mother. Within two hours of birth, the healthy foal is capable of joining its dam in flight if the need arises.

The factors that influence the mare's behavior during this important adjustment period are instinctive, hormonal, and learned. The mare's instinct to protect her foal is illustrated by the similarity of maternal behavior patterns between individuals and by the tendency for "good mothers" to produce daughters with strong maternal drives. Hormonal influence (e.g., oxytocin, estrogen, and progesterone) has been identified in association with instinctive maternal drive, and attempts to treat certain maternal behavioral problems with hormones has been somewhat successful. It is important to note, however, that the exact hormonal mechanisms that affect maternal behavior are not completely understood. Learned factors are sometimes difficult to differentiate from inherent influences, but the possibility that learning helps to shape maternal behavior cannot be ignored since most mares that reject their foals are maiden mares that have not witnessed maternal behavior patterns (i.e., in a herd situation).

REJECTING THE FOAL

In the wild, mares may reject or savage and kill their foals if they fail to respond favorably during the initial bonding period, but rejection of a healthy foal and failure to carry out the normal maternal routine is reportedly very rare under feral circumstances. In a domestic environment, however, the mare's maternal drive may be inhibited to some extent by isolation from herd relationships and by reconditioning of behavioral responses (i.e., training). In instances where the mare is simply not capable of utilizing the instinctive drive due to a lack of "education" (usually seen in first-foal mares), gentle restraint and encouragement is usually very effective in helping the mare realize her responsibilities. If the mare's udder is sore, she may refuse to allow her foal to nurse. Once again, gentle restraint is usually very effective, and once pressure within the udder is relieved, the mare is usually quite content and allows her foal to nurse without further problems. (If the mare's udder is infected, the foal may require bottle feedings until the infection can be corrected.)

There are a few instances, however, when the mare's disinterest or violent resistance is more difficult to correct. If she has suffered a difficult foaling, she may be too exhausted to carry out the stimuli-response routine that establishes the required bond with her foal. As a result, she may refuse to acknowledge the foal's presence or may reject it physically. (Excessive human interference in the foaling area may also result in rejection of the foal.) Techniques that have been used to help the mare identify the foal as her own include rubbing the foal with the mare's milk, urine, or feces and allowing the foal to nurse repeatedly while the mare is restrained physically or chemically (e.g., tranquilized).

If the mare repeatedly rejects each of her foals or if she rejects a foal after accepting it initially, she may have a hormone imbalance that inhibits normal maternal response. In such cases, the mare should be watched carefully after each foaling to prevent her from injuring her foal, especially if she has a history of savaging foals. Unfortunately, attempts to reunite the mare and foal are often futile. In some instances, hormonal therapy may prove helpful, but the use of hormones in treating maternal behavioral problems has not been studied extensively.

It should also be noted that failure of the mare to carry out her normal maternal responsibilities may be caused by the foal's failure to identify its true mother. This may occur when the newborn is greeted warmly and pampered by well-meaning bystanders. This interference in the normal mare-foal bonding period makes it very difficult for the foal to respond to the stimuli-response pattern that helps to establish a normal mare-foal relationship.

POSTFOALING AGGRESSION

The mare's instinct to discourage the approach of other horses during the postfoaling bonding period reflects her understanding of the foal's initial inability to determine who its mother is. After the mare's identity is imprinted on the foal's memory, she allows curious horses within the herd to investigate her youngster. In a domestic environment, this early instinctive aggression may be directed toward trusted handlers or bystanders. Through their acquired trust of humans, many mares learn to accept human assistance during and immediately after foaling. It is important to remember, however, that although aggression and nervousness may be considered maternal behavioral problems when directed toward the foal, they are not abnormal responses of the mare to human interference in the foaling process. ♘

18

LACTATION

A mare's pedigree, fertility, and her ability to carry a foal to term are obvious characteristics that affect her value as a broodmare. Once the mare has produced a foal, however, the importance of another characteristic, mothering ability, must be considered. Mothering ability involves both the mare's willingness to protect her foal and her physical capacity to provide the youngster with nutrients. The mare's ability to nourish her newborn foal depends on inherited and environmental factors. For example, even a heavy-milking mare may not be able to produce adequate amounts of milk if she does not receive enough energy or protein. Although it would be impractical to select mares solely on the basis of mothering ability (except when breeding nurse mares), the breeder must be able to recognize the mare's limitations and should understand how special management techniques can help the mare reach maximum production. As a result, the foal is more likely to attain optimal growth and development.

The mare's ability to nourish the foal through lactation is one of her most important maternal functions.

The mare's ability to produce milk for her foal is the focal point of this chapter. However, the reader should not forget that the mare must also exhibit a willingness to nurse the foal. In most instances, the mare encourages her newborn foal to nurse and seems relieved when the foal is finally successful in locating the udder. Occasionally, the mare may require some reassurance or gentle restraint while the foal has its first meal. In unusual cases, the mare totally rejects the foal so that the attendant must bottle feed the youngster or provide it with a nurse mare. (Refer to "Orphan Foals.")

As with many other technical subjects in this text, the following discussion on the physical and managerial aspects of lactation is introduced by a description of important anatomical characteristics. This preliminary material provides important background information for subsequent discussions on milk production, milk let-down, and abnormalities of the udder.

Mammary Gland

The mammary gland, or udder, is a collection of highly modified oil glands that are capable of collecting important nutrients and secreting those nutrients in the form of milk. Although this feature is characteristic of all mammals, tissue structure and milk composition vary significantly between species. The mammary gland is not actually a part of the reproductive tract, but it is considered an important accessory gland of the reproductive system. In fact, the size and productivity of this gland are governed by many of the same hormones that control the mare's reproductive activities (e.g., estrus, pregnancy maintenance, and parturition).

ANATOMY

The mare's udder is located in the inguinal region between her hindlegs. It is arranged in two halves which lie on either side of a median line. This separation is marked by two suspensory ligaments. Each half of the udder is composed of two glands, or quarters. The individual glands have separate drainage networks although they are supplied by common nerves and blood vessels.

The udder's structure undergoes significant physical changes as the mare's reproductive status changes. During lactation, each gland contains a collection of tiny, sac-like cavities (alveoli) which are lined with special milk-producing cells. The alveoli are organized into groups of about 100 each. The resulting milk-producing unit is surrounded by a connective tissue septum and forms what is referred to as a lobule. (In the non-lactating mare, these secretory units are represented by fat and connective tissue and are difficult to differentiate from surrounding tissue.) The lobules and their interconnecting ducts are further organized into lobes, which are surrounded by even more connective tissue. It should be noted that the ducts leading from the alveoli also contain specialized secretory cells which are believed to contribute to the production of milk. In addition, both the alveoli and interconnecting ducts contain myoepithelial cells beneath and surrounding each secretory cell. The muscle-like myoepithelial cells aid in the expulsion of milk from the glands, a process that will be examined later in this chapter.

Ducts from each lobule drain into the teat cisterns which supply each teat canal. The four glands, or quarters, are drained by separate teat canals and separate streak canals. The two canals draining one half of the udder (two glands) then emerge from a single teat. Thus, although the mare has four mammary glands and four separate duct systems, she has only two teats. The teat opening is closed by a sphincter muscle, which normally prevents milk from streaming out of the teat until milk ejection is stimulated.

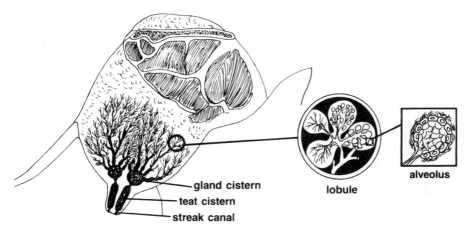

gland cistern
teat cistern
streak canal
lobule
alveolus

As can be seen when one half of the udder is viewed in cross-section, the alveoli are drained by a complex network of ducts leading to two streak canals. The ducts leading to each streak canal drain one quarter of the udder, and although the ducts from each side intermingle, they do not communicate with one another.

DEVELOPMENT

At birth, the filly has only slight mammary and teat development. Prior to puberty, an increase in the amount of fat and connective tissue causes a slight increase in the size of the filly's mammary glands, but the milk-producing system of alveoli and ducts remains incompletely developed. While the prepuberal increase in mammary size is a response to the same hormones that cause overall body growth, development of alveoli and ducts is caused by estrogen and progesterone. More specifically, estrogen causes the development of the duct system, while progesterone in the presence of estrogen stimulates development of the alveoli. Thus, alveoli and duct development is stimulated as the filly approaches puberty. The presence of insulin, a hormone produced by the pancreas, seems to be required for normal response of the mammary gland to estrogen, but the exact mechanism is unknown. During the mare's first pregnancy, development of the mammary gland proceeds rapidly. The organization of the alveoli into lobules becomes extensive, and milk production increases gradually.

MILK PRODUCTION

Although milk production is initiated during pregnancy, the release of milk from the alveoli and intralobular ducts is normally inhibited until just prior to or immediately after parturition. Researchers have suggested that the high levels of estrogen and progesterone during pregnancy stimulate the release of prolactin (a hormone that stimulates milk secretion) from the anterior pituitary gland. However, it is believed that high estrogen and progesterone levels also inhibit any action of prolactin on the mammary glands. Because other hormones are involved in stimulating milk production, some milk synthesis occurs during pregnancy. When estrogen and progesterone levels drop just after parturition, the effect of prolactin is no longer inhibited, and milk secretion increases significantly.

As the secretory cells become active, the collection ducts and teats fill with milk and the gland becomes warm, swollen, and firm. Retention of milk prior to parturition distends the alveoli and depresses milk production. Release of the hormone oxytocin from the neurohypophysis during parturition causes contractions of the muscle-like myoepithelial cells within the alveoli and intralobular sinus. This action causes milk ejection (milk let-down). As a result, pressure within the alveoli diminishes, and milk secretion is stimulated. For this reason, frequent suckling or manual milking is thought to enhance milk production.

Oxytocin release is triggered by stimulation of the uterus, vagina, and vulva (e.g., during parturition or breeding) and by tactile stimulation of nerve endings in the teats (e.g., when the udder is suckled or washed). Fear

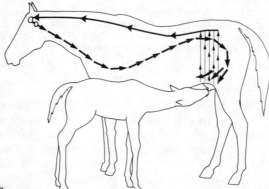

Oxytocin feedback

Nerve impulses originating from tactile stimulation of the teats (suckling) cause the hypothalamus to stimulate oxytocin release from the pituitary gland into the bloodstream. Oxytocin then stimulates milk let-down by causing the myoepithelial cells surrounding each alveolus to contract. This action forces the milk within the alveoli into the large ducts that supply the teat cistern. The milk is then retained in the teat cistern by the teat sphincter until it is removed by suckling.

and excitement in the mare can block these milk ejection stimuli. The exact mechanism for the inhibition of milk let-down is not known, but researchers suspect that the adrenal glands play an important role. In stressful situations the adrenal glands release the hormone epinephrine, also known as the "fight or flight" hormone. Epinephrine may act on the hypothalamus to block oxytocin release or it may cause blood vessels within the mammary gland to constrict, thereby preventing oxytocin from reaching the blood capillaries which supply the udder.

When the mammary gland produces milk but is not suckled or milked, as happens during late pregnancy, a yellowish, slightly syrupy milk is produced. This milk is called colostrum and contains high levels of fat, vitamins, minerals, and protein. It also contains elements which have a laxative effect on the foal's digestive tract and antibodies which establish passive immunity within the foal's bloodstream during the first few weeks of life. The importance of this concentrated milk is discussed in more detail in the next chapter, "The Neonatal Foal."

As long as the foal nurses, stimulation of the nerve endings within the mare's teats triggers the release of special hormones from the hypothalamus which, in turn, maintain lactation. Immediately after parturition, milk production increases dramatically and continues to increase at a slower rate until production peaks. Varying research reports have indicated that lactation in the mare peaks from one to three months after the foal's birth. After the period of increase, milk production gradually decreases until soon after weaning, when production halts completely. The quantity of milk produced varies significantly between individuals, depending on factors such as breed type, nutrition, environment, and health. For example, a heavy draft mare produces about 40 pounds (18 kilograms) of milk per day during peak lactation, while a light horse mare produces about 33 pounds (15 kilograms) per day and a pony mare produces about 27 pounds (12 kilograms) per day.

When determining the expected output from an individual, body weight must be considered. Light horses, for example, produce about 3 percent of their body weight per day in milk during the first 12 weeks of lactation

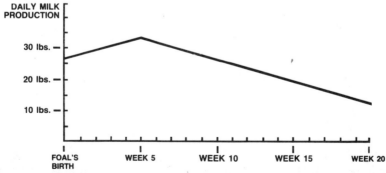

This graph represents a characteristic lactation pattern for an 1100 pound mare with milk production peaking at about five weeks. The mare still produces a significant quantity of milk after 20 weeks of lactation, although this milk is of low nutritional quality.

and 2 percent per day during the later stages of lactation. Pony mares, on the other hand, produce 4 percent of their body weight in milk per day during early lactation and about 3 percent per day after the first 12 weeks. A mare of average milking ability and weighing 1100 pounds (499 kilograms) can be expected to produce approximately 3000 pounds (1361 kilograms) of milk during a lactation of about five months. Her milk is usually composed of about 89 percent water, 1.6 percent fat, 2.7 percent protein, 6.1 percent lactose (sugar), and 0.5 percent ash (minerals). The foal's nutritional needs increase constantly, but the quality of the mare's milk decreases throughout lactation. For this reason, mare's milk gradually becomes an inadequate source of nutrition for the foal. Therefore, it is important to start a creep feeding program for the foal very early. (Refer to "The Foal's First Year.")

INVOLUTION

When the foal is weaned, milk accumulates in the mare's mammary gland and intramammary pressure discourages further milk production. The hormones mentioned earlier still act on the mammary gland, but increasing pressure prevents the alveolar cells from synthesizing additional milk. As the mammary gland returns to its nonlactating state (a process referred to as involution), milk in the gland is resorbed and milk-producing tissue degenerates as it is replaced by fat and connective tissue. Some of the residue from the degenerating tissue is found in the colostrum of the mare's next lactation. With each lactation, ducts and alveoli redevelop within the inactive, involuted gland.

Care of the Lactating Mare

Common-sense husbandry techniques are important to many aspects of equine management, including lactation. Proper care of the lactating mare from foaling until weaning benefits both the mare and her foal by minimizing the incidence of mammary infection or irritation and by maximizing milk production.

WASHING THE UDDER

Because the two halves of the mare's udder lie closely together, dirt, oily secretions, and dead skin accumulate within the fold of skin that divides the udder. Accumulated oil and debris are sources of irritation to the mare. In an attempt to relieve this discomfort, the mare may rub her tail against a stationary object. This accumulation is also a source of infection to an injured udder and may cause digestive problems in the suckling foal. Regular, gentle washing of the mare's udder with mild soap and warm water to remove built-up smegma prevents these problems and accustoms the mare to having her udder touched.

NUTRITION

With the possible exception of heavy exercise or work, no other equine function has higher nutritional demands than lactation. After parturition, the mare's nutrient requirements increase by at least 75 percent; most of this increase is due to the physical demand of milk production. The nutrient requirements for milk production are particularly high because the conversion of the mare's digested nutrients into milk is a very inefficient process. For example, the conversion of digestible energy to milk energy is only 60 percent efficient. In other words, 40 percent of the energy used by the mare for milk production is not present in the final product.

If the mare's nutrient needs are not met, both milk production and the mare's body weight and condition may be affected. When dietary protein levels are inadequate, milk production decreases. In contrast, dietary energy deficiencies cause weight loss in the mare due to the depletion of body energy for milk production. Some weight loss during peak lactation may be unavoidable, but an attempt should be made to maintain the mare's body weight. (Excessive weight loss may interfere with conception.) Similarly, a calcium-deficient diet causes the release of calcium from bone reserves to facilitate milk production and to maintain proper calcium levels in the blood. (A constant level of calcium in the blood must be maintained at all times to support important physiological processes.) Phosphorus levels are also important, not only for milk production but also for the support of important body functions and for proper calcium utilization. Other minerals are also required at higher levels during lactation than during gestation. Although mineral deficiencies rarely cause death, they may cause reproductive problems, inefficient feed utilization, and a reduction in overall performance and condition.

It is important to note that the mare cannot consume enough good-quality grass hay or grass forage to meet her protein, vitamin, and mineral requirements. However, if free-choice alfalfa or plenty of lush, green legume pasture is provided, she is usually capable of consuming enough roughage

Although most lactating mares require grain supplementation to meet their energy requirements, pasture can supply many of the nutrients that are required for milk production.

(approximately 2.5 to 3.0 pounds per 100 pounds of body weight, or 1.1 to 1.4 kilograms per 45 kilograms body weight) to meet her protein, calcium, and vitamin A needs. If legume roughage is not supplemented with grain, the mare's diet may be deficient in phosphorous, however. Thus, a lactating mare receiving no grain should have access to a free-choice, mineral and salt supplement that will provide adequate levels of phosphorus.

Energy is an extremely important component of the lactating mare's diet. She requires approximately 1.3 Mcal of digestible energy per pound (.45 kilogram) of feed during the first three months of lactation and about 1.2 Mcal of digestible energy thereafter.

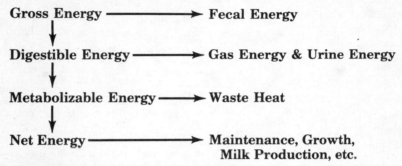

In the conversion to net energy, some gross energy is lost to byproducts such as feces, gas, urine, and waste heat. Thus, the mare is unable to use all of the energy she consumes for her maintenance and for milk production.

Because the conversion of dietary protein to milk protein is a very inefficient process, the mare should be fed a ration that contains at least 13 percent crude protein in order to insure proper milk production. This requirement can be met by feeding the mare a 14 percent crude protein grain supplement, a mineral and salt supplement, and high-quality roughage, containing approximately 12 percent crude protein. (The weight of the grain in the total ration should not exceed the weight of the roughage.) In addition to protein quantity, protein quality(the balance of amino acids in the ration) should be considered. Diets that are comprised entirely of grain and grass hay or pasture are notably deficient in lysine, an amino acid that is essential for growth and high-quality milk production. Fortunately, alfalfa hay or pasture and soybean meal are excellent sources of lysine and are easy to add to the mare's ration. Thus, a portion of the mare's total protein requirement during lactation should be supplied by one of these supplemental protein sources.

In order to insure that adequate amounts of each nutrient are provided to the mare, each ration should be analyzed for its protein, energy, calcium, phosphorus, and vitamin A content. The manufacturers of commercial rations provide analyses on their products, which can be found on feed tags. However, the analyses are seldom very detailed, and may list only crude protein and ash (a rough measure of mineral content). Thus, an analysis of the ration should be obtained from a testing laboratory. (State or county agricultural extension agents can usually provide addresses for testing laboratories.) The nutritional quality of hay can be estimated based on its variety and its physical condition, but it should also be analyzed in order to accurately determine nutritional value.

Most commercial feeds produced for the lactating mare not only meet the mare's protein and energy needs but also provide adequate amounts of calcium, phosphorus, and vitamins. The mare's vitamin A requirement can also be met by feeding high-quality hay or providing lush, green pasture. Her vitamin D needs are met by feeding sun-cured hay or allowing her to spend at least part of each day in the sunlight. A more detailed description of the mare's nutritional requirements during lactation is presented in the Appendix, but the reader should note that because of the great variability of milking ability and feed efficiency between mares, adjustments for supplemental feeding are often based on visual appraisal of the mare's condition. (Heavy milkers may require additional concentrate in their rations in order to maintain body condition.) To prevent obesity but maintain the mare in moderate condition, concentrate supplementation should be reduced gradually as milk production decreases. Regardless of the mare's stage of lactation or of her individual ability to convert digested nutrients into milk for her foal, one factor remains constant throughout lactation: the mare needs plenty of fresh, clean water since the milk she produces contains about 90 percent water.

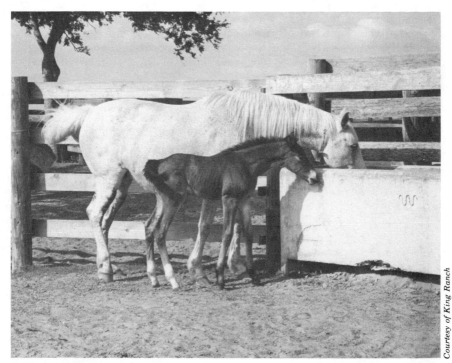

Courtesy of King Ranch

Water is essential for the lactating mare due to the demands of milk production. She should have access to a constant supply of clean water at all times.

WEANING

When the foal is weaned or otherwise prevented from nursing, milk accumulates in the mare's mammary gland. The resulting intramammary pressure stops milk production within the alveoli and intralobular ducts even though the hormones that encourage milk production are still present in the mare's bloodstream. The drying-up process, which results in involution of the mammary gland, often causes the mare significant discomfort. The mare becomes sensitive due to increased intramammary pressure and due to dry udder skin which may result during this period. To prevent the discomfort of dry skin, cocoa butter, camphorated oil, or a mixture of lard and spirits of camphor can be applied to the mare's udder. Although it is tempting to milk out a ballooning udder to relieve some of the mare's discomfort, this process only encourages further milk production and prolongs the drying-up process. In addition, excessive handling of the mare's sensitive teats may increase the mare's discomfort rather than relieve it. (In some instances, mares whose tight bags persist longer than four days after weaning are milked to check for the presence of infection. Refer to the discussion on mastitis within this chapter.)

The most successful technique for relieving painful intramammary pressure after weaning is to reduce grain intake and establish an active exercise program just prior to weaning. Decreased grain rations help to minimize further milk production, and exercise encourages involution of the mammary gland. Many breeders take their mares off grain completely after weaning to hasten the involution process. If a mare must be fed a concentrate ration after weaning to improve condition, her diet should be increased only after she has dried-up completely and then only increased sparingly at first.

Courtesy of Earl and Barbara Lang Quarter Horses.

Free exercise helps promote involution of the mare's mammary gland after the foal is weaned.

Abnormalities of the Udder

Although the most common causes of insufficient milk production are inadequate nutrition and inherited limitations, there are several abnormalities of the udder that may affect the mare's ability to transfer nutrients to her foal. Fortunately, however, abnormalities of the udder are uncommon in the mare.

FAILURE TO PRODUCE SUFFICIENT MILK

When the mare is not producing enough milk, her foal may be seen nursing vigorously at frequent intervals but never seems to be satisfied after feeding. The foal may be slow to develop and usually displays few of the behavioral patterns characteristic of a normal foal (e.g., spurts of energy and play with other foals). If in doubt about how much milk the mare is producing, separate the mare and foal for 30 to 60 minutes to prevent nursing and then milk the mare gently. If the mare is producing sufficient milk, the hand-milking process should secure about 6 ounces (177 milliliters) of milk.

Low milk production may be caused by a number of different circumstances. Most commonly, the factors mentioned earlier (inadequate nutrition and inherited limitations) are related to insufficient milk production. If the mare is not limited by inherited milk production handicaps, a high-protein feed containing salt (to increase the mare's water intake) may be helpful. If the mare is simply incapable of producing enough milk to support her foal, supplemental feeding of the foal (creep feeding) at an early age may be helpful.

AGALACTIA

Complete failure of milk flow after foaling is referred to as agalactia. This absence of milk may be caused by a failure of either the milk ejection or the milk production processes. Failure of milk to be ejected from the mammary gland occurs occasionally after parturition and is believed to be caused by the presence of adrenalin in the mare's bloodstream. Adrenalin, a hormone that inhibits the effect of oxytocin on the mammary gland, is released when the mare is frightened or stressed. Injections of oxytocin and applications of warm compresses to the udder seem to be very effective in stimulating milk ejection.

On the other hand, if agalactia is caused by failure of the mammary gland to produce milk, treatment is often futile. A breakdown in the production process may be caused by hormonal deficiencies, abnormal mammary development (i.e., absence of secretory tissue), or indigestion. Diseases such as acute metritis, mastitis, or fescue poisoning may also prevent milk production. In some instances, injections of special lactation-promoting hormones may be effective, but usually the foal must be maintained as an orphan (e.g., bottle fed or provided with a nurse mare).

TEAT ABNORMALITIES

Congenital teat abnormalities, such as inverted or conical nipples, do not necessarily affect milk production or ejection but do make it very difficult for the foal to nurse. In very rare instances, a filly may be born with an abnormal number of teats. The effect of extra, or supernumerary, teats on milk production or ejection has not been studied extensively. Although teat abnormalities are very unusual in the mare, a prepurchase or prebreeding examination should include careful evaluation of the mare's udder.

UDDER EDEMA

Occasionally, the udder and the lower abdomen become engorged with fluid due to circulatory problems. When extreme, this so-called udder edema is sometimes confused with a ruptured prepubic tendon, an umbilical hernia, a blood-filled swelling (hematoma), or an abscess. Udder edema, however, is usually characterized by an accumulation of fluid within the mammary tissue and around the lymph vessels located along the mare's belly. The condition, which occurs most frequently in heavy-milking mares during early lactation and after weaning, often causes the udder to become warm and extremely painful. Severe edema predisposes the udder to infection and may cause necrosis of the skin or prevent milk ejection. Although exercise may relieve mild cases, treatment of severe cases usually involves massage, hand milking, cold and hot applications, and the use of mild counterirritants. The veterinarian may administer diuretic drugs and recommend lowering the mare's salt intake in order to temporarily reduce fluid retention in body tissues. Reducing her grain intake may also help relieve swelling caused by fluid build-up within and around the mammary gland.

MASTITIS

Inflammation in one or more quarters of the mammary gland is referred to as mastitis. Although all mastitis cases are characterized by the presence of bacteria and numerous infection-fighting blood cells, infectious organisms do not necessarily initiate this condition. For example, a mare may be resistant to bacterial strains located in the mammary gland or on the udder until an injury or illness predisposes the mare to infection. In contrast to its high incidence in dairy cows, mastitis rarely occurs in mares. The low incidence of mastitis in mares is probably due to frequent nursing by the foal and to the mare's relatively low level of milk production.

Mastitis, which may occur during any stage of lactation or even in the nonlactating mare, is characterized by physical changes in the glandular tissue and by chemical and bacteriological changes in the milk. Milk from the infected gland looks curdled and may contain blood. Although the condition is also characterized by a swollen, warm, painful udder, the best diagnosis is laboratory analysis of the milk. Microscopic studies reveal how much of the gland is involved in the infection and indicate the stage of inflammation. If detected early, mastitis responds well to systemic antibiotics and medication infused into the drained gland via the teat.

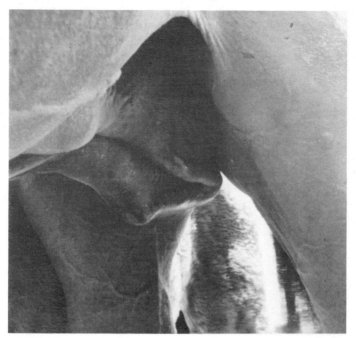

The right side of this mare's udder is swollen due to mastitis.

ABSCESSES

An abscess is a collection of pus in a localized area, or cavity. Udder abscesses are uncommon but may occur in association with mastitis. The infected area usually ruptures and drains without causing permanent tissue damage. Although the condition's cause and method of transmission are unknown, researchers believe that *Corynebacterium pseudotuberculosis* may cause one type of udder abscess.

CYSTIC UDDER

Occasionally, the non-pregnant mare's udder may fill with a clear fluid secretion or with milk. Like mastitis, this condition may be confined within one quarter or may affect several quarters. Cystic udder is relatively common in mares, and it requires immediate veterinary attention.

19

THE NEONATAL FOAL

The moment that foaling is completed, the attendant's responsibilities shift to observation and care of both the mare and her newborn foal. In fact, care of the mare and her foal during the period immediately after foaling is one of the greatest concerns of breeders. Evaluating the newborn foal's progress and recognizing abnormalities are very important during the foal's first three to four days of life, which is referred to as the neonatal period. This is a very critical time for the foal: life support systems such as respiration are tried for the first time; a bond is established with the mare; the antibody-rich milk (colostrum) is received; and physical problems that interfere with the foal's progress or threaten its life become apparent for the first time. Although this chapter is not intended to replace the diagnostic skills and experience of a veterinarian, the following information will aid the breeder in determining when veterinary supervision is important.

The foal's behavior is a good indicator of its well-being. A healthy foal rests for short but frequent periods and appears alert and comfortable.

Fetal Adjustments at Birth: The Adaptive Period

Certain anatomical and biochemical changes must take place in the foal's body immediately following birth if it is to survive in its new environment. These changes are reflected in the foal's behavior, pulse, temperature, and respiration. Because failure of any of these changes to take place is often life-threatening, the breeder should be familiar with the foal's normal vital signs and behavioral patterns. Any abnormalities should be recognized quickly and reported immediately to the farm veterinarian.

CIRCULATORY

The heart rate of a healthy foal immediately following birth ranges from 50 to 90 beats per minute. If the birth was difficult or if the mare foaled standing up, the foal's heart rate may exceed 90 beats per minute. The pulse jumps to approximately 150 beats per minute when the foal struggles to his feet for the first time, but by the time the foal is two or three hours old, its resting heart rate should stabilize between 70 and 95 beats per minute.

During the 5 to 10 minutes following birth, the foal receives approximately 10 to 30 percent of its blood from the placenta via the umbilical cord. For this reason, it is imperative that the umbilical cord not be severed prematurely. Premature rupture of the cord causes a drop in the foal's blood pressure and may interfere with cardiopulmonary function. The resulting oxygen deficiency may cause convulsions. Under normal circumstances, the cord ruptures at a point a little more than an inch (about 3 cm) from the foal's abdomen when the mare rises or the foal first struggles to stand.

The foal's hemoglobin (oxygen-carrying blood protein) and the number of red blood cells increase at birth but drop between 1 and 12 hours after foaling. This drop is accentuated when the foal's umbilical cord is broken prematurely. It is generally believed that high postpartum hematocrit values are due to the stress of foaling. As the red blood cell count decreases after foaling, the infection-fighting white blood cell count increases.

RESPIRATORY

Respiratory movements of the chest and abdomen begin within thirty seconds after birth. If the amnion (i.e., the innermost layer of the placenta) is still intact, it is usually torn at this time by struggling movements of the foal's forelegs and head. (If the amnion is not torn, it should be removed from the foal's muzzle as soon as possible.) A series of gasps may precede respiration, but that is no cause for alarm unless the foal fails to establish a respiratory rhythm promptly. Respiration is irregular at first but stabilizes within 60 seconds. Any foal that fails to breathe within 30 to 60 seconds after birth should be resuscitated. (Refer to 'Resuscitation of the Depressed Foal' within this chapter.)

EXCRETORY

Meconium is the brown, greenish-brown, or black substance that accumulates in the foal's intestinal tract during gestation. Prior to birth, digested amniotic fluid and cellular debris are propelled along the bowel by peristalsis and stored in the colon, cecum, and rectum. Under stressful conditions, such as asphyxia, the meconium may be expelled before or during birth. The foal passes the first bits of meconium within 12 hours after birth, and by the fourth postpartum day, the meconium should give way to the yellowish feces of the normal neonatal foal.

Dark-colored meconium is the first excretion of the foal's digestive tract.

NUTRITIONAL

Prior to birth, the fetus depends on the mare for all of its nutritional needs. When it is released from the protective uterine environment, the foal experiences a temporary period of nutritional independency; the placental relationship with the mare is severed, but the foal has not yet nursed. This period of nutritional independency is also a time of heavy physical exertion. To meet the energy needs required to stand, walk, and nurse for the first time, glycogen (an important carbohydrate source and, thus, an important energy source) is stored in large quantities in the liver during the last stage of gestation. Immediately after birth, this glycogen is mobilized in the form of glucose, a blood sugar that readily supplies the newborn foal's energy needs. The amount of glucose in the blood reflects the degree to which stored glycogen is available for cellular use. Prior to nursing, the foal's blood glucose level is between 50 and 70 mg/100 ml of blood. By the time the foal is 36 hours old, these levels normally reach 100 to 110 mg/100 ml of blood.

TEMPERATURE

Unlike the young of many other species, the neonatal foal is normally able to maintain its body temperature at the intrauterine level (about 100°F) even though the external environment may be much colder. The normal temperature range during the first four days of life is 99-101°F (37.2-38.3°C). The newborn foal is equipped with a layer of fat and a plentiful supply of hair, two important sources of insulation against heat loss. In addition, vigorous shivering during the first three hours after birth and muscular exertion involved in standing and following the mare generate heat.

IMMUNOLOGICAL

During gestation the placenta prevents large immunoglobulin molecules from being passed from the mare to her foal. Consequently, the foal is born with a very low level of antibodies in its circulatory system. The foal begins to make its own antibodies during gestation, but the immune system does not reach maximum efficiency until the foal is three to four months old. During most of this transitional period, the foal is protected by the antibodies (immune proteins) received from the dam's colostrum. For approximately 36 hours following birth, the foal has the ability to absorb immune proteins from the mare's colostrum through special epithelial cells lining the small intestine. (Afterward, the antibodies are digested rather than absorbed.) Each epithelial cell absorbs as much protein as it can hold before discharging it into the foal's circulatory system. The protein globules pass from these specialized cells, through the local lymphatics, and into the foal's circulatory system. Because the foal's intestine rapidly loses its ability to absorb colostral antibodies, the foal should receive colostrum within the first 24 hours of life (preferably within the first 12 hours).

AWARENESS

Within five minutes after birth, the foal becomes aware of its surroundings and responds to sound and physical stimuli. During this period the foal begins to move into an upright position resting on its sternum. This puts the foal into a position from which it can rise to its feet. At this time, blinking and pupillary reflexes are present, although visual competency in the newborn appears to be limited to the ability to avoid obstacles. Full visual competency and the ability to interpret the environment develop over a period of several days following birth.

PHYSIOLOGICAL PARAMETERS OF THE NEWBORN FOAL

Respiration Initiated	a few seconds after foaling
Normal Respiration	approximately 60 breaths/minute within 2 to 3 minutes after birth

Normal Heartrate	40 - 80 beats/minute a few minutes after birth
Heartrate as the Foal Stands	90 - 150 beats/minute
Stabilized Heartrate	70 - 95 beats/minute
Respiratory Distress	more or less than 60 breaths/minute
Abnormal Cardiac Function	above 120 beats/minute in a truly resting state
Blood Supply	10 - 30% passes through umbilical vessels during first 10 - 15 minutes postfoaling
Umbilical Cord Ruptures	within 30 minutes after foaling
Foal Stands Unassisted	within 1 to 2 hours after foaling (individual variation)
Foal Nurses	within 2 to 3 hours after foaling (individual variation)
Temperature of Newborn	may drop to 98.6° F (37° C) immediately after birth but should reach 100.4° F (38° C) within 1 hour
Normal Temperature Range	99 - 101° F (37.2 - 38.3° C)
Meconium Evacuation	completely voided 4 to 96 hours after birth

Treatment of the Newborn Foal

Although about 90 percent of all foals pass through the neonatal adjustment period without serious problems, there are several management techniques with which the attendant should be familar. Before they are examined, it is very important to note that the bond between the mare and her newborn foal is established and strengthened during the first few days after birth. Handling the newborn foal excessively or other physical interference during this important adjustment period may prevent a normal mare/foal relationship from forming. Any examinations, routine management procedures, or special treatments should be carried out quietly and quickly, leaving the mare alone with her foal as much as possible. This is not to imply, however, that periodic inspections of the newborn foal's progress are not important, only that these inspections should be carried out with as little noise and distraction as possible.

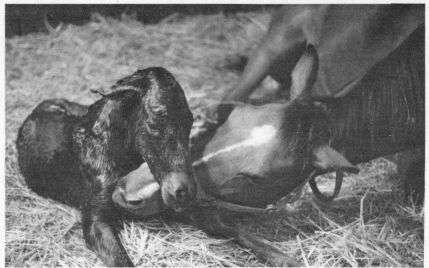

Courtesy of The Thoroughbred Record   Photo by Susan Rhodemyre

Even though they must be observed closely, the mare and her foal should be given as much undisturbed time together as possible during the first few hours and days following birth.

RESUSCITATING THE DEPRESSED FOAL

Any foal that fails to breathe within 30 to 60 seconds after birth should be resuscitated by one or more of the following techniques:

1. Extend the foal's head and clear its nostrils of mucus, using gauze or a clean rag. Rub the foal briskly with a clean, dry towel or burlap sack.
2. Lower the foal's head to drain any fluid from the respiratory tract and tickle the inside of the foal's nostril with a straw or other small instrument to stimulate the cough reflex.
3. If these techniques are not successful, kneel between the foal's head and forelegs with the foal lying on its side. Place the right hand under the foal's muzzle to seal the lower nostril. Inflate the foal's lungs by blowing gently into its open nostril, using the left hand to help seal your mouth around the foal's open nostril. Blow hard enough to expand the foal's chest every two seconds, releasing both nostrils momentarily between breaths to allow exhalation of carbon dioxide. Resuscitation activities should be continued at a rate of 25 breaths per minute, but the attendant should pause after several breaths to determine whether the foal is breathing on its own.
4. Some farms routinely use oxygen on depressed foals, but because misuse or overuse of oxygen can be harmful to foals, oxygen should only be administered by trained personnel.

(Note: Use these resuscitation methods only if the foal's heart is beating. A thump on the girth behind the foal's elbow may produce a heartbeat.)

TREATING THE NAVEL

The umbilical cord usually ruptures during postpartum movements of either the mare or the foal and, under normal circumstances, it should not be cut or broken by the attendant. Occasionally, however, the mare expels the placenta before the cord has broken. In these cases, the cord should be broken (not cut) 2 to 3 inches from the foal's abdomen after the transfer of blood from the placenta is complete. The cord should be held securely near the foal's abdomen so that pressure is not applied to this area when the cord is pulled apart.

After the umbilical cord ruptures, the navel stump should be immersed in Lugol's solution (5 percent iodine) or in tincture iodine (7 percent iodine) to help prevent the entry of pathogenic bacteria. (There are varying opinions on the best method of treating the navel, but immersion in an iodine solution remains the most commonly recommended treatment.) The object of navel treatment is not to treat an established infection but to prevent contamination, and for this reason, the navel should be treated as soon after birth as possible.

In order to thoroughly treat the umbilical cord, the newborn foal must be gently restrained in a recumbent position.

Photos courtesy of Earl and Barbara Lang Quarter Horses

The umbilicus can be immersed in a small container filled with an iodine solution.

When properly treated, the umbilicus should be completely saturated.

The umbilical cord contains one vein and two arteries. In some instances, bleeding from these vessels persists after the cord ruptures. Normally, bleeding can be checked by pinching the navel stump for several minutes, but if this technique is not successful, the veterinarian may decide to suture the cord temporarily. Because natural drainage from the cord is very important, the attendant should not attempt to stop bleeding by tying off the navel stump. However, in rare instances, when profuse bleeding persists, the veterinarian may choose to suture the cord.

ADMINISTERING ANTIBIOTICS AND TETANUS ANTITOXIN

Unless the mare has been vaccinated against tetanus within 30 days prior to foaling, the foal should be given tetanus antitoxin soon after birth. Some authorities advise that tetanus antitoxin be administered as a precautionary measure, even when the mare is on a tetanus vaccination program. Preventative vaccination of the foal is especially important when partial or total FPT is suspected. (Refer to 'Failure of Passive Transfer' within this chapter.)

Opinions vary on whether the healthy foal should be given antibiotics as a routine precaution following birth. Some veterinarians advocate a daily antibiotic injection for three to five days following birth to help prevent infection during the foal's first few vulnerable days, especially in areas where postnatal infection is common. Other veterinarians maintain that the administration of antibiotics to a healthy foal is unnecessary unless the farm has persistent infection problems and point out that routine use of antibiotics may encourage the development of antibiotic-resistant bacteria.

GIVING AN ENEMA

Most foals eliminate their meconium without assistance, but because meconium retention is not uncommon and because it can often be alleviated by administering an enema shortly after birth, many breeders routinely give enemas to all newborn foals. There are several effective enemas that can be used on the newborn foal. One consists of 2 ounces of glycerin delivered into the rectum very carefully from a dose syringe through a soft rubber tube. One or two pints of a mild, nondetergent soap solution can also be administered in the same manner. Many breeders prefer to use prepackaged enemas since they are easy to use and are readily available in most areas. Because the young foal's rectal tissues are very delicate, the enema tubing should be inserted no more than 4 or 5 inches into the foal's rectum, and the enema (regardless of type) should be administered gently without excessive pressure.

If the foal fails to defecate and shows signs of straining or colic, it may be suffering from a meconium impaction or from a congenital abnormality of the digestive tract. However, even if the foal does not appear colicky, failure to eliminate the meconium should be reported to a veterinarian. (Refer to 'Atresia Ani' and 'Atresia Coli' within this chapter.)

Prepackaged enemas come in small, soft plastic bottles that are convenient for use on the breeding farm.

CARE OF PREMATURE FOALS

The term *premature* is used to describe foals born following a gestation of 300 to 325 days that show lack of developmental maturity. (Delivery prior to 300 days is referred to as abortion.) Prematurity is characterized by general weakness, low birth weight, reduced sucking reflex, delay in standing, and inability to maintain body temperature or to establish a normal mare/foal relationship. The tongue of the premature foal is often deep red, and the skin is usually very silky. In many instances, premature foals stand within 48 hours of birth and develop normal behavioral patterns within a week, but the first three days are very critical. During this time, some foals die due to general collapse of neurological and cardiopulmonary functions.

Care of the premature foal includes maintaining the foal in the most supportive environment possible. The best environment is achieved through the use of an incubator, which allows the temperature, oxygen content, and humidity of the foal's surroundings to be controlled. Unfortunately, foal incubators are very expensive and are available at very few veterinary clinics or breeding farms. If an incubator cannot be obtained, the premature foal should be kept warm and dry by equipping a well-bedded stall with heat lamps. The foal should be fed mare's milk through a tube or by means of a bottle every one or two hours until it is able to rise and nurse without assistance. Most importantly, the mare's first milk, or colostrum, should be milked from the mare and given to the foal as soon as possible since the foal's intestine is unable to absorb colostral antibodies after the foal is about 36 hours old.

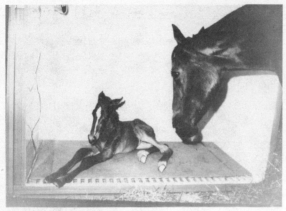

A foal incubator that
provides controlled temp-
erature, oxygen, and humidity is an excellent means
of regulating the premature
foal's environment.

Courtesy of Kansas State University, Department of Surgery and Medicine

RESTRAINT

The foal should not be restrained for treatment by holding or tying its
halter even if the youngster is halterbroken. A young foal is unlikely to
stand still under stressful circumstances and may escape or become
injured if a struggle occurs. When restraint is necessary, the foal should
either be cradled or tailed but should never be picked up with the
handler's arms wrapped around the rib cage as the foal's ribs are easily
broken. Similarly, the foal should not be picked up with the handler's
arms cradling the abdomen as injury to the internal abdominal organs
may result. Cradling, which involves placing one arm around the foal's
chest and the other arm around its hindquarters above the hocks, is an
effective and safe restraint technique for small foals. Tailing, usually
used on older foals, involves putting one arm around the foal's chest
while the opposite hand is used to gently but firmly grasp the tail and
lift it upward. (Note: Rough handling may injure the foal.) The effect of
tailing is similar to that of twitching a mature horse; if tailing is done
correctly, the foal usually remains stationary and quiet.

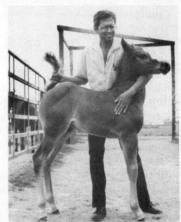

**Tailing the foal by placing one arm around
the foal's chest and gently lifting the tail is
an effective restraint method.**

*Courtesy of Texas A & M University
Department of Large Animal Medicine and Surgery*

POSTFOALING PROCEDURES: A REVIEW

1. Make sure that all fluid and mucus are removed from the foal's mouth, nose, and throat. If the foal does not begin breathing within 30 to 60 seconds, begin resuscitation procedures immediately. (Refer to 'Resuscitating the Depressed Foal' within this chapter.)

2. Allow the mare and foal to rest for 10 to 45 minutes. The foal's hind legs may remain inside the mare's vagina.

3. If the foal's breathing is shallow or irregular, massage its body briskly with a clean, dry cloth. (Complete removal of the membranes may change the foal's odor and cause the mare to reject it.)

4. If the foal's respiratory response is very poor (less than 50 to 60 breaths/minute), a veterinarian should be summoned. Oxygen should be administered only by a trained technician.

5. The navel cord should be allowed to break naturally. The foal's navel stump is then treated (e.g., soaked in an iodine solution such as Lugol's solution or tincture iodine). If the cord does not rupture naturally within about 20 minutes after birth, it should be broken by hand by holding the cord firmly at a point two inches from the foal's belly and pulling the portion of the cord which is attached to the placenta until it breaks. The cord should never be cut, and the open end should not be touched. The cord should be observed for two to three days to make sure that it dries normally.

6. The foal should be up and nursing within two to three hours after birth. If the mare's udder is sore, she may have to be restrained gently while the foal nurses for the first time. This problem is seen most frequently in mares that have never been suckled.

7. If the foal is too weak to stand and nurse within three to six hours, milk some colostrum from the mare and bottle feed the foal using a lamb nipple on a soda bottle. The foal's strength should improve quickly, but if it does not, a veterinarian should be summoned.

8. Administer antibiotics and tetanus antitoxin as recommended by the farm veterinarian. On some farms, enemas are given routinely to prevent meconium impaction.

9. Although it is important to minimize distractions and handling during the foal's first few days, the youngster should be observed closely to detect the presence of any congenital abnormalities or infections. For example, is the foal alert and responsive to the mare? Is its breathing labored or its nursing excessive? Is there any diarrhea, constipation, nasal discharge, or excessive watering from the eyes? Is it resting comfortably between meals? If the foal is dull, is its rectal temperature normal and are its mucous membranes a healthy pink color? If there is any reason to suspect that a problem exists, a veterinarian should be contacted immediately.

Behavior of the Newborn Foal

For several minutes following delivery, the foal usually lies behind the mare with its hind legs still inside the mare's vagina. After a short rest period, the foal extends its forelegs, withdraws its hind legs, rolls up on its chest, and begins to show some interest in its surroundings. Within 10 to 20 minutes of birth, the foal should make its first attempt to stand. (The foal is not capable of standing before it develops sufficient muscle tone, a condition that is usually present once the foal raises its ears.) As the foal struggles to stand and moves away from the mare, the umbilical cord should break if it is still intact.

The foal does not usually succeed in standing on its first attempt. Instead, a trial-and-error period of approximately an hour is necessary for the average foal to muster the strength and coordination to stand successfully. Once the foal stands, each subsequent attempt seems to require less effort. If the foal does not attempt to stand or if it is unable to stand unassisted within two hours of birth, an illness or abnormality should be suspected.

The foal should stand within two hours after birth. However, many foals begin their efforts to rise much sooner.

The normal foal is born with a sucking reflex and instinctively begins to search for the mare's udder as soon as it is able to stand and walk. As with standing, the foal's first attempt to nurse is a trial-and-error procedure. When searching for the mare's udder, the foal may attempt to nurse stationary objects such as the stall wall and feed bin or between the mare's front legs or along her belly. There is no cause for alarm, and human intervention is seldom necessary. Given time and a little assistance from the dam, the foal usually arrives at the proper location. After nursing and establishing an affinity for the mare, the foal usually takes a short nap. Any foal that fails to nurse within two or three hours of birth should be examined by a veterinarian.

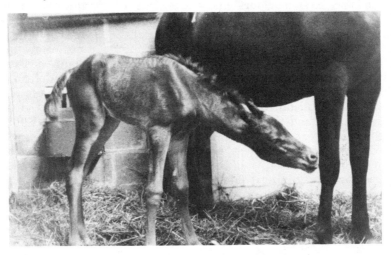

In the foal's early attempts to nurse it often searches the mare's forelegs and belly before locating the udder.

Noninfectious Complications

Although abnormalities do not appear with great frequency in foals, the attendant's ability to evaluate the foal's progress during its first few days of life can be very important to the foal's immediate survival and future performance. The following section on abnormalities is not complete, but it includes descriptions of the more common noninfectious complications along with examples of very rare abnormalities, such as cerebellar hypoplasia. The conditions have been categorized according to the system or organ affected, for use as a reference.

EXCRETORY

MECONIUM RETENTION

Meconium retention, a condition in which the fecal matter collected in the large intestine during gestation becomes painfully impacted, is a fairly common problem in neonatal foals. The cause of meconium retention is not clear, but workers have suggested that a vitamin A deficiency during the last trimester of gestation may be partially responsible. Foals that do not receive an adequate amount of colostrum, which is laxative, are also believed to be candidates for meconium retention. In addition, some veterinarians have reported a higher incidence in foals born after a gestation of more than 340 days and in male foals.

Signs include straining, tail switching, thrashing, rolling, and lying on the back or in other unusual positions. Affected foals may exhibit these signs as early as three hours postpartum, but the condition usually does not become apparent until 24 to 36 hours after birth and may occur despite the occurrence of one or more bowel movements. A gently administered enema usually relieves simple constipation, but if the meconium mass is out of the enema's reach, the foal will not respond to this treatment. The administration of a mixture of castor oil, mineral oil, and milk of magnesia via a stomach tube may help break up the impaction and stimulate bowel movement. In addition, the veterinarian may administer an analgesic (i.e., pain killer) to relieve the foal's discomfort and a sedative to prevent injury before the laxative takes effect. If correction of the impaction takes several days, further treatment may be required for associated conditions, such as dehydration, exhaustion, and/or bloat. If the foal's impaction fails to respond to chemical therapy, surgery may be indicated.

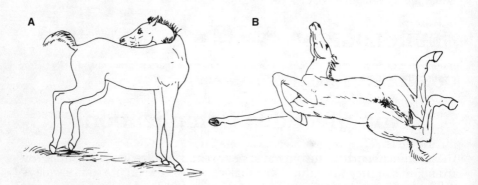

Foals affected by meconium retention may show evidence of colic, including (A) general signs of intestinal discomfort and (B) thrashing.

PATENT URACHUS

The urachus is a small vessel in the umbilical cord that connects the fetal bladder to the allantois, where fetal urine is stored until the placenta ruptures at birth and releases the allantoic fluid. (Refer to "Fetal Development" for a complete description of the allantois.) The urachus normally closes when the umbilical cord breaks and, consequently, the flow of urine is diverted through the urethra. If the urachus fails to close, a condition called patent urachus develops, and urine may be observed dribbling from the foal's navel. When the foal attempts to urinate, the dribble becomes a steady stream. Usually, the veterinarian will suture the opening to seal it and to prevent navel infection. Because the wet, soiled navel is an excellent environment for bacterial growth, the navel stump should be treated with a product such as silver nitrate, strong iodine, or 10 percent formalin, until it shrivels and dries. If this treatment is not successful, surgery is usually indicated.

RUPTURED BLADDER

Rupture of the bladder, a condition that is not uncommon in newborn foals, is believed to be the result of compression on a full bladder during delivery or of a sharp jerk on the umbilical cord following birth. Affected foals appear normal for 12 to 24 hours following birth but then become depressed and progressively weaker. Constant attempts at urination usually produce little or no urine, although depending on the position of the rupture, nearly normal amounts of urine may be voided. About 48 to 72 hours after the bladder ruptures, abdominal distension is noticeable, and the foal may become short of breath. If left untreated, the foal may die of uremic poisoning or of asphyxiation due to pressure on the diaphragm. Surgery, accompanied by antibiotics and other supportive therapy, is the only method of correcting this condition. If the problem is diagnosed and treated quickly, the foal's chances of survival are much better. In most cases, early surgery results in complete recovery.

NEUROLOGICAL

CONVULSIVE FOAL SYNDROME (CFS)

Convulsive foal syndrome (also called barker, wanderer, or dummy foal syndrome and sometimes referred to as neonatal maladjustment syndrome) is believed to be caused by decreased flow of oxygen to the brain prior to or during birth. This oxygen deficiency may be caused by early separation of the placenta from the uterine wall, or by premature rupture of the umbilical cord.

The foal usually appears normal at birth but begins to show signs of convulsive foal syndrome (CFS) within a few minutes or hours. The first stage (barker stage) of this syndrome is characterized by convulsions and, in severe cases, by pronounced respiratory distress. As a result of this respiratory difficulty, the foal may make a barking noise. In recumbency the foal makes galloping motions and suffers frequent, strong contractions of the jaw, back, and neck muscles. Pulse and respiratory rates increase greatly, and profuse sweating may occur. If the foal is able to stand, it may charge blindly around the stall, seemingly unaware of obstacles in its path. Some CFS foals appear to skip the barker stage, and although the point is disputed, some authorities have suggested that, in these cases, the convulsions may take place prior to birth.

If the foal survives the barker stage, it passes into the dummy stage. During this period, the foal does not react to light or sound and remains recumbent and inert. This stage may appear at birth, several hours after birth, or after a barker stage. Following the dummy stage, the foal becomes extremely active and walks about aimlessly, without regard to objects in its path. These wanderer foals seldom lie down voluntarily but may remain down for hours if placed in that position.

There is no specific treatment for CFS other than supportive nursing and sedation. Due to the lack of a sucking reflex, CFS foals must be tube fed every few hours. During the barker stage, sedation is indicated when convulsions are severe. With proper care, approximately 50 percent of the affected foals live through the barker stage. If the foal survives into the dummy and wanderer stages, the prognosis becomes increasingly favorable. When properly cared for, the wanderer foal usually begins to recover within two or three days, and complete recovery may be accomplished as early as the fourth or fifth day.

Infections of *Actinobacillus equuli* cause signs somewhat similar to those exhibited by foals in the dummy and wanderer stages of CFS. Foals affected by *A. equuli* are often called dummies or wanderers. However, these two conditions differ in several ways. For example, foals infected with *A. equuli* lack the barking characteristic of CFS foals. Unlike CFS foals born in the dummy stage, which then pass into the active wanderer stage, foals infected with *A. equuli* prenatally may never stand. A blood culture may detect the presence of *A. equuli* and should be run on any foal displaying dummy or wanderer behavior. (Refer to the discussion of *Actinobacillus equuli* infection in 'Infectious Complications' found in this chapter.)

MENINGEAL HEMORRHAGE

Hemorrhage and congestion of the membranes covering the brain and spinal cord (meninges) sometimes cause death in young foals. The meningeal lesions are considered an index of injury sustained by the fetal central nervous system due to trauma or oxygen deficiency during birth. Affected foals rarely live more than 10 days after birth, and those foals that do survive are usually abnormal.

HYDROCEPHALUS

Hydrocephalus is a rare congenital abnormality caused by the accumulation of excess cerebral spinal fluid within the cranium, or skull. As a result, the fetus develops an enlarged head, and its brain tissue becomes compressed. In severe cases, the cranium is very large and a Caesarian section or fetotomy may be required to deliver the foal. When hydrocephalus is severe, foals are usually stillborn or die soon after birth. In some cases, the foal is born with mild cerebral atrophy and is able to nurse with assistance. Some hydrocephalic foals may even live for several days despite cerebral tissue damage, but the mildly affected foal appears lethargic and unsteady on its feet and may suddenly stagger and lower its head while walking. Most hydrocephalic foals become progressively worse since there is no known treatment for this unusual condition.

CEREBELLAR HYPOPLASIA AND DEGENERATION

In the condition cerebellar hypoplasia the brain's cerebellum is incompletely developed and undergoes some deterioration during gestation.

Although a similar disease occurs in Gotland ponies and Oldenburg horses, cerebellar hypoplasia seems to be limited to pure and part-bred Arabian foals. It is generally believed that this condition is inherited, since affected horses have a close family relationship, and the disease appears most often in horses from a few heavily inbred lines. However, the possibility of a viral cause has not yet been ruled out.

Head tremors and incoordination are characteristic of this condition, but they may not appear for several months after birth. Regardless of when the signs of cerebellar hypoplasia appear, the affected foal becomes progressively worse, and an effective treatment for the condition has not yet been developed.

COLIC

Colic is not a disease; it is the expression of abdominal pain through tail switching, pawing, rolling, sweating, or straining to urinate or defecate. Retained meconium is the principal cause of colic in foals. Atresia coli, atresia ani, strangulated hernias, and other abnormalities which may cause colic are examined later in this chapter. Colic that cannot be relieved by the administration of an enema should be treated by a veterinarian.

RESPIRATORY

PNEUMOTHORAX

Pneumothorax is the introduction of air into the pleural cavity (the space surrounding the lung) usually through the puncture of a lung by a fractured rib. Because the presence of air in the pleural cavity destroys the vacuum that keeps the lung inflated, this condition results in partial or total collapse of the affected lung. This loss of vital capacity interferes with inspiration and pulmonary circulation. As a result, the foal's breathing becomes increasingly labored, and death occurs in a few hours. Surgery may be of benefit if the condition is diagnosed promptly.

ASPHYXIA

Asphyxia (hypercarbia and anoxia) is an impaired exchange of oxygen and carbon dioxide. It can be caused by a number of conditions, including prolonged or delayed birth, premature separation of the placenta from the uterine wall, or early rupture of the umbilical cord. Asphyxia can also occur when the foal is presented in a backward position during foaling. In this position, the umbilical cord is compressed against the maternal pelvis so that oxygenated blood cannot reach the foal from the placenta. Prolonged asphyxia (generally more than 2.5 minutes) may result in convulsions, cerebral edema and hemorrhage, or an inability to stand or nurse. Foals born in a state of asphyxiation must be resuscitated in the manner described earlier under 'Treatment of the Newborn Foal.' Some foals are resuscitated successfully but die later from residual damage.

CIRCULATORY

PATENT DUCTUS ARTERIOSUS

The ductus arteriosus is the short blood vessel that allows blood to by-pass the nonfunctional fetal lungs by connecting the pulmonary artery to the aorta. During gestation, oxygenated fetal blood flows from the placenta, through the veins, and into the fetal heart. With the help of two by-pass mechanisms (the ductus arteriosus and the foramen ovale), blood is diverted

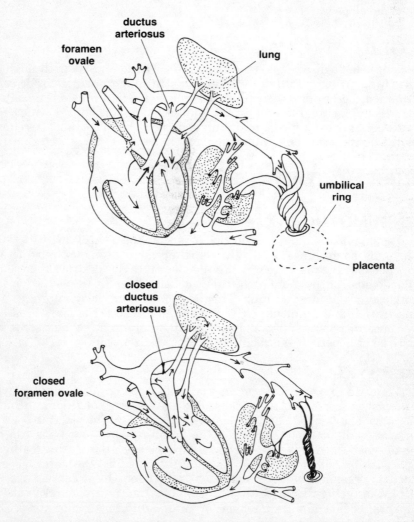

(A) In the fetal circulatory system blood passes through two openings: the foramen ovale in the atrium wall and the ductus arteriosus by-passing the lungs. **(B)** In the normal foal the foramen ovale and the ductus arteriosus close within a few days after birth.

past the pulmonary vessel so that it can be distributed throughout the fetal body without passing through the lungs. Within a few days of birth, the ductus arteriosus closes, and all of the foal's deoxygenated blood flows through the pulmonary artery to the lungs. Failure of the ductus arteriosus to close allows deoxygenated blood carried by the pulmonary artery to mix with the oxygenated blood carried by the aorta. Affected foals are usually weak and may have blue mucous membranes. Although the severity of the signs depends on how large the ductus arteriosus is, both cyanosis of the mucous membranes and weakness are especially pronounced after exercise. Very little research has been conducted on patent ductus arteriosus in horses. Therefore, there is little information on successful treatment of this condition or its long-term effects.

CONGENITAL HEART DEFECTS

Although approximately a dozen different heart defects have been reported, congenital heart defects are rare in the equine. Affected foals are weak and unthrifty and have a poor tolerance for exercise. Following exercise, a blue tinge of the foal's mucous membranes is often noticeable. A thorough cardiovascular examination is necessary before accurate diagnosis and prognosis can be made.

NEONATAL ISOERYTHROLYSIS (NI)

Neonatal isoerythrolysis (NI) is a condition that is characterized by destruction of the newborn foal's red blood cells. This potentially fatal disease is caused by a series of events:

1. The foal inherits a blood type from its sire that is incompatible with the dam's corresponding blood type (e.g., sire's blood type q^R vs. dam's blood type q^S).
2. Abnormal circumstances in the placenta allow the foal's blood to cross the placental barrier and enter the dam's circulatory system.
3. A normal defense mechanism, called isoimmunization, causes the mare to produce antibodies against the foal's sire-related blood factors.
4. Although these antibodies cannot cross the placenta, they do become concentrated in the mare's colostrum.
5. When the newborn foal ingests colostrum, the antibodies cross the intestinal wall and enter the foal's blood, where they recognize, combine with, and destroy the foal's red blood cells. (The isoimmunization process can also be induced by blood transfusions or by the use of vaccines containing horse erythrocytes.)

The NI foal appears normal at birth and remains healthy until after it ingests the mare's colostrum. At this time, acute cases are characterized by weakness, loss of appetite, dullness, normal or subnormal temperature, weak but rapid respiration, increased heart rate, repeated yawning, and pale yellow or pale blue mucous membranes. As the disease progresses, the foal's urine becomes reddish (due to the presence of pigment originating from destroyed red blood cells), and the foal becomes too weak to stand. Unless the disease is detected promptly, the foal usually dies on about the fourth or fifth day.

In less severe cases, neonatal isoerythrolysis may not be detected for several days. At that time, the only evidence of the condition is slight discoloration of the eyes and mucous membranes and slight anemia. However, if slightly affected foals exercise during this period of anemia, the absence of sufficient oxygen-carrying red blood cells may cause severe fatigue and possible onset of the signs described previously. It may take three to five days of strict inactivity before the slightly anemic foal's red blood cells reach normal levels.

Although early detection of neonatal isoerythrolysis is important, emphasis should be on the prevention of the disorder. Records should be kept of all sire and dam crosses that have produced an affected foal. Any mare that has produced an NI foal should be carefully tested during each subsequent pregnancy. The mare's serum can be cross-matched with the stallion's red blood cells by a special laboratory procedure to determine whether the mare is sensitized to the stallion's red blood cells and also to determine the extent of isoimmunization. If the mare has ever produced an NI foal it is recommended that the veterinarian serotype both the mare and stallion.

If the test results are positive, the newborn foal should be confined or muzzled and fed colostrum obtained from another mare and a milk substitute until the dam's colostrum has been milked out and the foal's digestive tract is no longer capable of absorbing the mare's antibodies. The foal's intestine can absorb colostral antibodies for 24 to 36 hours after birth, although mare's milk is often withheld for 48 hours. If the foal inadvertently receives colostrum from its isoimmunized dam, it should be muzzled, fed a milk substitute, and watched carefully until its red blood cell count approaches normal. (It is very important not to stress the foal physically during the recovery period.) If the foal receives a large quantity of colostrum from a sensitized mare or if it receives a small amount of colostrum that contains a high concentration of destructive antibodies, it will develop acute neonatal isoerythrolysis within minutes after ingestion. In such cases, massive blood transfusions may provide the only hope of saving the foal's life. Ideally, the red cell portion of the dam's blood, which has already been exposed to the antibody, is administered. However, facilities must be available for completely removing antibody-containing plasma from the erythrocyte fractions. Then only the erythrocytes are administered to the foal. Optionally, blood from a cross-matched donor can be given to the foal.

ANATOMICAL

CLEFT PALATE

Normally, the mouth and the nasal cavities are separated by the bony hard palate and the muscular soft palate, but occasionally a foal is born with a fissure, or opening, in the partition. This congenital abnormality ranges in severity from a cleft in both the hard and soft palate to a cleft in the soft palate only. Affected foals cough, choke, and ooze milk from the nostrils during and immediately after nursing. Although this disorder cannot usually be corrected, surgical intervention may be of benefit in some cases. If left untreated, affected foals become weaker and often develop pneumonia due to aspiration of milk into the lungs.

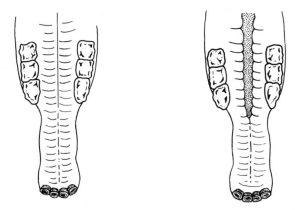

(A) The hard and soft palates in the roof of the mouth normally divide the nasal and oral cavities from one another. (B) However, a cleft in the hard and soft palates forms an opening between the two cavities.

TYMPANY OF THE GUTTURAL POUCH

The guttural pouch is the sac-like enlargement of the lower end of the Eustachian tube. In some foals the flap of tissue that covers the opening to the guttural pouch acts as a one-way valve, trapping air in the pouch and causing a swelling in the parotid area. Unlike an abscess, tympany of the guttural pouch is soft and cool to the touch. In most cases swelling occurs after strenuous coughing, panicky screaming, or some other circumstance that allows air to enter the guttural pouch. Affected foals show no signs of illness or discomfort. If the condition is not corrected spontaneously, surgical correction may be necessary. Occasionally, infections are associated with tympany of the guttural pouch, and in such cases, treatment should be accompanied by antibiotic therapy.

ENTROPION

Entropion is a relatively common condition characterized by profuse tearing of the affected eye or eyes. On close inspection, the lower eyelid is found to be inverted, causing the eyelashes to rub against the sensitive cornea of the eye. If allowed to go untreated, the corneal surface of the eye becomes dry and damaged, and permanent blindness may result. Entropion is easily corrected by everting the eyelid several times daily or through minor surgery. The veterinarian may recommend an ophthalmic ointment or solution to treat any irritation.

ECTROPION

Ectropion is the eversion of one or both eyelids away from the eye. Because tears cannot bathe the cornea properly when the eyelid is everted, the eye becomes dry. Ectropion can be corrected and should be dealt with as early as possible.

CONGENITAL NIGHT BLINDNESS

Congenital night blindness is a hereditary condition that occurs in Appaloosas. Day vision is seldom affected, but at night the foal stumbles, bumps into objects, acts confused, and is reluctant to move. A visual examination does not reveal any abnormalities, but a special diagnostic method called electroretinography can be used to identify the disorder. This diagnostic test is accomplished with an instrument that monitors electrical impulses from the eye and traces them on a sheet of paper in a continuous up and down pattern of varying amplitude (heights). When tested in the dark, the amplitude of the impulses from a horse with night blindness is higher than the amplitude of those from a horse with normal vision.

CONGENITAL MATURE CATARACTS

Congenital mature cortical cataracts are present at birth and are the most common cause of blindness in foals. These cataracts are dense, white opacities that cover the entire lens. This condition usually affects both eyes and may be accompanied by microphthalmia. Surgical removal is successful in some cases.

MICROPHTHALMIA

Microphthalmia is abnormal smallness of one or both eyes. This condition ranges in severity from a slightly smaller than normal globe to near anophthalmia (complete absence of the eye). Although the cause of this abnormality is not yet known, a genetic basis is suspected.

FRACTURED RIBS

Due to the compressive forces of delivery, fractured ribs are not uncommon in newborn foals. Rapid, shallow breathing and a characteristic grunt that occurs when air is inspired are indicative of this condition. In most cases, the foal is very sore and is unwilling to move. Rough handling must be avoided to minimize the chances of displacing the ends of the broken ribs. If the fractured end of a rib penetrates a lung, death usually results. When there is no displacement of the fractured ends, the condition is very difficult to detect. Therefore, if it is even suspected that a foal is suffering from fractured ribs (i.e., if he appears to be sore), a veterinarian should be summoned at once. The young foal should never be restrained or picked up by holding the rib cage.

ATRESIA COLI

Atresia coli is a congenital abnormality involving the absence of a section of the intestinal tract. Each of the two unconnected sections ends in a blind pouch. The foal appears normal at birth but begins to develop colic within 8 to 24 hours. After an enema is administered, very little of the liquid is expelled and no fecal material is passed. Because several feet of intestine may be missing, surgical correction is usually impossible.

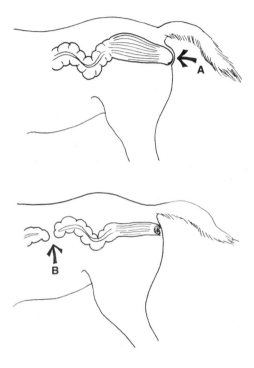

Two congenital abnormalities which occlude the intestinal tract are sometimes seen in foals. (A) The intestine ends blindly in foals affected by atresia coli, while (B) the rectum is closed by atresia ani.

ATRESIA ANI

Atresia ani is a rare, congenital defect in which the external rectal opening, or anus, is absent. This condition can be corrected with surgery unless the atresia is so severe that connection of the rectum to a surgically constructed anus is impossible. However, if atresia ani is not discovered promptly and treated immediately, surgery is seldom successful.

UMBILICAL HERNIA

The umbilical opening allows the passage of the urachus, the umbilical arteries, and the umbilical vein from the foal's body to the placenta. During the last stage of gestation, this opening normally closes around the umbilical cord. If it fails to close soon after birth, a loop of intestine may slip through the opening, forming an umbilical hernia, which appears as a small swelling protruding from the abdominal wall near the navel.

Small umbilical hernias usually disappear by the time the foal is twelve months old. But large, persistent, or strangulated hernias require surgical correction. Strangulation occurs when the ring closes around the entrapped intestine, cutting off circulation to the displaced tissue. If the hernia cannot be pressed gently back through the opening (i.e., manual reduction) or if it is firm and painful, strangulation should be suspected. If a strangulated hernia is not corrected immediately, it may become gangrenous, resulting in fatal peritonitis.

SCROTAL HERNIA

As explained in "The Stallion Reproductive System," descent of the testes from the abdominal cavity, through the inguinal canal, and into the scrotum normally occurs within a year after birth. If the testes have not descended at birth or if the inguinal rings do not close soon after the testes descend, a piece of intestine may become displaced within the inguinal canal and possibly within the scrotum. Due to their locations, many inguinal and scrotal hernias are not detected for several days or months. Although most of these hernias disappear without complications, affected colts should be watched closely for signs of bowel strangulation (e.g., colic), and if strangulation is suspected, a veterinarian should be called immediately.

LEG DEFORMITIES

Foals are commonly born with angular deviations of one or more limbs. In most instances, these abnormalities do not affect the foal's ability to stand or walk and improve with continued exercise and stretching of the leg muscles. If the foal is unable to support its weight on one or more legs or if the condition does not improve naturally, bracing, splinting, or casting may be required. When mechanical assistance is given, it is extremely important that the support device be adequately padded. If padding is not sufficient, pressure on the limb may irritate the skin and underlying tissue, possibly resulting in necrosis. (Refer to "The Foal's First Year" for more information on leg deformities.)

CONTRACTED FOAL SYNDROME

Contracted foal syndrome is characterized by paralyzing muscular contractions of the legs, asymmetrical formation of the skull, and curvature of the spine. The joints of affected foals are permanently and rigidly fixed into place. If carried to term, these foals must usually be removed by fetotomy or Caesarian section. If delivered alive, contracted foals are unable to rise and nurse, and because surgical correction is seldom successful, euthanasia is indicated in most cases.

MULTIPLE EXOSTOSIS

Multiple exostosis is a hereditary condition involving the formation of bony enlargements on the ribs, pelvis, and/or long bones. These enlargements increase in size and in number until the skeleton reaches maturity. Foals may be affected at birth or may appear normal for several months before the abnormal bone growth begins. Although the condition is unsightly, it seldom affects the horse's health or athletic ability.

IMMUNOLOGICAL

FAILURE OF PASSIVE TRANSFER (FPT)

Although the equine fetus is capable of producing some antibodies, it does not produce enough to provide adequate protection at birth. This protection is acquired through the ingestion of antibodies in the mare's colostrum during the first few hours of life when the foal's small intestine is still capable of absorbing the large antibody structures. Unfortunately, approximately 10 to 20 percent of all foals fail to receive adequate protection. Failure of the foal to receive or utilize antibodies in the mare's colostrum is termed failure of passive transfer (FPT).

Premature lactation is one of the most common causes of FPT since it is not unusual for a mare to run milk for several hours or days prior to parturition. This early flow does not necessarily exhaust the mare's colostrum but may markedly reduce the amount available to the foal. Another cause of FPT is the failure of some mares, particularly those that have never foaled before, to pull immune proteins from the blood and concentrate them in the colostrum. In some cases, the foal does not receive adequate colostrum because of weakness, deformity, or the lack of a sucking reflex. In other cases, the foal nurses, receives an adequate volume of colostrum, but fails to show useful levels of protective proteins in the blood. Finally, overproduction of corticosteroids by either the mare or the foal just prior to birth may reduce the foal's intestinal capacity for absorption, thus causing FPT.

Several methods can be used to assess the foal's immune status, but the most commonly used technique is radial immunodiffusion, a quantitative assay of antigen/antibody reactions. The foal's blood serum is tested against specific antigens (foreign compounds that stimulate the production of antibodies) to determine the concentration of antibodies in the serum. Because a specific antigen combines with only one antibody to form a precipitate in the test sample, information can be obtained by comparing the size of the precipitation ring with sizes for known standards. A 12 to 18 hour old foal should have blood protein IgG (immunoglobulin G) levels of 800 to 1000 mg/dl (milligrams/deciliter). This value is approximately equal to that found in the mare. If IgG levels are between 200 and 400 mg/dl, the foal has partial FPT. Levels less than 200 mg/dl indicate total FPT.

Another method of detecting failure of passive transfer is the zinc turbidity test, which can be purchased in a kit. The following steps are used to conduct the zinc turbidity test on the farm:

1. Prepare zinc sulfate solution by adding 250 milligrams of zinc sulfate to 1 liter of freshly boiled distilled water. (This solution can be mixed by a pharmacy.)

2. Add 6 milliliters of the prepared zinc sulfate solution to 0.1 milliliter of the mare's blood serum (or serum from a normal horse on the farm).

3. Add 6 milliliters of the prepared zinc sulfate solution to 0.1 milliliter of the foal's blood serum.

4. Agitate both samples and leave at room temperature for one hour. Read the test results using the mare's sample as a control for comparison. The foal's sample should be cloudy (turbid). A clear sample indicates FPT.

The age of the FPT foal must be considered when treatment methods are selected. If the foal is less than 24 hours old, for example, 2 to 3 liters of colostrum via bottle or stomach tube over a period of several hours is the preferred treatment. Fortunately, colostrum can be stored in a freezer for about five years, and many breeding farms collect and freeze colostrum from foaling mares to provide an adequate supply for foals that do not receive colostrum from their dams. The colostrum can be thawed and bottle-fed to the newborn to provide passive immunity. If the colostrum-deficient foal is over 24 hours old, however, intravenous administration of cell-free plasma (20 milliliters per kilogram of body weight) from a carefully screened donor is the treatment of choice.

A unique and potentially useful method of providing antibodies to the foal is currently being studied. In the initial research trials, blood serum was collected from a number of vaccinated donors and then combined and lyophilized, or freeze-dried. (Blood serum is the source of the antibodies that are concentrated in the mare's colostrum.) The freeze-dried serum was then reconstituted and given orally to a group of newborn foals. Blood samples taken from the foals during the first two weeks of life showed that the foals had absorbed adequate levels of antibodies. Although the use of freeze-dried serum is only being used experimentally, it shows great promise for practical use on the breeding farm. Some interesting applications of this technique might include intravenous administration of reconstituted serum to foals that fail to receive colostrum before 36 hours of age. The donor horses might also be hyperimmunized against certain diseases so that their blood serum would contain an even higher level of antibodies to these diseases than does the mare's colostrum.

COMBINED IMMUNODEFICIENCY (CID)

Although combined immunodeficiency, or the inability of the foal to produce sufficient antibodies, is a congenital defect (i.e., present at birth), the condition does not become apparent until the antibodies received in the mare's colostrum no longer provide passive immunity for the foal. For this reason, combined immunodeficiency is examined within "The Foal's First Year."

Infectious Disease

Normally, antibodies for many of the organisms present in the newborn foal's environment are concentrated in the mare's colostrum before foaling. After birth, the foal receives antibodies by suckling, thereby providing passive immunity against some diseases to which the foal may be exposed. However, for a number of reasons a foal may fail to receive passive immunity from its dam. For instance, if the mare is moved to a new location just before foaling, she may not have time to develop antibodies for the organisms found in the new environment. In other instances, the mare may lose antibody-rich colostrum before foaling. Thus, a foal which is healthy and normal at birth may be left susceptible to infectious disease after birth. Therefore, all foals should be observed closely during the first few weeks of life in order to promptly detect and treat disease.

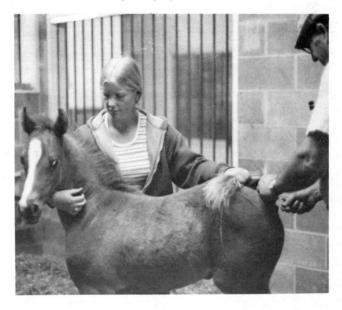

The foal's temperature should be monitored if the foal shows signs of illness, because temperature fluctuations may indicate the onset of disease. The normal temperature range for the neonatal foal is 99 - 101° F (37.22 - 38.33° C).

It is important to note that the fetus and newborn foal's ability to cope with foreign chemicals is minimal. For this reason, the use of drugs on mares during parturition and in patients under 30 days of age should be conservative. To minimize the stress on the neonatal foal's system, the veterinarian usually selects drugs that are easily broken down for elimination. In order to maximize the benefits and minimize the risks of drug therapy, the infectious organisms should be identified (when possible) so that a very specific antibiotic therapy can be used.

SEPTICEMIAS

Septicemias are infections that affect the entire body due to the presence of pathogenic bacteria or toxins produced by bacteria in the bloodstream. If these so-called systemic diseases develop during gestation, the foal may be born weak or comatose. The mortality rate for these foals is very high, but some cases respond to prompt treatment. After birth, exposure to pathogenic organisms may result in septicemia. The main paths of entry for these organisms are 1) through the open navel, 2) by inhalation, and 3) by ingestion. *Streptococcus pyogenes, Escherichia coli, Actinobacillus equuli, Corynebacterium equi, Salmonella typhimurium, Salmonella abortus equi*, and *Pseudomonas aeruginosa* are often responsible for foal septicemias.

STREPTOCOCCUS PYOGENES

Foals acquiring *Streptococcus pyogenes* infections in utero are usually aborted, stillborn, or born very weak and dehydrated. Prompt treatment and intensive care save some affected foals, but foals that are born semi-comatose and too weak to stand seldom recover. If an infected foal appears normal at birth, it usually becomes depressed and develops a fever within a few days. Constipation is followed by diarrhea, and the foal becomes lame due to swollen joints (joint ill). The infection may shift from one joint to another joint or may regress and then reappear in the same joint. (Refer to the section entitled 'Joint Ill' in this chapter.) If the navel is involved, it becomes hot and swollen and may exude a purulent discharge. Abscesses caused by secondary pathogenic invaders may develop in or around the navel.

S. pyogenes infections are extremely resistant to treatment and may remain in a quiescent form even though the foal appears to have completely recovered from the initial infection. In these cases, the young foal develops hot swelling in the stifle, knee, or hock joints. Treatment seldom yields satisfactory results, and even if the infection is eliminated, the youngster often suffers a life of chronic joint trouble.

ESCHERICHIA COLI

Normally an inhabitant of the intestinal tract, *Escherichia coli* may enter the bloodstream of a foal that is already weakened by other pathogens. *E. coli* invades the central nervous system so that the resulting infection is characterized by a gradual loss of equilibrium. The infected foal is usually depressed and may be unable to rise, but if the foal is lifted to its feet, it is usually able to walk or nurse in a normal manner. (In some instances, the foal may appear normal for as long as an hour before staggering and falling.) When the foal is recumbent, its legs move in a running or walking motion, and as the disease progresses, the foal's neck muscles may draw the head around to one side. In the advanced stages of the disease the foal is unable to rise and becomes comatose.

Complete recovery is possible if the infection is recognized and treated before the advanced stages. Treatment usually involves the administration of antibiotics and/or sulfonamides. Vitamins, corticosteroids, electrolyte solutions, and blood transfusions are also beneficial in many cases.

ACTINOBACILLUS EQUULI

Actinobacillus equuli is one of the organisms which causes the sleeper foal syndrome. Affected foals are also called dummies and wanderers (not to be confused with foals in the "dummy" or "wanderer" stages of the convulsive foal syndrome). The foal may be stillborn or born semicomatose and unable to rise. The prognosis for these weak foals is poor. In some instances, affected foals appear healthy at birth but begin to show signs of infection within two or three days. These signs include lack of a sucking reflex, increased pulse and respiration, and a subnormal temperature. Foals that are strong enough to rise wander around the stall but refuse to nurse. In subacute cases, the foal is depressed, has a poor appetite and a fever, and may have one or more swollen joints.

Treatment of an *A. equuli* infection consists of the administration of broad-spectrum antibiotics. Affected foals must be fed artificially, and care should be taken to prevent dehydration and resulting electrolyte imbalances. Even when diagnosis and treatment of this disease are prompt, the affected foal's chances of survival are very poor.

SALMONELLA TYPHIMURIUM

Salmonella typhimurium is the cause of an acute and usually fatal intestinal infection in foals. The foal is weak and has severe diarrhea and a fever. Fortunately, the judicious use of vaccines and proper farm management have reduced the occurrence of *S. typhimurium* infections significantly.

PSEUDOMONAS AERUGINOSA, SALMONELLA ABORTUS EQUI

Pseudomonas aeruginosa and *Salmonella abortus equi* are seldom primary causes of disease in foals, but they are known to be present as secondary invaders in many foal diseases.

SEPTIC ARTHRITIS

Septic (infectious) arthritis in the foal may be due to a septicemia settling in the joints, a laceration of the joint capsule, or the introduction of bacteria through improper use of a hypodermic needle. After pathogenic organisms enter the joint capsule, pus forms and the surrounding bone and articular cartilage erode, often leaving the foal with permanently damaged joints. Prompt treatment, which is necessary to effect a complete recovery, involves removing the purulent contents from the joint capsule, injecting antibiotics into the joint, and casting or bandaging the area.

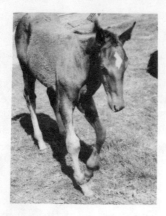

Even after recovery, a severe case of joint ill can leave the foal with arthritis and permanently enlarged joints that are incapable of adequately supporting the foal.

JOINT ILL

Joint ill is an infection in foals caused by organisms which may be ingested by the foal or which may enter through the navel cord shortly after birth. Abscesses within the umbilical cord may serve as sources of infection even after the umbilical cord has closed. Although foals as young as five days of age may develop joint ill, foals up to four months old are also susceptible. A number of opportunist bacteria may cause this disease, including *Actinobacillus equuli, Escherichia coli, Staphylococcus aureus*, and *Streptococcus zooepidemicus*.

The first definite signs of joint ill are a fever of 102-104°F and lameness due to one or more hot, swollen joints. The fetlock, stifle, and hock joints are most susceptible to this type of infection. The foal may be listless or lose its appetite, but in many cases lameness is the only external sign of infection. Synovial fluid in the affected joints appears cloudy and may contain pus. If joint ill is left untreated, abscesses may form in the infected joints and can cause permanent lameness.

Joint ill is very difficult to treat, and thus, the foal's navel should be treated (e.g., immersed in Lugol's solution or tincture iodine) as soon as the umbilical cord ruptures to prevent the entry of bacteria. Treatment for joint ill consists of massive doses of antibiotics and complete stall rest, and mildly affected foals recover if treated promptly. However, the prognosis is poor for severely afflicted foals, and even if the foal survives a severe case of joint ill, it may have chronically arthritic joints for the rest of its life.

DIARRHEA

There are many possible causes of diarrhea in the neonatal foal, but the most common type of diarrhea occurs in association with the mare's foal heat, which occurs approximately 7 to 12 days postpartum. This condition is characterized by soft to watery feces, normal appetite, and mild dehydration, and it usually lasts for only two or three days. Although the cause of foal heat scours is not clear, one suggestion is that high milk production in the mare overloads the foal's digestive tract. (In this instance, scours can often be controlled by reducing the mare's grain intake.) Although changes in the mare's milk composition during estrus and the presence of vaginal discharges on the udder have long been suspected as causes of foal heat scours, research studies have not been able to substantiate these theories.

Other causes of scours in the young foal include increased fiber in the diet, changes in intestinal flora, and the ingestion of miscellaneous tidbits when the foal's digestive tract is not yet ready for solids. The ingestion of manure causes a loose, foul-smelling bowel and may also cause impaction. Persistent diarrhea may be caused by strongyloides infestations or salmonella infections. *Strongyloides westeri* infestations in the foal can be controlled through deworming with thiabendazole or cambendazole. Because *Strongyloides westeri* are concentrated in the mare's milk, it also may be necessary to treat the mare.

Normally, foal heat diarrhea is not profuse, but it is very important to treat diarrhea promptly because the foal can dehydrate rapidly. If the condition is the result of disease rather than excessive milk consumption, the foal will often be weak, feverish, and reluctant to nurse. Some foals require antibiotics and/or fluid therapy, and in all cases the foal's hindquarters should be bathed daily with a mild soap solution and rinsed thoroughly to remove accumulated fecal matter. In addition, petroleum jelly applications around the foal's anus and on its hindquarters help to prevent loss of hair due to irritation.

TETANUS

Clostridium tetani is a spore-forming bacterium that is responsible for the production of a toxin (poisonous agent) that causes tetanus in animals and man. Because this bacterium changes from its spore stage to its toxin-producing form only in the presence of necrotic (dead) tissue and only in the absence of oxygen, tetanus is often associated with puncture wounds. When the bacteria become established within a wound, toxin is released and then absorbed by nerves and lymph vessels in the area surrounding the wound. (The incubation period is about 10 to 14 days.) The toxin affects the central nervous system and causes a characteristic progression of symptoms, resulting in death in 80 percent of all cases. Foals that develop tetanus require immediate veterinary attention and must be kept in a quiet, darkened box stall to minimize stimulation. When the disease is acute, the following signs may be observed:

1. Localized stiffness of muscles in the jaw, neck, hind limbs, and in the region of the wound is apparent.
2. General stiffness is apparent about a day later.
3. Muscle spasms become evident.
4. Reflexes become more intense.

5. Violent muscle spasms are stimulated by sudden movement or noise, and the nictitating membrane flips across the eye.
6. Spasms of the muscles in the jaw make eating very difficult.
7. Ears are erect.
8. Tail is stiff and held away from the body.
9. Walking and turning become very difficult.
10. Head and neck are extended, and legs adopt a sawhorse stance.
11. Sweating may be present.
12. Respiratory and heart rates increase.
13. In severe attacks, the temperature rises to slightly above normal. (Just before death, the temperature may rise to 108-110°F.)

Due to the high mortality rate associated with tetanus, a tetanus prevention program should be pursued. The newborn foal receives passive immunity through ingested colostrum if its dam has been vaccinated within 30 days prior to foaling. Some veterinarians administer tetanus antitoxin to the newborn foal as a precautionary measure regardless of whether the mare is on a regular vaccination program. Tetanus antitoxin can be given within three days after birth and again three weeks later to protect the foal. ◄

20

ORPHAN FOALS

An orphan is any foal whose mother is unwilling or unable to provide milk and/or a normal maternal relationship due to her death, illness, injury, or behavioral problems. Because the newborn foal depends on its dam for immunity against infection and for the development of normal behavioral patterns, loss of the mare during or immediately after foaling can be a serious set-back to the youngster's physical and psychological development and in some instances may even threaten its survival. Fortunately, research on the foal's immediate and long-term needs has helped to make the transition from orphan to independent weanling less traumatic than in the past. With proper supervision, a well-planned feeding program, and adequate discipline, the orphan foal has an excellent chance of reaching its inherent potential.

Immediate Needs

Because the orphan foal's immediate needs vary with individual circumstances, a brief review of situations which may lead to the separation of mare and foal precedes more detailed discussions on health care, feeding, and handling. As mentioned earlier, the foal may be orphaned by events ranging from death of the mare to absence of normal maternal behavioral patterns. In any instance, foals should be categorized according to their immunity status: foals that received colostrum within 48 hours of birth versus foals that did not. The immunological protection effected by the ingestion of colostrum lasts for 3-4 months and gives the orphan foal a marked advantage over those orphans that, for various reasons, do not receive colostrum. To help insure that every foal receives this important passive immunity, many farms maintain a colostrum bank—a supply of frozen colostrum. Frozen colostrum should only be stored for five years before being discarded and replaced with fresh colostrum. In some instances, the mare's colostrum can be milked and bottle-fed to her newborn foal. This procedure has also been carried out successfully soon after the death of the dam.

Rupture of the uterine arteries, uterine torsion, and twisted intestines are just a few of the conditions that can prove fatal to a small percentage of foaling and post-foaling mares. Depending on the circumstances surrounding birth (e.g., type of presentation, length of parturition), the orphan foal may be strong and healthy or may be so weak that it is unable to stand or nurse. A weak or depressed foal requires immediate attention to insure that it receives adequate immunity (colostrum via stomach tube or broad-spectrum antibiotics parenterally) and stimulus for the evacuation of its bowels. Stronger orphans may adapt readily to sucking a bottle and should be placed on a regular feeding program (colostrum for the first feeding) as soon as possible. It is important to note that the routine post-foaling procedures described in "The Neonatal Foal," are especially important to the orphan foal. Every possible precaution should be considered to help insure that the orphan makes normal progress.

In some instances, the mare's life is not lost, but her maternal role is minimized or eliminated by illness or injury. For example, a foal whose dam contracts mastitis may require bottle feeding until the infection can be eliminated. If the mare is isoimmunized against the foal's red blood cell type, the foal must be separated from its dam or muzzled to prevent the ingestion of lethal antibodies. (Refer to the discussion on neonatal isoerythrolysis within "The Neonatal Foal.") This is only a temporary measure, since the foal can safely ingest the mare's milk once the foal's intestine is no longer permeable to large antibodies. In other instances, the mare may not produce enough milk to meet the foal's nutritional requirements. Supplementary bottle feeding should leave the foal slightly hungry so that it will continue to suckle its dam and hopefully stimulate milk production. Similarly, if the mare is not producing any milk, the foal's bottle feedings should leave the youngster slightly hungry, so that suckling continues and milk let-down is stimulated. If milk secretion is not stimulated within a reasonable period (usually in aged or first-foal mares) or if the mare is simply not capable of producing enough milk to meet her foal's needs, the attendant should supplement the foal's diet with regular bottle feedings so that its nutrient requirements are met. In such cases, separation from the dam is not usually necessary.

The importance of the mare-foal bond, established soon after parturition, has been emphasized in previous chapters. When human intervention is minimal and when the mare's maternal behavior patterns are normal, the mare sniffs her newborn foal and licks the fetal fluids which serve as a type of identification marker. Normally, the mare continues to investigate and nuzzle her foal for several hours, and a strong bond develops between dam and foal. However, if parturition has been unusually stressful, the mare may be too exhausted to investigate and lick her foal. Fright or lack of experience in first-foal mares and human interference in the foaling stall also inhibit this important activity. Regardless of the cause, the absence of these important interactions between the mare and her foal results in a weak mare-foal bond and may cause the mare to reject her foal. In these cases, an adoption procedure similar to that used when uniting orphans with nurse mares may be very effective. (Refer to the discussion on foster mothers within this chapter.)

In rare instances, the mare may savage her foal to the point that the youngster must be removed from the mare's reach to prevent serious injury. Although some mares eventually accept their foals, other mares continue to refuse their foals violently and often repeat this abnormal behavioral pattern after subsequent foalings. If the mare has a history of savaging her foals, she should be watched closely after foaling. At the first sign of aggression, the foal should be removed for its own protection, and if the mare can be restrained without undue trauma, her colostrum should be milked and fed to the foal. If the mare cannot be encouraged to accept her foal, it should be placed on one of the orphan feeding programs described later in this chapter. If the mare has a history of delayed acceptance, she should be milked at regular intervals to prevent involution of her udder. The foal should receive this fresh milk from a bottle until acceptance is complete.

Health Care

Preventive medicine and good husbandry techniques are important on any animal operation, but the benefits of such programs are especially significant when the foal starts life with the handicap of temporary or permanent separation from its dam. The newborn orphan requires the same routine care and observation that any newborn foal receives. The navel should be soaked in iodine, and the foal's vital signs should be checked carefully. If the foal did not receive colostrum, tetanus antitoxin should be administered, and an enema should be given. A series of antibiotics and vitamins may be recommended by the veterinarian. It is also wise to check the orphan's vital signs twice a day for the first few weeks to insure that any illness or congenital abnormality is detected as quickly as possible. (Refer to 'Physiological Parameters of the Newborn Foal' within the chapter, "The Neonatal Foal.") The foal's temperature is the single most important indication of its health status. Any rise above normal should be considered an immediate danger signal. (One hour after birth the foal's temperature should be 100.4-101°F.) A subnormal temperature is also significant and requires that the foal be blanketed or placed under heat lamps until a veterinarian can be summoned.

The hand-raised foal is often somewhat more susceptible to many of the common foal illnesses than is the foal that is raised by its dam. Diarrhea, for example, frequently occurs in bottle-fed orphans due to artificial formulas and overfeeding. If diarrhea persists for more than 24 hours, or if it is accompanied by other signs of illness, such as colic, fever, or lethargy, a veterinarian should be called immediately. As emphasized in "The Neonatal Foal," diarrhea may also occur in association with several serious foal diseases, and persistent diarrhea can rapidly dehydrate the foal to the extent that its body defenses are lowered significantly. If the foal is alert and its vital signs are normal, diarrhea may be controlled by reducing milk consumption or by reducing the formula's fat content. Orphan foals also seem to be more susceptible to meconium retention and respiratory diseases, especially when the foal failed to receive colostrum. The treatment of meconium retention is also examined in "The Neonatal Foal." Because respiratory infections range in severity from a common cold to foal pneumonia, treatment depends on the veterinarian's diagnosis and recommendations. (For more information on respiratory diseases in foals, refer to the chapters "The Neonatal Foal" and "The Foal's First Year.")

Housing is an important consideration when raising an orphan. Because the orphan foal is particularly susceptible to health problems, it should be housed in a well-designed stall and turned out for periods of free exercise in good weather. The foal's stall or pen should be secure and free of safety hazards. All youngsters are inquisitive and prone to injury, but without its dam's supervision the orphan foal is even more susceptible to injury. Attention should be given to the temperature and humidity of the stall.

Although it is helpful to remove any chill or dampness from the foal's stall, overheating can damage the foal's sensitive nasal mucosa and respiratory tract. Infra-red heat lamps should be directed in such a way that the foal can move in and out of the light to remain comfortable. Ample sunlight and fresh air help to minimize dampness and the growth of harmful bacteria. However, windows or vents should be located high enough to prevent drafts at the foal's level.

Feeding the Orphan

There are several acceptable methods for feeding the orphan foal. Individual circumstances determine which method is most suitable for a particular orphan. For example, not all owners of orphan foals will be fortunate enough to locate an agreeable nurse mare, and some owners may not be able to provide the attention required to hand-raise a foal. The following discussion reviews important points of several different feeding methods, but the breeder should consult a knowledgeable professional (e.g., veterinarian, university horse expert, or extension agent) before selecting a technique.

FOSTER MOTHERS

The best possible alternative for an orphan foal is a nurse mare. Sometimes a mare whose foal has died or is old enough to be weaned may be secured and successfully introduced to an orphan foal. Cases of mares with exceptional milking abilities adopting orphans in addition to their own foals have been reported, but this is an unusual technique that would require close observation of each foal's progress.

In some parts of the country, particularly Kentucky, nurse mares can be leased from special farms that maintain lactating mares. Besides substituting for an unwilling or unable dam, nurse mares have an expanded role in the industry. They replace aged mares and mares with poor breeding records to allow them to channel all of their energies toward conception and gestation. In addition, to spare valuable foals from travel and exposure to unfamiliar disease organisms on other farms, nurse mares are sometimes employed after natural dams are shipped away to be rebred.

Mares which are maintained specifically as nurse mares are often draft-type mares. In order to be successful, they should be good milkers with large udders and be healthy, sound breeders. They are bred each year, often to teasers, and when their services are required, their foals are given away or sold. However, fillies may be retained as replacement nurse mares. Occasionally, foaling is induced several days before the mare's due date in order to provide a nurse mare for a well-bred orphan. However, gestation and foaling are usually allowed to proceed normally.

After a nurse mare has been obtained, the first stage in the adoption process involves introducing the orphan to the mare. The nurse mare is more likely to accept the foal and recognize it as her own if it smells familiar. Recognition has been accomplished by rubbing the foal with the nurse mare's urine, milk, or manure. Also, both the mare's nose and the foal can be wiped with peppermint spirits to mask the foal's unfamiliar odor.

At least three people are required to insure the foal's safety during the initial adoption period: one person to hold the mare's lead shank, one to restrain the foal, and another to hold one of the mare's forelegs in such a way that kicking is prevented. The mare should be allowed to smell the foal, but she should be watched carefully and pulled away if she attempts to bite. After the initial introduction, the foal should be led to a nursing position. At this point, all three handlers must remain alert and ready to separate the mare and the foal if a problem arises. (The foal should be allowed to nurse only if the mare stands quietly.) This procedure should be repeated every two to three hours, and if all goes well, the mare can be allowed to stand on all four legs after the third or fourth feeding. However, the handler should be prepared to turn the mare's hindquarters away from the foal in case she attempts to kick.

As an alternative to hand-restraint, the mare can be restrained in a special stall which has a partition between her and the foal. An opening should be provided in the partition through which the foal can reach the mare to nurse. If the mare is quiet, she and the foal can be left in the partitioned stall for several hours before putting them together under supervision. Protective behavior is a good sign that a mare has adopted a foal. Once the initial adoption takes place, it is unlikely that the mare will later reject the foal. After the mare has allowed the foal to nurse a few times without being restrained, they can be turned out together. They should be closely supervised for the first few days.

When nurse mares are unavailable, nurse goats can be used with good results. The composition of goat's milk is much closer to that of mare's milk than is cow's milk (which might be used in a hand-feeding program), and most foals readily adopt a nanny as a foster mother. The goat must be held or tied on a platform so that the foal can reach the goat's udder. If this feeding method is used, the foal should be weaned at two months to allow maximum growth and development.

MECHANICAL METHODS

Another possible technique for feeding the orphan foal is the use of an automatic feeder. The standard automatic calf feeder consists of a drum containing milk powder, a mixing vessel, and systems to control the temperature and the amount of milk available to the orphan. The foal drinks milk through a teat assembly on the wall of the stall, and one person working regular stable hours can care for six machine-fed foals. The feeders should be stripped down, and all removable parts should be washed in hot, soapy water each day. The use of a dirty feeder or old milk can lead to serious digestive problems. When this method is used to feed the orphan, animal companionship (e.g., another foal or goat) should be provided.

Courtesy of Manx Farms

The companionship of another animal, such as a goat, may prevent the orphan foal from becoming too dependent on human handlers and from developing vices due to boredom.

HAND-FEEDING

If a foster mother or an automatic feeder cannot be used, the foal must be raised by hand. A satisfactory feeding bottle can be made by fitting a clean soft-drink bottle with a lamb nipple. Lamb nipples are the right size and shape for foals and can be found at most farm supply stores. The foal should be fed only while it is standing up or lying on its chest. Attempts to feed the foal while it is lying on its side may result in the entry of milk into the lungs and, possibly, death.

To make feeding much easier on the attendant, the foal should be taught to drink milk from a bucket as soon as possible. Smearing the foal's muzzle with formula or allowing the foal to lick milk off a finger may help hasten the transition from bottle to bucket feeding. If the foal shows any interest in the milk, the handler should transfer the foal's attention by gently pushing the foal's muzzle into a bucket that contains the foal's formula.

Courtesy of Manx Farms

A large soft-drink bottle equipped with a lamb nipple can be used to hand-feed the foal. Later, the foal may be taught to drink milk from a bucket.

After a few attempts, the foal usually drinks readily from the bucket, but it should never be forced to investigate the contents of the bucket. Hanging the bucket at the foal's shoulder level and directing the foal toward it will facilitate the learning process. The bucket should be washed and rinsed thoroughly after each feeding and should only be used to feed that particular foal.

It should be mentioned in passing that premature foals or foals that are born with a physical disadvantage that prevents a normal sucking reflex must be fed through a stomach tube until sufficient strength to nurse is acquired. Tube feeding involves the passage of a soft, clean stomach tube through the foal's nostril, down its throat, and into its stomach. The administration of a special formula (or mare's milk if it is available) through this tube allows the foal to receive nutrients without effort. It should be noted, however, that this is a procedure that requires special training. (A veterinarian or trained technician can usually teach the orphan handler how to tube feed in 2-3 sessions.)

MILK REPLACERS

There are many formulas that can be used to replace mare's milk with some degree of equivalence. The commercial milk substitute Foal-Lac™ (Borden) is an excellent milk replacer for foals. If a commercial milk replacer is unavailable, there are several formulas that can be prepared on the farm. In these instances, the objective is to approximate the composition of mare's milk as closely as possible. For example, cow's milk is lower in carbohydrates than mare's milk but it is higher in fat and protein. By diluting the milk and adding special sugars, the fat and protein content are lowered, and the carbohydrate content is increased. The following formulas are examples of suitable milk replacers:

Formula A
   1 quart homogenized cow's milk
   8 ounces limewater (aqueous solution of calcium hydroxide)
   2 tablespoons dextrose
   2 tablespoons lactose

Formula B
   1 pint lowfat cow's milk
   4 ounces limewater

Formula C
   1 can evaporated cow's milk mixed with an equal part of water
   4 tablespoons limewater
   1 tablespoon sugar or corn syrup

If a particular formula does not agree with the foal (i.e., if diarrhea or poor development is observed), an equine nutritionist should be consulted, and another formula should be adopted.

FEEDING SCHEDULES

There are no firm rules for feeding the orphan foal, since individual needs vary with the foal's age, size, state of health, and amount of solids consumed. The following is an adequate feeding schedule for a normal, healthy foal that is orphaned at birth:

day 1-7	once every hour
day 8-14	once every 2 hours
day 15-21	once every 3 hours
day 22-28	once every 4 hours
day 29-weaning	4 times a day

(Large, robust foals may be fed at slightly longer intervals.)

In the beginning, the foal should be fed about 4-8 ounces of formula at each feeding. Each week, the amount should be increased to the maximum quantity that the foal willingly consumes without exhibiting diarrhea. When unaccompanied by fever or other signs of illness, diarrhea is often a sign that the foal is being overfed or that its formula is too rich. In these cases, the formula should be quickly adjusted.

CREEP FEED

Many sources recommend introducing the orphan foal to solid feed as early as 1 week of age. Although the foal will not consume significant amounts of grain or hay at this time, intake usually increases gradually. The use of milk-replacer pellets (e.g., Foal-Lac™ pellets) may encourage the transition from formula to grain. When the foal is eating grain readily, the amount of milk replacer fed and the frequency of feedings should be adjusted accordingly. Eventually, grain intake reaches an amount that allows the foal to be weaned, usually at about 2 months of age.

It is important to select a feed that is easily digestible, highly palatable, and suited to the foal's nutritional needs. The average foal requires a feed containing 16-18 percent protein, 0.8 percent calcium, and 0.6 percent phosphorus to realize its inherent growth potential. The diet should be balanced with other essential minerals and vitamins to insure proper development. (Refer to the nutrient requirement charts within the Appendix.)

When solid feed is introduced to the foal, the importance of high quality protein should not be overlooked. Most commercial feeds print the minimum crude protein content on the label, but it is important to remember that crude protein is not necessarily indicative of digestible protein. As a rule, only 1/2 to 3/4 of the crude protein in a particular feed is digestible. (This varies among feeds.) Among the numerous types of proteins (amino acids) that may be present, ten amino acids are essential to the horse: arginine, histidine, isoleucine, leucine, lysine, methionine, phenylalanine, threonine, tryptophan, and valine. Of these, lysine is most often deficient in grain and hay. Thus, protein feeds that are high in lysine are of the greatest value in equine nutrition, especially for growing foals, which require a high level of digestible protein. Both soybean meal and linseed meal are good sources of lysine and may be fed in limited amounts to raise the overall protein content of the diet.

Another major concern in preparing a balanced diet for the foal is the calcium-phosphorus ratio. A correct ratio (about 2:1) is essential for the development of healthy bones, joints, and tendons. Slight excesses of calcium are not detrimental, but gross excesses can cause the formation of abnormally dense, thick bone. Excess phosphorus can also cause bone problems. When excessive levels of phosphorus are present in the blood, calcium is pulled from bone to maintain the proper calcium-phosphorus ratio in the blood. This safety mechanism may weaken bones and may possibly cause epiphysitis. Because most grains, including corn and oats, are higher in phosphorus than in calcium, it may be necessary to provide a supplemental calcium source (e.g., bonemeal or alfalfa). Care should be taken to insure that a 2:1 calcium-phosphorus ratio is maintained in the total diet (i.e., hay, grain, and supplement).

Hay of good quality is also important to the growing foal. Both clover and alfalfa are high in calcium and digestible protein and can be incorporated into the foal's diet. Regardless of type, the quality of the selected hay is of utmost importance. Any hay that is dusty, hot, wet, moldy, or is characterized by an objectionable odor should be rejected.

WEANING

By 2-3 months of age, the orphan foal should be completely weaned. However, the foal's diet should not be switched abruptly from milk to solids. To prevent digestive and emotional upsets, the creep feeding program described previously should be implemented. This program should encourage the foal to eat more grain and hay and drink less milk over a period of several weeks. When the foal is weaned, it is provided free-choice grain for another 3 months and then placed on a regular feeding program. If the foal gains excessive weight during any stage of its feeding program, the diet should be adjusted immediately, since excess weight places a strain on the young horse's skeletal structure and predisposes its legs to future problems. (For a more detailed description of the growing foal's nutritional requirements, refer to the Appendix and to the text FEEDING TO WIN.)

Discipline

Due to the nature of the orphan foal's upbringing, its psychological needs are just as important as its physical requirements. If the foal is isolated from other horses, dependence on human companionship may become a deterrent to future handling and training. Ideally, the orphan should be placed in a paddock where other horses can be viewed. When the foal is several months old, direct contact with a gentle horse or pony might be considered. It is important that the companion horse does not bully the foal or consume the orphan's available feed. When the orphan is 4-6 months old, interaction with other weanlings of comparable age and size helps the foal to adapt to rejection, intimidation, and other normal behavioral patterns involved in establishing a social hierarchy. The sooner the orphan is allowed to become a "horse," the easier it will be to carry out normal discipline and training activities.

The orphan foal benefits from the play and companionship with other foals its age.

A major problem encountered with hand-raised foals is their lack of respect for human authority. If handled excessively and isolated from other horses, the orphan may identify only with humans, a situation that encourages play with handlers as if they were other foals or horses. This play takes the form of kicking, rearing, and biting. As the foal grows older, this behavior becomes increasingly dangerous. Therefore, it is important that the orphan be taught at an early age that such behavior is undesirable. (It is equally important that the orphan be allowed to romp with other foals.)

Biting and nipping may be observed when the foal is playing, investigating, or cutting teeth, but this behavior should not be inflicted on humans regardless of the foal's age or intent. When correcting the nipper, the handler should use discretion. Overly harsh treatment is unnecessary and may only succeed in frightening the foal or making it head shy. If the foal is very young, the handler should simply say "no" and push the foal's muzzle firmly away when an attempt to bite is made. An older foal that has not been disciplined in the past may require slightly more severe correction. If the foal does not respond to a "no" and a push, it should be slapped on the neck or shoulder when an attempt to bite is made. Hand-feeding tidbits and petting the foal's face and nose encourage biting and should be avoided.

Kicking and rearing are natural behavioral patterns in certain situations. The foal may kick when approached unexpectedly from the rear or when its back feet are handled. Foals often kick and rear during play, and the orphan foal frequently attempts to play with human caretakers in the same fashion. When a strike or rear is directed toward the handler, the foal should be pushed firmly away and told "no" in a harsh tone. If the foal kicks, it should be slapped on the rump. The best way to prevent serious behavioral problems from developing is to meet the foal's early half-hearted attempts with appropriate discipline.

21

THE FOAL'S FIRST YEAR

Whether the foal is still at its dam's side or whether it is an independent weanling, the first year of life is a critical time of growth and development. The dam is the primary influence on the very young foal. She provides security, companionship, and nutrition, and the foal learns many early lessons and behavior patterns from her. However, as the foal matures, its interactions with other foals, humans, and other aspects of its environment become increasingly important. Thus, during the first year, proper management of such factors as handling, health care, and nutrition can make a tremendous difference in the foal's quality as a yearling.

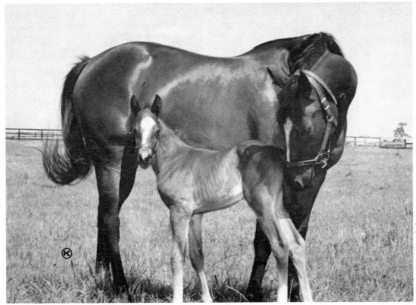

Photo by Art Kunkel

Behavior

Initially, the young foal establishes a close play relationship with its dam through behavior such as nibbling at her mane and tail and kicking. By the time the foal is two or three weeks of age, it begins to engage in mutual grooming with its dam. Another early play pattern involves galloping around the mare in small circles, which become larger as the foal grows older. As contact with other foals increases and as the foal becomes less dependent on its dam, mare/foal play patterns and solitary play become less frequent. Thus, an eight-week-old foal spends approximately 7 percent of its time playing alone or around its mother as compared to 56 percent in a foal under one week of age. Similarly, time spent playing with other foals increases from about 3 percent at one week to 50 percent at eight weeks of age. These figures are influenced by a number of conditions. Without exposure to other foals, for example, the foal continues to play with its mother far past the age when this behavior normally subsides. In addition, if the mare is nervous or highly protective, her foal may be shy and hesitant to venture from her side.

By the time the foal is a few weeks of age, it has usually formed relationships with other foals and is less dependent on its dam. These two foals are engaged in mutual grooming, a common behavior pattern of all horses.

Like other babies within the animal kingdom, the foal spends more time lying down and sleeping than do mature horses. Normally, the foal takes short naps separated by periods of play and nursing, but prolonged periods of sleep are unusual and may indicate illness or other physical abnormality.

Normally, very young foals spend much more time sleeping than do mature horses.

Lying upside down or in other unusual positions is definitely abnormal and may indicate colic. A healthy foal sleeps stretched out on its side or resting up on its chest, and the foal often remains in this position even when approached. (There is no cause for alarm unless the foal displays other signs of illness, such as rapid or shallow breathing, fever, or loss of appetite).

The young foal suckles for periods of several minutes separated by 30 to 60 minute intervals, but as it grows older, these feeding periods gradually decrease in frequency and duration until the foal is essentially weaned. In the wild, the foal is weaned a few days or weeks before the mare gives birth to her next foal, usually within one year after the first foal was born. Under these conditions, however, it is not unusual for a yearling or two-year-old to run back to its dam when threatened or frightened. In contrast,

Courtesy of Earl and Barbara Lang Quarter Horses

The day old foal nurses frequently for short periods.

under most breeding farm management programs the foal is supplied with hay and grain as a nutrient source and becomes less dependent on the mare's milk by the age of three to four months. Although the six-month-old foal may nurse lightly every one to two hours, mare's milk is only a minor part of its nutrient intake. The foal should be weaned by at least six months of age to give the mare an opportunity to gain condition lost during lactation.

Between two and five weeks of age, foals may often be seen nibbling at their dams' feces. This practice is termed coprophagy, and several explanations have been advanced to interpret the behavior. Manure ingestion may benefit the foal by establishing desirable bacteria in the intestine or by supplying needed vitamins or minerals, or coprophagy may even be a normal step in the growing foal's behavioral development. However, coprophagy can cause diarrhea, and it also allows parasites to infest the foal. Because almost all young foals engage in coprophagy, the mare should be on an effective deworming program to reduce the number of parasites in her feces. Also, the foal should be dewormed beginning at seven days of age and at two month intervals thereafter, using a product suitable for young foals. Droppings should be removed regularly from its stall or paddock.

Often a foal develops the habit of chewing on its dam's mane and tail. This habit usually appears when a mare and foal are confined for extended periods of time and, thus, is generally attributed to boredom. Attempts to correct this behavior pattern are usually futile, although some breeders report successful results when the mare's mane and tail are painted with various foul-tasting but nontoxic substances.

Managing the Young Foal

The first few months of life provide an excellent opportunity to influence the foal's development. During this period, the foal is impressionable and is usually very cooperative as long as its dam remains near. Therefore, basic training, such as leading and haltering, should begin at least within a few weeks after birth. As well as its mental development, the foal's growth and physical development can be influenced easily during early life.

EARLY HANDLING

The horse's natural instinct to flee when faced with unusual or dangerous circumstances is a normal behavioral pattern which protects the horse from predators. Under domestic conditions, however, the horse must be taught to accept restraint, resist the flight instinct, and trust the judgement of its handler. The horse's first experience with restraint can be an emotional and physical battle that ends with injury, especially if it is postponed until the horse is over six months of age or until the foal must be restrained for an upsetting procedure, such as medical treatment. Because these experiences can affect the horse's attitude toward future training, handling should begin when the foal is very young. If various forms of restraint and handling are calmly introduced during the suckling period, the foal should be very tractable by the time it is weaned.

Courtesy of Earl and Barbara Lang Quarter Horses

The mare's response to human handling is a part of the foal's early experience. Thus, if the mare is calm and cooperative, the foal's first impressions of early training are likely to be positive.

HALTERBREAKING

Most breeders prefer to fit the foal with a halter within a few days after birth and to begin halterbreaking the foal shortly thereafter. At this age, a foal is surprisingly strong but is small enough to be restrained by the handler. If the foal is several months old when halterbreaking begins, however, the process is usually more traumatic for both parties, since the foal is heavier and usually stronger than its handler. Regardless of the foal's age, lessons should not last longer than 15 to 30 minutes since the young horse has a very short attention span.

Initial haltering is best accomplished in the stall occupied by the foal and its dam. The foal will be much more relaxed in familiar surroundings and can usually be convinced to accept haltering with little difficulty. The handler should gradually move close to the foal's side with a calm, relaxed attitude. At the first sign of discomfort or fright in the foal, the handler should stop until the foal relaxes. This procedure should be repeated until the handler finally reaches the foal's side. Then the foal should be allowed to investigate the handler and make the initial contact but should not be allowed to bite at any time. Once the foal is confident that the handler is relatively harmless, the handler should touch the foal, starting with firm but gentle strokes at the shoulder and working toward the head. Then the halter should be presented and the foal allowed to investigate the unfamiliar object until its apprehension subsides. The halter can then be slipped over its nose and buckled quickly but gently.

To halter the foal, it must first be restrained. This step is usually carried out in a stall or other enclosure, since the foal can be caught more easily when closely confined.

The halter should be presented quietly.

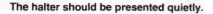

The noseband should be placed on the foal's nose and the halter drawn up into the proper position to be secured behind the foal's ears.

For the foal's comfort and for safe handling, the halter should be properly adjusted so that the nose band is slightly loose but not too low on the nose, and so that the top of the halter does not slip down the foal's neck.

If haltering is postponed until the foal is several weeks or months old, the session may be much more traumatic. To corner and catch the foal, the handler should use a stall or other small enclosure to confine both the foal and its dam. If the mare is very protective, she should be held while her foal is restrained, since such mares sometimes become very aggressive when their foals are handled. The handler should approach the foal quietly and slide one arm around its chest. If necessary, the handler can further restrain the foal by backing its rump into a hazard-free corner. A small foal can be restrained by cradling, which involves placing one arm behind the foal's rump and the other around its chest. Alternatively, the foal can be restrained by placing one arm around the foal's chest and holding the foal's tail straight over its back with the other hand. While scratching the foal gently and speaking to it in a calm tone, the handler or an assistant should gradually position the halter so that the foal is not startled by its appearance. After the halter is slipped carefully over the foal's head, it should be adjusted so that it fits snugly without being tight.

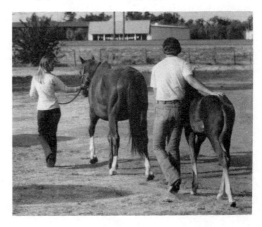

The young foal's natural inclination to follow its dam can be an aid to the handler when teaching the foal to lead in early training sessions.

LEADING

With proper handling, foals may be taught to lead by as early as two weeks of age. While the mare is being moved around the farm, her foal can be led at her side. Because the foal is naturally inclined to follow the dam, it usually accepts this activity. After the foal becomes accustomed to the process, its dam can be led away and the foal can then be led back to the mare. This is an excellent method for halter-breaking very young foals with minimum risk of trauma or injury since it takes advantage of the foal's natural willingness to follow its dam.

An older foal can be taught to lead by using patience and a sequence of cues that are reinforced with reward. Because the mare has already taught her youngster to follow her, the lesson should not be too difficult once the foal understands what is expected. The foal should first be fitted with a comfortable and properly adjusted halter, and a rump rope should be placed on its hindquarters. By standing beside the foal's shoulder, facing the same direction as the foal, and taking a few steps forward, the handler imitates the dam's leading cue to which the foal is already accustomed. At the same time, the handler should encourage the foal to move forward by tugging gently but firmly on the rump rope. If the foal does not step forward, the handler should step back to the foal's shoulder and try again, as several attempts are often required before the foal will respond. The handler should never face the foal while asking it to step forward, but by pushing the foal sideways or pulling on a rump rope while stepping forward, the handler may encourage the foal to take one or two steps. An extra handler is valuable at this point to encourage the foal. If the foal refuses to respond to cues, the second person can often help move the foal forward by placing one hand on the foal's rump while standing close enough to avoid being kicked. Any progress should be rewarded with praise or a gentle rub. Eventually, the foal will catch on and realize that it is actually very simple to please the handler. It is important to note that if a steady pull is placed on the rump rope instead of several short tugs, the foal may lean back into the rope and may even rear backwards, possibly injuring himself or the handler. If, on the other hand, the foal lunges forward, the handler should restrain the foal with the lead rope.

Gentle tugs on a rump rope placed around the foal's hindquarters encourage the foal to move forward.

TYING

Although many horses that have been handled and halter-broken at a very early age learn to stand tied without fighting the rope, learning to accept head restraint can be one of the most difficult lessons the young horse must undergo. One of the oldest techniques used to teach the horse not to fight head restraint is to simply fit the youngster with a sturdy halter, tie its lead shank to a post, and let the horse fight until it submits to restraint. This method is never suitable for young foals since serious injuries to the legs, head, vertebrae, and nerves frequently result, and permanent damage to the horse's neck is not unusual. However, the method can be modified to reduce the risk of injury to the head and neck by attaching the lead shank to a pliable but sturdy object, such as a tire or innertube which is, in turn, attached to an immovable structure, such as a pole or solid wall. For safety, the foal should be tied with a quick-release knot secured above the level of its withers. In addition, the area should be free of any hazards that might injure the foal.

Another method used to acquaint the horse with head restraint is the drag line. The horse is allowed to wander in a pen while dragging about 8 feet of 0.5 to 1 inch diameter soft cotton rope from its halter. When the foal steps on the rope it must stop and drop its head. Eventually the foal learns that when it drops its head, pressure is relieved. Although this method requires minimal labor, many trainers feel that it causes the foal to become depressed and hesitant to move and that it does not promote a good learning attitude.

A third method used to teach the foal to stand tied involves tying a soft cotton rope around the foal's girth in a nonslip bowline knot. The ends of the rope are then run between the foal's front legs, through its halter, and to the stationary tie structure. When the foal fights the rope, pressure is placed on its girth and not on its neck. This method is especially effective on spoiled foals and on older foals that have not been handled much.

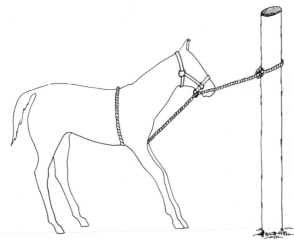

A soft cotton rope run through the halter ring and around the girth is an excellent tool for early attempts at teaching a foal to stand tied. In this arrangement, pressure is placed on the foal's girth, thus protecting the neck from injury.

HANDLING THE LEGS

The foal should be accustomed to having its legs handled as soon as possible in order to save a great deal of time and trouble when the foal is larger and to make the foal's first trimming much easier for the farrier. If these lessons are initiated during the foal's first or second week, the foal will probably submit quietly to having its legs handled by the time it is a month old. After haltering the foal, the handler should push the foal so that its weight is shifted off the leg to be handled. The foot should be lifted gently and then put back down before the foal resists. With daily repetition, the foal soon learns to comfortably support its weight on three legs while the fourth leg is being handled.

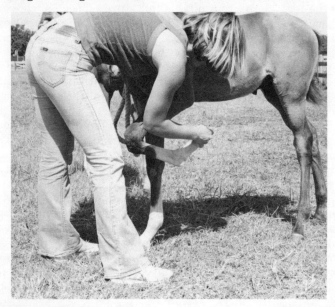

If the foal's feet are handled as a part of the daily routine, the foal will learn to pick up its feet readily.

TRAILERING

In most instances, it is much easier to load and trailer a suckling foal than a weanling. If its dam is cooperative, the suckling foal can usually be persuaded to follow her into a trailer with little physical force. If the foal does not follow its dam into the trailer, it can be lifted in, assuming that trailer lessons begin at an early age. Once the mare and foal have been loaded, they should be taken for a brief trailer ride. If this process is repeated two to three days each week, the foal usually loads readily and rides calmly within two to four weeks. When hauling a mare and foal, many people prefer to remove the trailer partition, tie the mare, and let the foal stand free. Whenever a foal is being hauled, it is important to cover the back of the trailer to prevent the foal from attempting to escape over the tailgate.

With a little calm encouragement, a foal will usually enter the trailer along with its dam.

NUTRITION

Studies have shown that the mare's milk production peaks when the foal is only about six to eight weeks old. Some mares are better milk producers than others, but unfortunately, the foal's increasing needs exceed the available nutrients in even the best mare's milk. When the mare is well fed and when quality pasture or hay is available, the young foal usually receives all of the necessary nutrients from a diet of milk and forage, but there is no rule of thumb applicable in all cases.

Some breeders begin feeding free-choice creep feed when the foal is one week old; however, each foal's progress should be evaluated separately, and a feeding program should be designed to fit that foal's needs. For example, if the mare is a poor milk producer, her foal may show significant improvement when creep feed is provided at an early age. Generally, the foal does not consume a significant quantity of grain at this time, but its intake gradually increases as its needs increase. (Note: Ingestion of grain without adequate roughage increases the risk of serious digestive upsets.)

When the mare's milk quality declines significantly (usually when the foal is about three to four months old), creep feed helps to insure optimum growth and development and also plays an important role in preparing the foal for the transition from suckling to weaning. It is important to note, however, that pushing the foal to achieve rapid gains and early maturity may be detrimental to its soundness and overall health. Foals may gain up to 4 pounds per day, but at one year of age the foal's weight should not exceed 80 percent of its expected mature weight.

Frequently, the foal may be seen nibbling at its dam's grain; however, as the foal matures and consumes more of the mare's grain, both the foal and the dam may be underfed. Thus, although the dam may willingly share her grain ration, a separate feeder should be provided, to insure that the foal's diet is supplemented and to determine how much feed the foal eats.

The mare should not be able to reach the foal's feeder. Some mares can be discouraged by simply nailing a board across one corner of the stall at a height that allows the foal to walk under while keeping the mare at a sufficient distance from the foal's feeder. Pastures can also be equipped with special foal feeders. A typical pasture feeder consists of a small pen with a feed trough in the center. A board should be nailed across each of the pen's entrances at a height that prevents the mares' entry. When foal feeders are used in pasture, they should be located near the mares' loafing shed or water trough, since foals are more likely to enter these feeding pens when their mothers are nearby.

A pasture creep feeder must be placed in an enclosure that permits foals to enter easily, while the feeder itself should be covered to protect the feed from moisture.

The foal's creep feed should consist of a uniform mixture of nutrients that meet the foal's immediate nutrient needs. (Refer to the nutrient requirement chart within the Appendix.) An excellent way to insure that the foal receives a uniform mixture is to feed a pelleted ration. Pelleting involves grinding the feed components into a homogenous mixture and then compressing this mixture into pellets. Because the pellets are homogenous, each bite of the ration is nutritionally balanced, and the foal cannot sift out certain ingredients. Pellets are also compact, easy to store, and relatively free of the spoilage problems associated with other feeds. (Pellets must be protected from moisture, however, or they swell and spoil.) Despite the advantages of pellets, good-quality grain is also suitable for the foal. Until the foal becomes accustomed to creep feed, it may leave significant amounts of uneaten feed, which should be removed each day and replaced with fresh feed to prevent spoilage.

Whether supplied by pasture or hay, roughage is an important part of the foal's diet. The very young foal consumes only small amounts of roughage, but as it consumes less milk and becomes more dependent on creep feed, its roughage consumption increases. A certain amount of roughage is required in the diet merely to prevent digestive upset. However, roughage also provides important nutrients for balancing the ration. One of the objectives of ration formulation is to select a roughage and a concentrate which complement one another nutritionally. For instance, legumes or legume-grass mixes are often used since they provide high levels of protein and calcium to the growing foal. If only grass hay or forage is available, the concentrate portion of the creep ration must contain higher levels of protein and calcium to compensate for the lower levels of these nutrients in grass hay and forage. Since roughage quality varies considerably due to maturity and season, this factor should be taken into account when feeding the foal.

Whenever possible, foals should be fed individually, since free-choice feeding of a highly palatable ration permits the foal to consume large quantities of concentrate. If free-choice feeding is unavoidable, 1 part chopped hay to 1 part concentrate can be mixed in the ration to limit intake. Because creep rations are difficult to balance and mix, most breeders feed high-quality commercial mixes designed specifically for creep feeding. A three-month-old foal weighing 340 pounds (154 kg) should have a total intake of approximately 9.2 pounds (4.2 kg) per day. At three months of age the foal should be provided with about 0.5 to 0.75 pounds of concentrate per 100 pounds of body weight daily (0.5 to 0.75 kg concentrate/ 100 kg body weight/day). A proper creep feed ration should contain 16 percent crude protein, 0.8 percent calcium, 0.55 percent phosphorus, and a calcium to phosphorus ratio of 1:1 to 3:1.

EXERCISE

Exercise is extremely important to the foal's development since it stimulates appetite and encourages proper muscle development. Muscles used for support develop at a faster rate than muscles used for movement when the foal does not get exercise. For this reason, a normal foal should not be

Courtesy of The Quarter Horse Journal

An ideal method of exercise for the foal is free exercise with its dam in a safe and spacious pasture or paddock.

stalled for prolonged periods. Ideally, the foal should have access to a pasture or a large paddock where it can exercise freely. (Unless a carefully supervised pasture breeding program is utilized, foals should not be pastured with stallions.)

If for some reason the foal must be forced to exercise, the handler should avoid overworking the young foal, and a rest period between each 5 to 10 minute exercise session is advisable. It is best to pony the youngster from another horse, preferably its mother. In a safe area such as a pasture, the foal can be allowed to follow its dam without being led while she is being ridden. Mechanical walkers and longe lines are not suitable for young foals as the constant circling associated with them places unequal strain on the foal's legs. The strain can prevent proper muscular development and may cause unsoundness.

Dentition

Dentition, or the eruption of teeth, is a gradual process that occurs with significant uniformity among individuals. For this reason, the appearance or loss of teeth can be used to determine a horse's age. By the time the foal is one year old, for example, it usually has 24 deciduous, or "baby," teeth. The foal's deciduous teeth are smaller, lighter in color, and more distinctly grooved than its permanent teeth. Unlike permanent teeth, the deciduous teeth have a well-defined constriction at the point at which they enter the gum.

The foal is usually born toothless but may have central incisors at birth. In most instances, the central incisors erupt within eight to nine days of birth; the middle incisors erupt at approximately eight weeks of age, and the corner incisors appear by the time the foal is eight months of age. The deciduous premolars erupt by the foal's third week, while the small pointed teeth, referred to as wolf teeth, appear in some foals by the eighth month. (Fillies seldom have wolf teeth.) These teeth are often removed before training begins since they serve no useful purpose and may interfere with the bit later in life.

| 1 week | 6 - 8 weeks | 6 - 9 months |

The foal's age can be estimated roughly by the eruption of its incisor teeth.

Hoof Care

One of the most important and, unfortunately, one of the most overlooked aspects of foal management is hoof care. A regular program of cleaning and trimming accustoms the foal to having its feet and legs handled, but more importantly, it provides frequent opportunities to inspect for injury and to periodically evaluate the foal's structural development.

NORMAL TRIMMING

Proper trimming during the first year of life is necessary to encourage correct structural development. The foal grows very rapidly during this period, and changes in hoof length and angle and uneven wear can place stresses on other parts of the foot and leg. As a result, existing conformational faults may be aggravated or deviations may develop in an otherwise correctly conformed horse. Therefore, regular trimming should be incorporated into the breeding farm management schedule by the time the foal is one month old.

After cleaning the foot, the farrier removes the dry, flaky portions of the frog and sole. Hoof nippers are then used to trim away the excess horn from the hoof wall, and a rasp is used to level the foot and to round off the edges of the wall. This procedure gives the hoof an attractive appearance and prevents chipping. In most instances, the young foal only needs to have the hoof wall slightly rasped and beveled.

Weanlings should be trimmed at least every four weeks, and those foals requiring corrective measures should be trimmed more often. Because the bones are still growing, some angular problems of the foot and leg can be corrected through trimming. However, the role of trimming in early hoof care is largely to maintain the normal shape and angle of the hoof, and corrective trimming should be done conservatively.

CORRECTIVE TRIMMING

The foal's conformation and "way of going" should be studied to determine whether corrective trimming is necessary. In general, corrective trimming is capable of modifying some conformational abnormalities involving the fetlock, pastern, and hoof. However, many leg problems in foals are self-correcting, and overzealous attempts to correct them may only lead to more serious problems. Although angular deviations involving the entire leg are best treated by a veterinarian, simple corrective trimming is frequently used to correct the following deviations in the young foal.

Hoof/pastern angle can be used to illustrate the limitations of corrective trimming. Ideally, a line drawn through the hoof and pastern should form a 45° angle to the ground when viewed from the side. This angle is a part of the foot's normal shock absorption mechanism, which reduces much of the stress that is placed on the foot and leg by concussion. Thus, when the hoof/pastern angle is significantly less than or greater than 45°, the horse becomes predisposed to foot and leg injuries.

To a degree, variations from the ideal hoof/pastern angle can be adjusted by changing the length of the horse's heel through trimming. However, each horse has a characteristic hoof/pastern angle which restricts the amount of correction that can be achieved through trimming. For instance, a foal with extremely upright pasterns cannot be trimmed so that the hoof forms a 45° angle to the ground without breaking the line which passes through the hoof and pastern. When this line is broken, the shock of concussion is largely concentrated at the navicular bone and lameness often results. Thus, corrective manipulation is often restricted by the horse's basic conformation.

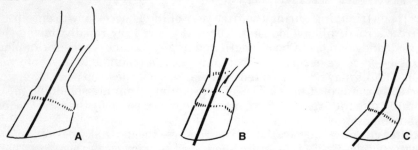

(A) In a correctly trimmed, normal foot the hoof and pastern are positioned at a 45° angle to the ground. However, (B) the angle of a hoof with an extremely upright pastern cannot be trimmed to form a 45° angle to the ground without (C) breaking the line that passes through the hoof and pastern.

TOEING OUT

Many foals have a tendency to toe out (stand or move with the toes pointing outward) to some degree. As the foal moves, the affected foot breaks over the inside wall of the hoof (instead of the center of the toe) and lands on the inside of the toe. As a result, the inside wall of the affected hoof is worn more rapidly than the outside wall, placing abnormal strain on the foal's legs. In addition, the leg may strike the opposite limb as it swings through an inward arc. To correct this condition, the hoof must be trimmed to compensate for uneven wear on the hoof wall. The inside hoof wall should be trimmed if the horse stands base-narrow with toes pointed outward. The outside hoof wall should be trimmed if the horse stands base-wide and toed-out. If this type of trimming is done very early, while the foal is developing, the toed-out condition can usually be corrected. (Depending on the extent of abnormal wear, the hoof wall may require gradual trimming over a period of several months.)

TOEING IN

Like toeing out, toeing in subjects the foal's front legs to strain due to a deviation in the normal shock absorption mechanism. If the hoof points inward, it lands on the outside hoof wall as the horse completes each stride. Foals that toe in are characterized by a paddling motion as the affected foot swings through an outward arc. This condition can frequently be corrected by trimming the inside wall to compensate for abnormal wear.

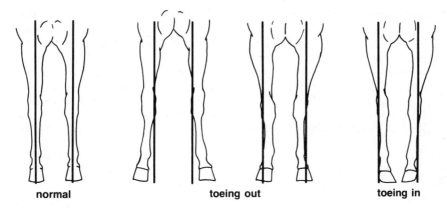

normal   toeing out   toeing in

EXESSIVE TOE

An excessively long toe can be the result of neglect. When the foal's feet are not trimmed for an extended period and when the terrain does not wear the hoof wall naturally, the toe may become long enough to strain the animal's legs and eventually cause serious leg problems. This condition is easily corrected by trimming the toe and leveling the heel at the proper height.

CLUB FOOT

The club foot is characterized by a short toe that changes the normal angle of the foot. If the upright foot is caused by a short, straight pastern, correction is difficult. However, if the foal's toe has been shortened by improper trimming or chipping of the hoof wall, the problem can be corrected by gradually trimming the heel and allowing the toe to grow out. In severe cases, light plates are required to protect the toe from wear.

club foot

Preventive Medicine

A carefully planned disease-prevention program is perhaps the most valuable insurance policy that the breeder can secure. Management practices that emphasize hygiene and disruption of parasite life cycles are important to the health and productivity of all horses, but there are several factors that increase the need for disease-prevention practices: 1) a high concentration of horses on the farm; 2) the presence of suckling foals, weanlings, and/or yearlings; and 3) the admittance of outside horses (e.g., for breeding or training purposes).

IMMUNIZATION

If the foal has received adequate passive immunity via the mare's colostrum, it is probably protected from most infectious organisms for about two to three months. At that point, immunization programs should be initiated. (Note: Tetanus antitoxin is frequently given immediately after birth as an added precaution, but this protection only lasts for two weeks.) Because foals are more susceptible than adults to a wide variety of infectious organisms, routine vaccinations should be used to help minimize complications and reduce the chances of epidemics on the breeding farm. The type of preventive medicine appropriate for a particular farm depends on the types of problems common to that area. For example, rabies vaccinations are usually administered only in areas where a rabies problem has been identified (e.g., in stray cats or skunks). The possibility of a strangles epidemic (and therefore the need for a strangles vaccination program) is much greater on farms that have not been exposed to the infectious organism for many years. Horses become more susceptible to the disease because their level of natural immunity is reduced by absence of exposure to the strangles organism. (For a more detailed discussion on immunization programs, refer to the Appendix.)

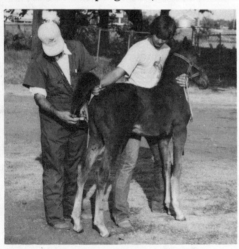

A well-planned immunization program protects the foal from disease after the immunity received via the mare's colostrum begins to decline.

PARASITE CONTROL

Fecal examinations reveal the type of parasites common on a particular farm so that specific deworming agents can be utilized efficiently. It is usually recommended that the foal be dewormed at seven days of age to prevent foal-heat diarrhea and every two months thereafter until it reaches two years of age. It is important to note that, unlike mature horses, foals frequently have infestations of *Strongyloides westeri* and ascarids. Therefore, the parasite control program for horses under one year of age should be effective against those parasites. Specific information on deworming agents and parasite control programs can be found in the Appendix.

Angular Limb Deformities

Because their bones are growing rapidly, young horses are particularly susceptible to angular limb deformities as well as other leg problems. Sometimes deformities can be prevented by feeding a balanced diet and by providing proper exercise and regular hoof trimming. On the other hand, some deformities have a genetic cause. When leg problems already exist in the foal, they may be accentuated by improper diet, strenuous forced exercise, and improperly trimmed or neglected hooves. Although the precise cause or causes of these disorders often cannot be identified, prompt recognition and treatment frequently help to minimize or even correct the abnormality.

BUCKED KNEES

Bucked knee is a deformity of the carpal joint that results in constant partial flexion of the knee. This condition is thought to be caused by abnormal positioning of the foal's limbs during gestation, by genetic predisposition, by a vitamin or mineral deficiency in the mare's diet during pregnancy, or by an imbalance or deficiency of calcium, phosphorus, vitamin A, or vitamin D in the young foal's diet. In mild cases, the condition can often be corrected by simply adjusting the diet. If the condition is severe or if other angular deviations of the limbs are also present, it may be necessary to cast the affected limbs. Casts should be left on for about 10 days but no longer than two weeks at one time. If a longer casting period is necessary, a minimum of two weeks should be allowed between castings. This interval allows the foal to overcome (at least partially) any muscle atrophy caused by the first casting.

KNOCK KNEES

The condition referred to as knock knees (inward deviation of one or both knees) may be caused by genetic factors, by abnormal intrauterine positioning, or by a mineral imbalance. In young horses, knock knees can often be corrected with complete stall rest and proper diet. In severe cases, however, casting may be required. Many cases respond favorably to casting, but if the foal is over two months old, epiphyseal stapling may be more effective. Epiphyseal stapling involves the insertion of a special staple into the convex side of the epiphysis. The staple limits growth on the convex side in such a way that the opposite side has a chance to increase in size and straighten the leg.

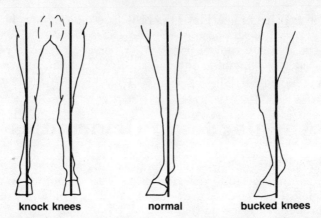

knock knees   normal   bucked knees

CONTRACTED DIGITAL FLEXOR TENDONS

The digital flexor tendons connect the coffin bone (the largest bone within the hoof) and the pastern bones to muscles within the upper portion of the forelimb. Normal flexion of the fetlock joint is caused by contraction of these muscles; however, foals are occasionally born with or develop shortened flexor tendons, which cause abnormal flexion of the fetlock joint.

Depending on the degree of contraction, shortened flexor tendons may prevent the heel from touching the ground or may even cause the fetlocks to knuckle over completely. In extreme cases, the foal may actually stand and walk on the front of the hoof and the fetlock. Genetic factors, abnormal intrauterine positioning, rapid growth rates, and nutritional imbalances (calcium, phosphorus, vitamin A, and vitamin D) have been suggested as causes of contracted flexor tendons in foals. However, the precise cause or causes of contracted flexor tendons are unclear. Mild cases are often corrected by simply allowing the foal to exercise freely and by adjusting any ration imbalances, although some cases require casting or surgical correction. In severe cases that do not improve, euthanasia may be recommended by the veterinarian.

Contraction of the digital flexor tendons causes the foal to stand with pasterns upright and with its weight on the toes of the forefeet.

WEAK FLEXOR TENDONS

In contrast to contracted tendons, weak digital flexor tendons cause the toe on the affected limb to tip upward. In severe cases, the fetlock joint touches the ground. This condition, which is usually present at birth, is most common in the rear legs but may occur in all four legs. The affected fetlock may be bandaged to prevent abrasion, and the condition usually improves naturally within a few days. If spontaneous correction does not occur, however, weak flexor tendons may be corrected by supporting the foal's affected foot with a trailer shoe for approximately 7 to 10 days.

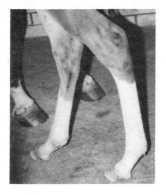

The hind legs are most often affected by weak flexor tendons, which cause the toe to point upwards.

If the flexor tendon is exceptionally weak, the foal may even walk with the pasterns and fetlocks touching the ground. To prevent abrasions until the tendon strengthens, the foal's foot can be fitted with a trailer shoe applied with surgical adhesive tape. The rear extension helps to support the foal's foot and forces the toe downward.

EPIPHYSITIS

The epiphyseal plate is a cartilaginous area located at each end of the long bones. This area is the site of bone growth in the young horse until the plate ossifies. If excessive pressure is placed on the epiphyseal plates before ossification occurs, the cartilage frequently becomes compressed, and inflammation and uneven growth rates result in epiphysitis. For example, increased compression of one side of the plate retards growth of that side, while growth continues on the opposite side. The result is seen in foals and weanlings as angular deformities and enlarged or knobby fetlocks or knees. During the active phase of epiphysitis, the horse may experience lameness, but generally, once the epiphysis ossifies the condition is no longer painful. However, the area may remain enlarged even after the epiphysis has closed. The two epiphyses most often affected by pressure are those of the lower cannon bone, which closes between 7 and 12 months of age, and of the radius, which closes between 24 and 30 months of age.

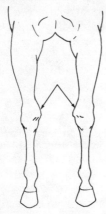

Enlarged areas on the knees may be seen externally in foals affected by epiphysitis.

Epiphysitis is thought to be caused by a number of conditions. The condition is commonly seen in youngsters that are fed heavily and that are gaining weight rapidly. It also seems to be associated with dietary imbalances and improper exercise. When excessive weight is placed on one leg due to injury of the opposite leg, epiphysitis sometimes develops in the supportive leg.

The prognosis in cases of epiphysitis varies according to the severity of the condition and the promptness with which it is treated. Epiphysitis is usually treated by reducing, or at least by maintaining, the horse's weight, balancing the ration, and allowing limited exercise. In the young horse,

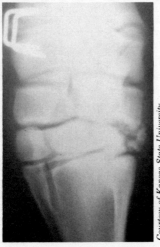

Courtesy of Kansas State University Department of Surgery and Medicine

Staples are sometimes inserted across the epiphysis, or growth plate, to correct angular limb deviations. The staples placed are placed on the more rapidly growing side of the epiphysis. Staples restrict growth on the stapled side, while still allowing growth to continue on the opposite side.

epiphyseal stapling is sometimes successful in correcting the growth irregularities associated with epiphysitis. Staples anchored across the epiphyseal plate on the convex side restrict growth in the stapled area while permitting growth on the opposite side to continue until the bone is straightened.

Disease

As mentioned earlier, the foal loses its passive immunity to infectious disease by about two months of age. Although the foal may succumb to an infection even during the period of passive immunity, the likelihood of illness increases during the transition from passive to active immunity. The neonatal foal is capable of producing the protein structures that provide active immunity, but adequate levels of these immune proteins are obtained gradually. As the foal's level of circulating immune proteins (active protection) rises, the level of maternal antibodies (passive protection) gradually declines. Because the foal's level of protection is borderline during this transitional period, a special disease-prevention program is particularly important.

Although descriptions of all possible primary and secondary infections in foals is beyond the scope of this text, several diseases of special interest with respect to the suckling and weanling foal are reviewed in the following discussion. Hopefully, awareness of these diseases will help the breeder recognize potentially serious illnesses and seek veterinary assistance before complications arise.

VIRAL DISEASES

EQUINE INFLUENZA

Equine influenza is a highly contagious respiratory disease caused by one of two different influenza strains: *Myxovirus influenza A-equi I* and *Myxovirus influenza A-equi II*. Due to the highly contagious nature of this disease, the airborne influenza virus is capable of invading an entire barn within hours. The A-equi I type of virus is the more prevalent of the two influenza viruses, but its effects are usually limited to the upper respiratory tract. If the foal is confined and rested, the infection does not usually become serious. However, the A-equi II type of virus has an affinity for lung tissue and often causes bronchitis and pneumonia in foals. Foals that have not been exposed to the organism previously are especially susceptible to serious infections, which may even result in death.

Influenza is characterized by depression, a dry hacking cough, fever, loss of appetite, and a clear, watery nasal discharge. Young foals are affected more severely by influenza than are older horses, and when secondary bacterial infections accompany influenza, they can result in the foal's death.

There is no specific treatment for influenza, but affected horses should be confined to a well-ventilated barn and given complete rest. Antibiotic therapy and supportive treatment may be required to prevent or combat secondary infections, and after recovery the animal should be rested an additional 10 days. Influenza may be prevented by vaccinating foals with a series of two injections; the first shot is given when the foal is approximately two months old, and the second is given 6 to 12 weeks later. In the past, annual doses have been recommended for continued protection against the disease, but recent reports suggest that booster vaccinations should be given more frequently to insure adequate protection.

EQUINE VIRAL ARTERITIS

Equine viral arteritis is an infectious disease caused by a virus that damages arterial walls. Originally described as pinkeye or a form of influenza, viral arteritis was recognized as a specific disease in 1957. This disease causes degeneration of the middle layer of the arterial walls, and affected mares frequently abort.

Viral arteritis may be spread through direct contact, inhalation of infected droplets (i.e., spread by coughing), or through the ingestion of contaminated material. Infected foals begin to show signs of illness 2 to 10 days following exposure to the virus. The first signs of the disease are fever (102-106°F) and a watery nasal discharge. Later, the foal may suffer from nasal congestion, inflammation of the lining of the eyelids, and teary eyes. Other signs include weakness, depression, diarrhea, colic, coughing, and edema of the legs and eyelids. If the disease is not complicated by secondary infection, it usually subsides in one to two weeks; however, pneumonia due to secondary bacterial infection is a common complication that may delay recovery or even cause death. Severely affected animals may continue to suffer from impaired blood circulation due to residual arterial damage.

The treatment for viral arteritis is absolute rest, plenty of fresh water, and high-quality feed. Veterinarians often recommend antibiotics to prevent or combat secondary bacterial infections and two to three weeks of rest after signs of the disease have subsided. Horses in areas affected by viral arteritis should be vaccinated against this disease.

RHINOPNEUMONITIS

Rhinopneumonitis is a viral infection caused by equine herpesvirus I. This agent causes viral abortion in mares, is frequently responsible for outbreaks of respiratory infections among young horses, and has been associated with a neurological disease that results in paralysis. Rhinopneumonitis usually runs its course without great consequence to the animal, although foals that are infected during gestation seldom respond to treatment. However, the possibility of secondary infections is a constant threat, especially in foals.

ADENOVIRAL INFECTION

Adenoviral infection is a potentially fatal respiratory disease that has been observed in foals less than three months old. This disease is caused by a group of viruses (adenoviruses) that affect the upper respiratory tract and the membranes lining the eyelid (conjunctiva).

Recent reports indicate that adenoviruses are opportunists which cause infections in foals suffering from combined immunodeficiency disease (an inherited deficiency of immune proteins that occurs in Arabian and part-Arabian foals). Equine adenoviral infections have been reported in the United States, Great Britain, and Australia.

Mucous discharges from the eyes and nose are the first signs of the disease, although the foal later shows signs of pneumonia, including coughing and labored breathing. The foal's pulse and respiratory rates become elevated. Additionally, its temperature may reach 106°F, and it is usually characterized by weight loss and a rough hair coat. Blood tests often reveal a lowered white blood cell count. Although vision does not seem to be impaired, the foal's eyes may have a glazed appearance.

At present, a vaccine against adenoviral infections has not been developed, and effective treatment has not been discovered. Attempts to treat the disease usually involve antibiotic therapy to control secondary bacterial and fungal infections.

PAPILLOMAS

Papillomas (warts) are benign tumors with fibrous centers. They develop on yearlings and two-year-olds more frequently than on older horses. They are believed to be caused by a virus and are transmitted by direct and indirect contact and flies. Papillomas usually disappear in two to four months without treatment, but they are unsightly and may cause discomfort if they are located on the horse's lips or in areas that are rubbed by the halter. Papillomas can be removed by surgery, vaccines, and applications of various substances. The use of nitrogen, which freezes warts, is fast and effective and is a popular means of removing papillomas.

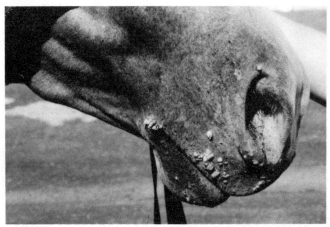

Papillomas, often found on young horses, usually disappear within a few months without treatment.

BACTERIAL DISEASES

FOAL PNEUMONIA

Corynebacterium equi is the cause of a serious and often fatal disease commonly referred to as foal pneumonia. Autopsies of infected foals reveal multiple abscesses on the lungs, liver, lymph glands, kidneys, and intestines. Foal pneumonia does not appear to spread directly between animals. Rather, it usually appears in isolated cases and seldom affects foals over four months of age.

In most cases, foal pneumonia is well established by the time external signs become apparent. Fever, rapid pulse, rapid respiration, and coughing are characteristic of this disease. Watery eyes, diarrhea, and a purulent (pus-filled) nasal discharge may also be present. Many foals continue to nurse and maintain the appearance of good condition until just prior to death, while in other cases, loss of appetite and condition are the first signs of illness. Death usually occurs within one or two weeks after the first signs of the disease appear.

Prompt diagnosis and treatment of foal pneumonia are necessary if the foal is to survive. Broad-spectrum antibiotics such as tetracycline and chloramphenicol may be effective in fighting *Corynebacterium equi* infections, but the bacteria are capable of developing resistance to these antibiotics very rapidly. Horses that recover from *Corynebacterium equi* infections are not thought to develop any immunity to future attacks, and attempts to develop a vaccine have not been successful. Yet, older horses are apparently not susceptible.

STRANGLES

Strangles is a contagious disease of the upper respiratory tract caused by a *Streptococcus equi* infection. The disease is spread through the nasal discharges of infected horses. It is important to note that horses are capable of contaminating their surroundings for about four weeks after recovering from the disease. Although horses of all ages are susceptible to the infectious organism, young horses are the most common victims.

Signs of the disease are usually apparent within 4 to 10 days after exposure. These signs include a fever (103-106°F) and a nasal discharge which becomes thick and full of pus. Later, lymph glands in the throat area become swollen, and the foal stands with its head and neck stiffly extended and may have difficulty swallowing.

Antibiotic therapy should be initiated in an attempt to prevent systemic complications, and warm compresses may be applied to the swollen lymph glands to encourage drainage. Strangles is seldom fatal, and an infection usually leaves the horse with some degree of immunity against future *Streptococcus equi* infections.

STREPTOCOCCUS ZOOEPIDEMICUS INFECTION

Streptococcus zooepidemicus is a bacterium which may be present on the skin and mucous membranes of healthy horses. It does not usually infect healthy horses but often attacks the upper respiratory tract tissues after they have been weakened by a respiratory virus. For example, streptococcal pneumonia is often a complication of influenza in foals and is usually the cause of death in fatal influenza cases.

Streptococcus zooepidemicus infections produce signs similar to those characteristic of strangles: a nasal discharge of pus and mucus, swollen lymph nodes, and abscesses in the region of the pharynx. Abscess formation caused by *Streptococcus zooepidemicus* occurs less rapidly than abscess formation due to *Streptococcus equi* (the strangles organism). Unlike horses recovering from strangles, however, a horse that recovers from a *Strep. zooepidemicus* infection does not develop immunity to future infection by the organism.

TYZZER'S DISEASE

Tyzzer's disease is a liver disease of foals under six weeks of age caused by *Bacillus piliformis* infection. Symptoms include diarrhea, collapse, shock, and death within two or three hours of the appearance of the first signs. It is not known how the disease is spread, but rats and mice are suspected carriers. The extremely rapid course of the disease makes diagnosis very difficult, but the disease can be identified during autopsy by characteristic lesions on the liver. An effective treatment for Tyzzer's disease is not known at this time.

DIARRHEA

The most common causes of foal diarrhea are excessive milk consumption, ingestion of manure or other solid debris, dietary incompatibilities, and infection. The first three causes have been examined in "The Neonatal Foal." Foals suffering from bacterial infections usually have elevated pulse and respiratory rates in addition to diarrhea. These foals often become listless and lose their appetite. When a systemic infection is present, the foal's temperature rises to 103 to 105°F, and its pulse and respiration rates may be very high. However, thermometer readings of rectal temperature may be inaccurate because the rectum is relaxed in foals affected by diarrhea.

While most cases of foal diarrhea are easily corrected, death may occur if diarrhea is accompanied by septicemia, dehydration, and/or shock. Foals are particularly susceptible to diarrhea and they weaken quickly when diarrhea is severe or prolonged. In any instance, the cause of diarrhea should be determined quickly to insure that the appropriate corrective measures are taken before the foal has weakened significantly. For the foal's comfort, soiled areas of the hindquarters should be cleaned regularly and a petroleum jelly product applied to prevent irritation and hair loss.

SALMONELLOSIS

Salmonellosis is classified as a zoonotic disease, meaning that the causative organism (usually *Salmonella typhimurium* and *Salmonella enteritides*) can be transmitted to humans, where it can cause significant digestive disturbance. Infection in humans might be caused by the use of contaminated farm utensils, while infection in horses is usually caused by the ingestion of feed or water that has been contaminated with feces. In horses, the salmonella organism causes fever of up to 107.6°F (42°C), increased respiratory rate, loss of appetite, decreased white blood cells, teeth grinding, and colic. In most cases, watery and blood-stained diarrhea is present, but intermittent periods of constipation have been reported.

Foals ranging from two weeks to four months of age seem to be affected more commonly and more severely. Although the disease may last for several days, younger foals sometimes die within 24 to 36 hours after the onset of the disease. In any instance, rapid diagnosis and immediate treatment with a suitable antibiotic is extremely important to the foal's survival. Treatment for salmonellosis consists of the administration of nitrofurans (i.e., antimicrobial agents which are effective against Gram-positive and Gram-negative organisms), neomycin sulfate compounds, or other chemical agents that are effective against Salmonella organisms. In addition, foals with severe diarrhea are very susceptible to dehydration and usually require fluid replacement therapy.

NONINFECTIOUS DISEASES

In addition to infectious disease, the foal may also be subject to physical abnormalities that affect its ability to survive or function normally. Although an attempt to identify all possible congenital and acquired abnormalities is not made in the following discussion, a few examples of noninfectious complications that may affect the foal are examined briefly. (For a more detailed discussion of inherited abnormalities, refer to EQUINE GENETICS & SELECTION PROCEDURES.)

SHAKER FOAL SYNDROME

Shaker foal syndrome is a sporadic, noncontagious, highly fatal condition that occurs in foals between three and five weeks of age. The onset of this disease is sudden. Without warning or a gradual change in condition, the foal suddenly becomes very weak and unable to rise. It may appear bright and alert, however, and may even rise and nurse with assistance only to tremble and drop to the ground within minutes. The muscular trembling progresses to severe shaking, peristalsis (gut movement) ceases, and the foal's pupils dilate, showing little reaction to light. Death usually occurs within 72 hours due to respiratory failure. Recent research indicates that this condition is of neurological origin.

Treatment is seldom successful, but the foal may be saved through dedicated nursing. Limited success in treating the foal has been achieved through the administration of neostigmine in 2 milligram doses every two hours until the foal is able to rise unassisted. This treatment is accompanied by the administration of mineral oil, steroids, and antibiotics. If the foal survives for five days, the prognosis is favorable.

WOBBLER SYNDROME

Wobbler syndrome is a neurological condition that affects foals and young adult horses, with those in the 9- to 18-month-old range showing the highest incidence of disease. This condition appears to be common in the Thoroughbred, and large, rapidly growing, long-necked males are most often affected. However, Wobbler syndrome has been reported in almost all other breeds, including draft horses and mules. The signs of Wobbler syndrome include lowered head carriage, hind leg incoordination, a swinging gait lacking impulsion, and difficulty in stopping and backing. Clinical signs may vary from day to day, but affected horses very seldom show any permanent improvement. The condition of an affected horse may stabilize for varying lengths of time or may rapidly worsen until the horse is unable to stand. In some cases, surgery involving cervical fusion may bring about a dramatic improvement in the animal's condition. However, surgery can only be done in cases where radiographs reveal compression of the spinal cord by joint abnormalities between adjacent cervical vertebrae. In rare cases, the horse may be returned to the show ring or race track.

HERNIAS

A hernia is the protrusion of an organ or tissue through the wall of a cavity that normally contains the organ or tissue. In foals, two types of hernias occur with some degree of frequency: umbilical and scrotal hernias.

Umbilical hernias result when a section of intestine protrudes through the umbilical opening in the abdominal wall. Normally, this opening constricts shortly after birth, but occasionally closure is delayed. If a piece of intestine becomes displaced through this opening, a lump is usually observed in the navel area. Although the bowel often becomes repositioned and the opening usually closes within one year, surgical closure is required in some cases. It is important to note that the opening may close around a piece of intestine, cutting off the blood supply to the displaced tissue. When bowel strangulation occurs, the foal initially becomes listless and only later shows signs of colic. Bowel strangulation requires surgical correction.

Scrotal hernias occur in colts when a piece of intestine is displaced through the inguinal canal (the pathway for the descending testes) and into the scrotum. (If the intestine is lodged in the inguinal canal, it is referred to as an inguinal hernia.) The affected colt may show signs of intestinal pain or altered gait and usually has scrotal swelling. Like the umbilical hernia, this condition is usually corrected naturally, but the possibility of bowel strangulation is always a concern. Scrotal hernias may be congenital or acquired; although when such a hernia is seen in very young foals, the condition is thought to be congenital. Because the condition is also thought to be heritable, these animals are usually castrated.

COMBINED IMMUNODEFICIENCY DISEASE (CID)

Combined immunodeficiency disease (CID) is a genetic disease seen in pure and part-bred Arabian foals. The term refers to the lack of both B and T lymphocytes (blood cells which are necessary to produce normal immune reactions). Due to this inherited weakness in the immune system, CID foals are extremely susceptible to disease and usually die by the age of five months despite extensive supportive treatment.

CID is inherited as an autosomal recessive trait, meaning that both the mare and stallion must be CID carriers to produce a CID foal. Theoretically, a cross between two carriers will produce the following results:

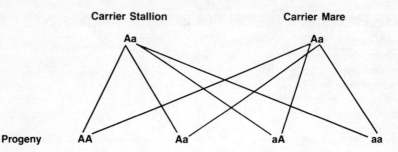

As can be seen in the illustration, there is a 25 percent chance of producing a CID (homozygous) foal when two carrier (heterozygous) horses are crossed.

Due to the temporary immunity that the foal receives from its dam's colostrum, CID foals usually appear healthy until they are about 2 to 10 weeks old. The first signs of illness may be a discharge from the eyes and nose. Early infections may respond to treatment, but repeated infections of increasing severity are characteristic of the disorder. As a result, the CID foal weakens and dies within about five months after its birth. Although death may be due to a variety of viral, bacterial, fungal, or protozoal infections, the most common cause of death is foal pneumonia.

Weaning

In the breeding industry weaning is necessary to safeguard the mare's health and to encourage maximum growth of the foal. Because most mares are rebred shortly after foaling, the mare is usually three to five months pregnant with her next foal at the time of weaning. If the suckling foal is removed, the mare is given time to regain any condition lost during lactation. In addition, after several months of lactation the mare's milk quality drops significantly; thus, supplemental nourishment must be provided to the growing foal. Weaning the foal from its dam and providing adequate supplemental feed encourages the foal to consume a proper balance of nutrients, which can be supplied by the ration. Despite the obvious advantages of weaning over leaving the foal with the mare indefinitely, weaning is a stressful event for the mare and her foal. For this reason, an important goal in all weaning situations is to minimize stress, making the separation as painless as possible. There are several methods of separating the mare from the foal, each having advantages and disadvantages. To determine which method is best suited to a particular breeding farm, the facilities, the condition of the mare and foal, and the number of foals to be weaned in a season should be considered. With this information, a weaning program can be selected and tailored to fit specific management requirements.

PREPARING THE FOAL FOR WEANING

At weaning, a foal's life changes drastically. Not only is it separated from its dam, but the foal also comes into increasing contact with human handlers who require certain standards of behavior. Ideally, foals should be accustomed to human handling before weaning, since it will be necessary to catch the foal, lead it, handle its feet, and in case of injury, administer medications without the calming influence of the foal's dam. Therefore, the foal should be taught to accept basic handling and discipline from humans before weaning.

Careful preparation for weaning may reduce the stressfulness of this period, and because the combined effects of disease and weaning stress can be devastating, the foal should be relatively healthy at the time of weaning. (In exception to this general rule, veterinarians occasionally recommend weaning unhealthy foals to facilitate medical treatment.) Also before weaning, the foal should be thoroughly accustomed to supplemental feed (creep feed), so that its digestive tract need not make a sudden adjustment from milk to solid feed. Most foals begin eating solid food by seven days of age, gradually increasing their intake, until, by the time that the foal is five months old, it depends on its dam primarily for security and companionship rather than nutrition.

Because weaning is very stressful, the foal should be in good health and condition before being separated from its dam.

AGE FOR WEANING

Because weaning is so stressful, the foal's age at weaning should be given some consideration. A number of factors, including the foal's health and even the dam's disposition should affect this decision. Usually, after examining farm conditions and experimenting with different weaning techniques for a few seasons, each manager develops a method that is well suited to the breeding farm's management schedule.

If necessary, foals can be weaned very early in life, but the usual age for weaning is around five months. By this time the foal is accustomed to eating creep feed and has developed active immunity to many organisms in the environment. In addition, the foal is usually less dependent on its dam for protection and emotional support by this age. However, on some farms, foals are routinely weaned as early as three months of age. Proponents of early weaning claim that foals weaned at three months of age and fed an adequate ration grow more rapidly than those weaned at six months of age and creep fed. Actually, however, by one year of age the foals weaned at six months catch up to early weaned foals and, thereafter, growth differences between the two groups are negligible. Thus, if foals are to be marketed or shown before one year of age, early weaning might be helpful in producing rapid growth in weanlings.

Because of the foal's relative lack of maturity at three months of age, early weaning should be undertaken cautiously. Young foals require the companionship of another animal, so if the early weaned foal cannot be placed with a compatible group of other weanlings, a substitute companion

To reduce the stress of weaning, a foal that cannot be weaned with a group of foals its own age should be provided with a gentle companion such as a pony.

such as a small, gentle donkey or goat should be provided. If no companion is available, it may be better to postpone weaning until the foal is older and can better cope with separation. Otherwise, the trauma caused by isolation could offset any benefits gained from early weaning. Gradual weaning, which is discussed later, may be useful for early weaning, since it allows the foal to remain close to its dam.

In practice, the actual age of weaning varies considerably. If either the mare or the foal is unhealthy, weaning earlier than three months of age is usually advisable, but if an individual foal is ill at the time of weaning or was born very late in the year, it may be left with its dam when other foals on the farm are weaned. The dam is often a significant factor in determining when the foal shoud be weaned, too. Because behavior patterns, such as aggression and vices, are learned from the dam, many breeders prefer to separate a mare with a poor disposition or stall vices from her foal as early as possible, with the hope that her behavior will have less influence over the foal's disposition.

On very large breeding farms, there may be enough foals to justify separating them into groups of similar age for weaning. (Separating the foals into small groups is also helpful when weaning facilities are limited.) On small farms, however, all the foals are usually weaned at once, when most of the foals are around five months of age. Because the age of the foals in such a group may vary from three to six months, care should be taken to insure that younger foals are eating well and that they are not being bullied by older, larger foals.

WEANING METHODS

Weaning procedures comprise a very important aspect of breeding farm management. Plans for separating mares and foals should be well thought out in advance of weaning time, and once a satisfactory method has been found, it should be followed each year as long as it proves successful. In this way, weaning can be accomplished efficiently since the breeding farm staff is already acquainted with the usual procedure.

Several weaning methods are commonly used on breeding farms. Traditionally, the mare and foal are abruptly separated, and one of the two is left in the stall or pasture while the other is moved to a location out of sight and hearing. This area could be a pasture or barn at a distant location on the same farm or on another farm. After three to six days of complete separation, the mare and foal become quiet and settle into a normal routine. Although abrupt weaning successfully ends the bond between the mare and her foal, the first few days of weaning are anxious for both horses and their handlers. During this period, the foal often ceases to eat and is more susceptible to injury and disease. For these reasons, other weaning methods are sometimes used which reduce the stressfulness of weaning.

One of the best ways of easing the newly weaned foal's anxiety is to maintain familiar surroundings. This can be accomplished by leaving the foal in the stall formerly shared with its dam or by weaning the foal with a group of familiar companions. When other elements of its environment remain the same, the foal is far less upset by the separation from its dam. For example, an established group of up to fifteen foals can be weaned in a safe pasture or paddock by removing one or two mares from the group at a time. The weaned foals may spend the first few hours looking for their dams, but they are generally occupied with their companions in play and are quieted by the presence of the other mares. At two- to three-day intervals additional mares are removed until the entire group of foals is weaned. In this weaning arrangement, nervous or ill-natured mares are removed first so that they do not disrupt the entire group, and a well-mannered gelding or an old mare is often placed with the group to calm the weanlings.

Recent research has explored the concept of separating the mare and foal in a more gradual manner. Some horse breeders believe that prolonging the weaning period only increases the foal's anxiety; however, one behavioral study indicates that gradual weaning is actually less stressful to the foal than is abrupt separation. In this study, one group of foals was weaned gradually by separating mares and foals and then placing them in adjoining pens. Obviously, this method prevents the foals from nursing but provides them with the reassuring presence of their dams. The study's results showed that gradually weaned foals were much quieter after weaning than foals in the control group, which were weaned in an abrupt manner. However, no follow-up research has yet been completed showing what, if any, long-term effect this difference may have on the foal's behavior and future performance.

Despite the seeming advantages of some gradual weaning programs, more traditional methods in which the mare and foal are abruptly and finally separated prevail on most breeding farms. If the farm's fencing can prevent the foal from reaching the mare's udder, a gradual weaning method could be used. The safest fence for gradual weaning is a small-weave mesh wire. This technique might be well suited to some small breeding farms where it is impossible to keep the mare out of the sight and hearing range of her foal.

No matter how weaning is accomplished, fences, gates, and other equipment should be checked regularly for safety hazards on which a panicky foal might injure itself. Under the stress of weaning, the foal may attempt to jump fences and fail to notice obstacles in its path. The best way to prevent injury to the foal is to provide it with adequate free exercise, since a tired foal is much less likely to become excited than one which has been confined. As a result, weaning methods that utilize stall confinement are frequently criticized. If the foal must be kept in a stall, it should be released in a moderately sized paddock to work off built-up energy before being placed in a larger pasture.

POSTWEANING MARE CARE

Immediately after weaning, the mare becomes a primary concern. She usually calms down more quickly than the foal, although the period of time required for her to resume normal behavior may vary from a few hours to a few days. If her foal is at least six months of age when weaned, the mare often appears relieved to be free of her responsibility.

Initially, the mare's milk supply must be dried up. A decreased energy and protein ration a week before and after weaning, applications of camphorated oil to her udder, and plenty of exercise will help to dry the mare's milk. Avoid milking the mare since a full bag decreases milk production, while milking promotes further milk production. When the mare's milk production has ceased, her ration can then be increased to improve her condition. Postweaning mare care is discussed further in "Lactation."

Developing the Weanling

Although all weanling management programs share some common characteristics, such as the need for a disease-prevention program and high-quality nutrition, the anticipated use of the individual must be taken into consideration. Generally, one is interested in developing a strong musculoskeletal system and overall soundness. Yet each breeding program is usually committed to producing horses for specific activities, some of which demand a high degree of physical development at a very young age. For instance, halter horses are often campaigned vigorously as weanlings and yearlings, and in order to be competitive, these horses must attain maximum growth and development during this period.

Unfortunately, managers often feel forced to choose between conditioning foals for maximum early performance and delaying conditioning until the foal is more mature. However, a well-balanced combination of good nutrition, sensible exercise, and individual attention can be used to achieve maximum results without sacrificing the horse's soundness as an adult.

HOUSING AND SAFETY

A young horse's nutritional and exercise requirements are regulated in part by whether it is kept in a stall or pasture. Keeping a youngster separated from other horses and housed in a stall helps to keep it free from scratches and blemishes and prevents its coat from being bleached by the sun. These considerations may be important if the horse is bound for a sale or show; however, a stalled youngster must be exercised daily, and improper forced exercise can easily put excessive strain on the youngster's legs. To provide adequate exercise, confined youngsters are usually turned out of their stalls into paddocks and pastures for a portion of their exercise. They can be turned out individually to lessen the chance of blemishes and turned out at night to avoid sun-bleached hair.

Whenever possible, weanlings and yearlings should be left at pasture to grow and develop, since free exercise develops agility, coordination, confidence, and good muscle tone. Natural exercise also greatly reduces the strain on growing bones. Additionally, the pasture-raised youngster develops a positive mental attitude and is not bored by inactive confinement. As a precaution, the health and eating habits of horses fed at pasture should be monitored closely. When several horses are fed together, a timid horse may not receive his share of the feed. If this happens, the horses should be fed separately. In addition, an extra feeder may be provided so that there is always an unoccupied feeder available to the least dominant animal.

The foal's stall and pasture should be free of sharp objects and projections, such as nails, protruding boards, and wire. Thus, periodic inspection of the facilities for safety hazards should be a regular part of the farm's maintenance program. Additionally, to prevent the horse from becoming caught on a projection, its halter should be removed whenever it is free in the stall or pasture.

NUTRITION

For the first two to three months of life the foal's nutritional needs are met primarily by the mare's milk. However, as the foal's nutrient requirements increase and the mare's milk quality decreases, the foal's diet must be supplemented. During this period, the foal's digestive tract adjusts gradually from a liquid diet to a diet composed of concentrates and roughage. Then at weaning, the foal must make the final transition to a solid diet for all of its nutrients.

NUTRIENT REQUIREMENTS

At various stages of its life, the horse has different nutritional needs. For instance, the growing horse requires more protein, more minerals, and more vitamins than does a mature horse. For this reason, rations designed for older horses are not suitable for weanlings. Because nutritional deficiencies are much more critical in growing horses than in mature horses, the weanling ration should be formulated carefully to meet the growing foal's requirements. All of the nutrient requirements in this discussion are based on 90 percent dry matter in the ration; however, further information on nutrient requirements can be found in the Appendix.

Energy

Energy is the most important factor in the older foal's diet, with the exception of water. It is especially critical in weanling rations, since an energy deficiency in an otherwise nutritionally balanced diet causes decreased growth rates. A growing horse requires a high level of energy, which is usually supplied by grain in the concentrate portion of the diet. For instance, a three-month-old weanling requires a minimum of 1.5 Mcal of digestible energy per pound of ration (3.25 Mcal/kg), while a six-month-old weanling needs 1.4 Mcal per pound (3.1 Mcal/kg) to achieve proper growth.

Despite the weanling's need for a high energy and, therefore, a high concentrate ration, an adequate amount of roughage must be included to prevent digestive disturbance. It is generally recommended that the weanling diet be made up of 65 percent concentrate and 35 percent roughage when the roughage being fed contains approximately 1.0 Mcal of digestible energy (D.E.) per pound (2.2 Mcal D.E./kg). Legumes, such as alfalfa or clover, and some grasses, such as high-quality bluegrass, brome, or timothy, usually contain at least 1.0 Mcal of D.E. per pound. However, if other types of roughage which contain significantly less than 1.0 Mcal of D.E. per pound are being fed to the weanling, the ration will be deficient in energy. When low energy roughages are used in the ration, the concentrate portion can be increased and the roughage portion decreased to provide more energy. However, the roughage portion probably should not fall below 25 percent in the three-month-old weanling's diet and 30 percent of the six-month-old weanling's diet. Alternatively, the energy density of the concentrate could be increased to compensate for energy-deficient roughage.

The foal's energy intake should always be adjusted according to its body condition. Whenever foals are being fed for maximum performance, the manager must remain constantly aware of the individual foal's progress and condition. However, there is generally a tendency among managers to follow one of two extreme feeding methods. On some farms, creep feeding is started too late, to the detriment of early growth. On other farms, however, young foals are fed excessively in order to prepare for sale or show. The obesity caused by excessive feeding places abnormal stress on the foal's legs and can result in temporary or even permanent unsoundness.

High energy intake has been implicated as a cause of structural problems such as epiphysitis and contracted tendons in the growing horse. However, several other factors, including nutritional imbalances, genetic predisposition, and periods of malnutrition followed by rapid growth may also be involved in these conditions. In fact, a combination of several factors actually may be necessary to cause structural unsoundness. Until more is known about the relationship between nutrition and development, a ration that is properly balanced according to current recommendations is probably the best management policy. (The phrase *properly balanced* is used often in nutrition discussions to refer to the relative amount of each nutrient in the ration. Because the level of each nutrient may affect the availability of another nutrient, the entire ration must be balanced for efficient utilization.)

Protein

Protein is necessary for proper bone and muscle growth and, as discussed in the previous section, for energy utilization. The current recommendations for feeding growing horses are 16 percent crude protein for the three-month-old weanling and 14.5 percent crude protein for the six-month-old weanling.

Although the total amount of crude protein in the diet is important, all proteins are not alike. Arrangements of certain chemical units called amino acids make up protein. While some amino acids can be manufactured by the horse's body, others, called essential amino acids, must be provided in the horse's diet. Whether a dietary protein source contains all of the essential amino acids determines its protein quality. For instance, soybean meal, fish meal, and milk protein are considered to be high-quality protein sources because of their amino acid compositions. On the other hand, cottonseed meal and linseed meal proteins are of lower quality and should not be used as sole protein sources for growing horses, although they are suitable for mature horses.

The amino acid lysine is most often deficient in rations for young horses. If the ration lacks adequate lysine, a protein deficiency can result even though the crude protein content of the ration is sufficient. Fortunately though, lysine deficiencies can be prevented easily by feeding a source of high-quality protein. Therefore, soybean meal is often added to rations for growing horses because it contains high levels of lysine and because it is palatable and is readily available in most areas.

Protein requirements can be influenced by the energy content of the ration. High energy levels increase the need for dietary protein, and protein deficiencies can hamper energy utilization. Although there is probably an optimum energy-to-protein ratio for every stage of the horse's life, little is known about this area of nutrition. Further research may uncover the appropriate ratio, but until then, protein can be fed at slightly higher than recommended levels to insure that energy is being utilized efficiently. Increased protein levels also provide a margin of safety in case protein quality in the ration is slightly lower than anticipated.

Minerals

Mineral deficiencies may occur due to the nature of the feedstuff being used in a ration. For instance, cereal grains naturally contain more phosphorus than calcium. Thus, if a grain mix comprises the majority of the ration and if a calcium supplement is not added, an imbalance in the calcium-to-phosphorus ratio will probably result. However, mineral imbalances can also occur due to mineral deficiencies in the feedstuff itself. Soils which are deficient in a specific mineral will sometimes produce crops that are also deficient in that nutrient. Unless the forage or grain is analyzed, the deficiency may be overlooked until developmental problems become apparent in the horses. For this reason, samples of hay and grain should be analyzed for mineral content, especially calcium, phosphorus, selenium, and iodine content.

Although premixed commercial mineral supplements can be used to correct ration deficiencies, indiscriminant supplementation may only create new dietary imbalances or aggravate existing imbalances. First, the entire ration must be examined to determine its mineral content. Then an appropriate supplement can be selected to correct the ration's mineral deficiencies.

Calcium and Phosphorus

Because calcium and phosphorus often occur together in the body and because the level of each affects the absorption of the other, these two minerals are often discussed together. A six-month-old weanling usually requires 0.65 percent calcium and 0.45 percent phosphorus in the ration, but because actual mineral requirements vary, it is wise to add very small amounts above the recommended levels to provide a margin of safety in the feeding program. Calcium at 0.70 percent and phosphorus at 0.55 percent of the total ration will provide adequate levels of these minerals for rapidly growing foals, but these levels are not toxic.

In addition to the total amount of calcium and phosphorus supplied by the ration, special attention should be paid to the ratio of calcium to phosphorus. If the level of one of these minerals is excessive, it can cause a deficiency of the other mineral. For instance, if calcium is in excess, it ties up phosphorus in the intestine so that the phosphorus cannot be absorbed. As a result, a phosphorus deficiency occurs even though analysis of the ration would show phosphorus to be adequate. Further, an inadequate supply of either calcium or phosphorus interferes with utilization of the other. Although the ideal calcium-to-phosphorus ratio for the weanling is 1.5:1, it can tolerate variation from 1:1 to 2:1. As the foal matures, it can tolerate wider variation in the calcium-to-phosphorus ratio, but even during the yearling year these minerals should be kept within the range from 1:1 to 3:1.

Salt

Salt is usually the only mineral other than calcium and phosphorus which needs to be supplemented in significant amounts. The horse requires about 0.9 percent salt in its total ration. Commercial concentrate mixtures usually contain added salt, but approximately 0.7 percent salt can be added to concentrates mixed on the farm if salt has not already been added to any portion of the concentrate. Salt intake varies considerably due to exercise and environmental temperature, but a horse will adjust salt intake to fit its needs if the salt is available. Therefore, horses should always be given free access to loose salt or salt blocks.

Trace Minerals

A number of minerals must be provided in the horse's ration in very small amounts. These minerals are called trace minerals and they are usually supplied in adequate amounts by the ration's feed ingredients. However, the soils in some regions of the world are deficient in one or more of the trace minerals. The feedstuffs grown in these areas are also deficient and must be supplemented.

One of the widely recognized regional mineral deficiencies is that of selenium in the Great Lakes area of the United States and in New Zealand. Selenium deficiency is also known to occur in other isolated areas, and where the deficiency is recognized, selenium is usually added to ration formulations for local use. In other areas, adequate selenium is provided by free access to trace-mineralized salt blocks. Other trace minerals that are essential to the horse's diet are iodine, iron, copper, cobalt, manganese, potassium, magnesium, molybdenum, sulfur, and fluorine. These minerals are usually supplied by feedstuffs without supplementation being required, and excesses may cause toxicity. Fortunately, trace minerals can be supplemented safely by giving the horse free access to trace-mineralized salt.

SAMPLE RATIONS

Although a variety of formulations can be fed to meet the nutritional needs of a six-month-old weanling foal, the following rations are examples of formulations that are well-balanced when fed with specific types of roughage.

The following ration is nutritionally balanced when the foal's diet is made up of 70 percent concentrate and 30 percent good-quality bermuda-grass hay each day. When fed in this manner, the total ration supplies 16 percent crude protein, 0.70 percent calcium, and 0.56 percent phosphorus, and the calcium-to-phosphorus ratio is 1.25:1, well within the acceptable range for a six-month-old horse.

corn	55%
oats	20%
soybean meal	22%
molasses	1%
bonemeal	1%
limestone	1%

When the six-month-old weanling's daily intake is comprised of 30 percent legume roughage and 70 percent concentrate, the following ration supplies 16 percent crude protein, 0.70 percent calcium, and 0.57 percent phosphorus. The calcium-to-phosphorus ratio is 1.22:1.

corn	45%
oats	39%
soybean meal	12%
dicalcium phosphate	3%
molasses	1%

HANDLING AND MANNERS

If foals are brought in from pasture to pens or individual stalls each day, this time presents an excellent opportunity to inspect each foal for injuries or illness. In addition, every foal should be handled and groomed during this period. These sessions should be a continuation of the gentling program which begins before the foal is weaned. However, on large farms where weanlings are seldom brought in from pasture, all of the foals should be halter-broken and taught to pick up their feet before being turned out. Although they may be rambunctious when gathered from the pasture, they will gradually recall their earlier lessons. Even if foals are not handled daily, they should be checked for injuries or signs of illness on a regular schedule, such as at daily feedings.

The foal that has been introduced to grooming, haltering, hoof care, clipping, and even trailering before it reaches three months of age will seldom object to these practices when it enters training. If properly conducted, these very early training sessions with the weanling build a foundation of confidence, which can be drawn on in future training sessions as a yearling and two-year-old. A foal that is relaxed and comfortable with humans is also far more likely to concentrate on training sessions and to accept new lessons quickly.

Safety should be of top priority when handling young horses. Care must be taken to prevent accidents, since most foals tend to be flighty and excitable. Sudden and loud noises should be avoided initially; however, the handler's actions should be confident rather than cautious or the foal may begin to suspect the handler's intentions. The foal can be gradually introduced to distracting situations as it begins to trust its handler, so that it will become accustomed to the activities that are common on every farm.

The foal should be taught early in its handling not to bump, push, or bite its handler, to let the handler go through gates and doors before following, and to stand quietly. Because the foal's patience is limited, these lessons are best taught as incidental parts of the youngster's routine handling, rather than within specific training periods. For example, if the youngster bumps the handler when being lead from pasture to stable, the handler should correct the foal firmly. If necessary, the foal can be taught to maintain proper distance from the handler by use of the blunt end of a crop or longe whip. Leading the horse on his right, and holding the whip in his left hand, the handler should bump the horse's upper arm or shoulder with the blunt end of the crop, causing the horse to move away. While the bump should be forceful enough to make an impression on the horse, it need not be so strong as to hurt or frighten him.

When a horse tries to charge through an opening such as a gate or stall before the handler, the horse should be told "whoa" and halted immediately. When halting the horse, the handler should use one brief pull on the lead rope. If the horse fails to halt, a stronger pull or series of pulls and releases should be used. A constant pull should be avoided as the horse can easily push into and evade it. After the horse has halted, the handler should have him stand for a few moments, pat him, and then lead him calmly through the opening. Several repetitions may be necessary before the horse can be

Courtesy of Earl and Barbara Lang Quarter Horses

If the foal is to be shown as a weanling, its early training must go beyond basic lessons in leading and tying. The foal must also learn how to stand with its weight distributed evenly on all four legs and to stand quietly in distracting situations.

relied upon to follow, rather than lead, through a gate. If the horse persists in trying to bolt ahead of the handler, it may be necessary to halt him and back him several steps before he is led through the gate.

Excitable youngsters that are constantly intrigued by their surroundings often refuse to stand still when halted. This behavior is unendearing, particularly as the horse grows older and is introduced to new situations. To be pleasant to handle, the horse must learn to be contained, and this quality can be instilled by teaching the horse the meaning of the word "whoa." The horse must learn to stand still while at halter and later when being mounted, and the lesson is taught more easily while the horse is young. If the horse becomes unsettled after being halted, the handler should tell him "whoa," and firmly steady him with the halter. If he is unusually nervous, exercise may help calm him and make him responsive to the handler.

GROOMING

Grooming is important for health and training as well as for appearance. It stimulates circulation and hair growth by massaging the skin and muscles, and it cleans pores and brings out the sebum (natural oil) in the horse's hair. It encourages close examination of the horse for cuts, foreign objects lodged in the skin, and disease. It is also a particularly pleasant way for the foal to become accustomed to being handled and to being touched. During the sessions, the foal can also be taught good manners. A horse kept in its stall should be groomed daily, while those kept in pastures should be brought in and groomed at least three times a week. Of course, all horses should be groomed before and after training sessions.

Courtesy of The Thoroughbred Record

Turning out weanlings during periods when the sun is not intense, such as early evening, prevents the foals' hair coats from becoming sun bleached.

If the foal is being prepared for sale or show, grooming efforts should be intensified to encourage shedding and good skin condition. To protect the hair coat from sun-bleaching, foals may be brought in from pasture during the day for grooming, handling, and feeding. They can then be turned back out to pasture during the late afternoon for free exercise when the sun is less intense.

Occasionally, the horse will need to have its bridle path trimmed and the long hairs clipped from its ears, muzzle, and fetlocks. Clippers are upsetting to most young horses, so the youngster should be introduced to them quietly and slowly. A pair of small, quiet clippers is best for this purpose. At least two people will be needed for the process: one to steady the youngster and the other to handle the clippers.

The handlers must exercise the utmost patience when introducing the youngster to the clippers and may find it wise to spread the training over a period of several days. First, the clippers should be run slowly over the horse while they are turned off. When the horse has learned to accept and even to enjoy this, the handler should run the clippers, still turned off, over the horse while making a buzzing or humming noise. When the foal is accustomed to the feel and the noise, the next step, actually clipping, usually can be accomplished with ease. The handler should take care to avoid overheating the clippers and oil them regularly. Should the horse become upset, the handler should stop clipping and steady the horse. If after several tries the horse is still upset, it may be necessary to postpone clipping until another day or to use a twitch or mild sedative to calm him. With persistence the horse can be taught to accept clipping, but an upsetting experience makes further clipping sessions even more difficult.

EXERCISE

In order to lay a foundation of physical stamina for later training, some form of conditioning should begin when the horse is a weanling. Adequate exercise is necessary for proper development of the musculoskeletal system and is essential for developing coordination in the young horse. In fact, the only time that the foal should be completely confined for more than a few hours is during convalescence for injury or illness. Despite the need for physical activity, the foal's growing skeletal structure is very susceptible to injury. Yet, inactivity is not an acceptable means of protecting the foal.

When considered from the viewpoint of conditioning, the foal's susceptibility to injury of bones, muscles, ligaments, and tendons can be interpreted as a strength when managed properly. For instance, immature bone is weak, but for this reason, it also responds readily to appropriate stress by rearranging its composition into a stronger structure. Therefore, regular stress actually creates more stress-resistant bone. It is only when stress increases beyond the bone's ability to adapt that injury occurs. On the other hand, mature bone is stronger but is also less adaptable to stress. Thus, stress for the purpose of conditioning is more effective when started while the horse is young. The same principles of managing bone stress also apply to muscles, ligaments, tendons, and to the hoof wall. Therefore, in order to condition the animal, the conditioning program must place an adequate level of stress on the animal to strengthen its musculoskeletal system, but the stress cannot be great enough to cause injury.

If the foal is destined for training as a two- to three-year-old, early conditioning should be moderate. In fact the best activity for this type of horse is free exercise in a safe, spacious pasture with a group of foals its own age. In this setting, the foal can run and play at will, and because it is not confined, the horse is unlikely to become so energetic as to injure himself. When the foal is destined for sale or show as a weanling or even as a yearling, though, a more controlled exercise program is required. At this point, the manager's ability to evaluate condition and conformation becomes extremely important.

Courtesy of The Thoroughbred Record

Free access to pasture encourages exercise for proper bone development.

Some confinement and forced exercise become part of the foal's schedule when it is necessary to condition the youngster for sale or show. Confinement should always be minimized whenever possible, but in order to bring the horse's hair coat into peak condition, sun exposure must be limited. Some breeders avoid this problem by turning out their foals in the early morning and late evening. This arrangement encourages free exercise but protects the hair coat.

Whenever a forced exercise program is implemented for the young horse, there is an accompanying risk of injuring its developing musculoskeletal system. However, good muscle tone and the appearance of athletic condition, which are sought by judges and buyers of young horses, cannot be achieved without such a program. Preparation for sale or show is also complicated by the fact that top condition cannot be maintained indefinitely in an animal. Therefore, the conditioning program must be designed to bring the horse's coat condition, muscle tone, and training to a peak at the proper time. It is generally agreed that the peak can only be maintained for approximately two weeks. However, whether or not this peak can be reached at the appropriate time is dependent upon how much stress the foal can withstand. Since foals are particularly susceptible to injury and disease, it can be seen that weanling preparation programs demand the utmost skill on the part of the manager.

The duration and level of a conditioning session should always be determined according to the foal's current condition. Therefore, the manager must develop a keen ability to evaluate individual weanlings, to determine the level of exercise that is appropriate for the foal's current level of condition, and to recognize the warning signs of overstress. If the foal is healthy and was free in pasture with its dam before weaning, it can be started on a moderate exercise program. However, conditions such as obesity, extremely poor condition, prolonged periods of inactivity, and conformation faults are reasons for restricting the initial exercise sessions. For instance, an obese foal is very likely to injure its legs during vigorous forced exercise, since excess weight places considerable strain on the foal's legs. Similarly, a foal in very soft condition is ill prepared to begin strenuous exercise, since it will very likely develop muscle fatigue early in the session. Both of these conditions predispose the foal to injury.

Courtesy of McDermott Ranch

Ponying is one of the best methods of providing suckling or weanling foals with supplemental exercise.

Although the weanling can be started on a forced exercise program, care must be taken to design the program with an immature individual in mind. Longeing, round pen work, and mechanical walkers are totally unsuitable for young foals. Each of these exercise techniques requires the foal to circle in a relatively small area, placing excessive strain on the foal's legs. With the first two of these methods, it is also difficult to control energetic foals, and if the young horse has not had sufficient free exercise, it is very likely to injure its legs by running uncontrollably on the longe line or in the round pen. If, for some reason, longeing, round pen work, or a walker is the only method of forced exercise available, the harmful effects can be minimized by limiting the length of the exercise session, by providing a soft but not deep working surface, and by providing plenty of free exercise to manage the foal's energy.

Treadmills are sometimes used for exercising foals, particularly when the primary goal is muscle development. Treadmills are beneficial to older horses, but some veterinarians claim to have noticed increased leg problems and particularly rear leg lamenesses in young foals conditioned on treadmills.

When conducted properly, ponying the foal from a calm, patient horse is a useful means of conditioning. This method is more time-consuming than those previously mentioned. However, the foal's speed can be controlled easily and ponying can be done in any safe area where the surface is not too hard. Ponying also has the advantage of not being as boring to the young horse as some other exercise activities. It also provides an opportunity to introduce the foal to unfamiliar surroundings before its training under saddle begins later in life.

The ideal method of conditioning foals for top condition has yet to be found, but it would likely incorporate some of the following characteristics:

1. slow, regular workouts of gradually increasing duration;
2. no tight circling;
3. a combination of forced exercise and free exercise to develop agility and provide mental stimulation. ♞

APPENDIX

APPENDIX 1

COMMON MEASUREMENTS AND CONVERSION FACTORS

The following information is provided to familiarize the reader with the most commonly used systems of measurement and to explain how values in one system can be converted to other systems.

Length

U. S. SYSTEM			METRIC SYSTEM		
12 inches (in)	=	1 foot (ft)	10 millimeters (mm)	=	1 centimeter (cm)
3 feet	=	1 yard (yd)	100 centimeters	=	1 meter (m)
1760 yards	=	1 mile (mi)	1000 meters	=	1 kilometer (km)
1 inch	=	2.54 centimeters	1 millimeter	=	0.03937 inch
1 foot	=	30.48 centimeters	1 centimeter	=	0.3937 inch
1 yard	=	91.44 centimeters	1 meter	=	39.37 inches
1 mile	=	1.609 kilometers	1 kilometer	=	0.6214 mile

To convert the U.S. system of length to the Metric system, multiply the number of U.S. units by the number of Metric units per U.S. unit.

Example: To change 12 inches to centimeters, multiply the number of inches by the number of centimeters per inch.

12 inches x 2.54 centimeters/inch = 30.48 centimeters

To convert the Metric system of length to the U.S. system, multiply the number of metric units by the number of U.S. units per Metric unit.

Example: To change 2 meters to inches, multiply the number of meters by the number of inches per meter.

2 meters x 39.37 inches/meter = 78.74 inches

Volume

U. S. SYSTEM			METRIC SYSTEM		
16 fluid ounces (oz)	=	1 pint (pt)	1000 milliliters (ml)	=	1 liter (l)
2 pints	=	1 quart (qt)			
4 quarts	=	1 gallon (gal)	1 milliliter	=	0.033 ounce
			1 liter	=	33.814 ounces
1 fluid ounce	=	29.57 milliliters	1 liter	=	2.113 pints
1 pint	=	473.16 milliliters	1 liter	=	1.056 quarts*
1 quart	=	946.33 milliliters	1 liter	=	0.264 gallon
1 quart	=	0.946 liter			

*For practical purposes, 1 quart is usually considered to be equivalent to 1 liter.

To convert the U.S. system of volume to the Metric system, multiply the number of U.S. units by the number of Metric units per U.S. unit.

Example: To change 8 fluid ounces to milliliters, multiply the number of fluid ounces by the number of milliliters per ounce.

8 ounces x 29.57 milliliters/ounce = 236.56 milliliters

To convert the Metric system of volume to the U.S. system, multiply the number of Metric units by the number of U.S. units per Metric unit.

Example: To change 350 milliliters to ounces, multiply the number of milliliters by the number of ounces per milliliter.

350 milliliters x .033 ounces/milliliter = 11.55 ounces

APPROXIMATE HOUSEHOLD MEASURES

1 teaspoon	=	5 milliliters	=	5 cubic centimeters (cc)
3 teaspoons	=	1 tablespoon		
1 tablespoon	=	½ fluid ounce	=	15 milliliters
1 jigger	=	1½ fluid ounces	=	45 milliliters
1 cup	=	8 fluid ounces	=	2365 milliliters

Weight

U. S. SYSTEM			METRIC SYSTEM		
16 ounces (oz)	=	1 pound (lb)	1000 milligrams (mg)	=	1 gram (g)
2000 pounds	=	1 ton (tn)	1000 grams	=	1 kilogram (kg)
1 ounce	=	28.35 grams	1 gram	=	0.03527 ounce
1 pound	=	453.6 grams	1 kilogram	=	2.2046 pounds
1 ton	=	907.18 kilograms	1 kilogram	=	0.0011 ton

To convert the U.S. system of weight to the Metric system, multiply the number of U.S. units by the number of Metric units per U.S. unit.

Example: To change 12 ounces to grams, multiply the number of ounces by the number of grams per ounce.

12 ounces x 28.3 grams/ounce = 339.6 grams

To convert the Metric system of weight to the U.S. system, multiply the number of Metric units by the number of U.S. units per Metric unit.

Example: To change 2.5 kilograms to pounds, multiply the number of kilograms by the number of pounds per kilogram.

2.5 kilograms x 2.2046 pounds/kilogram = 5.5115 kilograms

Temperature

There are two major systems that are commonly used to measure temperature: Fahrenheit and Centigrade. The Fahrenheit system is set up so that water freezes at 32°F and boils at 212°F, while the Centigrade system is based on a freezing point of 0°C and a boiling point of 100°C.

To convert Fahrenheit degrees to Centigrade degrees, subtract 32 from the number of Fahrenheit degrees, multiply the remainder by 5, and divide the result by 9.

Example: 72° F - 32 =   40
                40 x 5 =   200
             200 ÷ 9 =   22.2° C

To convert Centigrade degrees to Fahrenheit degrees, multiply the number of Centigrade degrees by 9, divide the result by 5, and then add 32.

Example: 38° C x 9 =   342
               342 ÷ 5 =   68.4
             68.4 + 32 =   100.4° F

TEMPERATURE CONVERSION TABLE

Fahrenheit	Centigrade	Fahrenheit	Centigrade
230°	110°	101.3°	38.5°
212	100	100.4	38
203	95	99.5	37.5
194	90	98.6	37
185	85	97.7	36.5
176	80	96.8	36
167	75	95.9	35.5
158	70	95	35
149	65	93.2	34
140	60	91.4	33
131	55	89.6	32
122	50	87.8	31
113	45	86	30
111.2	44	77	25
109.4	43	68	20
107.6	42	59	15
105.8	41	50	10
104.9	40.5	41	5
104	40	32	0
102.2	39	23	−5

APPENDIX 2   POINTS OF THE HORSE

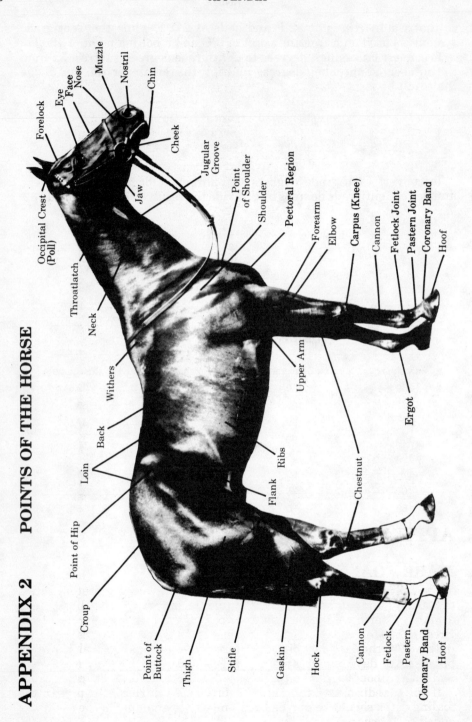

Forelock
Eye
Face
Nose
Muzzle
Nostril
Chin
Cheek
Jugular Groove
Point of Shoulder
Shoulder
Pectoral Region
Forearm
Elbow
Carpus (Knee)
Cannon
Fetlock Joint
Pastern Joint
Coronary Band
Hoof
Occipital Crest (Poll)
Jaw
Throatlatch
Neck
Withers
Upper Arm
Back
Loin
Ribs
Flank
Chestnut
Ergot
Point of Hip
Croup
Point of Buttock
Thigh
Stifle
Gaskin
Hock
Cannon
Fetlock
Pastern
Coronary Band
Hoof

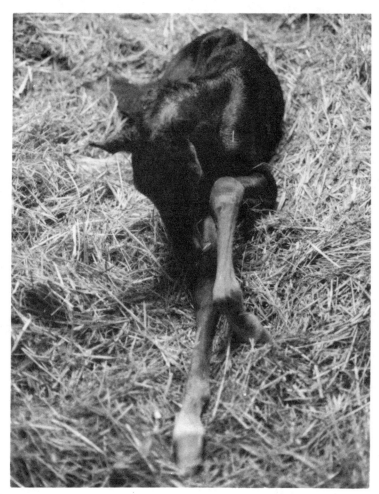

APPENDIX 3

MARE FOALING CALENDAR

The following foaling calendar is based on an average gestation period of 336 days and can be used to determine the approximate foaling date of a mare, if her breeding date is known. Use the columns marked "Date Bred" to locate the day on which the mare was last bred. Across from that date will be the date on which the mare can be expected to foal. However, because 336 days is only an average gestation length, many mares will foal within about ten days either before or after the date given in this table.

If the breeding date falls after February 28 during a leap year, the foaling dates should be moved back one day to account for the extra day added to February.

DATE BRED	DATE DUE TO FOAL	DATE BRED	DATE DUE TO FOAL	DATE BRED	DATE DUE TO FOAL
Jan. 1	Dec. 3	Mar. 1	Jan. 31	May 1	Apr. 2
2	4	2	Feb. 1	2	3
3	5	3	2	3	4
4	6	4	3	4	5
5	7	5	4	5	6
6	8	6	5	6	7
7	9	7	6	7	8
8	10	8	7	8	9
9	11	9	8	9	10
10	12	10	9	10	11
11	13	11	10	11	12
12	14	12	11	12	13
13	15	13	12	13	14
14	16	14	13	14	15
15	17	15	14	15	16
16	18	16	15	16	17
17	19	17	16	17	18
18	20	18	17	18	19
19	21	19	18	19	20
20	22	20	19	20	21
21	23	21	20	21	22
22	24	22	21	22	23
23	25	23	22	23	24
24	26	24	23	24	25
25	27	25	24	25	26
26	28	26	25	26	27
27	29	27	26	27	28
28	30	28	27	28	29
29	31	29	28	29	30
30	Jan. 1	30	Mar. 1	30	May 1
31	2	31	2	31	2
Feb. 1	3	Apr. 1	3	Jun. 1	3
2	4	2	4	2	4
3	5	3	5	3	5
4	6	4	6	4	6
5	7	5	7	5	7
6	8	6	8	6	8
7	9	7	9	7	9
8	10	8	10	8	10
9	11	9	11	9	11
10	12	10	12	10	12
11	13	11	13	11	13
12	14	12	14	12	14
13	15	13	15	13	15
14	16	14	16	14	16
15	17	15	17	15	17
16	18	16	18	16	18
17	19	17	19	17	19
18	20	18	20	18	20
19	21	19	21	19	21
20	22	20	22	20	22
21	23	21	23	21	23
22	24	22	24	22	24
23	25	23	25	23	25
24	26	24	26	24	26
25	27	25	27	25	27
26	28	26	28	26	28
27	29	27	29	27	29
28	30	28	30	28	30
		29	31	29	31
		30	Apr. 1	30	Jun. 1

DATE BRED	DATE DUE TO FOAL	DATE BRED	DATE DUE TO FOAL	DATE BRED	DATE DUE TO FOAL
Jul. 1	Jun. 2	Sep. 1	Aug. 3	Nov. 1	Oct. 3
2	3	2	4	2	4
3	4	3	5	3	5
4	5	4	6	4	6
5	6	5	7	5	7
6	7	6	8	6	8
7	8	7	9	7	9
8	9	8	10	8	10
9	10	9	11	9	11
10	11	10	12	10	12
11	12	11	13	11	13
12	13	12	14	12	14
13	14	13	15	13	15
14	15	14	16	14	16
15	16	15	17	15	17
16	17	16	18	16	18
17	18	17	19	17	19
18	19	18	20	18	20
19	20	19	21	19	21
20	21	20	22	20	22
21	22	21	23	21	23
22	23	22	24	22	24
23	24	23	25	23	25
24	25	24	26	24	26
25	26	25	27	25	27
26	27	26	28	26	28
27	28	27	29	27	29
28	29	28	30	28	30
29	30	29	31	29	31
30	Jul. 1	30	Sep. 1	30	Nov. 1
31	2	Oct. 1	2	Dec. 1	2
Aug. 1	3	2	3	2	3
2	4	3	4	3	4
3	5	4	5	4	5
4	6	5	6	5	6
5	7	6	7	6	7
6	8	7	8	7	8
7	9	8	9	8	9
8	10	9	10	9	10
9	11	10	11	10	11
10	12	11	12	11	12
11	13	12	13	12	13
12	14	13	14	13	14
13	15	14	15	14	15
14	16	15	16	15	16
15	17	16	17	16	17
16	18	17	18	17	18
17	19	18	19	18	19
18	20	19	20	19	20
19	21	20	21	20	21
20	22	21	22	21	22
21	23	22	23	22	23
22	24	23	24	23	24
23	25	24	25	24	25
24	26	25	26	25	26
25	27	26	27	26	27
26	28	27	28	27	28
27	29	28	29	28	29
28	30	29	30	29	30
29	31	30	Oct. 1	30	Dec. 1
30	Aug. 1	31	2	31	2
31	2				

APPENDIX 4　HORMONES ASSOCIATED WITH REPRODUCTION

HORMONE	PRODUCTION SITE	TARGET TISSUE	EFFECT	THERAPEUTIC USE
Estrogen	ovarian follicle placenta (mare)	genital tract	secondary sex characteristic development; uterine gland growth; uterine contraction; cyclic genital tract changes; expression of estrous behavior	enhances sexual receptivity of the mare during estrus
		mammary glands	mammary duct growth	
	Sertoli cells (stallion)	unknown in the stallion	unknown in the stallion	
Follicle Stimulating Hormone (FSH)	anterior pituitary	ovary (mare)	growth and maturation of follicles	
		seminiferous tubules (stallion)	spermatogenesis	
Gonadotropin Releasing Hormone (GnRH)	hypothalamus	anterior pituitary	LH and FSH release	
Luteinizing Hormone (LH)	anterior pituitary	ovary (mare)	follicle maturation; ovulation; estrogen secretion	
		Leydig cells of testis (stallion)	stimulates interstitial tissue and androgen secretion	
Oxytocin	posterior pituitary	uterine myometrium	contraction of uterine musculature	elimination of retained placenta; induced parturition
		contractile mammary tissue	smooth muscle contraction	milk let-down and ejection
Progesterone	corpus luteum of the ovary placenta	uterus	quiets smooth muscle, maintains pregnancy	prevent abortion; synchronize or suppress estrus
		mammary gland	growth of mammary and uterine glands	
Prolactin	anterior pituitary	alveoli of mammary gland	milk production	
Prostaglandin	uterus*	corpus luteum	corpus luteum regression	synchronize estrus; treat persistent corpus luteum
Relaxin	ovary placenta	pelvic tissues	relaxation of pelvic area for parturition	
Testosterone	interstitial cells of testis	accessory sex glands; seminiferous tubules	male sexual behavior; accessory sex gland growth; secondary sex characteristic development; spermatogenesis	treatment of infertility and low libido in stallion (not usually recommended due to possible long-term effect on fertility)

*Prostaglandins are produced by a number of tissues in the body. However, this chart deals specifically with prostaglandins produced by the uterus.

APPENDIX 5

NUTRITION TABLES

Tables reproduced from "Nutrient Requirements of Horses, 1978," with permission of the National Academy of Sciences, Washington, D. C.

TABLE 1   Dietary Minerals and Vitamins for Horses

	Adequate Levels		
	Maintenance of Mature Horses	Growth	Toxic Levels[a]
Calcium	—[b]	—[b]	
Phosphorus	—[b]	—[b]	
Sodium, %	0.35	0.35	
Potassium, %	0.4	0.5	
Magnesium, %	0.09	0.1	
Sulfur, %	0.15	0.15	
Iron, mg/kg	40	50	
Zinc, mg/kg	40	40	9,000
Manganese, mg/kg	40	40	*
Copper, mg/kg	9	9	*
Iodine, mg/kg	0.1	0.1	4.8
Cobalt, mg/kg	0.1	0.1	
Selenium, mg/kg	0.1	0.1	5.0
Fluorine, mg/kg	—	—	50+
Lead, mg/kg	—	—	80
Vitamin A	—[b]	—[b]	*
Vitamin D, IU/kg	275	275	150,000
Vitamin E, mg/kg	15	15	
Thiamin, mg/kg	3	3	
Riboflavin, mg/kg	2.2	2.2	
Pantothenic acid, mg/kg	15	15	

[a] *Nutrients known to be toxic to other species but without adequate information on the horse are indicated by *.*

[b] *See Tables 2-5.*

TABLE 2　Nutrient Requirements of Horses (Daily Nutrients per Horse), Ponies, 200 kg Mature Weight

	Weight		Daily Gain		Digestible Energy (Mcal)	TDN		Crude Protein		Digestible Protein		Calcium (g)	Phosphorus (g)	Vitamin A Activity (1,000 IU)	Daily Feed[a]	
	kg	lb	kg	lb		kg	lb	kg	lb	kg	lb				kg	lb
Mature ponies, maintenance	200	440	0.0		8.24	1.87	4.12	0.32	0.70	0.14	0.31	9	6	5.0	3.75	8.2
Mares, last 90 days gestation			0.27	0.594	9.23	2.10	4.62	0.39	0.86	0.20	0.44	14	9	10.0	3.70	8.1
Lactating mare, first 3 months (8 kg milk per day)			0.0		14.58	3.31	7.29	0.71	1.56	0.54	1.19	24	16	13.0	5.20	11.5
Lactating mare, 3 months to weaning (6 kg milk per day)			0.0		12.99	2.95	6.50	0.60	1.32	0.34	0.75	20	13	11.0	5.00	11.0
Nursing foal (3 months of age)	60	132	0.70	1.54	7.35	1.67	3.68	0.41	0.90	0.38	0.84	18	11	2.4	2.25	5.0
Requirements above milk					3.74	0.85	1.87	0.17	0.37	0.20	0.44	10	7	0.0	1.20	2.7
Weanling (6 months of age)	95	209	0.50	1.10	8.80	2.0	4.40	0.47	1.03	0.31	0.68	19	14	3.8	2.85	6.3
Yearling (12 months of age)	140	308	0.20	0.44	8.15	1.85	4.07	0.35	0.77	0.20	0.44	12	9	5.5	2.90	6.4
Long yearling (18 months of age)	170	374	0.10	0.22	8.10	1.84	4.05	0.32	0.70	0.17	0.37	11	7	6.0	3.10	6.8
Two year old (24 months of age)	185	407	0.05	0.11	8.10	1.84	4.05	0.30	0.66	0.15	0.33	10	7	5.5	3.10	6.8

[a] Dry matter basis.

TABLE 3 Nutrient Requirements of Horses (Daily Nutrients per Horse), 400 kg Mature Weight

	Weight		Daily Gain		Digestible Energy (Mcal)	TDN		Crude Protein		Digestible Protein		Calcium (g)	Phosphorus (g)	Vitamin A Activity (1,000 IU)	Daily Feed[a]	
	kg	lb	kg	lb		kg	lb	kg	lb	kg	lb				kg	lb
Mature horses, maintenance	400	880	0.0		13.86	3.15	6.93	0.54	1.19	0.24	0.53	18	11	10.0	6.30	13.9
Mares, last 90 days gestation			0.53	1.17	15.52	3.53	7.76	0.64	1.41	0.34	0.75	27	19	20.0	6.20	13.7
Lactating mare, first 3 months (12 kg milk per day)			0.0		23.36	5.31	11.68	1.12	2.46	0.68	1.50	40	27	22.0	8.35	18.4
Lactating mare, 3 months to weanling (8 kg milk per day)			0.0		20.20	4.59	10.10	0.91	2.00	0.51	1.12	33	22	18.0	7.75	17.1
Nursing foal (3 months of age)	125	275	1.00	2.2	11.51	2.62	5.76	0.65	1.43	0.50	1.10	27	17	5.0	3.55	7.8
Requirements above milk					6.10	1.39	3.05	0.40	0.88	0.30	0.66	15	12	0.0	1.95	4.3
Weanling (6 months of age)	185	407	0.65	1.43	13.03	2.96	6.51	0.66	1.45	0.43	0.95	27	20	7.4	4.20	9.2
Yearling (12 months of age)	265	583	0.40	0.88	13.80	3.14	6.91	0.60	1.32	0.35	0.77	24	17	10.0	4.95	10.9
Long yearling (18 months of age)	330	726	0.25	0.55	14.36	3.26	7.17	0.59	1.30	0.32	0.70	22	15	11.5	5.50	12.2
Two year old (24 months of age)	365	803	0.10	0.22	13.89	3.16	6.95	0.52	1.14	0.27	0.59	20	13	11.0	5.35	11.8

[a] Dry matter basis.

TABLE 4 Nutrient Requirements of Horses (Daily Nutrients per Horse), 500 kg Mature Weight

	Weight		Daily Gain		Digestible Energy (Mcal)	TDN		Crude Protein		Digestible Protein		Calcium (g)	Phosphorus (g)	Vitamin A Activity (1,000 IU)	Daily Feed[a]	
	kg	lb	kg	lb		kg	lb	kg	lb	kg	lb				kg	lb
Mature horses, maintenance	500	1,100	0.0		16.39	3.73	8.20	0.63	1.39	0.29	0.64	23	14	12.5	7.45	16.4
Mares, last 90 days gestation			0.55	1.21	18.36	4.17	9.18	0.75	1.65	0.39	0.86	34	23	25.0	7.35	16.2
Lactating mare, first 3 months (15 kg milk per day)			0.0		28.27	6.43	14.14	1.36	2.99	0.84	1.85	50	34	27.5	10.10	22.2
Lactating mare, 3 months to weaning (10 kg milk per day)			0.0		24.31	5.53	12.16	1.10	2.42	0.62	1.36	41	27	22.5	9.35	20.6
Nursing foal (3 months of age)	155	341	1.20	2.64	13.66	3.10	6.83	0.75	1.65	0.54	1.19	33	20	6.2	4.20	9.2
Requirements above milk					6.89	1.57	3.45	0.41	0.90	0.31	0.68	18	13	0.0	2.25	4.9
Weanling (6 months of age)	230	506	0.80	1.76	15.60	3.55	7.80	0.79	1.74	0.52	1.14	34	25	9.2	5.00	11.0
Yearling (12 months of age)	325	715	0.55	1.21	16.81	3.82	8.41	0.76	1.67	0.45	0.99	31	22	12.0	6.00	13.2
Long yearling (18 months of age)	400	880	0.35	0.77	17.00	3.90	8.58	0.71	1.56	0.39	0.86	28	19	14.0	6.50	14.3
Two year old (24 months of age)	450	990	0.15	0.33	16.45	3.74	8.23	0.63	1.39	0.33	0.72	25	17	13.0	6.60	14.5

[a] Dry matter basis.

TABLE 5 Nutrient Requirements of Horses (Daily Nutrients per Horse), 600 kg Mature Weight

	Weight		Daily Gain		Digestible Energy (Mcal)	TDN		Crude Protein		Digestible Protein		Calcium (g)	Phosphorus (g)	Vitamin A Activity (1,000 IU)	Daily Feed[a]	
	kg	lb	kg	lb		kg	lb	kg	lb	kg	lb				kg	lb
Mature horses, maintenance	600	1,320	0.0		18.79	4.27	9.40	0.73	1.61	0.33	0.73	27	17	15.0	8.50	18.8
Mares, last 90 days gestation			0.67	1.47	21.04	4.78	10.52	0.87	1.91	0.46	1.01	40	27	30.0	8.40	18.5
Lactating mare, first 3 months (18 kg milk per day)			0.0		33.05	7.51	16.53	1.60	3.52	0.99	2.18	60	40	33.0	11.80	26.0
Lactating mare, 3 months to weanling (12 kg milk per day)			0.0		28.29	6.43	14.15	1.29	2.84	0.73	1.61	49	30	27.0	10.90	23.9
Nursing foal (3 months of age)	170	374	1.40	3.08	15.05	3.42	7.53	0.84	1.85	0.78	1.72	36	23	6.8	4.65	10.2
Requirements above milk					6.93	1.58	3.47	0.51	1.12	0.38	0.84	18	15	0.0	2.25	4.9
Weanling (6 months of age)	265	583	0.85	1.87	16.92	3.85	8.47	0.86	1.89	0.57	1.25	37	27	10.6	5.45	12.0
Yearling (12 months of age)	385	847	0.60	1.32	18.85	4.28	9.42	0.90	1.98	0.50	1.10	35	25	14.0	6.75	14.8
Long yearling (18 months of age)	475	1,045	0.35	0.77	19.06	4.33	9.53	0.75	1.65	0.43	0.95	32	22	13.5	7.35	16.2
Two year old (24 months of age)	540	1,188	0.20	0.44	19.26	4.38	9.64	0.74	1.63	0.39	0.86	31	20	13.0	7.40	16.3

[a] Dry matter basis.

TABLE 6 Nutrient Concentrations in Diets for Horses and Ponies Expressed on 100 Percent Dry Matter Basis [a]

| | Digestible Energy | | Example Diet Proportions | | | | Crude Protein (%) | Calcium (%) | Phosphorus (%) | Vitamin A Activity | |
| | | | Hay Containing 2.2 Mcal/kg | | Hay Containing 2.0 Mcal/kg | | | | | | |
	Mcal/kg	Mcal/lb	Concentrate[b]	Roughage	Concentrate[b]	Roughage				iu/kg	iu/lb
Mature horses and ponies at maintenance	2.2	1.0	0	100	10	95	8.5	0.30	0.20	1,600	725
Mares, last 90 days of gestation	2.5	1.1	25	75	35	65	11.0	0.50	0.35	3,400	1,550
Lactating mare, first 3 months	2.8	1.3	45	55	55	45	14.0	0.50	0.35	2,800	1,275
Lactating mare, 3 months to weanling	2.6	1.2	30	70	40	60	12.0	0.45	0.30	2,450	1,150
Creep feed	3.5	1.6	100	0	100	0	18.0	0.85	0.60		
Foal (3 months of age)	3.25	1.5	75	25	80	20	18.0	0.85	0.60	2,000	900
Weanling (6 months of age)	3.1	1.4	65	35	70	30	16.0	0.70	0.50	2,000	900
Yearling (12 months of age)	2.8	1.3	45	55	55	45	13.5	0.55	0.40	2,000	900
Long yearling (18 months of age)	2.6	1.2	30	70	40	60	11.0	0.45	0.35	2,000	900
Two year old (light training)	2.6	1.3	30	70	40	60	10.0	0.45	0.35	2,000	900
Mature working horses (light work[c])	2.5	1.1	25	75	35	65	8.5	0.30	0.20	1,600	725
(moderate work[d])	2.9	1.3	50	50	60	40	8.5	0.30	0.20	1,600	725
(intense work[e])	3.1	1.4	65	35	70	30	8.5	0.30	0.20	1,600	725

[a] Values are rounded to account for differences among Tables 2-5 and for greater practical application.

[b] Concentrate containing 3.6 Mcal/kg.

[c] Examples are horses used in western pleasure, bridle path hack, equitation, etc.

[d] Examples are ranch work, roping, cutting, barrel racing, jumping, etc.

[e] Examples are race training, polo, etc.

TABLE 7 Nutrient Concentration in Diets for Horses and Ponies Expressed on 90 Percent Dry Matter Basis[a]

	Digestible Energy		Crude Protein (%)	Cal- cium (%)	Phos- phorus (%)	Vitamin A Activity	
	Mcal/kg	Mcal/lb				IU/kg	IU/lb
Mature horses and ponies at maintenance	2.0	0.9	7.7	0.27	0.18	1,450	650
Mares, last 90 days of gestation	2.25	1.0	10.0	0.45	0.30	3,000	1,400
Lactating mare, first 3 months	2.6	1.2	12.5	0.45	0.30	2,500	1,150
Lactating mare, 3 months to weanling	2.3	1.1	11.0	0.40	0.25	2,200	1,000
Creep feed	3.15	1.4	16.0	0.80	0.55		
Foal (3 months of age)	2.9	1.35	16.0	0.80	0.55	1,800	800
Weanling (6 months of age)	2.8	1.25	14.5	0.60	0.45	1,800	800
Yearling (12 months of age)	2.6	1.2	12.0	0.50	0.35	1,800	800
Long yearling (18 months of age)	2.3	1.1	10.0	0.40	0.30	1,800	800
Two year old (light training)	2.6	1.2	9.0	0.40	0.30	1,800	800
Mature working horses							
(light work[b])	2.25	1.0	7.7	0.27	0.18	1,450	650
(moderate work[c])	2.6	1.2	7.7	0.27	0.18	1,450	650
(intense work[d])	2.8	1.25	7.7	0.27	0.18	1,450	650

[a] Values are rounded to account for differences among Tables 2-5 and for greater practical application.
[b] Examples are horses used in western pleasure, bridle path hack, equitation, etc.
[c] Examples are ranch work, roping, cutting, barrel racing, jumping, etc.
[d] Examples are race training, polo, etc.

APPENDIX 6

PASTURE MANAGEMENT

Valuable benefits can be realized from the use of pastures on the breeding farm. For instance, well-managed pastures can reduce supplementary feed requirements — a major farm-operating cost. During the growing season, horses on maintenance diets which are grazing high-quality pastures may not require any supplemental grain. (They should, however, always have access to trace-mineralized salt and fresh, clean water.) Additionally, pastures promote self-exercise, which contributes to the horse's overall well-being. In order to provide these benefits, however, pasture programs must be tailored to fit the breeding farm's individual requirements.

Pasture Programs

Pasture programs on horse farms fall into two general categories: permanent or temporary. The most suitable program takes into consideration the needs and resources of the farm and factors such as geographic location, climate, and water availability. These factors influence pasture productivity and should be evaluated before designing a pasture management program.

Permanent Pastures

Permanent pastures consist of native grasses or introduced plant species which provide forage for livestock over an indefinite period of time (usually in excess of ten years). Through proper fertilization and weed control, permanent pastures provide high-quality forage at lower cost and with less labor than other forage production systems. In addition, the well-developed root systems in a permanent pasture are better able to tolerate moisture and temperature stresses and minimize erosion and water runoff. Permanent pastures are usually planted to warm-season, perennial grasses. However, they can be over-seeded with cool-season grasses or legumes to extend the grazing season.

Temporary Pastures

Temporary pastures have life spans of from one to several years. Historically, temporary pastures have been used for the production of hay or grain crops (e.g., oats, wheat, or rye) with animal products generated indirectly through grazing during a certain part of the growing season. However, temporary pastures are sometimes used exclusively for grazing. High management inputs (e.g., tillage, fertilization, and erosion control) are required because of the shorter life span of temporary pastures as compared to permanent pastures. An advantage of temporary pastures is that high-producing plant species can be selected in order to maximize forage production during certain seasons when permanent pastures are less productive. Cool-season grasses such as rye and wheat and cool-season legumes like alfalfa and clover are commonly planted in temporary pastures for grazing. Proper management of temporary pastures to supplement permanent pastures can provide nearly year-round grazing in some regions.

PASTURE GRASSES

Cool-season Grasses	Warm-season Grasses
Bulbous bluegrass	Dallisgrass
Canada bluegrass	Blue grama
Kentucky bluegrass	Sideoat grama
Bromegrass	Rothrock grama
Mounting bromegrass	Buffalograss
Reed canarygrass	Lehman lovegrass
Tall fescue	Boer lovegrass
Meadow fescue	Weeping lovegrass
Tall oatgrass	Galletagrass
Orchardgrass	Sand dropseed
Redtop	Hardinggrass
Perennial ryegrass	Carpetgrass
Russian wild rye	Bermudagrass
Crested wheatgrass	Big bluestem
Slender wheatgrass	

Cool-season Legumes	Warm-season Legumes
Alfalfa	Lespedeza
Ladino clover	Low hop clover
White clover	Persian clover
Red clover	California burclover
Alsike clover	Black medic
Strawberry clover	
Sweet clover	
Birdsfoot trefoil	

Management Practices

After choosing a pasture program and a forage species, the manager should select methods of tillage, seeding, fertilization, pH control, and weed management.

Tillage

The primary purposes of tillage are to prepare the pasture for seeding, to conserve water for plant growth, and to control weeds. The conventional method of tillage includes a primary step involving relatively deep cultivation (by mold-board plow, disc plow, chisel plow, or rotary tiller) followed by a secondary step to firm the soil (with a disc or spring-tooth harrow). The purposes of primary and secondary tillage are to improve soil condition, to incorporate surface residues from past crops, and to control weeds. Additionally, secondary tillage helps to control weed invasion in a previously prepared seedbed.

Alternative methods of cultivation, such as minimum tillage and no tillage, have had greater acceptance in recent years because high costs of fuel, labor, and equipment have increased tillage expenses. Minimum tillage and no tillage methods can be successful when they are managed properly. In comparison to conventional tillage methods, decreased tillage reduces soil compaction due to machine traffic. Furthermore, wind and water erosion are minimized because plant residue is left undisturbed on the soil surface. State and county agricultural extension agents can provide information on the suitability of these practices in specific areas.

Planting

The first step of planting is careful selection and handling of planting materials. Whether seeds or sprigs are used, they should be of high quality (i.e., free of noxious weeds and undamaged). Improper handling and storage after purchase can significantly decrease viability. Thus, seeds should be stored in a cool, dry place and handled carefully to avoid cracking the seed. Sprigs should be kept moist and out of direct sunlight.

Planting is accomplished by one of several methods. Most pasture grasses and legumes are either broadcast (sown over the ground surface) or drilled (planted in furrows). Broadcasting can be accomplished manually or with equipment which evenly scatters seeds or sprigs. Drilling, on the other hand requires specialized equipment which digs furrows, sows seeds in the open furrows, and covers the seeds. Some pasture grasses which produce very little seed, but which reproduce themselves from certain vegetative portions of the plant (e.g., some varieties of bermudagrass), are usually established by transplanting grass sprigs. In this process, sods, crowns, or stolons of the grass are placed in the soil and then covered. The sprigs are broadcast manually or distributed with a manure spreader. They are then disced lightly to cover them with soil and rolled to establish adequate soil contact. There are also specialized implements which distribute and cover the sprigs and firm the soil in one trip across the pasture.

Depending on the planting method used, several pieces of equipment, such as a tractor, disc, sprigging attachment, and pull-type broadcaster or drill, may be required. If the expense of owning this equipment cannot be justified, custom operators can be contacted to prepare the seedbed and plant. Alternatively, seeds can be broadcast by a fertilizer dealer simultaneously with fertilizer application. Although combining fertilizer with seeds is time-saving and economical, fertilizer can have a drying effect on seeds, which may cause lowered seed vigor.

Fertilization and pH Control

Fertilization and soil pH control are closely associated with planting and plant growth. Soil tests should be made before fertilization to determine the nutrient content and chemical character of a given pasture. Test results show the amounts of nitrogen, phosphorus, potassium, and the hydrogen ion concentration (pH) within a soil sample. These are the four major factors which influence plant growth. Based on soil tests, appropriate application rates for fertilizer and neutralizing amendments can be determined.

On many farms, fertilization is necessary because leaching and plant growth have caused gradual declines in soil fertility. Many organic ele-

ments are essential for plant growth and maintenance, but the macronu-
trients — nitrogen, phosphorus, and potassium — are of most concern.
Nitrogen is required for plant growth and it increases the quality of protein
available to grazing horses. Phosphorus aids in root development, while
potassium is necessary for plant drought resistance and disease tolerance.
(Legumes, which can fix their own nitrogen from root nodules, require
little nitrogen fertilization in comparison to grasses.)

Applications of fertilizer at the appropriate rate and at the appropriate
time increase plant yield and water use efficiency and discourage competition from weeds. Agricultural extension agents can recommend suitable
fertilization schedules and rates based on soil test results and knowledge
of local climate and crops. Fertilizer can be broadcast on new or previously
established pasture, and then incorporated by tillage or left on the surface
to gradually dissolve into the soil. Another method of fertilization, banding,
is usually done when seeds are planted by drilling. Banding places fertilizer
either to the side of or below seeds for greater availability to desirable
plants and less competition from weeds. (Fertilizer should not be placed in
direct contact with some seeds due to its drying effect.)

Not only should soil fertility be considered, but soil pH should also be
evaluated and, if necessary, adjusted in order to achieve maximum plant
growth. The acidity of low pH soils can interfere with mineral absorption.
Fortunately, however, many crops are tolerant of mild acidity. Lime, which
is used to raise soil pH, can be applied as a topdressing or incorporated
during primary tillage. The advantage of incorporating during tillage is
that lime is carried into the root zone (where it is absorbed) faster than by
topdressing. Depending on the leaching rate, a lime application may be
effective from 2 to 20 years.

The opposite of acidic soil is alkaline (saline) soil, which decreases the
root systems' ability to absorb water. Generally, alkalinity is easier to
correct than acidity, unless the soil is strongly alkaline (pH greater than
8.5). In fact, controlled irrigation and adequate water drainage are common
tools of soil salinity alteration. A chemical amendment, such as sulfuric
acid, may be required to alter the pH of strongly alkaline soils to a desirable
level.

Weed and Brush Control

Although adequate fertilization and pH control give desirable plants a
head start, weeds may also take advantage of improved growing conditions.
Strictly defined, a weed is any plant that is growing in the wrong place.
Obviously, then all weeds are not necessarily toxic, unpalatable, or of low
nutritional quality. However, uncontrolled weed growth lowers the quality
and performance of desirable plants because weeds compete for the same
nutrients, light, and water. Thus, control measures are necessary to increase the productivity of desirable plants in horse pastures.

The principal weed-control methods fall into three categories: burning,
mechanical, and chemical. Burning demands little time or labor and
provides some unique advantages. Besides controlling weeds, pasture fires
warm soils (in the spring) for faster regrowth, eliminate mature vegetation,

and kill some plant and animal parasites (e.g., worm larvae and ticks). Of course, safety precautions must be observed when burning pastures and, to be effective, burning must be timed properly with the weather and plant growth patterns. Therefore, agricultural agents, the forest service, a university range management specialist, or the Soil Conservation Service should be contacted for advice on controlled pasture burning.

The second type of weed control, mechanical control, effectively eliminates weeds and brush during land clearing or maintenance. Mowing is a familiar method of mechanical weed control and it can be used throughout the growing season to impair weed growth and reproduction and to prevent weeds from shading desirable pasture plants. Woody plants such as brush and small trees may require heavy-duty equipment such as bulldozers for initial clearing, but regular mowing usually controls this vegetation if it is not allowed to become too mature.

Finally, chemical control uses herbicides to create a toxic environment for a selected group of undesirable plants without damaging desirable species. Special safety measures are required when using herbicides. For instance, it may be necessary to remove livestock from sprayed pastures. Furthermore, proper handling techniques should prevent contamination of nearby pastures or water sources. (It is very important to **read the labels** on all herbicides to insure correct product usage.)

Irrigation

Irrigation is used to supplement the natural water supply of plants. Irrigated pastures require careful management and land preparation and are significantly more expensive than other pasture programs. However, water supplementation can be used profitably to produce high-quality pasture in low-rainfall areas and in areas where rainfall is irregular during certain seasons.

Pasture irrigation is commonly accomplished by one of three methods: flooding, furrows, or sprinkling. The suitability of a particular system is determined by a number of factors, including terrain, water availability, and soil type. For instance, sprinkling is useful for rolling terrain. Because more moisture is lost to evaporation by sprinkling as compared to other irrigation methods, sprinkling is also better adapted to areas of relatively high humidity. Beyond a few general principles, irrigation is a highly specialized management practice, which requires considerable knowledge of soil characteristics and water quality. In addition, in many areas irrigation is regulated by local water-use agencies. Thus, irrigation practices should be tailored to meet local conditions.

Grazing Systems

Even though forage production varies from year-to-year and from season-to-season, careful management of grazing animals can help to maintain efficient forage production. Pastures can be continuously grazed or rotated. A continuously grazed pasture is stocked for an entire season or year, whereas, rotational pastures are divided into small acreages and stocked at higher rates over shorter periods of time. Generally, continuous grazing systems demand less labor and fewer fences and watering areas than rotational grazing. On the other hand, a disadvantage of continuous systems in which only one plant species is grazed is that seasonal forage shortages may occur.

Generally, either grazing system can be modified to meet specific farm and management requirements. Seasonal pasture shortages can be partially overcome by planting forage species that can be grazed during specific seasons. For instance, in some parts of the United States, cool-season grasses such as rye and wheat are planted in the fall to provide high-quality supplemental winter grazing. Good forage can be obtained from cool-season grasses by either overseeding permanent pastures or by planting temporary pastures. One variation of the rotational system is to divide a large pasture into several pastures of approximately equal size. During peak production, animals can be restricted to one or two lots while the other lots are being harvested for hay. Then as plant growth slows or enters dormancy, additional lots can be opened to provide additional forage for grazing.

It is important to determine the optimum stocking rate for each pasture to prevent over- or under-utilization. Stocking rates are calculated from the estimated consumption per horse and the estimated total forage production of the pasture. Very productive pastures may be able to carry one mature horse per acre; however, two to three acres of average pasture per horse are usually required. A continuously grazed pasture under low or moderate stocking rates is not as fully utilized as a properly managed rotational pasture. Because horses selectively graze young and tender growth, they avoid mature parasite- and feces-contaminated areas of a continuously grazed pasture. As the ungrazed grass matures, its nutritional quality and digestibility decrease. However, by increasing the stocking rate (forcing consumption of less palatable growth) or harvesting the excess growth, forage remains actively growing and maintains high nutritional quality, digestibility, and palatability.

APPENDIX 7

INTERNAL PARASITES

Internal parasites are of continual concern to the farm manager because of the damage that can occur from an infestation of even small numbers. These parasites can become costly because they live at the expense of the host animal, causing irritation and physiological damage from their feeding, migration, and attachment habits. The following section presents information on the life cycles and control of internal parasites and their effects on horses.

Large Strongyles

Three species of large strongyles are important causes of parasitic disease in the horse: *Strongylus vulgaris, Strongylus edentatus* and *Stron-*

gylus equinus. During the adult stage, all three species suck blood from the intestine. During the migratory stage, however, each species is characterized by different patterns.

S. vulgaris is the most damaging of the large strongyles and is commonly found in any pasture where horses have grazed within one year. After the immature larvae are ingested by the horse, they pass through the digestive tract and burrow into the walls of the cecum and colon. Approximately eight days after ingestion, the larvae resume migration and penetrate the intima, or inner layer, of the small arteries of the intestine. Within the arterial walls, *S. vulgaris* larvae travel upstream into progressively larger arteries. Along their migratory paths, they cause inflammation of the arterial wall, a condition which may cause the development of thrombi, or clots. Pieces of these thrombi often break away from the walls and block small arteries, thus reducing blood flow to the horse's intestinal tract. Inflammation caused by *S. vulgaris* migration can also weaken the vessels so that blood pressure produces a sac-like protrusion, or aneurysm, in the arterial wall. Arterial damage resulting from blockages and aneurysms can cause severe and even fatal colic in the horse if circulation to the intestinal tract is reduced significantly. Fortunately, the horse's circulatory system is capable of bypassing the blocked arteries and repairing some of the damage caused by migratory larvae.

After migrating in the arterial walls for two to four months, *S. vulgaris* larvae enter the bloodstream, which carries them into the small arteries that supply the intestine. At this site they cause inflammation and arterial damage before emerging in the lumen of the cecum and colon as immature adults. After maturing, the adult females lay eggs which pass from the digestive tract with the feces. If conditions are appropriate, *S. vulgaris* eggs develop into infective larvae which act as a source of contamination in pastures and stalls, initiating the next parasitic life cycle. The life cycle of *Strongylus vulgaris* is completed in approximately 6 months.

Unlike *Strongylus vulgaris, Strongylus edentatus* and *Strongylus equinus* migrate into the liver rather than the arteries. Ingested *S. edentatus* larvae enter the intestinal wall and reach the liver by way of the portal vein (i.e., the vein which carries blood from the gastrointestinal tract to the liver). In contrast, *S. equinus* larvae burrow into the cecum and migrate into the liver across the peritoneal space. Both *S. edentatus* and *S. equinus* wander within liver tissue for approximately two months.

S. edentatus larvae leave the liver by way of ligaments which maintain the liver's position in the body, and they make their way through peritoneal tissue to the cecal wall. They finally emerge as adults in the intestinal lumen to mate and lay eggs. After a migratory period in the liver, *S. equinus* larvae burrow into the pancreas or abdominal cavity. Approximately four months after infecting the horse, *S. equinus* adults migrate to the intestinal wall and emerge in the gut. *S. edentatus* and *S. equinus* complete their life cycles in 11 months and 9 months, respectively. Although these parasites may cause scar tissue formation within the liver and result in decreased liver function during their migration, they are less harmful to the horse than *S. vulgaris*.

Small Strongyles

Small strongyles, of which there are more than 40 species, begin their life cycle in much the same manner as large strongyles. However, small strongyles inhabit the large intestine and become attached to the intestinal wall rather than migrate along the blood vessels. In the intestine, the parasites produce eggs which are eliminated with the feces. Although small strongyles are less debilitating than large strongyles, heavy infestations can cause chronic diarrhea or colon ulcers.

Research has shown that some small strongyle species may develop resistance to benzimidazole drugs (e.g., cambendazole, fenbendazole, mebendazole, oxfendazole, thiabendazole). Thus, when benzimidazole drugs are used continuously, small strongyle species may remain unaffected. Fortunately, drug resistance may be broken by using a nonbenzimidazole drug (e.g., carbon disulfide, dichlorvos, phenothiazine, piperazine, pyrantel pamoate, pyrantel tartrate) which is also effective against small strongyles.

Strongyloides westeri

Strongyloides westeri has a very unique life cycle in that the adult form is only found in horses less than 6 months of age. This parasite is only transmitted to foals through mare's milk. It is not known, however, how infective larvae get in the mare's mammary gland. The foal ingests larvae when it nurses from the carrier dam and it begins to shed eggs at 10 days to 2 weeks of age. The load of *Strongyloides westeri* within the small intestine peaks when the foal is about 6 to 10 weeks of age but then begins to drop as the foal develops an immunity to this particular parasite. When the foal is approximately 5 months of age, the infestation is nearly indetectable.

Ascarids

Ascarid (*Parascaris equorum*) infections are common in horses under six months of age and are particularly damaging to the general health and condition of foals. Horses are infected when they ingest egg-contaminated water, grass, hay, or grain. The larvae hatch and burrow into the wall of the small intestine and are then carried by the circulatory system to the liver, heart, and lungs. From the lungs, larvae migrate up the bronchial passages and are swallowed by their host. Once they are back in the intestines, the larvae mature and produce eggs which are eliminated with the feces. In general, ascarids stimulate inflammation of the liver and lung tissue and may cause malnutrition in the foal. In addition, a heavy load of ascarids may cause intestinal rupture.

Pinworms

Although the horse can be infested with two types of pinworms, only the species *Oxyuris equi* is of significance. The horse ingests pinworm eggs that are present on mangers, stall walls, or other animals. Adult pinworms live and mate in the horse's large intestine and the female migrates out of the intestinal tract to deposit sticky masses of eggs on the external surface of the anus. This substance causes itching which the horse attempts to relieve by rubbing its tail against stationary objects. As a result, surfaces become contaminated with parasite eggs, and horses in the area are exposed to infestation by the next generation of pinworms. In addition to anal irritation, a heavy load of pinworms can cause the horse discomfort due to inflammation of the cecum and colon.

Tapeworms

Although tapeworms are seldom a problem in horses, significant infestations of adult tapeworms are found occasionally. Tapeworm larvae are transmitted by mites that are ingested by the horse while grazing. Ingested larvae develop into adults within about 2 months and then tend to accumulate at the junction of the large and small intestines or in the bile and pancreatic ducts. Heavy tapeworm infestations can cause intestinal damage and diarrhea. Eggs produced by adult females pass out with feces and are ingested by mites, thereby completing the tapeworm life cycle.

Stomach Worms (Habronema)

Adult stomach worms, or Habronema, reside in the horse's stomach and lay eggs which are passed from the digestive tract with the feces. Although Habronema cause little internal damage to the horse, flies feeding on Habronema-contaminated manure can consume Habronema eggs. The eggs hatch and develop into infective larvae within the fly. The fly then transmits the larvae to moist areas, such as the lips, wounds or abrasions on the horse. Inflammation caused by the presence of Habronema in a moist, open wound causes the formation of excess granulation tissue, and these irritated areas are sometimes referred to as summer sores. Flies may also deposit Habronema larvae in the conjunctival membrane of the eye, causing inflammation and excessive tearing. Habronema infections of the skin and eye respond slowly to treatment and easily become chronic.

Bots

Unlike the other internal equine parasites examined in this discussion, the bot is not a worm. Rather, it is the larval stage of the botfly, which lives outside the horse during its adult stage. There are three major species of bots: *Gasterophilus intestinalis, Gasterophilus nasalis*, and *Gasterophilus hemorrhoidalis*. Only *G. intestinalis* and *G. nasalis* are commonly found in the United States. The adults of all three botfly species lay eggs on the hair of horses. However, each has unique egg-laying and migratory habits.

Yellow-white *G. intestinalis* eggs are deposited on the forelegs and shoulders of the horse. As the horse grooms its forequarters with its teeth and lips, its warm breath causes the eggs to hatch. The newly-hatched larvae enter the horse's mouth and burrow into its tongue. Approximately one month later, the larvae emerge from the back of the tongue and are swallowed. The immature bots anchor themselves to the stomach wall until spring when they release their hold, pass out with the horse's feces, and develop into adult botflies.

In contrast to *G. intestinalis,* the adult botfly of *G. nasalis* deposits its yellow eggs on hair located between the horse's jawbones, while the *G. hemorrhoidalis* adult lays black eggs on hair surrounding the horse's lips. *G. nasalis* eggs hatch after one week of incubation, but *G. hemorrhoidalis* eggs require moisture to hatch. Although the hatching larvae of both species enter the horse's mouth, *G. nasalis* larvae burrow between the lips and inner surface of the mouth. Several weeks later, the larvae emerge and are swallowed by the host. Thereafter, the life cycles of both bot species follow that of *G. intestinalis.*

All three botfly species produce one generation per year and survive the winter protected inside the horse's stomach. Colic, stomach rupture, and other digestive problems have been attributed to heavy bot infestations.

Management Tips

Because proper management can significantly reduce the number of parasites that horses are exposed to, it should be considered an integral part of a parasite control program. The following management practices help reduce the number of parasite eggs or larvae available to infect horses:

1. Feed horses from mangers or racks rather than on the ground where they can easily ingest parasite larvae or eggs.

2. Fence off or drain manure-contaminated water sources such as ponds and streams.

3. Compost manure before spreading it on horse pastures. The heat generated by manure decomposition helps kill parasite eggs.

4. Remove feces from stalls and paddocks frequently. If possible, clean stall walls, feed troughs, and other facilities with pressurized steam to kill parasites.

5. Drag pastures weekly or monthly (depending on the stocking rate) to break up manure piles and expose parasite eggs to the sun. When possible, remove manure droppings from pastures for composting.

6. Rotate horses onto rested pastures or pastures that have been grazed by other species. By temporarily removing a parasite's host (the horse) it is possible to reduce pasture contamination through interruption of the parasite's life cycle.

Dewormers

Another place to disrupt parasite life cycles is within the host. Horses can be dewormed with a number of products, many of which are detailed on the following chart. Each product is effective against specific parasites, and dewormers should be chosen to complement the total parasite control program. Dewormers must also be selected carefully when treating young horses and pregnant mares. Precautions for these groups are included in the dewormer chart.

DRUG	PRODUCT NAME	COMPANY	EFFECTIVE AGAINST	FORM/METHOD OF ADMINISTRATION	SAFETY FOR FOALS	SAFETY FOR PREGNANT MARES	ALSO CONTAINS
BUTONATE	Vet-Kem T-113*	Vet-Kem	bots, ascarids	liquid/stomach tube	not for use under 12 months of age		
CAMBENDAZOLE	Camvet suspension* Camvet pellets Camvet paste	Merck	large strongyles, small strongyles, ascarids, pinworms, strongyloides	suspension/stomach tube or drench pellets/feed paste/oral syringe		not for use during first trimester	
CARBON DISULFIDE	Parvex*	Upjohn	bots, habronema, ascarids	liquid/stomach tube or dose syringe		not for use during last 2 months	phenothiazine
	Parvex Plus*	Upjohn	bots, habronema, ascarids, large strongyles, small strongyles, pinworms	liquid/stomach tube or dose syringe		not for use during last 2 months	piperazine
DICHLORVOS	Equigard Equigel*	Shell	large strongyles, small strongyles, ascarids, pinworms, bots	pellets/feed gel/feed	not for use in sucklings or young weanlings		
	Shell Horse Wormer	Shell	large strongyles, small strongyles, ascarids, pinworms, bots	pellets/feed	not for use in sucklings or young weanlings		
	Tridex Paste*	Fort Dodge	ascarids, bots	paste/oral syringe			
	Tridex Pellets	Fort Dodge	large strongyles, small strongyles, ascarids, pinworms, bots	pellets/feed			
FEBANTEL	Cutter Paste Wormer	Cutter	large strongyles, small strongyles, ascarids, pinworms	paste/oral			
	Rintal	Haver-Lockhart	large strongyles, small strongyles, ascarids, pinworms	paste/oral			
FENBENDAZOLE	Panacur*	National	large strongyles, small strongyles, ascarids, pinworms	suspension/stomach tube			
LEVAMISOLE	Ripercol L-Piperazine*	American Cyanamid	large strongyles, small strongyles, ascarids, pinworms	liquid/stomach tube or drench			piperazine
MEBENDAZOLE	Telmin*	Pitman-Moore	large strongyles, small strongyles, ascarids, pinworms	powder/feed			
	Telmin SF*	Pitman-Moore	large strongyles, small strongyles, ascarids, pinworms	paste/oral syringe			

DRUG	PRODUCT NAME	COMPANY	EFFECTIVE AGAINST	FORM/METHOD OF ADMINISTRATION	SAFETY FOR FOALS	SAFETY FOR PREGNANT MARES	ALSO CONTAINS
OXFENDAZOLE	Benzelmin Equine Anthelmintic Powder*	Diamond	large strongyles, small strongyles, ascarids, pinworms	powder/stomach tube			
	Benzelmin Equine Anthelmintic Pellets	Diamond	large strongyles, small strongyles, ascarids, pinworms	pellets/feed			
PHENOTHIAZINE	Dyrex T. F.*	Diamond	large strongyles, small strongyles, ascarids, pinworms, bots	liquid/stomach tube	not for use before 3 months of age	not for use during last month	
	Parvex Plus*	SEE INFORMATION LISTED UNDER CARBON DISULFIDE					
	Pheno Sweet	Farnum	large strongyles, small strongyles	granules/feed	not for use in very young foals	not for use during last month	
PIPERAZINE	Alfalfa Pellet	Farnum	large strongyles, small strongyles, ascarids, pinworms	pellets/feed	not for use before 3 months of age		
	Dyrex T. F.*	SEE INFORMATION LISTED UNDER PHENOTHIAZINE					
	Equivet Jr.	Farnum	large strongyles, small strongyles, ascarids	paste/oral syringe			
	Equizole A*	Merck	large strongyles, ascarids, pinworms, strongyloides	liquid, powder/stomach tube, feed or drench			thiabendazole
	Foal Wormer	Farnum	large strongyles, small strongyles, ascarids, pinworms	granules/feed	for use in weanlings and yearlings		
	Parvex*	SEE INFORMATION LISTED UNDER CARBON DISULFIDE					
	Parvex Plus*	SEE INFORMATION LISTED UNDER CARBON DISULFIDE					
	Pipzine-34	Affiliated Labs	large strongyles, small strongyles, ascarids, pinworms	liquid/feed or dose syringe			
	Ripercol L-Piperazine*	SEE INFORMATION LISTED UNDER LEVAMISOLE					
	Wonder Wormer	Farnum	large strongyles, small strongyles, ascarids, pinworms	granules/feed	not for use in very young foals	not for use during last month	
PYRANTEL TARTRATE	Banminth	Pfizer	large strongyles, small strongyles, ascarids, pinworms	pellets/feed			

DRUG	PRODUCT NAME	COMPANY	EFFECTIVE AGAINST	FORM/METHOD OF ADMINISTRATION	SAFETY FOR FOALS	SAFETY FOR PREGNANT MARES	ALSO CONTAINS
PYRANTEL PAMOATE	Pyraminth	Beecham	large strongyles, small strongyles, ascarids, pinworms	liquid/feed, stomach tube, or dose syringe			
	Strongid T*	Pfizer	large strongyles, small strongyles, ascarids, pinworms	liquid/feed, stomach tube, or dose syringe			
THIABENDAZOLE	Equivet TZ	Farnum	large strongyles, small strongyles, ascarids, pinworms, strongyloides	paste/oral syringe			
	Equivet 14	Farnum	large strongyles, small strongyles, ascarids, pinworms, strongyloides, bots	paste/oral syringe pellets/feed	not for use before 4 months of age	not for use during last month	trichlorfon
	Equizole Horse Wormer*	Merck	large strongyles, small strongyles, pinworms, strongyloides	powder/feed, dose syringe, or stomach tube			
	Equizole Horse Wormer Pellets	Merck	large strongyles, small strongyles, ascarids, pinworms, strongyloides	pellets/feed			
	Equizole Suspension*	Merck	large strongyles, small strongyles, ascarids, pinworms, strongyloides	liquid/stomach tube or drench			
	Equizole A*	SEE INFORMATION LISTED UNDER PIPERAZINE					
	Equizole B*	Merck	large strongyles, small strongyles, ascarids, pinworms, strongyloides, bots	powder/feed, stomach tube, or drench	not for use before 4 months of age	not for use during last month	trichlorfon
	Omnizole Suspension*	Merck	large strongyles, small strongyles, ascarids, pinworms, strongyloides	liquid/stomach tube or drench			
TRICHLORFON	Combat Liquid*	Haver-Lockhart	ascarids, pinworms, bots	liquid/stomach tube	not for use before 4 months of age	not for use during last month	
	Combat Paste**	Haver-Lockhart	large strongyles, ascarids, pinworms, bots	paste/oral syringe		only upon veterinary recommendation	
	Dyrex Cap-tab*	Fort Dodge	large strongyles, small strongyles, ascarids, pinworms, bots	bolus/balling gun		not recommended during late pregnancy	
	Dyrex Granules*	Fort Dodge	ascarids, pinworms, bots	liquid/stomach tube		not recommended during late pregnancy	
	Dyrex T. F.*	SEE INFORMATION LISTED UNDER PHENOTHIAZINE					
	Equibot TC	Farnum	ascarids, pinworms, bots	paste/oral syringe			

DRUG	PRODUCT NAME	COMPANY	EFFECTIVE AGAINST	FORM/METHOD OF ADMINISTRATION	SAFETY FOR FOALS	SAFETY FOR PREGNANT MARES	ALSO CONTAINS
	Equivet 14	SEE INFORMATION LISTED UNDER THIABENDAZOLE					
	Equizole B*	SEE INFORMATION LISTED UNDER THIABENDAZOLE					
	Negabot	Cutter	ascarids, pinworms, bots	paste/oral syringe		not for use during last month	
	P/M Trichlorfon	Pitman-Moore	ascarids, pinworms, bots	powder/feed	not for use under 4 months of age	not for use during last month	
	Top Form Two	SEE INFORMATION LISTED UNDER THIABENDAZOLE					

*Federal law restricts this drug to use by or on the order of a licensed veterinarian.
**This drug has not been proven safe for use in stallions used for breeding purposes.

APPENDIX 8

EXTERNAL PARASITES

The following descriptions of parasites and appropriate control methods are preparatory aids for the task of protecting horses and stable areas from external parasites.

Flies and Mosquitos

Members of the order Diptera can cause considerable irritation and annoyance to man and horse because of their vicious biting and feeding habits. These two-winged external parasites are commonly grouped into two categories according to their feeding behavior, either biting or nonbiting.

Nonbiting Flies

The house fly (*Musca domestica*) and the face fly (*Musca autumnalis*), both of which are nonbiting flies, feed on secretions from the eyes, nose, and mouth as well as from wounds caused by injury or insect bites. These flies not only create a nuisance but are capable of transmitting pinkeye (contagious conjunctivitis), eyeworms, and other infectious diseases common to man and livestock.

The general behavior of both the house and face fly is the same in that they both feed primarily during the day. Nights are spent on nearby vegetation or on the rafters of barns or roofs. The egg-laying area for the face fly is livestock manure while the house fly deposits eggs in decomposing fruit or vegetation as well as animal manure.

Biting Flies

The biting flies, stable fly (*Stomoxys calcitrans*), horn fly (*Haematobia irritans*), horse fly (Tabanus sp.), mosquitos and gnats are blood-sucking parasites. These flies, like the nonbiting flies, spread infectious diseases as they feed on various hosts and lay eggs in decaying organic matter and manure. Other behavioral patterns of biting flies are much the same as nonbiting flies. Feeding usually occurs in the morning or late afternoon with activity decreasing at nightfall. However, mosquitos feed at night, and horn flies spend virtually twenty-four hours a day on the host, leaving only to deposit eggs in fresh manure. Egg-laying habits vary; the stable fly is a manure egg-layer, but the mosquito and horse fly lay eggs in aquatic or semi-aquatic environments. Mosquitos deposit eggs directly into stagnant or slow-moving water, while horse flies lay eggs on vegetation directly above water. Hatching horse fly larvae drop into the water below but later pupate on dry ground.

Ticks

Some tick species are specific to a particular region of the country, while others are widespread. Aside from blood loss, ticks can also cause allergic reactions, transmit disease, and annoy their hosts. Pastured horses are continually susceptible to ticks, which are grouped into two families according to their outer shell hardness.

Hard ticks have a hard outer covering on part or all of their backs and can be further classified according to the number of hosts their lifecycles involve. As a larvae (seed tick), the three-host tick engorges on an animal, drops to the ground and molts. In the nymph stage, it attaches to a second host and repeats the sequence. Finally, as an adult, the tick claims a third host on which it will mate and engorge itself with blood. The female then drops off to lay eggs. In comparison, the one-host tick exhibits the same lifecycle sequence but on a single host animal.

Soft ticks, on the other hand, are characterized by the wrinkled, leathery texture of their outer covering. One soft tick, the spinose ear tick (*Otobius megnini*), lives deep in the ears of the host animal, causing evident discomfort. Only the larval and nymph stages parasitize livestock, while the adults are free-living.

Lice

There are only two species of lice that attack horses; the horse biting louse (*Bovicola equi*) and the horse sucking louse (*Haematopinus asini*). Lice are blood-sucking parasites which attach eggs, called nits, to the host's body hairs. Because lice are almost constant companions of their hosts, major infestations can create significant irritation and loss of hair.

EXTERNAL PARASITE CONTROL

External parasite control is a demanding task. Because methods of complete eradication are unknown at present, the primary objective of external parasite management is population control. The following discussion of control systems and management tips can provide a basis for developing effective pest control practices.

Chemical Control Methods

Many different chemical compounds are used as the active ingredients in pesticides labeled for horses or stable areas. Many companies in the United States market pesticides for use against external parasites of horses, but a smaller number actually manufacture the chemical ingredients found in those products. Commercially, these pesticides appear in various forms including sprays, dusts, ointments, and baits.

Prior to purchase, product labels should be carefully studied to determine the product's effectiveness and safety. The label may state that the product should not be used on foals and sick or stressed animals. Other labels may indicate that the pesticide is intended for use on premises only. Many of the products should not be inhaled, ingested, or come in contact with skin or eyes. Because of these special restrictions, all label instructions must be followed to achieve maximum results without causing undesirable side effects.

Biological Control Method

The only biological control system for stable areas which seems practical (besides natural insect predators such as birds and reptiles) employs insects which parasitize noxious insects and their eggs. One insect which can be useful to the horseman is a tiny wasp-like insect, called a pteromalid, which lays eggs inside fly pupa (immature flies in a resting stage).

These fly predators are ready to mate and lay eggs immediately upon emerging from their pupal stage. Upon mating, the female seeks out fly pupa, drills a hole through the pupal casing and lays one or more eggs. The female is capable of laying up to 400 eggs in a lifetime. The incubation period of the pteromalid is one month, during which it will feed off the fly larvae and emerge as an adult to mate and start the cycle again. Although pteromalids may not completely eradicate a fly population, they can be integrated with other control methods to significantly reduce fly numbers.

Mechanical Control Methods

Most horsemen are familiar with fly control by some type of mechanical device or system. Pest management programs commonly utilize such devices as fly paper, electronic bugkillers, and well-planned manure disposal systems. Fly paper strips, suspended in the barn beyond the reach of animals, attract and ensnare flies but they seldom affect the fly population significantly.

Electronic bugkillers also control flying insects. A fluorescent light which attracts insects is attached to the inside of a metal screen which electrocutes bugs upon contact. However, a major disadvantage of these lights is that they are most effective at night, attracting only mosquitos and gnats while flies are roosting. Any type of electrical appliance, such as a bug killer, should be installed carefully and placed out of reach of horses.

Finally, careful manure disposal is essential to an effective pest control program. Manure is very attractive to pests; thus, if it must be stockpiled before disposal, manure should not be stored close to animal facilities.

APPENDIX 9

IMMUNOLOGY AND VACCINATION

Most bacteria and viruses carry a characteristic substance, or antigen, that sensitizes certain cells in an infected individual. This sensitivity stimulates the production of antibodies, which are agents that can neutralize or destroy the antigen of a specific disease. By artificially stimulating antibody production (i.e., active immunity) or by introducing protective antibodies into a horse (i.e., passive immunity), the individual can be protected from specific diseases.

Inside a horse's body, antigens reach lymph organs through blood vessels and lymph vessels. Concentrated in the lymph organs are antibody-producing cells, called plasma cells, which when exposed to antigens coming into lymph organs manufacture antibodies to combat the antigens. Plasma cells continue to produce antibodies for several months or even for years after initial exposure to a specific antigen. This long-lasting protection is called active immunity. Immunity to most diseases gradually wears off. Therefore, in a regular vaccination program, periodic boosters are given to keep an ample supply of antibodies functioning.

The products used to promote active immunity are called vaccines. Several types of vaccines are listed below, all of which stimulate antibody production by plasma cells.

Toxoid: A toxoid is a toxin that has been rendered nonpoisonous without impairing its ability to stimulate antibody formation (e.g., tetanus toxoid).

Bacterin: Bacterin vaccines are suspensions of bacteria killed by heat or chemical means (e.g., strangles vaccine). They are not usually as effective as live vaccines.

Live bacterial suspension: This vaccine contains viable bacteria that can cause disease when administered. However, it produces a stronger and longer-lasting immunity than does a bacterin.

Live unmodified virus: This type of vaccine can cause a mild form of the disease, but it produces a very effective immunity.

Modified live virus: These vaccines are prepared from suspensions of live virus particles. The disease-causing capacity of these particles is reduced but their antigen character is unaltered (e.g., rhinopneumonitis internasal vaccine). Live virus vaccines stimulate antibody formation by causing a mild case of the disease. The horse seldom becomes noticeably ill but may develop a swelling at the injection site.

Killed virus: These products are prepared from a suspension of chemically killed virus particles (e.g., eastern and western encephalomyelitis and influenza vaccines).

In contrast with active immunity, passive immunity is instant, although temporary, disease-resistance resulting from the administration of preformed antibodies from another individual. This type of immunity lasts only as long as the antibodies are effective, but it is useful for protecting an individual against an immediate threat of disease until it can develop its own immunity. Tetanus antitoxin, which contains antibodies to the toxin that causes tetanus, is an example of a product that protects by passive immunity. After an injury or before surgery, horses are frequently given both tetanus antitoxin and tetanus toxoid. The antitoxin temporarily protects against tetanus infection (i.e., for approximately two weeks) while the toxoid causes the body to produce its own antibodies for active immunity.

BREEDING FARM VACCINATION SCHEDULE

BIOLOGICALS	FOALS	YEARLINGS	BROODMARES	TWO-YEAR-OLDS THROUGH MATURITY
TETANUS ANTITOXIN	give at birth when dam has not been vaccinated one month before foaling	give in case of injury if vaccination history is unknown or if tetanus vaccination is outdated	give in case of injury if vaccination history is unknown or if tetanus vaccination is outdated	give in case of injury if vaccination history is unknown or if tetanus vaccination is outdated
TETANUS TOXOID	1st dose: 12 weeks of age; 2nd dose: 14-16 weeks of age	booster 6-12 months after initial doses	booster annually, 30 days prior to foaling	booster annually
EQUINE INFLUENZA	1st dose: 8 weeks of age; 2nd dose: 14-20 weeks of age	booster annually	booster annually, 30 days prior to foaling	booster annually
RHINOPNEUMONITIS INJECTABLE (modified live virus)	1st dose: 16 weeks of age; 2nd dose: 20 weeks of age	booster every 6 months	1st booster: prior to breeding; 2nd booster: at least 60 days pregnant; 3rd booster: between 5 and 7 months pregnant	booster every 6 months
RHINOPNEUMONITIS INTRANASAL*	Can be used to induce "planned infection" in all horses on a farm; 1st dose: July, 2nd dose: October			
STRANGLES	1st dose: 12 weeks of age; 2nd dose: 13 weeks of age; 3rd dose: 14 weeks of age	booster 12 months after immunization or previous strangles infection	do not use in pregnant mares	booster 12 months after immunization or previous strangles infection
EASTERN AND WESTERN ENCEPHALOMYELITIS (EEE/WEE)	1st dose: 8 weeks of age; 2nd dose: 9 weeks of age	2-dose series before mosquito season	2-dose series before mosquito season	2-dose series before mosquito season
VENEZUELAN ENCEPHALOMYELITIS (VEE)	24 weeks of age	vaccinate every other year	vaccinate every other year	vaccinate every other year
RABIES	Administer to all horses in areas where skunks, loose dogs, and other potential carriers of rabies are common			

*1. Use only on farms with planned (February to June) breeding when mares are less than five months pregnant.

2. Do not move horses for three weeks since vaccine virus can be transmitted between horses in close contact.

3. Vaccine is restricted in certain states and some countries other than the U.S.

4. Use only if infection is present in the area.

5. Do not expose non-vaccinated pregnant mares to animals vaccinated less than three weeks previously.

APPENDIX 10

MANAGEMENT CALENDAR

The following breeding farm management calendar incorporates much of the material from preceding sections on parasite control, immunizations, and reproductive management. It is provided as an example of a well-planned management program. Specific dates will vary on different farms depending on the beginning of the breeding season, regional vaccination requirements, and other factors.

SAMPLE BREEDING FARM CALENDAR[a]

MONTH	PREGNANT MARES	BARREN MARES	FOALS	YEARLINGS	STALLIONS
JANUARY		begin teasing, veterinary exam, treat if necessary		deworm 10 mo.	
FEBRUARY	tetanus toxoid, booster 1 month before foaling, influenza booster, deworm	tetanus toxoid, influenza booster, deworm, rhino-pneumonitis booster, strangles		tetanus toxoid, influenza booster, rhinopneumonitis booster 11 mo.	deworm, tetanus toxoid, influenza booster, rhinopneumonitis booster, strangles
MARCH	FOALING	begin breeding	BIRTH deworm at 7 days of age	deworm[b] 12 mo.	
APRIL	begin rebreeding	pregnancy test bred mares	1 mo.	13 mo.	
MAY	deworm, pregnancy test, EEE/WEE, VEE	deworm, EEE/WEE, VEE	deworm[b], 1st: influenza, EEE/WEE, 7th: EEE/WEE 2 mo.	deworm, EEE/WEE, VEE 14 mo.	deworm, EEE/WEE, VEE
JUNE	pregnancy test, rhinopneumonitis (killed virus vaccine during 5th month of pregnancy)	pregnancy test	early weaning, tetanus toxoid 3 mo.	15 mo.	
JULY			tetanus toxoid, influenza, rhino-pneumonitis, deworm 4 mo.	deworm 16 mo.	
AUGUST	deworm, rhinopneumonitis (killed virus vaccine during 7th month of pregnancy)	deworm, rhinopneumonitis	rhinopneumonitis 5 mo.	rhinopneumonitis 17 mo.	deworm, rhinopneumonitis
SEPTEMBER			VEE, deworm 6 mo.	deworm 18 mo.	
OCTOBER	rhinopneumonitis (killed virus vaccine during 9th month of pregnancy)		7 mo.	19 mo.	
NOVEMBER	deworm, check teeth	deworm, check teeth, 15th: start artificial lighting, flushing[c]	deworm 8 mo.	deworm 20 mo.	deworm, check teeth, reproductive exam, fertility check
DECEMBER	balance ration for last trimester		9 mo.	21 mo.	

[a] specific dates are applicable to the Northern Hemisphere
[b] deworm every 2 months until age two; deworm every 3 months thereafter
[c] if on an early breeding schedule with season beginning on March 1 or before

APPENDIX 11

COST ANALYSIS OF BREEDING FARM EXPENSES (SAMPLE YEAR - 1982)

This section is provided to supplement the 'Cost of Raising a Foal' charts in the chapter "Breeding Farm Economics."

Feed Expenses

Based on:
Whole oats	$ 9.80/cwt
Commercial foal feed	9.50/cwt
Grass hay	100.00/ton
Alfalfa hay	167.00/ton

Mare

If the mare is boarded at a breeding farm for 2 months of each year while lactating, the total feed cost drops to $123.48 for grain and $266.33 for hay, or a total of $389.81 feed cost for 10 months on the farm. During those 2 months, her feed is paid for in the form of mare care charged by the breeding farm.

	Estimated Daily Intake		Estimated Yearly Intake	
	grain (lb)	hay (lb)	grain (lb)	hay (lb)
first 90 days of lactation	12	10	1080	900
180 days, maintenance	2	15	360	2700
last 90 days of gestation	6	11	540	990
	TOTAL INTAKE		1980	4590

1980 lb grain	@ $9.80/cwt	=	$194.04
4590 lb hay (50% grass, 50% alfalfa)			
2295 lb grass hay	@ $100/ton	=	114.75
2295 lb alfalfa hay	@ $167/ton	=	191.63
Vitamin-mineral supplement	$ 25 per year	=	25.00
	TOTAL YEARLY FEED COST		$525.42

Stallion

The 1100 pound stallion in this example consumes approximately 16.5 pounds of feed per day, comprised of 1.65 pounds of grain per day and 14.85 pounds of hay per day. Therefore, he requires about 602 pounds of grain per year and 5420 pounds of hay per year.

602 lb grain	@	$9.80/cwt	=	$ 59.00
5420 lb hay (50% grass, 50% alfalfa)				
2710 lb grass hay	@	$100/ton	=	135.50
2710 lb alfalfa hay	@	$167/ton	=	226.29
Vitamin-mineral supplement		$ 25 per year	=	25.00
		TOTAL YEARLY FEED COST		$445.79

Weanling Foal

Assuming that the weanling will weigh approximately 1000 lb at maturity, it will consume an estimated 546 lb of a commercial grain mix formulated for foals and about 565 lb of alfalfa hay from 3 to 6 months of age.

546 lb commercial foal feed	@	$ 9.50/cwt	=	$ 51.87
565 lb alfalfa hay	@	$167/ton	=	47.18
Vitamin-mineral supplement				10.00
		TOTAL FEED COST		$109.05

Mare Board

Board for 10 months is estimated at $150.00 per month, or $1,500.00 per year. This value is used only in the table 'Cost to Raise One Foal: Nonbusiness Enterprise (Hobby).' It is assumed that the mare will spend the other two months of the year at a breeding farm.

Mare Care

Mare care expense is estimated for 60 days on the breeding farm at $4.00 per day or approximately $240.00 per year. This figure will vary, depending on the length of time that the mare must be kept at the breeding farm. Mare care expenses are only applicable if the mare is shipped to an outside stallion for breeding. Therefore, this expense appears only in the table 'Cost to Raise One Foal: Nonbusiness Enterprise (Hobby).'

Farrier Service

All horse's hooves are trimmed every six weeks at a cost of $12.00 per horse. The cost for the foal is figured on the basis of trimming every four weeks for the first six months of life.

	trims/yr		cost/horse		cost/horse/yr				cost/yr
Mare	8	x	$12.00	=	$96.00	x	50 mares	=	$4,800.00
Stallion	8	x	12.00	=	96.00	x	1 stallion	=	96.00
Foal	6	x	10.00	=	60.00	x	40 foals	=	2,400.00
			TOTAL FARRIER EXPENSES PER YEAR						$7,296.00

Veterinary Services

MARE	cost per unit	cost per no. of specified units
5 vaccinations	$ 4.00	$ 20.00
4 dewormings	10.00	40.00
2 pregnancy examinations	8.00	16.00
2 uterine cultures	15.00	30.00
1 dental examination	15.00	15.00
7 trip charges	8.00	56.00
miscellaneous expenses		50.00
		$227.00
		x 50 mares
		$11,350.00

STALLION		
5 vaccinations	$ 4.00	$ 20.00
4 dewormings	10.00	40.00
1 dental examination	15.00	15.00
1 reproductive examination and physical	50.00	50.00
5 trip charges	8.00	40.00
miscellaneous expenses		50.00
		$ 215.00

FOAL		
1 examination at birth	$10.00	$ 10.00
4 dewormings	10.00	40.00
9 vaccinations	4.00	36.00
6 trip charges	8.00	48.00
miscellaneous expenses		50.00
		$184.00
		x 40 foals
		$ 7,360.00

TOTAL EXPENDITURES FOR VETERINARY SERVICES $18,925.00

Pasture Expenses

Actual pasture expenses may differ considerably from those cited in the following example because fertilization, stocking rates, and specific management methods vary, depending on a number of factors (e.g., pasture quality, soil type, and climate). However, the figures that follow are representative of a well-established Coastal bermudagrass pasture that is located in an area that receives adequate annual rainfall for grass growth. (Supplemental irrigation is a very costly item in pasture grass production.) The stocking rate in the following example is one mare and foal per two acres.

Fertilizer: 16-20-0 (Nitrogen-Phosphorus-Potassium) is applied in April (200 lb/acre) and ammonia nitrate (150 lb/acre) is applied in August, at a total cost of $30 per acre.

Herbicide: 2,4-D is applied in April to control broadleaf weeds at a cost of $3 per acre.

Clipping: Mechanical weed control is often required during the growing season to reduce weed growth. Clipping costs are approximately $2.50 per acre.

PASTURE MAINTENANCE COSTS

	cost/acre	cost/100 acres
Fertilizer	$30.00	$3000
Herbicide	3.00	300
Clipping	2.50	250
TOTAL PASTURE MAINTENANCE COST		$3550

Farm Labor Expense

	salary
1 full-time employee	$12,000/year
1 full-time employee	9,600
1 part-time employee	4,000
TOTAL LABOR COST	$25,600/year

Depreciation Expense

Breeding Stock Depreciation*

Mare

Original value = $10,000; acquired in 1982 at 4 years of age; placed on the 5-year depreciation schedule below:

Year	Year Property Was Placed in Service 1981-1984
1	15%
2	22
3	21
4	21
5	21
	100%

In the first year of service (1982) the depreciation on property obtained in 1982 is 15%. Thus:

$10,000 original value x .15 = $1,500 depreciation

*See "Breeding Farm Economics" for a discussion of depreciation schedules.

Stallion

Original value = $100,000; acquired in 1981 at 6 years of age; placed on the 5-year depreciation schedule below:

	Year Property Was Placed in Service
Year	1981-1984
1	15%
2	22
3	21
4	21
5	21
	100%

In the second year of service (1982) the depreciation on property obtained in 1981 is 22%. Thus:

$100,000 original value x .22 = $22,000 depreciation

Foal

Home-grown livestock is not depreciable.

*Facilities Depreciation**

These calculations are made on a straight-line depreciation schedule over a 25-year write-off period at 4 percent annual depreciation because property was placed in service before the Economic Recovery Tax Act of 1981 took affect.

Loafing Sheds

In 1970, 10 loafing sheds were constructed at a cost of $1000 each.

$1000 original value - $30 salvage value = $970
$970 x .04 = $38.80 annual depreciation
$38.80 x 10 sheds = $388 annual depreciation on 10 loafing sheds

Individual Sheds

In 1972, 10 individual sheds were constructed at a cost of $700 each.

$700 original value - $20 salvage value = $680
$680 x .04 = $27.20 annual depreciation
$27.20 x 10 sheds = $272 annual depreciation on 10 individual sheds

Horse Barn

In 1977, a 10 stall horse barn was constructed at a cost of $30,000.

$30,000 - $1,500 salvage value = $28,500
$28,500 x .04 = $1,140 annual depreciation

*If new or used depreciable property is purchased within the current tax year and it has a useful life of 3 years or more, it may qualify for investment tax credit. Refer to the discussion on tax credit within the chapter, "Breeding Farm Economics."

Total Facilities Depreciation

$ 388 (depreciation on loafing sheds)
272 (depreciation on individual sheds)
+1,140 (depreciation on horse barn)
$1,800 total annual facilities depreciation*

*Other significant facilities expenses that are not included in this simplified example are: fencing, arenas, and improvements.

APPENDIX 12

BASIC GENETIC DEFINITIONS

An understanding of genetic principles is an important foundation for any breeding program. Because genetics and breeding are so closely related and yet each is an extensive subject in its own right, Equine Research, Inc. has published separate volumes on these topics so that each may be dealt with in detail. However, genetic principles are unavoidably mentioned throughout this book, and for this reason, a review of basic genetics is included in the Appendix for easy reference. The following description of basic genetic concepts is a compilation of excerpts from EQUINE GENETICS & SELECTION PROCEDURES, the companion volume to this text.

The Cell

The cell is the basic unit of all plant and animal life. Every organism is composed of one or more cells, from the single-celled amoeba to the multi-billion celled horse. Even a complicated organism, such as the horse, begins life as a single cell — the fertilized egg, or zygote. The zygote divides and multiplies, developing into an embryo. As the embryo develops, each of its cells divides into two "daughter" cells. Some of the resulting "daughter" cells differentiate (i.e., change their function). Some become brain cells, while others become red blood cells, etc. Cell division and differentiation continue as the embryo becomes a fetus and finally a foal. Even in the adult horse, new cells (e.g., blood cells, bone cells, skin cells) are constantly produced as needed.

Growth and maintenance of the entire body is the result of cell activity. There are hundreds of different types of cells in the body, each with a genetically determined function that is inherited in the genetic material. This genetic information is contained in each cell of the body.

Chromosomes

The chromosomes are rod-like structures that contain the genetic information carried by an individual. They are located within each cell's nucleus, which is an area of highly concentrated cellular material. The number of chromosomes in each cell is the same for all members of a species, but it differs between species. For example, man has 46 paired chromosomes in each cell, while the horse has 64 paired chromosomes. The normal number of chromosomes is found in all of the somatic cells (i.e., all body cells). However, the sperm and ovum, sometimes referred to as the sex cells, or gametes, contain only half the normal number of chromosomes, since they receive only one chromosome from each of the original pairs during gametogenesis. For instance, normal human gametes have 23 chromosomes, while normal equine gametes have 32 chromosomes. The equine sperm and ovum unite during conception to form a new individual that carries 64 paired chromosomes (32 from the sperm plus 32 from the ovum = 32 chromosome pairs = 64 chromosomes).

Whether the offspring is male or female is determined by the sex chromosomes, which are known as the X and Y chromosomes. An individual that has both an X and a Y chromosome is a male, while the presence of two X chromosomes results in the development of female characteristics. All of the other chromosomes are collectively called the autosomes, or autosomal chromosomes.

Genes

The genetic information carried by the chromosomes is divided into basic units known as genes. Although the exact number of genes per chromosome is not yet known, there are believed to be thousands of genes located on each chromosome. Each gene is responsible for determining one or more characteristics, or traits.

An individual's total genetic makeup is known as its genotype — the complete blueprint that is passed on to the next generation. The visible and measurable characteristics of an individual are known as its phenotype. The difference between genotype (inherited potential) and phenotype (outward appearance) can be illustrated with the trait height. A horse may have the genotype for a height of 16 hands, but due to poor early nutrition, he may only attain 15-2 hands at maturity. In this case, his phenotype (15-2 hands) is not the same as his genotype (16 hands) due to environmental conditions (poor nutrition).

Heredity

Heredity is the transmission of genetic information from parents to offspring through the sex cells (i.e., sperm and ovum). Half of the offspring's chromosomes are received from its sire, and half are received from its dam. Therefore, the offspring has a genotype that differs from either parent. This genotype usually remains unchanged throughout the individual's life, even if the phenotype changes somewhat due to environmental conditions. On rare occasions, a change may occur in the chemical composition of a gene or in the structure of a chromosome. Such a change is known as a mutation. If the change occurs in a somatic cell, an alteration in phenotype may result but the characteristic will not be heritable. On the other hand, if the mutation occurs in a gamete, it can be passed on to the individual's offspring. The mutation then becomes a part of the offspring's genotype and can be passed to its progeny. Although mutations are usually harmful, they may (in rare instances) cause a desirable change and subsequent improvement within a species.

Environment

The surrounding conditions which influence the growth and development of an individual are known as its environment. Both external factors (e.g., living conditions) and internal factors (e.g., hormones) are a part of the individual's environment. The environment is capable of drastically changing the phenotype of an individual. For example, a foal with the genotype for straight legs may receive a nutritionally imbalanced diet and consequently develop crooked legs. On the other hand, a foal with the genotype for crooked legs will not develop straight legs even with a sound diet (unless surgery or corrective shoeing can compensate for the inherited disorder).

Proliferation of Genetic Information - Part I

Although some cells function throughout the life of the individual (e.g., nerve cells), many are short-lived (e.g., skin cells). For this reason, many cells must be constantly duplicated without losing their genetically encoded function. The chromosome's ability to duplicate its entire structure and the cell's ability to divide into two identical sister cells are attributed to a process known as mitosis. Intricate cellular changes that occur during mitosis (e.g., replication and movement of chromosomes) can be observed through an electron microscope, and in this manner five distinct steps of mitosis have been identified: interphase, prophase, metaphase, anaphase, and telophase.

Interphase

Interphase is the resting, or dormant, stage between cell divisions. During interphase, the chromosomes are not visible through the electron microscope (except as tiny granules located throughout the nucleus). During this period, the chromosomes are believed to be in the diploid state. (Diploid refers to the normal chromosome number, such as 64 chromosomes in the horse.) Although it is generally accepted that interphase is a resting period, some sources think that the undetectable chromosomes begin to replicate near the end of this phase.

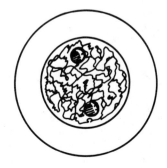

Interphase - Chromosomes are not visible during this dormant period.

Prophase

During prophase, the cell undergoes several changes. Small granular particles within the nucleus begin to concentrate into long strands, forming the visible chromosomes. A small cytoplasmic organelle (a specialized structure within the cellular fluid), called the centriole, divides into two similar structures. Each centriole appears to be surrounded by several ray-like extensions. Together, the centriole and its rays are called the astral body. Each astral body migrates to opposite ends of the cell, and long protein molecules form a spindle (three-dimensional arch) between the two bodies.

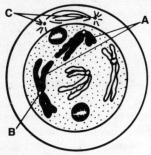

Prophase - Appearance of homo-logous chromosome pairs (A). A precise replication of each chro-mosome is connected to the original by the centromere (B). Division of the centriole to form two migrating astral bodies (C), which later become poles for the protein spindle.

As the chromosome strands enlarge, each completes a complicated replication process to form another identical chromosome strand or chromatid. Gradually, the identical chromosome strands separate partially, remaining attached at one point of constriction called the centromere. The chromosome number is now temporarily doubled (e.g., 128 chromosomes in the horse) and is referred to as the tetraploid chromosome number. Next, the nuclear membrane disappears, and the chromatids enter the cytoplasm, migrating toward the spindle fibers. The centromere of each chromatid pair moves to the spindle's equator.

Metaphase

Metaphase is a relatively short stage involving, simply, the presence of the chromatids at the spindle's equator.

Metaphase - Chromatids are located at the spindle's equator (A).

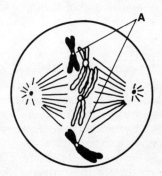

Anaphase

During this period, the sister chromatids should separate (the centromere divides) and move along the spindle fibers to opposite poles (astral bodies) of the cell. Occasionally, two chromatids do not part, and the segregation of chromosomes is unbalanced. This nondisjunction of chromatids results in the formation of daughter cells with missing or duplicated chromosomes.

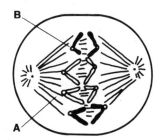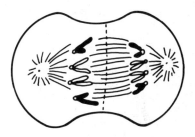

Anaphase - Attached to the spindle fibers (A) by their centromeres (B), the chromatids separate and migrate to opposite ends of the cell.

Telophase

The last stage of mitosis involves the collection of a complete set of chromosomes (one chromatid from each pair of sister chromatids) at opposite ends of the cell and the division of cytoplasm into two identical cells. The nuclear membrane reappears in both cells, separating the chromosomes from the cytoplasm. Telophase completes the cell division and results in the transition of one cell into two genetically identical cells.

Telophase - Separation of the cytoplasm to form cells with identical chromosomes.

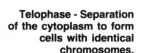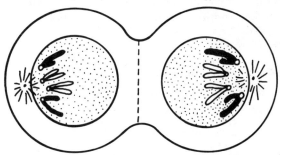

Proliferation of Genetic Material - Part II

The transfer of genetic material from parent to offspring is the basis of inheritance. This process is made possible by the formation of gametes (the ova and sperm), which unite during conception to form a new individual. Each gamete provides the offspring with a sample of genetic material from the respective parents. To insure that the characteristic chromosome number (e.g., 64 chromosomes in every body cell of the horse) remains the same from generation to generation, the sperm and ovum carry only half the normal chromosome number. This reduced chromosome number is referred to as the haploid number (e.g., 32 chromosomes in the horse).

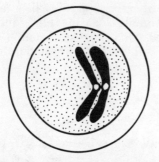 

The diploid chromosome number (left) occurs when each chromosome has one corresponding homologue. The haploid chromosome number (right) is found in the ovum and sperm; each chromosome is without its homologue.

The process by which the ovum and sperm are formed with only half the normal chromosome number involves mitosis and two specialized cell divisions, called meiosis. This chromosome reduction/cell division process can only occur within special cells of the ovary and testes. Because the overall results of meiosis are slightly different in the male and female, gamete formation is divided into two categories: spermatogenesis and oogenesis.

Spermatogenesis

Spermatogenesis occurs within the small twisting seminiferous tubules of the testes. Special cells (spermatogonium) along the inside walls of the tubules divide by mitosis to produce a second layer of similar cells, or

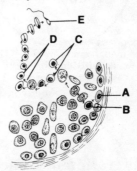

Spermatogenesis. Within the testicular tubules, spermatogonium (A) may divide into two primary spermatocytes (B). Each primary spermatocyte performs the first meiotic division to form two secondary spermatocytes (C). The second meiotic division forms two spermatids (D) from each secondary spermatocyte. These divisions are followed by maturation of the spermatid to form motile sperm (E).

primary spermatocytes. When the primary spermatocytes divide to form secondary spermatocytes, a more complicated process is involved. Designated as the first half of meiotic cell division, the formation of secondary spermatocytes is characterized by the following steps:

1. During prophase, the chromosomes become apparent and begin to duplicate in such a way that each chromosome is attached (via the

centromere) to its sister chromatid. During this period, the chromosomes that form each pair migrate toward each other. Because each chromosome is duplicated and is closely associated with its corresponding chromosome, the resulting structure (tetrad) appears to have four strands.

Prophase - Attraction between homologous chromosomes.

2. As prophase progresses, the astral bodies form and migrate to opposite ends of the cell. The spindle also develops, and the nuclear membrane disappears.
3. During metaphase, the corresponding chromosomes are completely separated. The sister chromatids are still attached by their centromeres which, in turn, are attached to the spindle's equator.

Metaphase - Tetrads are attached to the spindle's equator by their centromeres.

4. Anaphase involves the movement of corresponding chromosomes along the spindle fibers to opposite astral bodies, without separation of the sister chromatids.

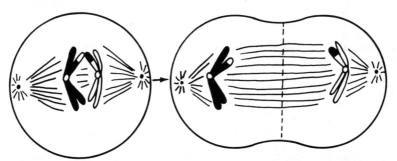

Anaphase - Homologous chromosomes migrate to opposite ends of the cell.

5. Telophase is characterized by the separation of cytoplasm to form two sister cells. Nuclear membranes form within each cell to enclose the respective set of chromosomes. Note, however, that each nucleus contains only half the original chromosome types (one chromosome from each corresponding pair) and that each type is doubled, so that the chromosome number is not altered. (The migration of corresponding chromosomes to opposite daughter cells allows sperm cells from the same individual to carry different chromosome combinations — an important cause of genetic variation.

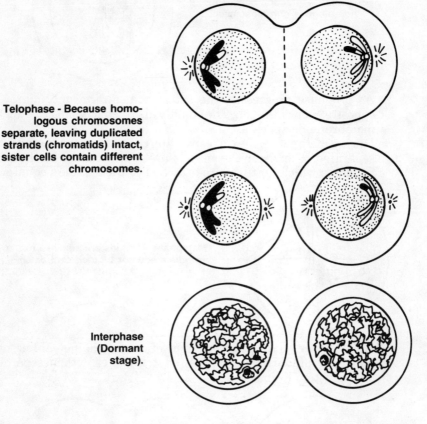

Telophase - Because homologous chromosomes separate, leaving duplicated strands (chromatids) intact, sister cells contain different chromosomes.

Interphase (Dormant stage).

After the first meiotic division is completed, the secondary spermatocytes enter a period of dormancy before the second meiotic division begins. The second division, which results in the formation of spermatids (immature sperm), resembles mitosis.

1. Prophase is characterized by the appearance of astral bodies, the formation of a spindle, and the disappearance of the nuclear membrane. Unlike mitosis, the second meiotic prophase is not characterized by duplication of chromosome strands. The chromosomes are already duplicated and attached by their centromeres.

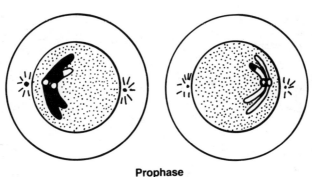

Prophase

2. During metaphase, the chromatid pairs are attached to the spindle equator by their centromeres.

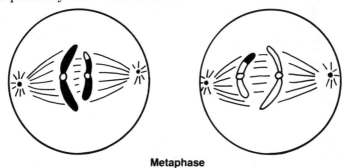

Metaphase

3. As the cell enters anaphase, each centromere splits, and the sister chromatids separate and move along the spindle toward opposite ends of the cell.

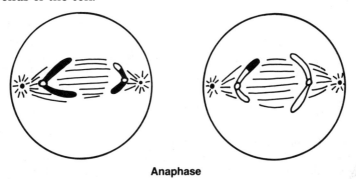

Anaphase

4. Telophase involves the division of cytoplasm and the appearance of nuclear membranes to form, once again, two sister cells. These cells are the spermatids which mature to form motile sperm. Each cell now carries only one chromosome from each chromosome pair, a condition that is necessary for normal recombination of chromosomes during fertilization.

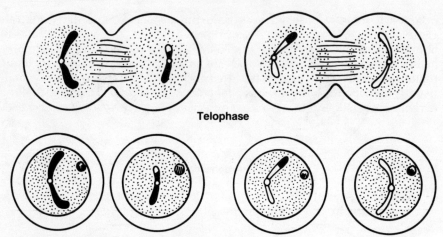

Telophase

The resulting cells carry one half the normal chromosome number (e.g., there are no homologous pairs). Some of the cells will become mature sperm.

Through meiotic division, one primary spermatocyte forms four sperm cells, each of approximately equal size. This rapid propagation of viable sperm (four-fold increase) is the major difference between spermatogenesis and oogenesis.

Oogenesis

The purpose of oogenesis is to produce female gametes, or ova, which contain the haploid number of chromosomes (32 chromosomes) and carry unusually large amounts of cellular fluid, or cytoplasm. This cytoplasm provides nourishment for the early embryo until the transmission of nutrients directly from outside of the embryo begins. To form this specialized structure, oogenesis is somewhat modified but organized into the same basic steps: mitosis, first meiotic division, and second meiotic division.

In the female fetus, special cells within the fetal ovary divide by mitosis, producing primitive egg cells, or primary oocytes. Because this process occurs before birth, the female is born with a specific number of primary oocytes. (Only a small number of these will ever develop into mature ova, however.)

After the female reaches sexual maturity, some of her primary oocytes divide by meiosis (first meiotic division) to form secondary oocytes in response to hormone changes that initiate the mare's estrus period. During this important cell division process, each oocyte is protected and nourished by a surrounding follicle. As the primary oocyte completes the first meiotic division, its follicle bursts and expels the partially developed ovum (secondary oocyte) into the reproductive tract.

The secondary oocyte does not divide again until after fertilization. If a sperm penetrates the egg, its head rests in the cytoplasm while the ovum's second meiotic division forms the ootid (ovum). Then, the nuclei of the ootid and the sperm combine, the chromosomes from each are paired, and embryonic development begins.

As each meiotic division occurs during oogenesis, one daughter cell receives most of the cytoplasm, while the other is left with its share of chromosomes and very little cytoplasm. The small cells are called polar bodies and are usually retained within the larger cell just below its outer membrane. The first polar body (product of the first meiotic division) usually disappears, but the second polar body (product of the second meiotic division) persists and may, in unusual cases, be fertilized by a sperm cell.

BIBLIOGRAPHY

Adams, O.R. **Lameness in Horses.** 2nd ed. Philadelphia: Lea and Febiger, 1972

American Horseman. Vol. 7 No. 3 March 1977
Wicke, B.S. "Taking Care of the Foaling Mare and Newborn Foal," pp. 22-23, 44-49.
Thomas, H. "Foal Diseases," pp. 28-29, 49-50.

American Horseman. Vol. 8 No. 4 April 1978
Hintz, H.F. "Equine Clinic," pp. 8-9.

American Journal of Veterinary Research. Vol. 33 No. 10 1972
Ginther, O.J., H.L. Whitmore and E.L. Squires. "Characteristics of Estrus, Diestrus, and Ovulation in Mares and Effects of Season and Nursing," pp. 1935-1939.

American Journal of Veterinary Research. Vol. 33 No. 12 1972
Steffenhagen, W.P., M.D. Pineda and O.J. Ginther. "Retention of Unfertilized Ova in Uterine Tubes of Mares," pp. 2391-2398.

American Journal of Veterinary Research. Vol. 35 No.9 1974
Ginther, O. J. "Occurrence of Anestrus, Estrus, Diestrus and Ovulation over a 12 Month Period in Mares," pp.1173-1179.

American Journal of Veterinary Research. Vol. 36 No. 8 August 1975
Engle, C.E., C.W. Foley, D.M. Witherspoon, R.D. Scarth and D.D. Goetsch. "Influence of Mare Uterine Tubal Fluids on the Metabolism of Stallion Sperm," pp. 1149-1152.

American Journal of Veterinary Research. Vol. 36 No. 10 October 1975
Kooestra, L.H. and O.J. Ginther. "Effect of Photoperiod on Reproductive Activity and Hair in Mares," pp. 1413-1419.

American Journal of Veterinary Research. Vol. 39 No. 12 December 1978
Hemeida, NA., W.O. Sack and K. McEntee. "Ductuli Efferentse in the Epididymis of the Boar, Goat, Ram, Bull and Stallion," pp. 1892-1900.

American Journal of Veterinary Research. Vol. 41 No. 2 February 1980.
Perryman, L.E. "Graft Versus Host Reactions in Foals with Combined Immunodeficiency," pp. 187-192-.

Appaloosa News. Vol. 36 No. 2 February 1979
Springer, B. "Points to Consider at Breeding Time," pp. 30-33.

Arabian Horse World. Vol. 18 No. 12 December 1978
"Pneumonia in Foals," pp. 123, 126.

Aronson, C.E., ed. **The Complete Desk Reference of Veterinary Pharmaceuticals and Biologicals 78/79.**
Medio, Pennsylvania: Harwal Publishing Company, 1978.

Arthur, G.H. **Veterinary Reproduction and Obstetrics.** New York: Macmillan Publishing Co., 1975.

Auburn Veterinarian. Vol. 16 No. 3 1968
Vaughan, J.T. "An Introduction to Equine Restraint."

Audio Veterinary Medicine-Equine Practice. Vol. 1
Purvis, A.D. and J.C. Bryant. "Clinical Induction of Labor in the Mare as a Routine Breeding Farm Procedure," No. 4, Jan. 1978

Audio Veterinary Medicine - Equine Practice. Vol. 2
Johnson, J.H. "Abnormal Tendon Development in Growing Foals," No. 1, Oct. 1978.
Knight, H.D. "Neonatal Septicemias," No. 7, April 1979
Rumbaugh, G.E. "Passive Immunity in the Foal," No. 8, May 1979.
Perryman, L.E. "Arabian Foal Diseases," No. 10, July 1979.
Perryman, L.E. "Immunologic Evaluation and Management of Foals," No. 12, Sept 1979.

Audio Veterinary Medicine - Equine Practice. Vol. 3
Lewis, L.D. "Nutrition for the Broodmare and Foal and Its Role in Epiphysitis," No. 4, Jan. 1980.

Audio Veterinary Medicine - Equine Practice. Vol. 4
Bello, T.R. "Equine Anthelmintic Therapy," No. 2, 1981

Australian Veterinary Journal. Vol. 39 Nov. 1963
Bain, A.M. "Common Bacterial Infections of Foetuses and Foals and Association of the Infection with the Dam," pp. 414-422.

Australian Veterinary Journal. Vol. 42 No. 12 1966
Thain, R.I. "Cystic Endometrium in Mares," p. 484.

Australian Veterinary Journal. Vol. 55 January 1979
Maxwell, J.A.L. "The Correction of Uterine Torsion in a Mare by Cesarean Section," pp. 33-34.

Backstretch. Vol. 15 No. 2 1976
Ensminger, M.E. "Foaling and Care of the Newborn Foal," pp. 66-72.

Backstretch. Vol. 16 No. 1 Jan/Feb/March 1977
Swanstrom, O.G. "All About 'Bute'" pp. 35-37, 40, 42, 44.

Backstretch. Vol. 16 No. 4 Oct/Nov/Dec 1977
"Stallion Promotion," pp. 36, 38, 40-42.

Backstretch. Vol. 17 No. 1 Jan/Feb/March 1978
Mayo, J.A. "Foaling the Mare," pp. 40, 42, 44, 46.

Backstretch. Vol. 14 No. 2 April/May/June 1978
"Risks Associated with Foaling," pp. 45-50.

Backstretch. July 1978
Swanstrom, O.G. "What's in a Bloodcount?," pp. 24-27.

Pfizer, Inc., Department of Veterinary Medicine. July, 1979. **Banminth, Strongid T.** New York.

Biochemic Review. Vol. 32 No. 4
Knappenberger, R.E. "Equine Pediatrics - Health and Development of the Foal."

Biochemic Review. Vol. 33 No. 3
   "Indwelling Uterine Infuser," pp. 8-10.
Blood, D.C. and J.A. Henderson. **Veterinary Medicine.** 4th ed. Baltimore: William and Wilkins Co., 1974.
Blood-Horse. March 8, 1971
   McGee, W.R. "Care of Newborn Foals," pp. 813-825.
Blood-Horse. May 1, 1972
   Farley, D. and R.H. Hagin. "Care of Foals," pp. 1486-1487.
Blood-Horse. May 21, 1973
   McGee, W.R. "The Foaling Mare," pp 1742-1748.
Blood-Horse. April 21, 1975
   Loy, R. "Problem Mares," pp. 1548-1549.
Blood-Horse. December 15, 1975
   Toby, M.C. "Palpation as an Assist to Mare Owners," pp. 5541.
   Kaufman, W.C. "Effect of Palpation on Reproduction in Mares," pp. 5538-5541.
Blood-Horse. February 14, 1977
   AAEP Paper: "Diagnosis of Equine Abortion Cause," p. 3498.
   Mearns, D. "Live Foal Percentages," pp. 732-733.
Blood-Horse. July 11, 1977
   Toby, M.C. "Veterinary Topics," p. 3014.
Blood-Horse. August 15, 1977
   Pons, J.P. "Weaning Time," pp. 3634-3636.
Blood-Horse. October 31, 1977
   Toby, M.C. "Bowed Tendons - Treatments Old and New," pp. 5082-5085.
Blood-Horse. February 3, 1978
   Pons, J.P. "Problems of the New Foal," pp. 764-765.
Blood-Horse. February 20, 1978
   Hughes, J.P. "Assessment of Breeding Potential in the Mare," pp. 834-838.
Blood-Horse. April 3, 1978
   Toby, M.C. "Artificial Insemination Mandate," pp. 1540-1543.
Blood-Horse. April 10, 1978
   Toby, M.C. "CEM After a Month," pp. 1622-1625.
Blood-Horse. April 17, 1978
   Toby, M.C. "Getting Into High Gear," pp. 1707-1709.
Blood-Horse. May 1, 1978
   Toby, M.C. "CEM Blood Test," pp. 1987-1989.
   "Breeding on Foal Heat," pp. 2017-2018.
Blood-Horse. August 8, 1978
   Franco, O.J. "Diagnosis of Equine Abortion Cause," pp. 3498-3499.
Blood-Horse. August 15, 1978
   Vandeplassche, M., R. Bouters, J. Spincemaille and P. Bonte. 'Caesarean Section in the Mare," pp. 3694-3695.
Blood-Horse. September 28, 1978
   Toby, M.C. "CEM: Guidelines for Interstate Shipment," pp. 4266-4267.
Blood-Horse. November 6, 1978
   Sturgill, D. "New Considerations for Old Stallion Syndicate Agreements," pp. 5250-5255.
Blood-Horse. November 20, 1978
   Toby, M.C. "National Breeder's Conference," pp. 5685-5687.
Blood-Horse. February 19, 1979
   Pons, J.P. "Nurse Mares: Not Only for Orphan Foals," pp. 908-909.
Blood-Horse. April 16, 1979
   Toby, M.C. "CEM in England," pp. 1846-1848.
Blood-Horse. February 9, 1980
   Sager, F.S. "Post-Partum Problems of the Foal," pp. 738-741.
Blood-Horse. February 23, 1980
   Toby, M.C. "The Breeding Shed," pp. 1022-1025.
Blood-Horse. November 1, 1980
   Toby, M.C. "Preventing Fires in the Horse Barn," pp. 6368-6371.
Bogart, R. **Improvement of Livestock.** New York: The MacMillian Co., 1959.
Breazile, J.E. **Textbook of Veterinary Physiology.** Philadelphia: Lea and Febiger, 1971.
British Veterinary Journal. Vol. 105 1949
   Haq, I. "Causes of Sterility in Bulls in Southern England," pp. 3-6, 71, 114, 143, 200.
British Veterinary Journal. Vol. 120 No. 9 1964
   "Equine Infertility," pp. 395-441.
British Veterinary Journal. Vol. 123 No. 11 November 1967
   Rossdale, P.D. "Clinical Studies on the Newborn Thoroughbred Foal," pp. 470-480.
British Veterinary Journal. Vol. 126 No. 10 October 1970
   Gluhovschi, N., M. Bistriceanu, A. Sucia and M. Bratu. "A Case of Intersexuality in the Horse with Type 2 A XXXY Chromosome Formula," pp. 522-527.
Britt, S.H. ed. **Marketing Manager's Handbook.** Chicago: The Dartnell Corporation, 1973.
Brumbaugh, J.E. **Heating, Ventilation, and Air Conditioning Library.** Indianapolis, Indiana: Howard W. Sams and Co., Inc. 1978
Bryans, J.T. and H. Gerber, eds. **Equine Infectious Diseases IV.** Princeton, New Jersey: Veterinary Publications, Inc., 1978.
Burns, S.J. and W.C. McMullan. **Junior Clinics: Equine Section. Clinic Handbook.** College Station, Texas: College of Veterinary Medicine, Texas A & M University.

California Horse Review. Vol. 15 No. 3 March 1978
"Fly Control on the Horse Ranch," pp. 172-180.
Medina, L.E. "Epiphysitis: A Serious Problem in the Young Horse," pp. 88-89.
California Horse Review. Vol. 15 No. 7 July 1978
Davis, B. "Whenever Horses Congregate Rhinopneumonitis Threatens," p. 44.
California Horse Review. Vol. 17 No. 5 May 1980
Bloom, L. "Teasing, Breeding and Artificial Insemination," pp. 64-65.
California Horseman's News. Vol. 7 No. 6 March 19, 1979
"CID is Mysterious Foal Killer," pp. 11-12.
California Horseman's News. Vol. 7 No. 7 April 16, 1979
Wood, K.A. "Law and the Horse Owner: Stallion Lease Should Benefit Both Parties," pp. 15, 36, 39.
California Horseman's News. Vol. 7 No. 10 June 11, 1979
"Twinning is Main Cause of Mare Abortion," p. 40.
Campbell, J.R. and J.F. Lasley. **The Science of Animals That Serve Mankind.** New York: McGraw-Hill Publications, 1969.
Canadian Journal of Comparative Medicine. Vol. 33 October 1969
Basrur, P.K., H. Knangawa, and J.P.W. Gilman. "An Equine Intersex with Unilateral Gonadal Agenesis," pp. 297-306.
Canadian Quarter Horse Journal. Vol. 5 No. 4 May 1978
Jones, W.E. "A Home Pregnancy Test for Pregnancy in Mares," pp. 12, 25.
Canadian Rider. Vol. 8 No. 9 March 1978
Burwash, L. "Stallion Syndication," p. 26.
Canadian Rider. Vol. 8 No. 10 April 1978
Burwash, L. "Handling the Broodmare," pp. 14, 25.
Catcott, E.J. and J.F. Smithcors. **Equine Medicine and Surgery.** 2nd ed. Wheaton, Illinois: American Veterinary Publications, Inc., 1972.
Clarke, E.G.C. and M.L. Clarke. **Veterinary Toxicology.** New York: Macmillan Publishing Co., Inc., 1975.
Cornell Veterinarian. Vol. 35 No. 4 October 1945
Britton, J.W. "An Equine Hermaphrodite," pp. 373-375.
Cornell Veterinarian. Vol. 54 No. 3 1964
"Uterine Dilation as a Cause of Sterility," pp. 423-438.
Cornell Veterinarian. Vol. 65 No. 3 July 1975
Whitlock, R.H., R.W. Dellers, and J.N. Shively. "Adenoviral Pneumonia in a Foal," pp. 393-401.
Davis, Thomas A. **Horse Owners and Breeders Tax Manual.** Washington, D.C.: The American Horse Council, 1979.
Davis, T.A. **American Horse Council Tax Reference Bulletins.**
"Tax Reform 1978," No. 96, July 1978.
"Hobby vs. Business," No. 97, August 1978.
"Revenue Act of 1978 - Tax Credit for Breeding Horses," No. 98, September 1978.
"New Capital Gains Rates; Alternative Mimimum Tax Computation," No. 100, November 1978.
"IRC Sec 3121, 3306, 3401...Federal Employment Taxes, Independent Contractor, Rev. Procedures," No. 101, December 1978.
"Tax Credit, Single Purpose Structure," No. 102, January 1979.
"Capital Gain Deduction - Effective Date," No. 103, February 1979.
"I.R.S. Procedures for Tax Disputes," No. 104, March 1979.
"Tax Legislation...Repel of Carryover," No. 104, March 1979.
"Partnership Returns, Penalty for Failure to File," No. 105, April 1979.
"Two-Out-Seven Presumption Period, Termination on Death," No. 108, July 1979.
"Business vs. Hobby, American Saddlebred Farm," No. 109, August 1979.
Days, J.M. **The Horse.** New York: Arco Publishing Co., 1977.
Dilts, R.V. **Analytical Chemistry Methods of Separation.** New York: D. Van Nostrand Co., 1974
Donahue, R. L., J. C. Schickluna and L. S. Robertson. **Soils: An Introduction to Soils and Plant Growth.** 3rd Ed. Prentice-Hall, Inc: Englewood Cliffs New Jersey, 1971.
Drudge, J.H. **Internal Parasites of Horses.** Somerville, New Jersey: American Hoechst Corporation, Animal Health Division, 1978.
Dukes, H.H. **Duke's Physiology of Domestic Animals.** 8th ed. New York: Melvin J. Swenson, ed. Cornell University Press, 1970.
Dyke, B. and B. Jones. **Horse Business: An Investor's Guide.** Colorado City, Colorado: Caballus Publishers, 1975.
Edwards, E.H. and C. Geddes. **The Complete Book of the Horse.** New York: Arco Publishing Co., 1973.
Emery, A.E.H. **Elements of Medical Genetics.** 4th ed. Berkeley: University of California Press, 1976.
Emery, L., J. Miller and N. Van Hoosen. **Horseshoeing Theory and Hoof Care.** Philadelphia: Lea and Febiger, 1977.
Ensminger, M.E. **Horses and Horsemanship.** 4th ed. Danville, Illinois: The Interstate Printers and Publishers, 1969.
Ensminger, M.E. **Horses and Tack.** Boston: Houghton Mifflin Co., 1977.
Ensminger, M.E. and C.G. Olentine. **Feeds and Nutrition.** Clovis, California: Ensminger Publishing Co., 1978.
Equine Medicine and Surgery. Vol. 3 No. 3 March 1979
Carson, R.L. and F.N. Thompson. "Effects of an Anabolic Steroid on the Reproductive Tract in the Young Stallion," pp. 221-224.
Hawkins, D. "A Breeding Platform for Stallion-Mare Height Differences," pp. 115-117.
Equine Practice. Vol. 1 No. 1 Jan/Feb 1979
Houpt, K.A. and T.R. Wolski. "Equine Maternal Behavior and Its Aberrations," pp. 7-20.
Equine Practice. Vol. 1 No. 2 1979
Adams, S.B. "Rupture of the Prepubic Tendon in a Horse," pp. 17-19.
Equine Practice. Vol. 1 No. 4 1979
Rumbaugh, G.E. and A.A. Ardans. "Field Determination of the Immune Status of the Newborn Foal," pp. 37-42.
Equine Practitioner. Vol. 60 No. 5 May 1979
Rooney, J.R. "Rupture of the Aorta," pp. 391-392.

Equine Veterinary Journal. Vol. 97 1965
   von Lepel, and J. Frhr. "Maintenance of Fertility in the Horse Including Artificial Insemination," p. 97.
Equine Veterinary Journal. Vol. 1 No. 65 1968
   Bowen, J.M. "Artificial Insemination in the Horse," pp. 98-110.
   Rossdale, P.D. "Modern Stud Management in Relation to the Oestrus Cycle and Fertility of Thoroughbred Mares," pp. 65-72.
Equine Veterinary Journal. Vol. 6 No. 1 January 1974
   Allen, W.E. "The Palpability of the Corpus Luteum in Welch Pony Mares," pp. 25-27.
Equine Veterinary Journal. Vol. 6 No. 2 April 1974
   Allen, W.E. "Palpable Development of the Conceptus and Fetus in Welsh Pony Mares," pp. 69-73.
Equine Veterinary Journal. Vol. 6 No. 3 July 1974
   Edwards, G.B., W.E. Allery, and J.R. Newcombe. "Elective Caesarean Section in the Mare for Production of Gnotobiotic Foals," pp. 122-126.
   Jeffcott, L.B. "Some Practical Aspects of the Transfer of Passive Immunity to Newborn Foals," pp. 109-115.
Equine Veterinary Journal. Vol. 6 No. 4 October 1974
   Stabenfeldt, G.H., J.P. Hughes, J.W. Evans and D.P. McNeely. "Spontaneous Prolongation of Luteal Activity in the Mare," pp. 158-163.
   Collery, L. "Observations of Equine Arrivals Under Farm and Feral Conditions," pp. 170-173.
Equine Veterinary Journal. Vol. 7 No. 2 April 1975
   Hughes, J.P., P.C. Kennedy and K. Benirschke. "XO Gonadal Dysgenesis in the Mare: Report of Two Cases," pp. 109-112.
   Ricketts, S.W. "The Technique and Clinical Application of Endometrial Biopsy in the Mare," pp. 102-108.
Equine Veterinary Journal. Vol. 8 No. 1 January 1976
   Kieffer, N.M., S.J. Burns and N.G. Judge. "Male Pseudohermaphroditism of the Testicular Feminizing Type," pp. 38-41.
Equine Veterinary Journal. Vol. 8 No. 3 April 1976
   Fretz, P.B. and W.C.D. Hare. "A Male Pseudohermaphrodite Horse with 63XO?/64XX/65XXY Mixoploidy," pp. 130-132.
Equine Veterinary Journal. Vol. 8 No. 4 1976
   McIlwraith, C.W., R.A.R. Owen and P.K. Basrur. "An Equine Cryptorchid with Testicular and Ovarian Tissues," pp. 156-160.
Equine Veterinary Journal. Vol. 9 No. 1 January 1977
   Dawson, R.L.M. "Recent Advances in Equine Reproduction," pp. 4-11.
Equine Veterinary Journal. Vol. 10 No. 2 April 1978
   Mayhew, I.G., A.G. Watson and J.A. Heissan. "Congenital Occipitoatlantoaxial Malformations in the Horse," pp. 103-113.
Equine Veterinary Journal. Vol. 11 No. 1 January 1979
   Pascoe, R.R. "A Possible New Treatment for Twin Pregnancy in the Mare," pp. 64-65.
   Smyth, G.B. "Testicular Teratoma in an Equine Cryptorchid," pp. 21-23.
Equine Veterinary Research. Vol. 46 1979
   Jansen, B.C. and P.C. Knoetze. "The Immune Response of Horses to Tetanus Toxoid," pp. 211-216.
Evans, J.W., A. Borton, H. F. Hintz and L.D. Van Vleck. **The Horse.** San Francisco: W.H. Freeman and Company, 1977.
Feirer, J.L. and G.R. Hutchins. **Carpentry and Building Construction.** Peoria, Illinois: Charles A. Bennett Company, Inc., 1976.
Fertility and Sterility. Vol. 15 No. 6 1964
   Blandau, R.J. and R.E. Rumery. "The Relationship of Swimming Movements of Epididymal Spermatozoa to Their Fertilizing Capacity," pp. 516-522.
Fort Dodge Laboratories. **Tridex and Index Pellets.** Fort Dodge, Iowa, 1979.
Fortune. Vol. 99 No. 8 April 23, 1979
   Rowan, R. "Those Business Hunches Are More Than Blind Faith," pp. 110-114.
Frandson, R. D. **Anatomy and Physiology of Farm Animals.** Philadelphia: Lea and Febiger, 1972.
Geary, D. **Roofs and Siding: A Practical Guide.** Reston, Virginia: Reston Publishing Co., Inc. 1978
Georgi, J.R. and V.J. Theodorides. **Parasitology for Veterinarians.** 3rd ed. Philadelphia: W.B. Saunders Co., 1980
Gibbons, W.J. **Clinical Diagnosis of Diseases in Large Animals.** Philadelphia: Lea and Febiger, 1966.
Ginther, O.J. **Reproductive Biology of the Mare: Basic Applied Aspects.** Ann Arbor, Michigan: McNaughton and Gunn Inc. 1979
Goody, P.C. **Horse Anatomy - A Pictorial Approach to Equine Structure.** London: J.A. Allen, 1976.
Hackney Journal. Vol. 4 Part 5 Dec/Jan 1979
   Jones, W.E. "Pinpointing Breeding Problems," pp. 17-18.
Hafez, E.S.E. and I.A. Dyer. **Animal Growth and Nutrition.** Philadelphia: Lea and Febiger, 1969.
Harris, S.E. **Grooming to Win.** New York: Charles Scribner's Sons, 1977.
Harrison, J.D. **Care and Training of the Trotter and Pacer.** Columbus, Ohio: The United States Trotting Association, 1968.
Hayes, M.H. **Points of the Horse.** New York: Arco Publishing Co., 1969
Hayes, M.H. **Stable Management and Exercise.** New York: Arco Publishing Co., 1970
Herman, H.A. and E.W. Swanson. "Variations in Dairy Bull Semen with Respect to Its Use in Artificial Insemination," Missouri Agricultural Experiment Station, Bulletin No. 326.
Holland, J. **Decision Making.** June 24, 1979
Hoofbeats. Vol. 46 No. 12 February 1979
   "Best Buy in Racing," p. 24.
Hoofbeats. Vol. 48 No. 3 May 1980
   Kemen, M.J. "Equine Rhinopneumonitis," pp. 104-105.
Horse and Rider Vol. 19 No. 9 September 1979
   Jones, W.E. "Debunking Deworming," pp. 16-22.

Hub Rail. Vol. 7 No. 1 Spring 1979
  "Syndication: A Game of Pool?," pp. 65-76.
International Trotter & Pacer. Vol. 13 No. 2 February/March 1978
  "Discussion on Metritis," p. 8.
  Greengard, K. "Pros and Cons of Early Breeding," pp. 6-7.
Irish Veterinary Journal. Vol. 30 1976
  "Off Season Treatment of Endometritis in Thoroughbred Mares with Ampicillin," pp. 100-102.
Journal of Animal Science. Vol. 31 No. 3 1970
  Johnston, R.H., L.D. Kamstra, and P.H. Kohler. "Mares' Milk Composition as Related to 'Foal Heat' Scours," pp. 549-553.
Journal of Animal Science. Vol. 31 No. 4 October 1970
  Pickett, B.W., L.C. Faulkner and T.M. Sutherland. "Effect of Month and Stallion on Seminal Characteristics and Sexual Behavior," pp. 713-727.
Journal of Animal Science. Vol. 41 No. 3 1975
  Voss, J.L., B.W. Pickett, D.G. Back and L.D. Burwash. "Effect of Rectal Palpation on Pregnancy Rate of Nonlactating Normally Cycling Mares," pp. 829-834.
Journal of Animal Science. Vol. 41 No. 5 1975
  Sharp, D.C. and O.J. Ginther. "Stimulation of Follicular Activity and Estrous Behavior in Anestrous Mares with Light and Temperature," pp. 1369-1372.
Journal of the AVMA. Vol. 119 September 1951
  Stocking, G.G. "Observations Concerning Conception in the Mare," pp. 190-192.
Journal of the AVMA. Vol. 151 No. 12 1967
  Morrow, G.L. "Uterine Curettage in the Mare," pp. 1615-1617.
Journal of the AVMA. Vol. 153 No. 12 1968
  Laufenstein-Duffy, H. "Uterine Curettage," pp. 1570-1573.
  Sager, F.C. "Management of Uterine Disease," pp. 1567-1569.
Journal of the AVMA. Vol. 153 No. 24 December 15, 1968
  McFeely, R.A. "Chromosomes and Infertility," pp. 1672-1675.
Journal of the AVMA. Vol. 156 No. 3 February 1, 1970
  Solomon, W.J., R.A. McFeely and F.B. Peterson. "Effects of Uterine Curettage in the Mare," pp. 333-338.
Journal of the AVMA. Vol. 157 No. 11 June 1, 1970
  Witherspoon, D.M. and R.B. Talbot. "Ovulation Site in the Mare," p. 1452.
  Lieux, P., R.H. Baker, A. Degroote, H.H. Laskey, R.E. Raynor, J.G. Simpson and E. Tobler. "Results of a Survey on Bacteriologic Culturing of Broodmares," pp. 1460-1464.
  Prickett, M.E. "Abortion and Placental Lesions in the Mare," pp. 1465-1470.
Journal of the AVMA. Vol. 160 No. 6 March 15, 1972
  Wheat, J.D. and D.M. Meagher. "Uterine Torsion and Rupture in Mares," pp. 881-884.
Journal of the AVMA. Vol. 161 No. 11 December 1, 1972
  Hughes, J.P. G.H. Stabenfeldt and J.W. Evans. "Estrous Cycle and Ovulation in the Mare," pp. 1367-1374.
  Witherspoon, D.M., R.T. Goldston and M.E. Adsit. "Uterine Culture and Biopsy in the Mare," pp. 1365-1366.
Journal of the AVMA. Vol. 165 1974
  Back, et. al. "Teasing With Two Stallions," p. 717.
Journal of the AVMA. Vol. 170 No. 2 January 15, 1977
  Davis, L.E. and A.B. Abbitt. "Clinical Pharmacology of Antibacterial Drugs in the Uterus of the Mare," pp. 204-206.
Journal of the AVMA. Vol. 172 No. 3 February 1, 1978
  Kenney, R.M. "Cyclic and Pathologic Changes of the Mare Endometrium as Detected by Biopsy, with a Note on Early Embryonic Death," pp. 241-262.
  Stickle, R.L. and J.F. Fessler. "Retrospective Study of 350 Cases of Equine Cryptorchidism," pp. 343-346.
Journal of the AVMA. Vol. 173 No. 7 October 1, 1978
  Eugster, A.K., H.W. Whitford, and L.E. Mehr. "Concurrent Rotavirus and Salmonella Infections in Foals," pp. 857-858.
Journal of the AVMA. Vol. 173 1978
  Brown, M.P., P.T. Colahan and D.L. Hawkins. "Urethral Extension for Treatment of Urine Pooling in Mares," pp. 1005-1007.
Journal of the AVMA. Vol. 174 No. 3 1979
  Rumbaugh, G.E., A.A. Ardans, D. Ginno, and A. Trommershausen-Smith. "Identification and Treatment of Colostrum-Deficient Foals," pp. 273-276.
Journal of the AVMA. Vol. 174 No. 10 May 15, 1979
  Valdez, H., T.S. Taylor, S.A. McLaughlin and M.T. Martin. "Abdominal Cryptorchidectomy in the Horse, Using Inguinal Extension of the Gubernaculum Testis," pp. 1110-1112.
Journal of the AVMA. Vol. 176 No. 11
  Liu, I.K.U. "Management and Treatment of Selected Conditions in Newborn Foals," pp. 1247-1249.
Journal of Equine Medicine and Surgery. Vol. 1 No. 4 April 1977
  Huston, R., G. Saperstein and H.W. Leipold. "Congenital Defects in Foals," pp. 146-161.
  "A Case Report of Inguinal Hernionhaphy in a Stallion," pp. 391-394.
Journal of Equine Medicine and Surgery. Vol. 2 No. 1 1978
  Dubielzig, R.R. "Streptococcal Septicemia in the Neonatal Foal," pp. 28-30.
Journal of Equine Medicine and Surgery. Vol. 2 No. 10 1978
  Meuten, D.J. and V. Rendano. "Hypertrophic Osteopathy in a Mare with a Dysgerminoma," pp. 445-450.
Journal of Equine Medicine and Surgery. Vol. 3 No. 2 February 1979
  Carter, M.E., H.G. Dewes, and O.V. Griffiths. "Salmonellosis in Foals," pp. 78-83.
Journal of Equine Medicine and Surgery. Vol. 13 No. 4 April 1979
  Davis, L.E. "Drug Therapy in the Neonatal Foal," p. 175.
  Hamm, D.H. and E.W. Jones. "Gentamycin in the Treatment of Foal Diarrhea," pp. 159-162.
  Knight, H.D. "Infectious Diseases of the Newborn Foal," p. 175.

Journal of Equine Medicine and Surgery. Vol. 3 No. 5 May 1979
   Cahill, C. "Contagious Equine Metritis in the Stallion," p. 229.
   Carson, R.L. and F.N. Thompson. "Effects of an Anabolic Steriod on the Reproductive Tract in the Young Stallion," pp. 221-224.
   Conboy, S.H. "The Diagnosis and Treatment of Endometritis," pp. 229-236.
   Terry, J.M. "Artificial Insemination, Its Use as a Managment Tool," p. 236.
   Vaillancourt, D., P. Fretz and J.P. Orr. "Seminoma in the Horse: Report of Two Cases," pp. 213-218.
Journal of Equine Medicine and Surgery Supplement. Equine Infectious Diseases IV: Proceedings of the Fourth International Conference on Equine Infectious Disease. No. 1, 1978
Journal of Heredity. Vol. 31 No. 2 December 1940
   Cole, L.J., E. Waletzky and M. Shackelford. "A Test of Sex Control by Modification of the Acid-Alkaline Balance," pp. 501-502.
Journal of Reproduction and Fertility. Supplement Vol. 23 1975
   Allen, W.R. "The Influence of Fetal Genotype Upon Endometrial Cup Development and PMSG and Progestagen Production in Equids," pp. 405-413.
   Arthur, G.H. "Influence of Intrauterine Saline Infusion Upon the Oestrous Cycle of the Mare," pp. 231-234.
   Bain, A.M. and W.P. Howey. "Observations of the Time of Foaling in Thoroughbred Mares in Australia," pp. 545-546.
   Bain, A.M. and W.P. Howey. "Ovulation and Transuterine Migration of the Conceptus in Thoroughbred Mares," pp. 541-543.
   Belonje, P.C. and C.H. van Niekerk. "A Review of the Influence of Nutrition Upon Oestrus Cycle and Early Pregnancy in the Mare," pp. 167-169.
   Bouters, R., M. Vanderplassche and A. Demoor. "An Intersex (Male Pseudohermaphrodite) Horse with 64XX/XXY Mosaicism," pp. 375-376.
   Chandley, A.C., J. Fletcher, P.D. Rossdale, C.K. Peace, S.W. Ricketts, R.J. McEnery, J.P. Thorne, R.V. Short and W.R. Allen. "Chromosome Abnormalities as a Cause of Infertility in Mares," pp. 177-383.
   Douglass, R.H. and O.J. Ginther. "Effects of Prostaglandin F2 on the Oestrous Cycle and Pregnancy in Mares," pp. 257-261.
   Evans, J.W., D.A. Faria, J.P. Hughes, G.H. Stabenfeldt, and P.T. Cupps. "Relationship Between Luteal Function and Metabolic Clearance and Production Rates of Progesterone in the Mare," pp. 177-182.
   Evans, M.J., and C.H.G. Irvine. "Serum Concentrations of FSH, LH, and Progesterone During the Oestrous Cycle and Early Pregnancy in the Mare," pp. 193-200.
   Geschwind, I.I., R. Dewey, J.P. Hughes, J.W. Evans, and G.H. Stabenfeldt. "Plasma LH Levels in the Mare During the Oestrus Cycle," pp. 207-212.
   Heinze, H. and E. Klug. "The Use of GnRH for Controlling the Oestrous Cycle of the Mare (preliminary report)," pp. 275-277.
   Hillman, R.B., "Induction of Parturition in Mares," pp 641-644.
   Hillman, R.B., and R.G. Loy. "Oestrogen Excretion in Mares in Relation to Various Reproductive Stages," pp. 223-230.
   Hughes, J.P., G.H. Stabenfeldt, and J.W. Evans. "The Oestrus Cycle of the Mare," pp. 161-166.
   Hughes, J.P., K. Benirschke, P.C. Kennedy and A. Trommershausen-Smith. "Gonadal Cysgenesis in the Mare," pp. 385-390.
   Irvine, D.S., B.R. Downey, W.G. Parker, and J.J. Sullivan. "Duration of the Oestrus and Time of Ovulation in Mares Treated with Synthetic GnRH (AY-24,031)," pp. 279-283.
   Jeffcott, L.B. "The Transfer of Passive Immunity to the Foal and Its Relation to Immune Status After Birth," pp. 727-733.
   Laing, J.A. and F.B. Leach. "The Frequency of Infertility in Thoroughbred Mares," pp. 307-310.
   Mitchell, D. and W.R. Allen. "Observations of Reproductive Performance in the Yearling Mare," pp. 531-536.
   Moberg, R. "The Occurrence of Early Embryonic Death in the Mares in Relation to Natural Service and Artificial Insemination with Fresh or Deep-Frozen Semen," pp. 537-539.
   Moor, R.M., W.R. Allen, and D.W. Hamilton. "Origin and Histogenisis of Equine Endometrial Cups," pp. 391-396.
   Neely, D.P., J.P. Hughes, G.H. Stabenfeldt, and J.W. Evans. "The Influence of Intrauterine Saline Infusion on Luteal Function and Cyclical Activity in the Mare," pp. 235-239.
   Noden, P.A., W.D. Oxender, and H.D. Hafs. "The Cycle of Oestrus, Ovulation and Plasma Levels of Hormones in the Mare," pp. 189-192.
   Palmer, E. and B. Jousset. "Synchronization of Oestrus in Mares with a Prostaglandin Analogue and HCG," pp. 269-274.
   Palmer, E. and B. Jousset. "Urinary Estrogen and Plasma Progesterone Levels in Non-pregnant Mares," pp. 213-221.
   Pickett, B.W. and J.L. Voss. "Abnormalities of Mating Behavior in Domestic Stallions," pp. 129-134.
   Sharp, D.C., L. Kooistra, and O.J. Ginther. "Effects of Artificial Light on the Oestrous Cycle of the Mare," pp. 241-246.
   Squires, E.L. and O.J. Ginther. "Follicular and Luteal Development in Pregnant Mares," pp. 429-433.
   Stabenfeldt, G.H., J.P. Hughes, J.W. Evans, and I.I. Geschwind. "Unique Aspects of the Reproductive Cycle of the Mare," pp. 155-160.
   Steven, D.H. and C.A. Samuel. "Anatomy of the Placental Barrier in the Mare," pp. 579-582.
   van Niekerk, C.H. and W.R. Allen. "Early Embryonic Development in the Horse," pp. 495-498.
   van Niekerk, C.H., J.C. Morgenthal, and W.H. Gerneke. "Relationship betweeen the Morphology of and Progesterone Production by the Corpus Luteum of the Mare," pp. 171-175.
   Varadin, M. "Endometritis, A Common Cause of Infertility in Mares," pp. 353-356.
   Voss, J.L. and B.W. Pickett. "Diagnosis and Treatment of Haemospermia in the Stallion," pp. 151-154.
Journal of Reproduction and Fertility. Supplement No. 27 1979
   Betterige, K.J., M.D. Eaglesome, and P.F. Flood. "Embryo Transport Through the Mare's Oviduct Depends Upon Cleavage and Is Independent of the Ipsilateral Corpus Luteum," pp. 387-394.
   Burns, S.J., C.H.G. Irvine, and M.S. Amoss. "Fertility of Prostaglandin-Induced Oestrus Compared to Normal Post-Partum Oestrus," pp. 245-250.
   Evans, M.J. and C.H.G. Irvine. "Induction of Follicular Development and Ovulation in Seasonally Acyclic Mares Using Gonadotrophin-Releasing Hormones and Progesterone," pp. 113-121.
   Evans, J.W., et. al. "Episodic LH Secretion Patterns in the Mare During the Oestrus Cycle," pp. 143-150.

Freedman, L.J., J.C. Garcia, and O.J. Ginther. "Influence of Ovaries and Photoperiod on Reproductive Function in the Mare," pp. 79-86.

Garcia, M.C., L.J. Freedman, and O.J. Ginther. "Interaction of Seasonal and Ovulatory Factors in the Regulation of LH and FSH Secretion in the Mare," pp. 103-111.

Hurtgen, J.P. and V.K. Ganjam. "The Effect of Intrauterine and Cervical Manipulation on the Equine Oestrus Cycle and Hormone Profiles," pp. 191-197.

Hyland, J.H. and F. Bristol. "Synchronization of Oestrus and Timed Insemination of Mares," pp. 251-255.

Kenny, R.M., W. Condon, V.K. Ganjan, and C. Channing. "Morphological and Biochemical Correlates of Equine Ovarian Follicles as a Function of Their State of Viability or Atresia," pp. 163-171.

Loy, R.G., J.R. Buell, W. Stevenson, and D. Hamm. "Sources of Variation in Response Intervals After Prostaglandin Treatment in Mares with Functional Corpora Lutea," pp. 229-235.

Munro, C.D., J.P. Renton, and R. Butcher. "The Control of Oestrus Behavior in the Mare," pp. 217-227.

Neely, D.P., H. Kindahl, G.H. Stabenfeldt, L.E. Edqvist, and J.P. Hughes. "Prostaglandin Release Patterns in the Mare: Physiological, Pathophysiological, and Theraputic Responses," pp.181-189.

Roser, J.F., B.L. Kiefer, J.W. Evans, D.P. Neely, and C.A. Pacheco. "The Development of Antibodies to Human Chorionic Gonadaotrophin Following Its Repeated Injection in the Cyclic Mare," pp. 173-179.

Snyder, D.A., et. al. "Follicular and Gonadotrophic Changes During Transition from Ovulatory to Anovulatory Season," pp. 95101.

Strauss, S.S., C.L. Chen, S.P. Kalva, and D.C. Sharp. "Localization of Gonadotrophinreleasing Hormone (GnRH) in the Hypothalmus of Ovariectomized Pony Mares in Season," pp. 123-129.

Vandeplassche, M., M. Henry, and M. Coryn. "The Mature Midcycle Follicle in the Mare," pp. 157-162.

Vivrette, S.L. and C.H.G. Irvine. "Interaction of Oestradiol and Gonadotropin Releasing Hormone on LH Release in the Mare," pp. 151-155.

Voss, J.L., R.A. Wallace, E.L. Squires, B.W. Pickett, and R.K. Shideler. "Effects of Synchronization and Frequency of Insemination on Fertility," pp. 257-261.

Journal of the South African Veterinary Association. Vol. 43 No. 4 1972.

van Niekerk, C.H. and J.S. van Heerden. "Nutrition and Ovarian Activity of Mares Early in the Breeding Season," pp. 351-360.

Journal of the South African Veterinary Association. Vol. 44 No. 1 1973

van Niekerk, C.H., R.I.Coubrough and H.W.H. Doms. "Progesterone Treatment of Mares with Abnormal Oestrous Cycles Early in the Breeding Season," pp. 37-45.

Kays, D.J. and J.M. Kays. **The Horse.** New York: Arco Publishing Company, 1977.

Levine, L. **Biology of the Gene.** Saint Louis: The C.V. Mosby Co., 1969.

Lewis, L.D. **Feeding and Care of the Horse.** Philadelphia: Led & Febiger, 1982.

Looney, J.W. **Estate Planning for Farmers.** St. Louis: Doane Agricultural Service, Inc. 1978.

Lose, P. **Blessed are the Broodmares.** New York: McMillan Publishing Co., Inc., 1978.

Louisiana Veterinarian. Vol. 4 No. 2 1961

Roberts, S.J. "Infertility in Mares," pp. 4-9.

Lovelace, D.A. **"Pastures for Horses."** Vernon, Texas: Texas Agricultural Extension Service, Texas A&M University, 1979

Lynn, R.J. **Introduction to Estate Planning.** 2nd ed. St.Paul, Minnesota: West Publishing Co., 1978.

Lytle, R.J. **Farm Builder's Handbook.** Farmington, Michigan: Structures Publishing Co., 1978.

Martens, R.J. "Foal Diseases." Texas A&M University. (Paper.)

Martin, J.H., W.H. Leonard, D.L. Stamp. **Principles of Field Crop Production.** 3rd ed. MacMillan Publishing Co., Inc: New York, 1976.

Maynard, L.A. **Animal Nutrition.** New York: McGraw-Hill Book Co.. 1969

McDonald, L.E. **Veterinary Endocrinology and Reproduction.** 2nd. ed. Philadelphia: Lea and Febiger, 1975.

Merchant, I.A. and R.A. Packer. **Veterinary Bacteriology and Virology.** 7th ed. Ames, Iowa: The Iowa University Press, 1967.

Merck and Company, Inc., Animal Health Division. **Product Information.** Rahway, New Jersey 1981.

Merrell, D.J. **An Introduction to Genetics.** New York: W.W. Norton and Co., 1975

Midwest Horseman. January 1979.

McWilliams, J. "What You Should Expect at the Breeding Farm?," pp. 12-14.

Midwest Plan Service. **Farmstead Planning Handbook.** Ames, Iowa: Iowa State University, 1974.

Midwest Plan Service. **Horse Handbook Housing and Equipment.** Ames, Iowa: Iowa State University, 1971.

Midwest Plan Service. **Structures and Environment Handbook.** Ames, Iowa: Iowa State University, 1977.

Miller, W.C. **Practical Essentials in the Care and Management of Horses on Thoroughbred Studs.** London: The Thoroughbred Breeder's Association, 1965.

MIP-Test Instruction Bulletin, Diamond Laboratories, Inc: Des Moine, Iowa.

Modern Veterinary Practice. Vol. 43 No. 10 1962.

Lieux, P. "Anestrus and Infertility in the Mare," pp. 52-55.

Modern Veterinary Practice. Vol. 48 No. 7 1967

Gibbons, W.J. "Mucometra: Case Report," p. 64.

Modern Veterinary Practice. Vol. 49 No. 2 1968

"Causes and Treatment of Infertility in Mares," pp. 66-67.

Morris Animal Foundation Newsletter. July 1981

"Fight Colic, Plow Deep," p. 1.

Morrow, David A. **Current Therapy in Theriogenology.** Philadelphia, Pa.: W.B. Saunders Company, 1980.

Moscove, S.A. **Accounting Fundamentals: A Self-Instructional Approach.** Reston, Va.: Reston Publishing Co., 1977.

Moselle, G. ed. **National Construction Estimator, 1980.** Solana Beach, Calif.: Craftsman Book Company, 1979.

The National Research Council. **Nutrient Requirements in Horses.** 4th ed. Washington, D.C., 1978.

Naviaux, J.L. **Horses in Health and Disease.** New York: Arco Publishing Company, Inc., 1976.

New Zealand Veterinary Journal. Vol. 19 1971

Elliott, R.E.W., E.J. Callaghan and B.L. Smith. "The Microflora of the Cervix of the Thoroughbred Mare: A Clinical and Bacteriological Survey in a Large-Animal Practice in Hastings," pp. 291-302.

New Zealand Veterinary Journal. Vol. 23 1975
  Thornbury, R.S. "Diseases of the Vulva, Vagina, and Cervix of the Thoroughbred Mare," pp. 244-280.
Norden News. Vol. 41 No. 3 Summer 1966
  Owen, D. "Breeding Problems in Mares," pp. 8-11.
Norden News. Vol. 46 No. 3 Summer 1971
  Heinze, C.D. "Physeal Stapling in the Horse for Correction of Angular Leg Deformities," p. 24.
Norden News Vol. 48 No. 2 Spring 1973
  Bass, E.P., R.L. Sharpee and W.H. Beckenhauer. "Equine Viral Rhinopneumonitis: Pathology and Prophylaxis," pp. 5-9, 34.
Novosad, A.C. and J.N. Pratt. **Keys to Profitable Summer Annual Forage Production.** United States Department of Agriculture, Texas Agricultural Extension Service.
Novosad, A.C. and J.N. Pratt. **Keys to Profitable Winter Annual Forage Production.** United States Department of Agriculture, Texas Agricultural Extension Agency.
Oehme, F.W. and J.E. Prior. **Textbook of Large Animal Surgery** Baltimore: The Williams and Wilkins Company, 1974.
Paint Horse. Vol. 12 No. 5 May 1978
  Hughes, J.P. "Breeding Hygiene," pp. 42-43.
Paint Horse Vol. 12 No. 7 July 1978
  Wood, K.A. "Law, Taxes, and the Horseman: Entertainment Deductions for the Horseman," pp. 66-67.
Paint Horse Vol. 12 No. 10 September 1978
  Harper, F. "Win with Early Foal Management," pp. 137-138.
Paint Horse Vol. 12 No. 11 November 1978
  Ensminger, M.E. "Horses, Horses, Horses," p. 36.
Paint Horse Journal. Vol. 14 No. 5 May 1980
  "Colostrum Substitute Developed," p. 113.
Palomino Horse Journal. Vol. 36 No. 9 December 1978
  Beaver, B.G. "Cleaning the Sheath," p. 11.
Pattern, B.M. and B.M. Carlson. **Foundations of Embryology.** New York: McGraw-Hill Book Co., 1974.
Phipps, L.J., H.F. McColly, L.L. Scranton and G.C. Cook. **Farm Mechanics Text and Handbook.** Danville, Illinois: The Interstate Printers and Publishers, 1955.
Pickett, B.W. and D.G. Back. **Procedures for Preparation, Collection, Evaluation, and Insemination of Stallion Semen.** Colorado State University Experimental Station. General Series 935, December 1973.
Pittenger, P.J. **The Back-Yard Foal.** Hollywood, California: Wilshire Book Company, 1965.
Planning Machinery Protection. American Association for Vocational Instructional Materials. Planning Machinery Protection Athens, Georgia, 1974.
Pony Journal. Vol. 34 No. 1 February 1979
  Wood, K.A. "Law, Taxes, and the Horseman: Agreement for Lease of Broodmare," pp. 7-9.
Practical Horseman. Vol. 5 No. 1 December 1974
  "Breeders Research Report," pp. 53-56.
Practical Horseman. Vol. 6 No. 1 December 1975
  "Artificial Insemination," pp. 22-38.
Practical Horseman. Vol. 6 No. 3 April 1976.(ql) Hughes, J.P. "Breeding Problem Mares," pp. 25-28.
Practical Horseman. Vol. 5 No. 2 January 1977
  "Construct Your Own Quarantine Stall," pp. 50-53.
Practical Horseman. June 1977
  Witwer, J. "Your Horse's Immunity: How It Works," pp. 25-30, 40.
Practical Horseman. Vol. 5 No. 8 July 1977
  Abbott, R.D. "Seller Beware," pp. 23-26.
Practical Horseman. Vol. 7 No. 6 June 1979
  "Saddle Ways and Bridle Whys," pp. 60-61.
Practical Horseman. Vol. 9 No. 5 April 1981
  Carmichael, A.B. "War on Flies," pp. 23-28.
Pratt, J.N., A.C. Novosad, and G.D. Alston. **Keys to Profitable Permanent Pasture Production in East Texas,** United States Department of Agriculture, Texas Agricultural Extension Service.
Price, J.D., R.D. Palmer, and G.O. Hoffman. **Suggestions for Weed Control with Chemicals,** United States Department of Agriculture, Texas Agricultural Extension Service.
Proceedings of the Agricultural Conference. Texas A&M University. 1979
  Householder, D.D. "Broodmare Management." Pickett, B.W. "Stallion Seminal Evaluation and Artificial Insemination,"
  Pickett, B.W. "Management of the Breeding Stallion."
Proceedings of the 17th Annual Convention of the American Association of Equine Practitioners. F.J. Milne, ed. 1971.
  Kaufman, W.C., H.M.J. Aldous, W.R. Brawner and J.R. Witmer. "Panel - Broodmare - Questions and Answers," pp. 97-109.
  Kingston, R.S., A.H. Rajamannon and C.F. Ramberg,Jr. "Stallion Semen Characteristics for Predicting Fertility."
  Phillups, T.N. "Equine Vaccinations."
  Solomon, W.J. "Rectal Examination of the Cervix and its Significance in Early Pregnancy Evaluation in the Mare," pp. 73-80.
  Stuart, Wm. R. "Methods of Restraint."
  Wearly, W.K., P.W. Murdock and J.D. Hensel. "A Five Year Study of the Use of Post-Breeding Treatment in Mares in a Standardbred Stud," pp. 89-96.
Proceedings of the 18th Annual Convention of the American Association of Equine Practitioners. F.J. Milne, ed., 1972.

Asbury, A.C. "Management of the Foaling Mare," pp. 487-490.

Ellsworth, K.S. "A Practical Approach to the Treatment of Endometritis," pp. 490-494.

Hughes, J.P., G.H. Stabenfeldt and J.W. Evans. "Clinical and Endocrine Aspects of the Estrous Cycle of the Mare," pp. 119-152.

Keeran, R.J. "Equine Ovariectomy: Anesthesia and Positioning," pp. 41-42.

Lyall, W.L. "Broodmare Panel: The 21-Day Pregnancy Examination," pp. 483-486.

Mitchell, D. "The Diagnosis of Pregnancy in the Mare by Doppler Ultrasound," pp. 499-500.

Monin, T. "Vaginoplasty: A Surgical Treatment for Urine Pooling in the Mare," pp. 99-102.

Morrow, G.L., R.S. Jackson and S.M. Teeter. "Gentamicin Therapy for Endometritis in the Mare," pp. 411-416.

Pickett, B.W. and J.L. Voss. "Reproductive Management of the Stallion," pp. 501-531.

Purvis, A.D. "Elective Induction of Labor and Parturition in the Mare," pp. 113-116.

Rossdale, P.D. "Pulmonary Function in the Newborn Foal: A Clinician's Approach to the Diagnosis and Therapy of Neonatal Respiratory Problems," pp. 69-97.

Rossdale, P.D. "The Estrous Cycle in the Context of Stud Management and Routine Veterinary Examinations," pp. 495-497.

Solomon, W.J., R.H. Schultz and M.L. Fahning, "A Study of Chronic Infertility in the Mare Utilizing Uterine Biopsy, Cytology and Cultural Methods," pp. 59-62.

Vaughan, J.T. "Surgery of the Prepuce and Penis," pp. 19-40.

Voss, J.L. and J. L. Wotowey. "Hemospermia," pp. 103-112.

Witherspoon, D.M. "Technique and Evaluation of Uterine Curettage," pp. 51-53.

Proceedings of the 19th Annual Convention of the American Association of Equine Practitioners. F.J. Milne, ed., 1973.

Bergin, W.C. and M.A. Collier. "Powder Gun Worming of Horses."

Hoffman, P.E. "Castration of Normal and Cryptorchid Horses by a Primary Closure Method," pp. 219-223.

Pierson, R.H. "Clinical Observations on Planned Infection and Preventive Vaccination for Control of Equine Rhinopneumontitis."

Swanson, T.D. "Restraint in Treatment," pp. 147-151.

Voss, J.L. and B.W. Pickett. "The Effect of Nutritional Supplement on Conception Rate in Mares," pp. 49-54.

Voss, J.L., B.W. Pickett and D.G. Back. "The Effect of Rectal Palpation on Fertility in the Mare," pp. 33-37.

Proceedings of the 20th Annual Convention of the American Association of Equine Practitioners. F. J. Milne, ed., 1974.

Bello, T.R., et. al. "Practical Equine Parasitology Based Upon Recent Research."

Pickett, B.W. "Stallion Seminal Extenders," pp. 155-174.

Voss, J.L., B.W. Pickett, R.K. Shideler and T.M. Nett. "Controlling the Estrous Cycle of the Mare," pp. 145-154.

Proceedings of the 21st Annual Convention of the American Association of Equine Practitioners. F.J. Milne, ed., 1975.

Bergman, R.V. and R.M. Kenney. "Representativeness of a Uterine Biopsy in the Mare," pp. 355-361.

Cooper, W.L. and N.E. Wert. "Wintertime Breeding of Mares Using Artificial Light and Insemination: Six Years Experience," pp. 245-253.

Gadd, J.D. "The Relationship of Bacterial Cultures, Microscopic Smear Examination and Medical Treatment to Surgical Correction of Barren Mares," pp. 362-368.

Ganjam, V.K. and R.M. Kenney. "Peripheral Blood Plasma Levels and Some Unique Metabolic Aspects of Progesterone in Pregnant and Non-Pregnant Mares," pp. 263-276.

Hurtgen, J.P. "Alteration of the Equine Estrous Cycle Following Uterine and/or Cervical Manipulations," pp. 368-377.

Kaufman, W.C. "The Effect of Rectal Palpation on Reproduction in Mares," pp. 229-233.

Kenney, R.M. "Minimal Contamination Techniques for Breeding Mares: Technique and Preliminary Findings," pp. 327-335.

Kenny, R.M. "Clinical Fertility Evaluation of the Stallion."

Lauderdale, J.W. and P.A. Miller. "Regulation of Reproduction in Mares with Prostaglandins," pp. 235-244.

Simpson, R.B., S.J. Burns and J.R. Snell. "Microflora in Stallion Semen and Their Control with a Semen Extender," pp. 225-261.

Proceedings of the 23rd Annual Convention of the American Association of Equine Practitioners. F.J. Milne, ed. 1977.

Burns, S.J., N.G. Judge, J.E. Martin and L.G. Adams. "Management of Retained Placenta in Mares," pp. 381-390

Ellsworth, K.S. "A Practical Approach to the Treatment of Endometritis," pp. 490-494.

Ganjam, V.K. "An Inexpensive, Yet Precise, Laboratory Diagnostic Method to Confirm Cryptorchidism in the Horse," pp. 245-249.

Hughes, J.P. "Assessment of Breeding Potential in the Mare," pp. 83-87.

Hughes, J.P. and G.H. Stabenfeldt. "Anestrus in the Mare," pp. 86-89.

Johnson, J.H., J.N. Moore, J.R. Coffman, H.E. Garner, L.G. Tritschler, and D.S. Traver. "Selection and Care of Endoscopic Equipment," pp. 239-243.

Kenney, R.M. "Clinical Aspects of Endometrial Biopsy in Fertility Evaluation of the Mare," pp. 105-122.

Knowles, R.C. "Contagious Equine Metritis in the Stallions," p. 229.

Meagher, D.M., J.D. Wheat, J.P. Hughes, G.H. Stabenfeldt and B.A. Harris. "Granulosa Cell Tumors in Mares: A Review of 78 Cases," pp. 133-141.

Powell, D.G. "Contagious Equine Metritis, 1977," pp. 233-237.

Purvis, A.D. "The Induction of Labor in Mares as a Routine Breeding Farm Procedure," pp. 145-160.

Shideler, R.K., A.E. McChesney, G.L. Morrow, J.L. Voss and J.G. Nash. "Endometrial Biopsy in the Mare," pp. 97-104.

Snow, D.H., C.D. Munro and M. Nimmo. "Anabolic Steroids in Equine Practice," pp. 411-418.

Vandeplassche, M. and M. Henry. "Salpingitis in the Mare," pp. 123-131.

Witherspoon, D.M. "Assessment of Breeding Potential in the Mare," pp. 81-82.

Proceedings of the 24th Annual Convention of the American Association of Equine Practitioners. 1978.
Cahill, C. "Contagious Equine Metritis."
Proceedings of the 115th Annual Convention of the AVMA. July 17, 1978
Stratton, L.G. and R.B. Hollett. "Breeding Soundness Evaluation in the Mare."
Proceedings of the British Equine Veterinary Association. Newmarket, 1974
Hughes, J.P. and R.G. Loy. "The Relation to Infertility in the Mare and Stallion."
Proceedings of the Second Annual Cornell Equine Conference. November 16-18, 1972. Cornell University, New York.
Carr, R.M. and H.E. Gill. "Practitioner's Approach to Disease Control."
Davis, T.A. "Outline: Taxation of Horse Farming."
Proceedings of the 6th Annual Horse Short Course. 1966. Texas A&M University, College Station, Texas.
Lockridge, B., C. McDonald and J.K. Northway. "Settling Mares," pp. 62-72.
Romane, W.M. "Immunization for Disease Prevention," pp. 73-75.
Smith, J.P. "Controlling Internal Horse Parasites Through Management," pp. 76-77.
Sorensen, A.M. "Fundamentals of Reproduction," pp. 57-61
Proceedings of the 7th Annual Horse Short Course. 1967. Texas A&M University, College Station, Texas
Kieffer, N.M. "Guidelines to Consider When Choosing a Mate for Your Mare."
Proceedings of the 10th Annual Horse Short Course. 1970. Texas A&M University, College Station, Texas.
Breuer, L.H., T.L. Bullard, B.F. Yeates. "A Suggested Schedule for Management of a Horse Breeding Farm," pp. 17-21.
Burns, S.J. "Why She Won't Breed," pp. 39-41.
Proceedings of the 11th Annual Horse Short Course. 1971. Texas A&M University, College Station, Texas.
Bullard, T.L. "Immunizations," pp. 44-46.
Romane, W.M. "Looking Back at Vee," pp. 62-64.
Proceedings of the 12th Annual Horse Short Course. 1972. Texas A&M University, College Station, Texas.
Burns, S.J. "Husbandry Considerations for the Stallion Owner," pp. 33-38.
Proceedings of the 16th Annual Horse Short Course. April 8-9, 1976. Texas A&M University, College Station, Texas.
Adkins, D.T. "The Mare's Reproductive Cycle," pp. 5-11.
Burns, S.J. "Broodmare Management," pp. 14-16.
Hughes, H.A. "Setting Up the Horse Production Facility," pp. 93-97.
Kieffer, Nat M. "Guidelines to Consider When Choosing a Mate for Your Mare," pp. 1-4.
McMullan, W.C. "Foaling Time to Three Months," pp. 32-42.
Umstadtler, L.W. "Horse Farm Facilities," pp. 90-92.
Proceedings of the Horsemen's Short Course. March 9-11, 1978. VPI and SU, Blacksburg, Virginia.
Bibb, T.L. "Preventive Medicine," pp. 77-84.
Conner, J.T. "Breeding Management and Artificial Insemination," pp. 38-44.
Hughes, H.A. "Planning Facilities for Horses," pp. 67-72.
Hutton, C.A. "Reducing the Cost of Horse Production," pp. 54-62.
Kohler, C.F. "Horse Mortality Insurance," pp. 89-94.
Mather, E.C. "Preparing a Mare for Breeding," pp. 73-76.
Meacham, T.N. "Current Topics in Equine Reproduction Research," pp. 17-19.
Proceedings of the 2nd International Horse Identification Seminar. Dec. 9-10, 1977. University of Arizona, Tuscon, Arizona.
Drayton, S.J. "Tattoo Branding of Thoroughbreds," pp. 1-13.
Arabian Horse Registry. "Freeze Mark Branding for Purebred Arabians," pp. 14-24.
Farrell, R.K. and B.P. Farrell. "Program of International Horse Identification," pp. 46-64.
Garner, L.S. "Permanent Identification Cards for Horses," pp. 78-79.
Proceedings of the 1st National Horsemens Seminar. 1976. Virginia Horse Council, Inc., Fredericksburg, Virginia.
Bello, T.R. "Practical Equine Parasitology," pp. 85-87.
Conner, J.T. "Record Systems for Horse Farms," pp. 105-111.
Farrell, R.K. "Horse Identification," pp. 146-151.
Hughes, H.A. "Setting Up the Horse Production Facility," pp. 93-97.
Kenney, R.M. "Mare Reproduction: Some Good and Some Bad Features," pp. 118-123.
Kohler, C.F. "Livestock Insurance—All Breeds of Horses," pp. 68-71.
Lauderdale, J.W. "Prostin F2 Alpha—Effective in Management of Estrus in the Mare," pp. 52-57.
Pickett, B.W. "Stallion Management with Special Reference to Semen Collection, Evaluation and Artificial Insemination," pp. 37-47.
Stear, R.L. "Diagnosis and Control of Equine Rhinopneumonitis," pp. 175-179.
White, W.E. "Establishing and Maintaining Horse Pastures in East Central US," pp. 112-117.
Williams, P.J. "Do's and Don'ts for Horse Facilities: Do's and Don'ts in the Actual Construction of the Farm Facilities," pp. 99-100.
Proceedings of the 2nd National Horsemen's Seminar. 1977. Virginia Horse Council, Inc. Fredericksburg, Virginia.
Loring, M. "Your Horse and the Law," pp. 108-112.
Rhulen, P. "Liability in Respect to the Horse Business," pp. 131-132.
Squires, E.L. "Reproductive Physiology of the Mare," pp. 66-73.
White, H.E. "Establishing and Managing Horse Pastures," pp. 32-36.
Proceedings of the 3rd National Horsemen's Seminar. January 19-21, 1978. Virginia Horse Industry Yearbook. Williamsburg, Virginia.
Evans, J.W. "Critical Information for Breeding Mares," pp. 73-78.
Evans, J.W. "Nutrition for Optimal Reproductive Efficiency."
Rackley, A. "Advertising and Public Relations: You Can't Have One without the Other."
Proceedings of the 1970 Stud Manager's Course. Lexington, Kentucky.
Schmidt, H.L. "Breeding Farm Stallion and Mare Records," pp. 186-191.

Proctor, D.L., "Anatomy, Care, and Trimming of Feet," pp. 203-219.
Drudge, J.H. "The Control of Internal Parasites" pp. 131-138.
Proceedings of the 1973 Stud Manager's Course. Lexington, Kentucky.
McGee, W.R. "Management of Broodmares," pp. 71-85.
Worthington, W.E. "Management of Stallions, Considerations of Feeding, Health, and Fertility."
Proceedings of the 1978 Stud Managers Course. Lexington, Kentucky.
Bryan's, J.T. "Vaccines and Vaccinations," pp. 165-175.
Proceedings from the Water Use Seminar. Damacus, 1971.
Zein El-Abdin, A.N. and Yiannakis Nicolaou. "Irrigation Application and Water Distribution," pp. 144-148.
"Water Management and Irrigation Practices," pp. 160-166.
Progress in Equine Practice, Vol. 1 E.J. Catcott and J.F. Smithcors, eds. Wheaton, Illinois: American Veterinary Publications, 1966.
Wurster, A.C. "Ovarian Granulosa Cell Tumor," pp. 217-218.
Grant, D.L. "Uterine Leiomyoma," p. 218.
Roberts, S.J. "Infertility in Mares," pp. 362-364.
Lieux, P. "Anestrus and Infertility in the Mare," pp. 365-366.
Knudsen, O. "Uterine Dialation as a Cause of Sterility," p. 366.
Various authors. "Equine Infertility," p. 367.
Collins, S.M. "Cervical and Uterine Infection in Thoroughbred Mares," p. 367.
Progress in Equine Practice, Vol. 2 E.J. Catcott and J.F. Smithcors, eds. Wheaton, Illinois: American Veterinary Publications, 1970.
Sager, F.C. "Management of Uterine Disease," pp. 374-375.
Thain, R.I. "Cystic Endometrium in Mares," p. 376.
Gibbons, W.J. "Mucometra: A Case Report," p. 377.
Finocchio, E.J. and J.H. Johnson. "Ovarian Granulosa Cell Tumor," pp. 378-379.
"Causes and Treatment of Infertility in Mares," pp. 430-431.
Laufenstein-Duffy, H. "Uterine Curettage," p. 374.
Morrow, G.L. "Uterine Curettage," p. 375.
Quarter Horse Journal. Vol. 28 No. 4 January 1975
Boss, R.L. "The Do's and Don't's of Foaling," pp. 156-58, 252
Quarter Horse Journal. Vol. 29 No. 3 December 1976
Evans, J.W. "New Techniques for Breeding Horses," pp. 228-232, 234, 236, 238, 240.
Jennings, J. and B. Wood. "Semen Handling Techniques," pp. 308-316.
Quarter Horse Journal. Vol. 30 No. 3 December 1977
Biasatti, H. "When to Castrate Your Colt," pp. 114-118.
Evans, J.W. "Evaluation of the Breeding Farm," p. 38.
Hughes, J.P. "Breeding Hygiene, Pregnancy, and Parturition," pp. 156-158.
Quarter Horse Journal. Vol. 30 No. 4 January 1978
Gordon, L.R. and E.M. Sartin. "Predicting Future Reproductive Performance by Uterine Biopsy," pp. 220-222, 224.
Quarter Horse Journal. March 1978
Albaugh, R. "Horse Behavior," pp. 186-188.
Quarter Horse Journal. Vol. 31 No. 1 October 1978
Oxender, W.D. and P.A. Noden. "Can Hormones Be Used to Cause Ovulation in Mares?," pp. 36-37.
Quarter Horse Journal. Vol. 31 No. 4 January 1979
Travis, B. "Parentage Determination by Blood Typing," pp. 138-141.
"CSU Short Course - Full of Breeder Information," pp. 356-359.
Travis, B. "Impotence in Stallions Cured at CSU Lab," pp. 236-238.
Quarter Horse Journal. February 1979
McCall, J.P. "Stallion Breeding Management," pp. 122-123.
Quarter Horse Journal. Vol. 31 No. 6 March 1979
Travis, B. "Dr. Marvin Beeman Answers Questions About Bowed Tendons," pp. 92-94.
Quarter Horse Journal. Vol. 31 No. 9 June 1979
Kester, W.O. "Equine Piroplasmosis: Old Disease, New Threat."
Squires, E.L. "Embryo Transfer in Horses," pp. 122-125.
Quarter Horse Journal. Vol. 31 No. 10 July 1979
Berndston, W.E. "Anabolic Steroids Impair the Reproductive Potential of Stallions," pp. 152-158.
Quarter Horse Journal. Vol. 32 No. 6 March 1980
Evans, J.W. "Prostaglandins: Some Answers," pp. 290-292.
Quarter Horse Track. Vol. 1 No. 4 December 1975
Self, L. "The Teasing Stud and How to Use Him," pp. 88-89.
Quarter Horse Track. Vol. 3 No. 11 November 1977
McAdams, J. "How to Prepare for the Spring Breeding Season This Fall," pp. 122-123.
Quarter Horse Track. Vol. 4 No. 6 June 1978
McAdams, J. "Mare Care After Breeding," pp. 86-87.
Quarter Horse Track. Vol. 4 No. 10 October 1978
Kennedy, B. "One Season, Two Foals," pp. 120-121.
Quarter Horse Track. Vol. 5 No. 3 April 1979
Yates, B. "What is an Extender and How Do You Use It?," pp. 70-71.
Quarter Horse Track. Vol. 6 No. 3 March 1980
McAdams, J. "Halter Breaking Those New Foals," pp. 159-160.
Quarter Horse World. Vol. 1 No. 6 January 1975
Edison, K. "...to save a foal...," pp. 34-35.
Quarter Horse World. Vol. 1 No. 8 March 1975
Moore, G. "Artificial Breeding," pp. 36-37.

Quarter Horse World. Vol. 4 No. 3 March 1976
  Kenney, R.M. "Your Mare's Heat Cycle and Causes of Seasonal Variation," pp. 30-32.
Quarter Horse World. Vol. 5 No. 9 September 1977
  Alwan, D. "Walter Merrick Talks About the Mare's Contribution," pp. 49-50.
Quarter Horse World. Vol. 5 No. 11 November 1977
  Lawrence, R.G. "So You Want to Get into the Business of Horses," pp. 40-45.
Quarter Horse World. Vol. 6 No. 5 May 1978
  Alverson, D.W. "Contagious Equine Metritis," pp. 10-14.
Quarter Horse World. Vol. 7 No. 3 March 1979
  Evans, J.W. "Erratic Behavior of Mares," pp. 48-51.
Quarter Horse World. May 1979
  Ensminger, M.E. "All About Horses," pp. 40-42.
  Kirk, M.D. "Broodmares and Reproduction," pp. 44-46.
Quarter Horse World. Vol. 7 No. 6 June 1979
  Ensminger, M.E. "All about horses," pp. 52-54.
Quarter Horse World. Vol. 7 No. 7 July 1979
  Squires, E.L. "What About Embryo Transfers in Horses?," pp. 12-14.
Quarter Horse World. Vol. 8 No. 1 January 1980
  Douglas, R.H. "Use of Increased Photoperiod in Reproductive Management of Mares," pp. 70-71.
Quarter Horse World. Vol. 8 No. 5 May 1980
  Wallace, G. "The Fly Predator: Getting Rid of Flies Nature's Way," pp. 38-40.
Quarter Horse World. Vol. 8 No. 8 August 1980
  Di Pietro, J.A. "Eliminating Internal Parasites," pp. 44, 46-47.
Quarter Racing Record. Vol. 15 No. 2 February 15, 1975
  Tobias, O.G. "Care of the Young Foal," pp. 196-202.
Quarter Racing Record. June 15, 1978
  Bailey, E.S. "Controlled Environment Leads to Successful Breeding Operations," pp. 42-46.
Quarter Racing Record. Vol. 18 No. 12 December 15, 1978
  Treharne, S. "A Nanny for Bars High Gear," pp. 182-183.
Quarter Racing Record. Vol. 19 No. 12 December 15, 1979
  Baker, J.P. "Nutrition of Young Horses," pp. 196-378.
Ragsdale, B.J., D.L. Huss, and G.O. Hoffman. **Grazing Systems for Profitable Ranching,** United States Department of Agriculture, Texas Agricultural Extension Service.
Rapidan River Farm Digest. Vol. 1 No. 3 1975
  Davis, T.A. "Taxation of Horse Farming," pp. 85-90.
  Evans, J.W. "Variability of a Mare's Estrus," pp. 95-100.
  Flynn, D.V. "Semen Production, Collection, and Evaluation," pp. 242-245.
Rapidian River Farm Digest. Winter 1976
  Ricketts, S.W. "The Diagnosis of Infertility in Mares," pp. 54-60.
Rhoades, J.D. and C.W. Foley. "Cryptorchidism and Intersexuality." In **Veterinary Clinics of North America: Symposium on Reproductive Problems.** Vol. 7 No. 4 pp. 789-794. Philadelphia: W.B. Saunders Company, 1977.
Rice, V.A. **Breeding and Improvement of Farm Animals.** New York: McGraw-Hill Publications, 1970.
Roberts, S.J. **Veterinary Obstetrics and Genital Diseases (Theriogenology).** Ithaca, New York: Stephan J. Roberts publisher. 1971
Rogers, T. ed. **Mare Owner's Handbook.** Houston, Texas: Cordovan Corporation, 1975.
Rose, D.E. **Depreciation for the Horse Industry.** Lexington, Ohio: Equine Publications, Ltd., 1979
Rossdale, P.D. **Seeing Equine Practice.** London: William Heinemann Medical Books, Ltd., 1976.
Rossdale, P.D. **The Horse.** Arcadia, California: The California Thoroughbred Breeders Association, 1972.
Rossdale, P.D. and S.W. Ricketts. **The Practice of Equine Stud Medicine.** Baltimore: The Williams and Wilkins Co., 1974.
Rossdale, P.D. and S.W. Ricketts. **Equine Stud Farm Medicine** 2nd ed. Philadelphia: Lea & Febiger, 1980
Schalm, O.W., N.C. Jan and E.J. Carrol. **Veterinary Hematology** 3rd ed. Lea and Febiger: Philadelphia: 1975
Scher, L. **Finding and Buying Your House in the Country.** New York: Collier Books, 1974.
Scoggins, R.V. **Professional Topics: Horse-Stallion Evaluation and Management.** University of Illinois, Public Service Extension, 1978.
Seiden, R. **The Handbook of Feedstuffs.** New York: Springer Publishing Co., 1957.
Seminal SH and Fertility. Vol. 7 No. 6 1956
  Haag, F.M. and N.T. Werthessen. "Relationship Between Fertility and the Non-protein Sulfhydryl Concentration Seminal Fluid in the Thoroughbred Stallion."
Shive, C. Pasture Breeding.(Personal Communication).
Siegmund, O.H., ed. **The Merck Veterinary Manual,** 4th ed. Rahway, New Jersey: Merck and Co., Inc., 1973.
Sisson, S. and J.D. Grossman. **The Anatomy of the Domestic Animals.** Philadelphia: W.B. Saunders Comp., 1953.
Smith, H.A., T.C. Jones and R.O. Hunt. **Veterinary Pathology.** 4th ed. Philadelphia: Lea and Febiger, 1972.
Sorensen, A.M. **A Laboratory Manual for Animal Reproduction.** 2nd ed. Dubuque, Iowa: Kendall/Hunt Publishing Co., 1973
South American Journal of Veterinary Research. Vol. 36 No. 4 April 1975.
  Dutta, S.K. and W.D. Shipley. "Immunity and the Level of Neutralization Antibodies in Foals and Mares Vaccinated with a Modified Live-Virus Rhinopneumonitis Vaccine," pp. 445-448.
Southern Horseman. July 1979
  Davis, T.A. "Taxation on Horse Owners," pp. 35-38.
  "Embryo Transfer Foals Eligible for AQHA Registration," pp. 48-49.
Southern Horseman. Vol. 19 No. 5 May 1980
  Harper, F. "Foals Can be Weaned Early," p. 117.
Southwestern Veterinarian. Vol. 17 No. 2 1964
  Wurster, A.C. "Ovarian Granulosa Cell Tumor," pp. 149-151.

Spalding Laboratories, **Product Information,** Arroyo Grande, California 1981.
Speedhorse. Vol. 9 No. 5 January 1977
  McGee, W.R. "Management of Broodmares," pp. 91-94, 96-97.
  Sager, F.C. "Care of the Foaling Mare," pp. 29-31.
Speedhorse. Vol. 9 No. 6 February 1977
  Wyant, T. "Where Should My Stallion Stand?," pp. 36-40.
Speedhorse. Vol. 10 No. 3 November 1977
  Hughes, J.P., G.H. Stabenfeldt and J.W. Evans. "Utero-Ovarian Relationship in the Mare," pp. 147-153.
Speedhorse. Vol. 11 No. 5 January 1979
  Gralla, S. "Horseman's Architect," pp. 54-55, 220-221.
Speedhorse. Vol. 11 No. 6 February 1979
  Gralla, S. "Horseman's Architect," pp. 98-99.
Speedhorse. Vol. 2 No. 8 April 1979
  Harper, F. "Nutrition for the Broodmare," pp. 158-160, 170.
Speedhorse. Vol. 12 No. 5 January 1980
  McGee, W.R. "Evaluation and Care of Foals," pp. 196-197.
Sperry, O.E., J.W. Dollahite, G.O. Hoffman, and B.J. Camp. "Texas Plants Poisonous to Livestock," United States Department of Agriculture, Texas Agricultural Extension Service.
Spotted Horse. Vol. 3 No. 3 February/March 1979
  Davis, J.H. "More Profit Through Proper Marketing," pp. 15-17.
Standardbred. Vol. 8 No. 24 November 22, 1978
  Ontario Ministry of Agriculture and Food. "Artificial Lighting for Mares," pp. 20, 26.
Stud Manager's Handbook. M.E. Ensminger, ed. Clovis, California: Agriservices Foundation, 1974. Vol. 10
  Adams, M. "Do You Furnish Your Banker with All the Facts?," pp. 168-175.
  Cooper, W.L. "Conditioning and Preparation of Sales Yearlings," pp. 106-108.
  Harding, N.D. "1973 Developments in Farm Taxation," pp. 164-167.
  Killian, C.M. "The Syndication of Stallions," pp. 186-190.
  Kurth, S.P. "Estate Tax—The High Cost of Dying," pp. 160-163.
  Kurth, S.P. "Income Tax—The High Cost of Living," pp. 156-159.
  Putnam, H.D. "Equine Pediatrics," p. 23.
  Rossdale, P.D. "Equine Reproduction-Infertility," pp. 15-17.
Stud Manager's Handbook. M.E. Ensminger, ed. Clovis, California: Agriservices Foundation, 1975. Vol. 11
  Heinemann, E.A. "Marketing Horses," pp. 147-149.
  Maxwell, D.A. "Do Horses Have a Place in the Tax Law?," pp. 151-155.
  McDonough, R.L. "Practical Marketing for the Small Breeder," pp. 142-145.
  Myers, V.S., Jr. "Important Equine Reproductive Anatomy," pp. 34-35.
  Ricketts, S.W. "Conception and Gestation in the Mare," pp. 1-4.
  Ricketts, S.W. "The Diagnosis of Infertility in the Mare," pp. 10-14.
Stud Manager's Handbook. M.E. Ensminger, ed. Clovis, California: Agriservices Foundation, 1975. Vol. 14
  Kaufman, W.C. "A Complete Equine Breeding Program," pp. 121-122.
  Kirk, M.D. "Broodmares and Reproduction," pp. 126-129.
  Mackay, A. "Pregnancy Diagnosis in the Mare," pp. 130-132.
  Mackay, A. "Preparing the Yearling for Sale," pp. 266-269.
  Wood, K.A. "Estate Planning and the Horseman," pp. 94-99.
  Wood, K.A. "Income Tax and the Horse," pp. 85-93.
Stud Manager's Handbook. M.E. Ensminger, ed. Clovis, California: Agriservices Foundation, 1977. Vol. 13
  Conner, J.T. "Developing a Sales Program for Horses," pp. 174-176.
  Poppie, M. "Congenital Abnormalities of the Foal," pp. 31-33.
  Taysom, E.D. "More Efficiency in Horse Breeding," pp. 1-4.
Swenson, M.J. **Duke's Physiology of Domestic Animals.** 8th ed. Ithaca, New York: Comstock Publishing Associates, 1970.
Symposium on Reproductive Problems: The Veterinary Clinics of North America. Vol. 7 No. 4 November 1977
  Rhoades, J.D. and C.W. Foley, "Cryptorchidism and Intersexuality," pp. 789-794.
Texas Southern Quarter Horse Journal. March 1978
  Russell, S. "Vet Talk: Horse Preventative Health Care," pp. 31, 33.
Texas Thoroughbred. Vol. 4 No. 13 January 1979
  Cahill, C. "Breeding Shed Procedures," pp. 66-67.
Texas Thoroughbred. April 1979
  Gayle, L.G. "Contagious Equine Metritis: New Threat to Texas Horsemen."
Texas Thoroughbred. Vol. 5 No. 17 January 1980
  Gayle, L.G. "Passive Immunity in the Foal," pp. 32-33.
Texas Thoroughbred. Vol. 5 No. 18 February 1980
  Martens, R.J. "Foaling and Care of the Newborn," pp. 68-80.
Theriogenology. Vol. 10 No. 5 November 1978
  Kreider, J.L., V.C. Murrell, L.C. Lonwell and R.A. Godke. "Control of Estrus in the Lactating Post-partum Mare with Fluprostenol (ICI-81,008)a," p. 371.
Theriogenology. Vol. 11 No. 1 January 1979
  Douglas, R.H. "Review of Induction of Superovulation and Embryo Transplant in the Equine," pp. 33-45.
  Squires, E.L., W.B. Stevens, D.E. McGlothlin and B.W. Pickett. "Use of an Oral Progestin for Estrus Synchronization in Mares," p. 110.
Thomas, H.S. **Horses: Their Breeding, Care and Training.** New York: A.S. Barnes and Co., 1974.
Thoroughbred Record. Vol. 203 No. 5 February 2, 1975
  Zent, W.W. "Breeding the Foaling Mare," pp. 397-398.
Thoroughbred Record. Vol. 205 No. 2 January 12, 1977
  Sager, F.C. "Foaling the Mare," pp. 136-150.

Thoroughbred Record. Vol. 205 No. 4 January 26, 1977
McGee, W.R. "The Care of Foals," pp. 334-342.
Thoroughbred Record. Vol. 207 No. 1 January 4, 1978
Walsh, B. "Selling the In-foal Mare," pp. 27-29, 42.
Thoroughbred Record. Vol. 207 No. 12 March 22, 1978
Knowles, R.C. and C.G. Mason. "Contagious Equine Metritis," pp. 911-911B, 944B.
Thoroughbred Record. Vol. 207 No. 14 April 5, 1978
"Text of the Kentucky Department of Agriculture's Regulation on Artificial Insemination, and Remarks by the Commissioner of Agriculture, Thomas Harris," p. 1059.
Thoroughbred Record. Vol. 208 No. 19 November 8, 1978
Simon, M. "Stallion Syndication," pp. 1738-1739.
Thoroughbred Record. Vol. 208 No. 20 November 15, 1978
Lohman, J. "Investing in Thoroughbreds: Serious Business for Fun and Profit," pp. 1871-1873.
Thoroughbred Record. Vol. 209 No. 5 January 31, 1979
"The Success of Weatherby's Blood Typing Scheme," pp. 398-407.
Thoroughbred Record. Vol. 209 No. 14 April 4, 1979
Rhodemyre, S. "Conception Problems in Maiden and Barren Mares," pp. 1099-1101, 1142.
Thoroughbred Record. Vol. 209 No. 16 April 18, 1979
Wilson, G.L. "Problems with Foaling Mares," pp. 1262-1267.
Thoroughbred Record. Vol. 209 No. 18 May 2, 1979
Wilson, G.L. "Not Just a Little Horse," pp. 1428-1430.
Thomas, H.S. "Hemolytic Foals - The Importance of Early Discovery," p. 1431.
Thoroughbred Record. Vol. 209 No. 27 July 4, 1979
Rhodemyre, S. "You Can Lead a Horse to Water," pp. 36-39.
Thoroughbred Record. Vol. 210 No. 3 July 18, 1979
Wilson, G.L. "Big Knees, Open Knees, Knock Knees, Knobby Fetlocks," pp. 298-299.
Trot. Vol. 5 No. 6 June 1978.
Morley, T.L. "Good Foal Care is Like Money in the Bank," p. 13.
Trot. Vol. 6 No. 6 June 1979.
Fretz, P.B. "Special Delivery," pp. 6, 8, 12.
Tyler, G. **Making Money on Western Horses.** Houston: Cordovan Corp., 1970.
Ulrey, H.F. **Carpentry and Building.** Indianapolis, Ind.: Howard W. Sams & Co., Inc. 1978
Utility Buildings. 2nd ed. American Association for Vocational Instructional Materials. Utility Buildings, Athens, Georgia, 1974.
Vallentine, J.F. **'Range Developement and Improvements.** Provo, Utah: Brigham Young University Press, 1980
Veterinary Digest. Vol. 1 No. 6 Nov/Dec 1976
Reid, M.M., D.R. Jeffrey and G.E. Kaiser. "Rare Maduromycosis of Equine Uterus," p. 13.
Veterinary Digest. January-February 1977
Kieffer, N.M., S.J. Burns, and N.G. Judge. "Male Psuedohermaphroditism of Testicular Feminizing Type," pp. 22, 26.
Veterinary Medicine/Small Animal Clinician. Vol. 64 No. 4 1969
Finocchio, E.J. and J.H. Johnson. "Ovarian Granulosa Cell Tumor," pp. 322-327.
Veterinary Medicine/Small Animal Clinician. February 1971
Brown, J.M. and J.R. Coffman. "A Modified Technic for Episioplasty in the Mare," pp. 103-107.
Veterinary Medicine/Small Animal Clinician. Vol. 71 No. 4 1976
Shideler, R.K. "Adenoviral Infection in a Foal," pp. 448-449.
Veterinary Medicine/Small Animal Clinician. Vol. 72 No. 9 Sept 1977
Purdy, C.W. "Safety of RhinoquinTM, a Modified-Life Virus Equine Rhinopneumonitis Vaccine in Foals and Pregnant Mares," pp. 1478-1480.
Veterinary Medicine/Small Animal Clinician. Vol. 74 No. 10 1979
Coffman, J. "Immunity: Autoimmunity, Isoimmunity, and Immunodeficiency in the Foal," pp. 1430-1433.
Veterinary Record. Vol. 16 No. 8 February 22, 1936
Buckingham, J. "Hermaphrodite Horse," p. 218.
Veterinary Record. Vol. 76 No. 17 1964
Grant, D.L. "Uterine Leiomyoma," pp. 474-475.
Veterinary Record. Vol. 76 No. 25 1964
Collins, S.M. "A Study of the Incidence of Cervical and Uterine Infection in Thoroughbred Mares in Ireland," pp. 673-675.
Veterinary Record. Vol. 85 No. 7 August 16, 1969
White, D.J. and D.A. Farebrother. "A Case of Intersex in a Horse," pp. 203-204.
Veterinary Record. Vol. 86 1970
Arthur, G.H. "The Induction of Oestrus in Mares by Uterine Infusion of Saline," pp. 584-586.
Burgess, D. "Horse Pox."
Veterinary Record. Vol. 88 1971
Scott, P., P. Daley, G.G. Baird, S. Sturgess and A.J. Frost. "The Aerobic Bacterial Flora of the Reproductive Tract of the Mare," pp. 58-61.
Veterinary Record. Vol. 89 December 1971
Cox, J.E. "Urine Tests for Pregnancy in the Mare," pp. 606-607.
Veterinary Record. Vol. 90 No. 2 1972
Crouch, J.R. J.G. Atherton and Platt. "Venereal Transmission of Klebsiella Aeruginosa in a Thoroughbred Stud from a Persistently Infected Stallion," pp. 21-24.
Veterinary Record. Vol. 91 No. 24 December 9, 1972
Rossdale, P.D. "Differential Diagnosis and Treatment of Equine Neonatal Disease," pp. 581-587.
Veterinary Record. Vol. 92 1973
Fraser, A.F., N.W. Keith and H. Hastie. "Summarised Observations on the Ultrasonic Detection of Pregnancy and Foetal Life in the Mare," pp. 20-21.

Veterinary Record. Vol. 97 1975
  Rossdale, P.D. and L.B. Jeffcott. "Problems Encountered During Induced Foaling in Pony Mares," pp. 371-372.
Veterinary Record. Vol. 98 1976
  Cooper, M.J. "Fluprostenol in Mares: Clinical Trials for Treatment of Infertility," pp. 523-525.
Veterinary Record. Vol. 98 No. 26 1976
  Haughey, K.G. "Meningeal Haemorrhage and Congestion Associated with the Perinatal Mortality of Foals," pp. 519-522.
Veterinary Record. Vol. 99 No. 11 September 11, 1976
  Bowen, J.M. "Intrauterine Use of Prostaglandin F2 Alpha in Mares," pp. 212-213.
Veterinary Record. Vol. 99 No. 18 October 30, 1976
  Pashen, R.L. and W.R. Allen. "Genuine Anoestrus in Mares," pp. 362-363.
Veterinary Record. Vol. 99 1976
  Merkt, H. "Equine Artificial Insemination," pp. 69-71.
Veterinary Record. Vol. 101 No. 3 July 16, 1977
  Ricketts, S.W., P.D. Rossdale, N.J. Wingfield-Digby, M.M. Falk, F. Hopes, M.D.N. Hunt and C.K. Peace. "Genital Infection in Mares," p. 65.
Veterinary Record. Vol. 102 No. 11 March 18, 1978
  Timoney, P.J. "CEM and the Foaling Mare," pp. 246-247.
Veterinary Record. Oct. 7 1978
  Akerejola, O.O., M.D. Ayivor, and E.W. Adams. "Equine Squamous Cell Carcinoma in Northern Nigeria."
Veterinary Record. Vol. 103 February 1979
  Allen, W.E. "Abnormalities in the Oestrus Cycle in the Mare," pp. 166-167.
Veterinary Scope. Vol. 19 No. 1 1975
  Kenny, R.M., V.K. Ganjam, and R.V. Bergman. "Non-Infectious Breeding Problems in Mares," pp.1-11.
Vietor, D.M. "Pasture Management." 1980. (Lecture Notes.)
Voice of the Tennessee Walking Horse. Vol. 8 No. 6 June 1979
  Adcock, N. "Clean Up Your Act pp. 36, 37, 39, 53.
Voss, J.L. and B.W. Pickett. **'Reproductive Management of the Broodmare,''** Colorado State University Experiment Station, General Series 961.
Wagner, W.H. **Modern Carpentry.** South Holland, Illinois: The Goodheart-Wilcox Co., Inc., 1976.
Wagoner, D.M. ed. **Conditioning to Win.** Equine Research Publications, 1974.
Wagoner, D.M. ed. **Equine Genetics and Selection Procedures.** Equine Research Publications, 1978.
Wagoner, D.M. ed. **The Illustrated Veterinary Encyclopedia for Horsemen.** Equine Research Publications, 1975.
Wagoner, D.M. ed. **Veterinary Treatments and Medications for Horsemen.** Equine Research Publications, 1977.
Way, R.F. and D.G. Lee. **The Anatomy of the Horse - A Pictorial Approach.** Philadelphia: J.B. Lippincott Company, 1965.
Welch, C.D. and C. Grey. "Soil Acidity and Liming," U.S. Department of Agriculture, Texas Agricultural Extension Service.
Western Horseman. June 1979
  Davis, R. "Building Fence," pp. 8-12.
Willis, L.C. **The Horse Breeding Farm.** New York: A.S. Barnes and Co., 1976
Wiseman, R.F. **The Complete Horseshoeing Guide.** 2nd ed. University of Oklahoma Press, 1973.
Wood, K.A. **The Business of Horses.** Rancho Santa Fe, California: Wood Publications, 1973.
Wynmalen, H. **Horse Breeding and Stud Management.** London: J.A. Allen & Co. Ltd., 1971.

GLOSSARY

ABDOMINAL CAVITY: The cavity that lies between the diaphragm and the pelvis. The horse's abdomen contains the digestive organs as well as the liver, pancreas, kidneys, and spleen.

ABDOMINAL CRYPTORCHIDISM: Retention of one or both testes in the abdominal cavity.

ABORTIFACIENT: An agent that causes abortion.

ABORTION: Expulsion of the products of conception from the uterus between day 30 and day 300 of gestation.

ABSCESS: Cavity containing pus.

ABSORPTION: Also resorption. Process by which the uterus takes up or absorbs a dead fetus.

ACCESSORY SEX GLANDS: Glands of the stallion's reproductive tract that contribute secretions to the total seminal fluid volume.

ACROSOME: A cap-like membrane which covers the head of the sperm cell.

ACRS: Accelerated Cost Recovery System.

ACUTE: Having short and relatively severe duration; not chronic.

ADHESION: Abnormal firm, fibrous attachment between two structures; often formed as a result of inflammation.

AFTERBIRTH: The expelled placenta.

AGALACTIA: Absence of milk in the udder after foaling.

AGAR: Culture media used for growing microorganisms.

AI: Artificial insemination.

ALLANTOCHORION: The extraembryonic membrane formed by fusion of the allantois and chorion.

ALLANTOCHORIONIC POUCH: Pocket-like structure in the allantochorion formed at the site of a sloughed endometrial cup.

ALLANTOIS: The inner sac of the placenta, formed when the ectoderm folds over the embryo beginning around day 16 after conception.

ALVEOLI: Plural of alveolus.

ALVEOLUS: A tiny, sac-like dilation in the mammary gland; each cavity is lined with milk-producing cells.

AMINO ACIDS: Organic building blocks for protein.

AMNION: The innermost of the membranes enveloping the embryo. It is filled with amniotic fluid in which the embryo is free to move and is cushioned from mechanical injury.

AMPULLA: The enlarged portion of the vas deferens in the stallion. The enlarged tubular portion of the fallopian tube near the uterine horn; the site of fertilization.

ANABOLIC STEROID: Chemical substance derived from testostorone; encourages protein-building.

ANATOMY: The science of organism structure. The structure of an organism.

ANDROGENS: Agents that stimulate the accessory sex glands, and encourage development of male sex characteristics.

ANEMIA: Condition in which the number or volume of red blood cells, hemoglobin, and packed red blood cells is below normal.

ANESTRUS: Period of sexual quiescence during which there is an absence of observable heat.

ANEUPLOIDY: State of having an abnormal number of chromosomes; not a multiple of the haploid number.

ANOPTHALMIA: Complete absence of the eye.

ANOREXIA: Lack of appetite for food.

ANOXIA: Absence of oxygen in arterial blood or tissues.

ANTERIOR: Pertaining to the front part of the body or of any structure.

ANTERIOR PITUITARY GLAND: A gland which is located at the base of the brain and which secretes several hormones, including FSH and LH.

ANTHELMINTIC: Agent administered to destroy parasitic larvae and/or parasite eggs.

ANTIBODY: Complex protein molecule which combines with molecules of antigen. Antibodies participate in the immune response which protects an animal against disease.

ANTIGEN: A substance that stimulates the formation of antibodies.

ANTITOXIN: An antibody to a biological poison.

ANUS: The external opening of the rectum.

AORTA: The major artery that carries blood leaving the left ventricle of the heart.

ARTERY: Blood vessel which carries blood away from the heart.

ARTHRITIS: Inflammation of a joint.

ARTIFICIAL INSEMINATION: The process of depositing semen into the female reproductive tract by artificial rather than by natural means.

ARTIFICIAL VAGINA (AV): Device used to collect semen for evaluation or artificial insemination.

ASCHHEIM-ZONDEK TEST: Test for detecting PMSG in the mare's blood; used for pregnancy diagnosis.

ASEPSIS: Freedom from infective organisms.

ASPERGILLUS: A genus of fungi of the class Ascomycetes.

ASPIRATION: Removal by suction of air or fluid from a body cavity or region.

ASSAY: Analysis; to analyze or examine.

ASYMPTOMATIC: Without symptoms.

ATRESIA: Absence of a normal opening.

ATRIUM: A chamber or cavity connecting several chambers or passageways. Specifically, each of the two upper chambers of the heart.

ATROPHY: Wasting of tissues, organs, or the entire body.

AUTOCLAVE: A sterilization device which uses pressurized steam. To sterilize in an autoclave.

AUTOLYSIS: Enzymatic self-digestion of cells.

AV: See artificial vagina.

BACTERICIDE: An agent that destroys bacteria.

BACTERIUM: Any microorganism in the class Schizomycetes. (plural: bacteria).

BALLOTTEMENT: A method of pregnancy diagnosis in which the uterine wall is tapped through the rectum and the fetus can be felt to bounce back against the wall.

BARKER FOAL: A foal suffering from a convulsive syndrome thought to be caused by lack of oxygen at birth. The foal is characterized by convulsions, blindness, and weakness. It also makes a barking noise.

BARREN: A mare (other than a maiden mare) that did not become pregnant during the last breeding season.

BARTHOLIN'S GLANDS: Glands that lubricate the posterior portion of the mare's reproductive tract.

BENIGN: Mild or nonmalignant in character.

BILATERAL CRYPTORCHIDISM: Retention of both testes in the body cavity.

BIOPSY: The process of removing tissue from living patients for diagnostic examination. A sample obtained by biopsy.

BIRTH CANAL: The organs, including the uterus, cervix, vagina, and vulva, which form the passageway of the reproductive tract.

BLASTOCYST: Early stage of embryonic development in which the embryo is hollow and spherical and contains the inner cell mass at one pole.

BREAKING WATER: Common term for the expulsion of allantoic fluid during the first stages of parturition.

BREECH BIRTH: Parturition in which the hindquarters are presented first.

BREEDER'S BAG: See "condom."

BREEDING CONTRACT: Legal and binding agreement between the stallion owner and mare owner stating the payment due for services and the contractual obligations of each party.

BREEDING ROLL: Cylindrical, padded instrument placed between the stallion and mare and above the penis during breeding to prevent the stallion from penetrating the mare too deeply.

BREEDING STITCH: Suture placed at the lower end of a Caslick's suture line to reinforce the healed incision.

BRIGHTFIELD: A type of microscope illumination.

BROAD LIGAMENTS: Fibrous bands of tissue which suspend the reproductive tract of the mare from the upper wall of the abdominal cavity.

BROAD-SPECTRUM ANTIBIOTIC: An agent having a broad range of antibiotic activity against a variety of microorganisms.

BRUCELLA ABORTUS: A species of bacteria which causes abortion in mares, Bang's in cattle, and undulent fever in man.

BULBOURETHRAL GLANDS: A pair of ovoid, lobulated glands located near the ischial arch which act as secreting glands by adding volume and nutrients to the ejaculate.

BULL POLE: A sturdy pole which can be fastened to the halter to keep an agressive stallion from crowding the handler.

CAESAREAN SECTION: Surgical removal of the fetus through an incision in the abdominal and uterine walls.

CAPITAL EXPENDITURE: Money spent for expanding and improving a business, but not including operating expenses.

CAPITAL GAIN: Profit resulting from the sale of a capital asset.

CARBOHYDRATE: Organic compound consisting of carbon, hydrogen, and oxygen. Includes sugars, starches, and lignins.

CARCINOMA: Malignant growth.

CARDIOVASCULAR SYSTEM: The system formed by the heart, arteries, and veins; the circulatory system.

CASLICK'S OPERATION: The process of stitching a section of the vulval lips together to prevent air and contaminants from entering the reproductive tract.

CATHETER: A hollow cylinder designed for withdrawing fluid from or introducing fluid into a body cavity.

CAUDAL: Denoting a position more towards the tail than some specified point of reference. Depending on context, the term is also synonymous with the terms *inferior* or *posterior*.

CELL: The structural unit that is the living basis of all plant and animal life. It consists of a mass of cytoplasm, a cell membrane, and a nucleus.

CELLS OF LEYDIG: See "interstitial cells."

CENTRAL NERVOUS SYSTEM: The brain, spinal cord, cranial nerves, and spinal nerves; carries messages to and from the brain.

CENTRIFUGE: The process of separating particles in suspension by centrifugal force. An instrument used for centrifuging.

CEREBELLUM: The large posterior brain mass lying above the pons and medulla and beneath the posterior portion of the cerebrum.

CEREBRUM: The largest portion of the brain, including the two hemispheres but not the medulla, pons, or cerebellum.

CERVICAL OS: Opening of the cervix.

CERVICAL STAR: The irregular, nonvillous portion of the placenta covering the cervical opening of the uterus.

CERVICITIS: Inflammation of the cervix.

CERVIX: The muscular, neck-like structure that separates the uterus from the vagina.

CHAIN SHANK: Length of chain approximately 12 to 18 inches long with flattened links. Can be attached to a lead rope and placed over a horse's nose or under its lower jaw for restraint.

CHESTNUTS: The horny growths found on the inner side of the horse's leg. They are located above the knee on the foreleg and below the hock on the hind leg.

CHORIOALLANTOIS: The placental membrane formed from the fusion of the chorion and the allantois.

CHORION: The outermost of the three placental membranes.

CHORIONIC GIRDLE: Shallow band of folds located on the allantochorion around day 25 of gestation.

CHORIONIC VILLI: Minute projections on the chorion which attach the placenta to the uterus. They are the sites of gas, nutrient, and waste exchange between the dam and the fetus.

CHROMOSOME: Protein strands within the cell nucleus. The chromosomes carry genetic information.

CHRONIC: Long-term; continued; not acute.

CID: See "combined immunodeficiency disease."

CILIA: Minute finger-like projections whose whipping actions help maintain fluid transport over certain membranes.

CLEFT PALATE: Congenital defect of the roof of the mouth involving a fissure in the soft and/or hard palates.

CLITORIS: A small, cylindrical, erectile body, situated in the lower portion of the vulva. It is the homologue of the penis.

COITUS: Copulation.

COLIFORMS: Enterobacteria other than Salmonella,Shigella, and Proteus.

COLORIMETER TUBE: A glass vial specially designed for use in a spectrophotometer.

COLOSTRUM: The first milk of the mare; it provides the newborn foal with protective antibodies and also acts as a laxative.

COLOSTRUM BANK: A supply of frozen colostrum collected from many mares and held for emergency use in colostrum-deprived foals.

COMATOSE: In the state of a coma.

COMBINED IMMUNODEFICIENCY DISEASE: Congenital defect which prevents the foal from producing sufficient antibodies.

COMPOUND MICROSCOPE: A microscope which contains two or more lenses.

CONCEPTION: Combination of genetic material of the sperm and ovum at ovulation.

CONCEPTUS: The products of conception (i.e., the embryo and the fetal membranes).

CONDOM: Rubber bag which is fitted on the stallion's penis for semen collection.

CONGENITAL: Existing at and usually before birth, referring to conditions that may or may not be inherited.

CONNECTIVE TISSUE: The fibrous tissue that binds and supports various body structures.

CONTAGIOUS EQUINE METRITIS: Highly contagious disease which causes inflammation of uterine, cervical, and vaginal membranes. Caused by *Haemophilus equigenitalis*.

CONVULSIVE FOAL SYNDROME: Also called barker, wanderer, dummy foal, or neonatal maladjustment syndrome. The condition is thought to be caued by oxygen deficiency of the brain before or during birth.

COPROPHAGY: Manure ingestion.

COPULATION: Sexual union; coitus.

CORPUS HEMORRHAGICUM: The blood-filled cavity formed at the site of ovulation immediately after the ovum is released.

CORPUS LUTEUM: The mass of endocrine cells formed after ovulation in the ovary at the site of a ruptured ovarian follicle. Also called a yellow body.

CORTICOSTEROID: The glucocorticoids produced by the adrenal glands and their synthetic counterparts.

CORTISOL: A steroid hormone produced by the adrenal gland. Also called hydrocortisone.

CORYNEBACTERIUM EQUI: A gram-positive bacteria found in the soil; causes foal pneumonia.

COVERSLIP: A square, thin piece of glass or plastic used to cover a sample on a microscope slide.

CRUDE PROTEIN: Refers to the total nitrogen content of a feed.

CRYPTORCHID: Male horse in which one or both testicles are retained in the body cavity.

CRYPTORCHIDISM: A condition in which one or both testicles are retained in the body cavity.

CUBONI TEST: An analytical technique which detects estrogen in mare's urine. Used as a pregnancy test between 120 and 290 days of gestation.

CULLING: Removal of selected animals from a herd.

CULTURE: The propagation of microorganisms on or in a medium. A mass of microorganisms on or in a medium.

CUMULUS OOPHORUS: A mass of cells surrounding the ovum inside a developing Graafian follicle.

CURETTAGE: A scraping of the interior of a cavity.

CUVETTE: See "colorimeter tube."

CYANOSIS: Blue or purple coloring of the mucous membranes in horses with reduced blood oxygenation.

CYST: A closed cavity or sac, usually containing fluid or semisolid material.

CYTOPLASM: The substance of a cell exclusive of the nucleus. It is surrounded by the cellular membrane and contains various organelles.

DAM: Female parent.

DAUGHTER CELL: The multiple cells that result from the division of a single cell during mitosis and meiosis.

DEGENERATION: Progressive deterioration.

DEHYDRATION: An abnormal depletion of body fluids.

DENSITY: The ration of mass to volume; compactness.

DENTITION: The development and cutting of teeth.

DEPRECIATION: Reducton in value.

DERMOID CYSTS: A tumor with an epidermal wall, sometimes containing teeth and hair.

DETERMINATION: The process by which undifferentiated cells become committed to developing into specific tissues or structures.

DEXAMETHASONE: A synthetic analogue of the hormone, cortisol.

DIAPHRAGM: A partition that is composed of muscle and membrane and which separates the abdominal and thoracic cavities.

DIESTRUS: The quiescent period between metestrus and proestrus, lasting approximately 12 to 13 days.

DIETHYLSTILBESTROL: A synthetic, nonsteroidal compound possessing estrogenic activity.

DIFFERENTIATION: Process by which general cells develop into specialized structures during embryonic development.

DIGESTIBLE ENERGY: The portion of the gross energy in a feedstuff that the animal is able to digest and absorb.

DISMOUNT SAMPLE: The portion of the stallion's ejaculate which trickles from the penis immediately following dismount from coitus. It is sometimes collected for semen evaluation.

DISTILL: To extract a substance by volatilization and condensation.

DIURETIC: A substance that increases the excretion of urine.

DOMINANT: A trait which is expressed in individuals who are hetorozygous for a particular gene.

DONOR MARE: The female which contributes the embryo for embryo transfer.

DOPPLER PRINCIPLE: An apparent change in wave frequency produced by motion of the source toward or away from the stationary observer, or by motion of the observer toward or away from the stationary source.

DORSAL: Pertaining to the back or denoting a position more toward the back surface; opposite of ventral.

DUCT: A tubular passage which conveys fluid.

DUCTUS ARTERIOSUS: A fetal vessel that connects the left pulmonary artery with the descending aorta. During the first two months after birth it normally becomes a fibrous cord called the ligamentus arteriosum.

DUCTUS DEFERENS: The secretory duct of the testicle running from the epididymis to the prostatic urethra where it terminates as the ejaculatory duct.

DUMMY FOAL: See "barker foal."

DUMMY MARE: See "phantom mare."

DYSGERMINOMA: A rare, malignant neoplasm of the ovary, that is a counterpart of seminoma of the testis.

DYSTOCIA: Abnormal or difficult birth.

EARLY EMBRYONIC DEATH: Death of the embryo prior to day 30 of gestation.

ECTODERM: The primary germ layer comprising the outer layer of cells in the developing conceptus.

ECTROPION: Eversion of the eyelid.

EDEMA: Excessive fluid accumulation in tissues, cells, or cavities.

EDEMATOUS: Affected by edema.

EFFERENT DUCTS: Vessels which transport material away from an organ or site (e.g., efferent ducts of the testes).

EJACULATION: Emission of seminal fluid.

ELECTRORETINOGRAPHY: Diagnostic method used to identify congenital night blindness.

EMACIATE: To lose flesh.

EMBRYO: In the horse, the conceptus up to approximately 30 or 40 days of gestation.

EMBRYOLOGY: The study of an individual's development from conception to birth.

EMBRYO TRANSFER: A method whereby a developing embryo is removed from its natural mother and implanted in the uterus of a host mother for the remainder of gestation.

EMISSION: A discharge.

ENCEPHALOMYELITIS: An acute viral disease of horses. Three strains affect the horse: Eastern equine e. (EEE), Western equine e. (WEE), and Venezuelan equine e (VEE).

ENCAPSULATED: Enclosed in a sheath or capsule.

ENDOCRINE: The characteristic of being secreted internally; pertaining to a ductless gland that produces an internal secretion.

ENDOCRINOLOGY: The study of internal secretions, their actions, and their internal relationships.

ENDODERM: The innermost of the three primary germ cell layers of the embryo. It gives rise to the primitive gut.

ENDOMETRIAL CUPS: Raised structures formed on about day 36 of pregnancy in the gravid uterine horn.

ENDOMETRITIS: Inflammation of the endometrium.

ENDOMETRIUM: The membrane that comprises the lining of the uterine wall. It consists of a thin epithelial lining, a glandular layer, and a layer of connective tissue.

ENDOSCOPE: An instrument used for the examination of the inerior of a hollow organ or canal.

ENTROPION: Infolding of the eyelid.

ENVIRONMENT: All of the external conditions that influence an organism.

EPIDIDYMAL DUCT: The vessel that receives sperm from the efferent ducts. It is thought to reabsorb fluid, thereby concentrating the sperm.

EPIDIDYMIS: A U-shaped, tubular structure attached to the long axis of each testis, comprised of a head, body, and tail.

EPIDURAL ANESTHESIA: Spinal anesthesia achieved by injection of anesthetic into the spinal canal, but outside the dura mater.

EPINEPHRINE: An adrenal gland hormone that raises blood pressure.

EPIPHYSEAL GROWTH PLATE: Cartilaginous area that lies between the metaphysis and the epiphysis; site of bone elongation.

EPIPHYSIS: End of a long bone, which is separated from the bone by a plate of cartilage in the immature animal.

EPIPHYSITIS: Swelling of the epiphyseal plate or plates.

EPISIOTOMY: Incision of the sutured vulva.

EPITHELIAL INCLUSION CYSTS: A cyst formed when a small portion of the ectoderm is surrounded by the mesoderm.

EPITHELIUM: The avascular cell layer that covers all free surfaces of the body.

ERECTILE TISSUE: Tissue containing vascular spaces that become engorged with blood.

ERECTION: The state of erectile tissue when blood-filled; particularly the condition of the external genital organs.

ERGOT: An elongated, black-purple mass of fungus which replaces the grain of rye. Ingestion of the fungus causes uterine muscle contractions, thereby precipitating abortion.

ERGOTHIONEINE: A derivative of histidine; present in ergot.

ERYTHROCYTE: Red blood cell.

ESCHERICHIA COLI: A bacterium that is found normally in the intestines of vertebrates and is widely distributed in nature.

ESTROGEN: General term for any substance that exerts an estrogenic effect; formed naturally by the ovary, placenta, and testis.

ESTROUS: Pertaining to estrus.

ESTROUS CYCLE: The cyclic series of periods consisting of estrus, metestrus, diestrus, and proestrus.

ESTROUS SYNCHRONIZATION: Controlling the time of estrus by advancing or delaying the progress of the estrous cycle.

ESTRUS: Period during which the mare is receptive to the stallion.

EXTENDER: Formula used for semen dilution.

EXTRAEMBRYONIC MEMBRANES: Membranes that surround, nourish, and protect the embryo during gestation, but which are discarded after birth. The yolk sac, amnion, and allantois.

EXTRAUTERINE: Outside the uterus.

EXUDATE: Fluid formed in response to tissue injury or infection.

EXUDE: To gradually seep through tissue.

FAILURE OF PASSIVE TRANSFER: Condition in which the foal does not receive or utilize antibodies from the mare's colostrum. Also FPT.

FALLOPIAN TUBE: Oviduct; the tube leading from the ovary to the fallopian attachment to the end of the uterine horn.

FEBRILE: Feverish.

FERAL: Not domesticated.

FERTILE: Able to produce offspring.

FERTILITY: The quality or state of being fertile.

FERTILIZATION: The union of the sperm cell and the ovum.

FESCUE: Grass of the genus Festuca.

FETOTOMY: Dispartment of the fetus for removal from the uterus.

FETUS: The unborn foal from approximately day 40 of gestation to birth.

FIBRIN: An elastic protein found in the blood; aids in blood clotting. Formed from fibrinogen under the action of thrombin.

FIBRIN TAGS: Scarred areas of the ovary caused by strongyle migration that may interfere with normal ovarian function.

FIMBRIAL CYSTS: Cysts occurring in the region of the fimbria or oviducts.

FIBROSIS: An abnormal increase in the amount of fibrous connective tissue in an organ, part, or tissue.

FLACCID: Hanging in loose folds or wrinkles.

FLAGELLA: Plural of flagellum.

FLAGELLUM: A lash or whip-like appendage.

FLEHMAN: Common behavior pattern consisting of curling and raising the upper lip.

FLOATING: Removal of uneven edges of the teeth by filing.

FLORA: Various microscopic organisms inhabiting an individual.

FOAL HEAT: The first postfoaling estrus period.

FOAL HEAT SCOURS: Diarrhea that commonly occurs in the foal seven to nine days after foaling.

FOLEY CATHETER: A catheter equipped with an inflatable device used to retain the catheter within the body.

FOLLICLE: A small sac or cavity.

FOLLICLE STIMULATING HORMONE (FSH): A hormone secreted by the anterior pituitary gland. In the mare, FSH stimulates follicular growth and controls estrogen secretion by the ovary. In the stallion, FSH stimulates sperm production.

FOLLICULAR STAGE: Portion of the estrous cycle including proestrus and estrus.

FORAMEN OVALE: An opening in the fetal heart which closes after birth.

FPT: See "failure of passive transfer."

FREEZE MARKING: The application of a permanent identification mark using an instrument which has been cooled in liquid nitrogen.

FRIEDMAN TEST: An analytical technique for detecting pregnant mare serum gonadotropin in the mare's blood. Used as a pregnancy test.

FROG: Wedge-shaped mass of tissue which lies between the bars of the horse's foot. The frog absorbs shock upon the hoof's impact with the ground.

FSH: See "follicle stimulating hormone."

FULL TERM: Refers to the mature state of the fetus at birth.

FUNGUS: A plantlike organism that feeds on organic matter.

GAMETE: A germ cell; sperm or ovum.

GANGRENE: Tissue decay due to obstruction of the blood supply.

GASTROINTESTINAL: Of the stomach and the intestines.

GEL FRACTION: The viscous, gelatinous fraction of the ejaculate.

GELDING: A castrated male horse.

GEL-FREE VOLUME: The sperm-rich fraction of the ejaculate which remains after the gel-fraction has been removed by filtration.

GENE: The functional unit of heredity carried by the chromosomes.

GENERATION INTERVAL: The period from birth to reproductive age.

GENETIC POOL: All of the genes in a population.

GENETIC VARIATION: Heterozygosity within a population.

GENITAL RIDGE: A thickened area in the middle of the urogenital ridge; gives rise to the gonads.

GENITAL TRACT: Reproductive tract; general term referring to the reproductive passageway and all of its associated organs.

GENITALIA: The genitals.

GENITALS: The organs of reproduction.

GERM CELL LAYERS: The three primary cell layers that arise during early embryonic development.

GERMINAL CELLS: Cells which give rise to gametes.

GESTATION: Pregnancy.

GLAND: A secreting organ.

GNRH: See "gonadotropin releasing hormone."

GOBLET CELLS: Mucus-producing secretory cells found within the cervix.

GOLDEN SLIPPERS: Soft pads sometimes found covering the foal's hooves at birth.

GONAD: An organ that produces gametes; the testis of a male or the ovary of a female.

GONADOTROPIC: Promoting gonadal growth or function.

GONADOTROPIN: A hormone that promotes gonadal growth or function.

GONADOTROPIN RELEASING HORMONE (GNRH): A hormone secreted by the hypothalamus; stimulates or initiates pituitary activity.

GRAAFIAN FOLLICLE: A mature ovarian follicle.

GRANULOSA CELL TUMOR: A tumor of the ovary arising from the membrana granulosa of the Graafian follicle.

GRANULOSA CELLS: Cells which line the inside of the ovarian follicle and which support the developing ovum.

GRAVID: Pregnant.

HABRONEMA: Stomach worm.

HARROW: To disturb the surface of the soil by means of a toothed drag.

HCG: See "human chorionic gonadotropin."

HEAT SYNCHRONIZATION: See "synchronization."

HEMACYTOMETER: An instrument used to count microscopic items (e.g., blood cells, sperm) in a measured volume of fluid.

HEMATOCRIT: The percent of the volume of a blood sample occupied by cells.

HEMATOMA: A localized mass of blood that is restricted to a definite space or within a tissue.

HEMOGLOBIN: The oxygen-carrying protein pigment of red blood cells.

HEMOLYSIS: The alteration, dissolution, or destruction of red blood cells in such a manner that hemoglobin is released.

HEMORRHAGE: Bleeding; release of blood, usually profuse.

HEMOSPERMIA: The presence of blood in the semen.

HERPESVIRUS: A group of viruses, one of which causes equine rhinopneumonitis.

HETEROZYGOUS: Possessing two different alleles at one particular locus on a pair of homologous chromosomes.

HINNEY: The offspring resulting from a cross between a horse stallion and a jennet (female donkey).

HIPPOMANES: Brown, free-floating objects sometimes found in the placental tissues after birth and thouht to be formed from mineral and protein deposits in the allantoic cavity.

HISTOLOGY: The study of tissues.

HOBBLES: A device used to restrain an animal by restricting its leg movement.

HOMOZYGOUS: Possessing two identical alleles at one particula locus on a pair of homologous chromosomes.

HORMONE: A chemical which is formed in one organ and carried by the blood to other organs to elicit a specific response.

HOT IRON BRAND: An identification mark made on the body of an animal through the use of a heated element.

HUMAN CHORIONIC GONADOTROPIN (HCG): A hormone found in the urine of women during the first 50 days of pregnancy. Its activity is similar to that of pregnant mare serum gonadotropin and is sometimes used to stimulate ovulation in mares.

HYDROCEPHALUS: A condition marked by an excessive accumulation of fluid dilating the cerebral ventricles, thinning of the brain, and separation of cranial bones; congenital hydrocephaus is due to developmental defect of the brain.

HYMEN: The thin membrane which partially covers the external vaginal opening; sometimes observed in young mares.

HYPERCARBIA: The presence of excessive amounts of carbon dioxide in circulating blood.

HYPERPLASIA: An increase in the number of cells in a tissue or organs, excluding tumor formation.

HYPERTROPHY: Overgrowth; general increase in bulk of a part or organ not due to tumor formation. Use of the term may be restricted to denote greater bulk through increase in size, but not in number, of the individual tissue elements.

HYPOCALCEMIA: Abnormally low calcium levels in the circulating blood.

HYPOPLASIA: Underdevelopment of tissue or an organ, usually due to a decrease in the number of cells. Atrophy due to destruction of some of the elements and not merely to their general reduction in size.

HYPOTHALAMUS: A gland which is prominently involved in the functions of the autonomic nervous system and in endocrine mechanisms.

HYPOTHYROIDISM: Diminished production of thyroid hormone, leading to conditions associated with thyroid insufficiency.

IMMUNITY: Power to resist infection.

IMPERFECT TESTICULAR DESCENT: Condition in which the testes are near the body at the external inguinal ring.

IMPERFORATE: Atretic; closed; without opening.

IMPLANTATION: Also placentation. Attachment of the placenta to the uterine endometrium between days 45 and 150 of gestation.

IMPOTENCY: In the stallion, the reduced ability, inability, or lack of desire to breed.

INCUBATOR: A container in which a controlled environment (e.g., temperature, humidity) can be maintained.

INCUBATOR STAGE: A device which attaches to a microscope stage to maintain the temperature of microscopic samples.

INDUCED LABOR: To cause the mare to begin parturition or labor through the use of hormones or drugs.

INDWELLING CATHETER: A hollow cylinder designed to be passed into an organ or passageway and left in place for the purpose of drainage or infusion.

INFANTILE: Very young or undeveloped.

INFARCT: An area of necrosis resulting from the arrest of or sudden insufficiency of local arterial or venous blood supply.

INFECTION: Abnormal multiplication of microorganisms in the body.

INFERTILITY: Diminished or absent fertility. In the stallion, the reduced ability of the sperm to fertilize the mare's ovum, or the reduced ability of the stallion to produce sperm.

INFLAMMATION: The response of tissue to injury or abnormal stimulation, usually involving a reaction which leads to healing.

INFUNDIBULUM: The funnel-shaped extremity of the fallopian tube near the ovary.

INFUSION: The introduction of fluid other than blood.

INGUINAL: Relating to the groin.

INGUINAL CANAL: The passageway from the internal to the external abdominal ring.

INGUINAL CRYPTORHIDISM: Retention of one or both testes in the inguinal canal.

INGUINAL HERNIA: Condition in which a loop of intestine passes into the inguinal canal.

INHERENT: Natural; inborn; innate.

INHERITED LETHAL: A gene passed from parent to offspring which is lethal to the embryo, fetus, or newborn.

INNER CELL MASS: The cluster of cells at one pole of a blastocyst from which the embryo develops.

INSULIN: A hormone produced by the pancreas.

INTERSEX: Individual having both male and female characteristics.

INTERSTITIAL: Relating to spaces or interstices in any structure.

INTRAUTERINE: Within the uterus.

INTROMISSION: Insertion.

IN UTERO: In the uterus; prenatal.

IN VIVO: In the living body.

INVERT: To turn inside out.

INVOLUTION: The process by which an organ returns to its former size and cellular state. Specifically, the process by which the uterus and mammary gland return to a normal state after pregnancy or lactation.

ION: Electrically charged atom.

ISOIMMUNIZATION: The development of antibodies as a result of antigenic stimulation by material contained in or on the red blood cells of another individual of the same species.

ISTHMUS (FALLOPIAN TUBE): The long, slender section of the fallopian tube.

JAUNDICE: Icterus; yellow discoloration.

JENNET: A female donkey. Also jenny.

KICKING BOOTS: Padded boots sometimes placed on the mare's back feet to reduce any blow sustained by the stallion during breeding.

KURASAWA METHOD: Pregnancy test which uses cervical mucus samples for diagnosis.

LABIA: The vulval lips arranged vertically on either side of the vulval opening.

LACTATION: The production of milk. Th period of milk production.

LACTATIONAL ANESTRUS: Failure to show estrus before or after foal heat, probably caused by prolonged corpus luteum function.

LAMINITIS: Inflammation of the laminae which is caused by either infectious or noninfectious agents.

LATERAL: Denoting a position away from the median plane of the body; towards the side; on the side.

LEG STRAP: A strap (usually leather) used to restrain an animal by securing the front leg.

LENS PAPER: A soft, lintless paper specially prepared for cleaning lenses, filters, and other highly polished optical glass surfaces.

LETHAL GENE: A gene whose phenotypic effect results in the death of the embryo, fetus, newborn, or mature individual.

LH: See "luteinizing hormone."

LIBIDO: Sexual drive or desire.

LIGAMENT: Any tough, fibrous band which connects or supports bones, cartilage, or muscles.

LIMEWATER: A solution of calcium hydroxide and water in the proportion of 1:700; used as an antacid.

LIQUOR FOLLICULI: An albuminous fluid contained in the Graafian follicle.

LITMUS PAPER: Paper which is sensitive to pH levels.

LIVE-DEAD PERCENTAGE: The ratio of live to dead sperm cells found in a semen sample.

LIVE FOAL GUARANTEE: Stallion contract provision which guarantees the mare owner a live foal as a result of the purchased breeding. The guarantee usually gives the mare owner the right to rebreed the next season to the same stallion.

LOAFING SHED: A small structure, usually consisting of a roof and three sides, used to shelter pastured horses.

LOBE: A flap or projection that is imperfectly connected with other parts.

LOBULES: Small lobes or primary divisions of a lobe.

LOCOISM: Poisoning by loco weed consumption.

LUMEN: The cavity or potential opening within a tube or tubular organ.

LUTEAL CELLS: Pertaining to the progesterone-producing cells of the corpus luteum.

LUTEAL STAGE: Portion of the estrous cycle including metestrus and diestrus.

LUTEINIZATION: The transformation of the mature ovarian follicle and its theca interna into a corpus luteum after ovulation.

LUTEINIZING HORMONE: A hormone produced by the anterior pituitary that stimulates the development of corpora lutea in the female, and the development of interstitial tissue in the male. Also LH.

LUTEOLYTIC: Causing the degeneration of luteal tissue.

LYMPHOCYTES: White blood cells formed in lymphoid tissue.

LYOPHILIZED: Freeze-dried.

LYSINE: An amino acid that is essential in the diet of the growing horse.

MACROVILLI: Tuft-like tissue folds in the placental chorion which become established in maternal endometrial pockets to form microcotyledons.

MAIDEN: A mare that has not been bred.

MALIGNANT: Resistant to treatment; occurring in severe form, and frequently fatal; tending to become worse. In the case of a neoplasm, having the property of uncontrollable growth and/or recurrence after removal.

MAMMARY GLAND: The udder; a collection of highly modified oil glands in the mare's inguinal region that collect nutrients and synthesize and secrete milk.

MANDIBLES: Jaws.

MASTITIS: Inflammation of the mammary gland. A disease which involves acute, gangrenous, chronic, and subclinical forms of inflammation of the udder, and is due to a variety of infectious agents.

MECONIUM: The first intestinal discharge of the newborn.

MEDIAL: Near or towards the median plane (or middle) of the body; relating to the middle or center.

MEIOSIS: The process of cell division which produces gametes with half the normal chromosome number.

MELANOMA: A malignant neoplasm derived from melanin-forming cells. It may occur in the skin of any part of the body, in the eye, or in the mucous membranes of the genitalia, anus, oral cavity, or other sites.

MENINGEAL HEMORRHAGE: Fluid accumulation in the membranes that surround the brain and spinal cord.

MENINGES: The membranous coverings of the brain and spinal cord.

MESODERM: The middle layer of the three primary layers of the embryo, which lies between the endoderm and ectoderm.

METESTRUS: The early luteal period of the estrous cycle which occurs between estrus and diestrus.

METRITIS: Inflammation of the uterus.

MICROBIOLOGY: The study of microorganisms.

MICROCOTYLEDON: The attachment of the fetal placenta to the maternal endometrium by the association of macrovilli and endometrial pockets.

MICRON: A unit of linear measurement in the metric system (10^{-3}mm or 10^{-6}M).

MICRONUTRIENTS: Any substance that plays a nutritional role in the body and which is required only in minute quantities in the diet.

MICROPTHALMIA: Abnormal smallness of one or both eyes.

MICROVILLI: Tiny projections on the allantochorion which become embedded in the uterine epithelium during placentation.

MILK LINE FILTER: A filter used to trap foreign material in milk as it passes through a pipeline leading to a bulk storage tank. It may also be used to filter the gel fraction from the ejaculate as it runs through the artificial vagina during semen collection.

MILK REPLACER: A liquid formulation usually fed to orphan foals.

MILLIMICRON: One thousandth of a micron.

MIP-TEST: Mare Immunological Pregnancy Test. Commercial test which diagnoses pregnancy by detecting pregnant mare serum gonadotropin in the mare's blood.

MITOSIS: The type of cell division which occurs in somatic cells. Mitosis results in the production of genetically identical daughter cells.

MONORCHID: A horse affected by monorchidism.

MONORCHIDISM: Complete absence of one or both testes.

MORPHOLOGY: In semen evaluation, the examination of semen samples for normal sperm cell structure.

MORULA: The tight mass of cells formed at around day 3 to 4 by repeated cleavage of the original ovum.

MOSAIC: An abnormality of chromosome division resulting in two or more types of cells containing different numbers of chromosomes.

MOSAICISM: The condition of being mosaic.

MOTHERING ABILITY: Willingness of the mare to protect her foal and her physical capacity to provide the young foal with nutrients.

MOTILITY: The ability of a sperm cell to move in a normal, forward manner. The measure of the percentage of sperm in a sample that are able to move in a normal manner.

MUCIN TEST: See "Kurasawa method."

MUCOSA: A mucous membrane.

MUCOSAL FOLDS: Folds of mucous tissue.

MICRON: A unit of length equal to one thousandth of a millimeter.

MUCOUS: Relating to mucus or a mucous membrane.

MUCUS: The clear, viscous secretion of the mucous membranes, consisting of mucin, epithelial cells, leukocytes, and various inorganic salts suspended in water.

MULE: Hybrid offspring resulting from the cross of a male donkey (jack) and a female horse (mare).

MULLERIAN DUCTS: Duct system in the early embryo which in the female develops into the fallopian tubes, uterine horn, uterine body, cervix and vagina; in the male these ducts degenerate.

MUTATION: A change in the character of a gene that is perpetuated in subsequent divisions of the cell in which it occurs. Manipulation of the foal's position during parturition.

MYCOTIC INFECTION: Any disease caused by fungal infection.

MYOEPITHELIAL: Contractile epithelial tissue.

MYOMETRIAL: Pertaining to the myometrium.

MYOMETRIUM: The muscular center layer of the uterine wall.

NATURAL BREEDING SEASON: Period of year when conception rates are highest under natural conditions.

NAVICULAR DISEASE: Chronic inflammation of the navicular bone and its associated structures.

NECROSIS: The pathologic death of one or more cells, or of a portion of a tissue or an organ. Often characterized by atrophy of the affected organ or tissue.

NECROTIC: Pertaining to or affected by necrosis.

NEONATAL ISOERYTHROLYSIS: Condition in which the blood types of the mare and foal are incompatible, and the mare produces antibodies that destroy the foal's red blood cells. These antibodies are concentrated in the mare's colostrum and are transmitted to the foal when it suckles after birth.

NEOPLASM: New growth; tumor; an abnormal tissue that grows by cellular proliferation more rapidly than normal and continues to grow after the stimuli that initiated the new growth ceases.

NEONATE: Foal less than three to four days of age.

NEUROHYPOPHYSIS: Main part of the posterior lobe of the pituitary gland.

NI: See "neonatal isoerythrolysis."

NUCLEUS: In cytology, the mass of protoplasm within the cytoplasm of a cell that encloses the cell's chromosomes as well as a number of other structures.

NURSE MARE: A mare selected for mothering ability and bred each year for the purpose of raising orphan foals.

NYMPHOMANIA: A mare that is constantly receptive to the stallion.

OOCYTE: Immature ovum.

OOGENESIS: A series of cell divisions starting with the primary germ cells in the ovary and resulting in the production of ova.

OPEN: Nonpregnant.

OPPORTUNIST: Living at the expense of or on the destruction of another.

ORCHITIS: Inflammation of testicular tissue.

ORGANELLES: A specialized part of a cell having a specific function.

ORGANOPHOSPHATE: One of the phosphorus-containing chemicals which are used as drugs or pesticides.

OSSIFICATION: To form bone or change into bone.

OSTIUM ABDOMINALE: The opening in the infundibulum that leads into the fallopian tube.

OSTIUM UTERINUM: The fallopian entrance in the uterine horn.

OUTSIDE MARE: Mares that are not owned by the farm where they are being boarded or bred.

OVA: Plural of ovum.

OVARIAN HYPOPLASIA: Underdeveloped or immature ovary.

OVARIECTOMY: Removal of one or both ovaries.

OVARY: One of the paired gonads in the female which contain germinal cells.

OVIDUCT: Fallopian tube; the tube that leads from the ovary to the uterine horn.

OVULATION: The release of an ovum from a mature follicle.

OVULATION FOSSA: An anatomical feature of the equine ovary located on the pinched-in face of the organ; site at which ovulation occurs in the horse.

OVUM: Germinal cell produced in the ovary.

OXYTOCIN: A hormone produced by the posterior pituitary which stimulates myometrial contraction.

PAIR BONDING: The formation of a strong social attachment between two animals.

PALPATION: Feeling or perceiving by the sense of touch.

PANCREAS: An elongated gland which secretes pancreatic juice into the small intestine and produces the hormone, insulin.

PAPILLOMA: Wart. A benign tumor with a fibrous center.

PAPULE: A small, circumscribed, solid elevation on the skin.

PAROOPHORON CYSTS: Rudimentary tubules in the broad ligament between the epoophoron and the uterus; remnants of the tubules and glomeruli of the lower part of the Wolffian body.

PAROTID: Situated near the ear.

PAROVARIAN CYSTS: Relating to the paroophoron; situated beside the ovaries.

PARTURITION: Process of giving birth.

PASSIVE IMMUNITY: Disease resistance acquired though the transfer of antibodies from another individual.

PASTURE BREEDING: Program of breeding under semiferal or natural conditions.

PATHOGEN: Any virus, microorganism, or other substance that causes a disease.

PATHOLOGIST: An individual who studies the causes, development, and results of disease processes.

PATHOLOGY: The study of the causes, development, and results of disease processes.

PELVIC BRIM: The most anterior portion of the floor of the pelvis.

PELVIC CAVITY: The orifice bounded by the bones of the pelvis.

PELVIS: The ring of bone, along with its ligaments, formed by the os coxae (the pubic bone, illium, ischium), the sacrum, and the occyx.

PENDULOUS: Hanging freely or loosely.

PENIS: The male organ of copulation. In mammals it is composed of veins and arteries, spongy erectile tissue, and a urethra.

PERINEAL REGION: See "perineum."

PERINEUM: The area between the thighs, originating at the anus; in the male it terminates at the scrotum, in the female at the mammary glands.

PERISTALSIS: Wave-like contractions of the walls of a tubular structure.

PERITONITIS: Inflammation of the peritoneum.

PERITONIUM: The serous sac which lines the abdominal cavity and covers most of the viscera contained therein.

PHARYNX: The cavity which joins the mouth and nasal passages to the esophagus and larynx.

PHANTOM MARE: A raised, padded support that is mounted by the stallion during semen collection.

PHASE CONTRAST MICROSCOPE: A microscope which converts phase differences in light rays reflected by or passed through an object into differences in intensity, thus producing an image in which the details are distinct despite their actual lack of contrast. Frequently used for examination of nearly transparent specimens, such as sperm cells.

PHOTOPERIOD: Length of time an organism is exposed to light, either natural or artificial. Photoperiods control many biological activities, such as the reproductive activity of some animals, including the horse.

PHYSIOLOGICAL SALINE: A 0.9 percent sodium chloride solution.

PHYSIOLOGY: The science that deals with living things, with the normal vital processes of animal and vegetable organisms.

PILOERECTION: Erection of hair, which in animals forms an insulating air space around the body.

PIPETTE: A tube used to transport small amounts of a gas or liquid in laboratory work; usually marked to deliver a particular volume of fluid with quantitative accuracy.

PLACENTITIS: Inflammation of the placenta.

PLACENTA: A vascular organ that surrounds the fetus during gestation. It is connected to the fetus by the umbilical cord and serves as the structure through which the fetus receives nourishment from, and eliminates waste matter into, the maternal circulatory system.

PLACENTATION: Attachment of the placenta to the uterine wall.

PLACENTITIS: Inflammation of the placenta.

PLEURAL CAVITY: Cavity which contains the lungs and thorax.

PMSG: See "pregnant mare serum gonadotropin."

POLAR BODY: A cell formed during oogenesis that contains a complete set of chromosomes but very little cytoplasm.

POLYMYXIN B: A mixture of antibiotic substances which are effective against Gram-negative bacteria, such as pseudomonas.

POSTERIOR: Behind or after in place. In veterinary anatomy, the term is used to denote some structures of the head. Also "caudal."

POSTNATAL: After birth.

POSTPARTUM: Pertaining to, or occurring during, the period following parturition.

POSTPUBERAL: Subsequent to the period of puberty.

POSTSPERM FRACTION: Sticky gel secreted by the seminal vesicles.

POTABLE: Suitable to drink.

PRECIPITATE: A solid separated from a solution or suspension.

PREGNANT MARE SERUM GONADOTROPIN (PMSG): Hormone produced by the endometrial cups between days 40 and 180 of gestation. Its activity in animals is similar to that of the follicle-stimulating hormone.

PREMATURE: Pertaining to a foal born between day 300 and day 325 of gestation.

PRENATAL: Before birth.

PREPUBERAL: Before puberty.

PREPUBIC TENDON: The tendon that supports the uterus in the abdominal cavity.

PREPUCE: The fold of skin that covers the glans penis.

PREPUTIAL ORIFICE: The opening to the prepuce.

PRESPERM FRACTION: That fraction of the ejaculate which precedes the sperm-rich fraction, cleaning and lubricating the penis.

PRIMARY GERM CELLS: Cells in the testes which give rise to spermatogonium.

PRIMARY SPERMATOCYTE: A germ cell in the testis which gives rise to a secondary spermatocyte.

PRIMARY OOCYTE: A germ cell in the ovary which gives rise to a secondary oocyte.

PROESTRUS: The two-day period of increasing receptivity in the mare which precedes estrus.

PROGESTERONE: A hormone produced by the corpus luteum which quiets uterine muscle contractions.

PROLACTIN: A hormone of the anterior pituitary that stimulates the secretion of milk and, possibly, mammary growth.

PROSTAGLANDIN: A class of physiologically active substances which are present in many tissues and which have a variety of functions. In reproduction, the term is used to indicate the specific prostaglandin, PGF_2 alpha.

PROSTATE: A gland consisting of two bodies situated on either side of the urethra. It produces secretions which are added to the seminal fluid.

PSYCHIC STIMULATION: Mental excitement or arousal.

PUBERTY: Age at which gametogenesis and the secretion of gonadal hormones begins.

PUPILLARY REFLEX: Reaction of the pupil to stimuli.

PURULENT: Consisting of, containing, or forming pus.

PYOMETRA: An accumulation of pus in the uterus.

QUIESCENT: Resting; in repose; not moving.

RAW SEMEN: Semen which has not been treated by filtering or by additions of antibiotics, extenders, or other ingretients. Refers primarily to semen before processing for artificial insemination.

REAGENT: Any substance that is added to another substance to participate in a chemical reaction.

RECESSIVE: A trait which is expressed when an individual is homozygous for a particular gene but not when an individual is heterozygous for that gene.

RECIPIENT MARE: The mare receiving a transferred embryo.

RECTAL PALPATION: Examination of organs or structures adjacent to the rectum by feeling for size, texture, and other characteristics through the rectal wall.

RECTUM: The terminal portion of the alimentary tract.

RECUMBENT: Position of lying down.

REPRODUCTIVE HISTORY: Cumulative record of a mare's cyclic activity, gestations, parturitions, infections, and number of foals or abortions.

REPRODUCTIVE STATUS: The current condition of a mare in terms of cyclic ovarian activities, stage of gestation, or time of parturition.

RESERVATION SHEET: Form or sheet listing all mares that are scheduled to be bred by a specific stallion during the breeding season.

RESORPTION: The removal of an exudate, blood clot, pus, etc., by absorption. A loss of substance by lysis.

RESTRAINT: Immobilization of an animal.

RESUSCITATION: The act of restoring the heartbeat and/or oxygen supply.

RETAINED PLACENTA: Afterbirth or placenta which is not expelled within 6 hours after foaling.

RETE CYST: A pouch or sac containing a network of blood vessels or nerve fibers.

RINGER'S SALINE: A clear, colorless liquid containing sodium chloride (NaCl), potassium chloride (KCl), and calcium chloride (CaCl) in boiled, purified water.

ROUGHAGES: Pertains to feedstuffs that are high in cellulose and fibers. These feeds, usually grass and hay, provide bulk to the digestive system vital to proper digestion.

SALINE: Relating to, of the nature of, or containing salt. A solution containing sodium chloride. See "physiological saline."

SALPINGITIS: Inflammation of the fallopian tube.

SCHISTOSOMA REFLEXUM: A congenital malformation that results in an opening in the ventral midline.

SCOLIOSIS: Curvature of the spine.

SCROTAL HERNIA: Condition in which a portion of the intestine enters the scrotum.

SCROTUM: The pouch which contains the testicles.

SECONDARY OOCYTES: Germ cells which give rise to ova.

SECONDARY SPERMATOCYTES: Germ cells in the testis which give rise to mature sperm cells.

SELECTION INDEX: A system in which traits are given a specific emphasis. The index is then used in culling breeding stock.

SEMEN: The ejaculate. The fluid comprised of sperm cells and secretions of the testes, seminal vesicles, prostate, and bulbourethral glands.

SEMEN EVALUATION: Examinaton of semen for characteristics which indicate its capacity for fertilization.

SEMINAL FLUID: See "semen."

SEMINAL VESICLES: Pear-shaped accessory sex glands which add secretions to the volume of the ejaculate and aid in the nourishment of sperm cells.

SEMINIFEROUS TUBULES: Tubules located in the testes in which sperm cell production occurs.

SEPSIS: The presence of various pus-forming and other pathogenic organisms, or their toxins, in the blood or tissues.

SEPTIC: Infectious. Of or characteristic of septicemia.

SEPTICEMIA: Systemic disease caused by the presence of pathogenic microorganisms and their toxic products in the blood.

SEPTUM: A thin wall that divides cavities or masses of tissue.

SEROTYPE: The type of microorganism as determined by the kinds and combinations of constituent antigens present in the cells.

SERTOLI CELLS: Elongated cells in the tubules of the testes to the ends of which spermatids become attached apparently for the purpose of nutrition until they are mature spermatozoa. Also called nurse cells.

SEX CHROMOSOMES: Chromosomes designated "X" and "Y" which exert genetic control over sex determination.

SHEATH: A tubular structure enclosing or surrounding an organ or part.

SIGNALMENT: A system of recording external markings and body conformation for the purpose of identification.

SILENT HEAT: Estrus period in which the mare ovulates but fails to show behavioral signs of heat.

SLIDE WARMER: A device which is used to regulate the temperature of microscope slides.

SMEGMA: Secretions of sebaceous glands. Often used to refer to the accumulated material in the preputial area of a stallion's sheath.

SOW MOUTH: Condition in which the upper jaw protrudes beyond the lower jaw, causing malalignment of the upper and lower teeth. Also overshot jaw or parrot mouth.

SPECTROPHOTOMETER: An instrument used to measure the transmission of light through a liquid sample.

SPECULUM: An instrument for enlarging the opening of a canal or cavity in order to permit examination.

SPERMATIC CORD: The cord-like structure that suspends the testicle within the scrotum and which contains the vas deferens, blood vessels, and nerves supplying the testicle.

SPERMATID: A cell derived from a secondary spermatocyte.

SPERMATOGENESIS: Development of mature sperm cells from primitive germ cells.

SPERMATOGENIC WAVE: The phenomenon in which sperm cells of varying maturity are found in the seminiferous tubules.

SPERMATOGONIUM: A primitive germ cell which develops through mitosis and meiosis into a sperm cell.

SPERMATOZOA: Male gamete or sex cell which is composed of a head, neck, and tailpiece. The head contains the genetic information which is transmitted during fertilization, and the tail propels the cell.

SPERMICIDAL: Having the characteristics of a spermicide.

SPERMICIDE: An agent which kills sperm, such as water, soap, antiseptics, and germicides (except antibiotics). An agent destructive to sperm.

SPERM-RICH FRACTION: The portion of the stallion's ejaculate which contains 80 to 90 percent of the ejaculate's total number of sperm.

SPHINCTER MUSCLE: Circular muscle fibers or specially arranged oblique fibers which partially or totally reduce the lumen of a tube or the interior of an organ.

STALLION CONDOM: Rubber receptacle used to cover the stallion's penis during coitus for semen collection.

STALLION RING: Rubber or plastic ring secured at the base of the glans penis of the stallion which discourages penile erection.

STEREOSCOPE: An instrument which permits a view of a sample in a magnified, three-dimensional field.

STERILE: Infertile or barren. Aseptic; not producing microorganisms.

STERILITY: The inability to produce progeny. Complete or permanent loss of the sperm's fertilizing capacity or the stallion's ability to produce sperm.

STEROID: Any member of the class of compounds that includes sterols, bile acids, and sex hormones.

STILLBIRTH: The birth of a dead fetus.

STOCK: Restraining chute.

STREAK CANAL: The terminal opening of the mammary through which milk is expressed.

SUBACUTE: Disease course between acute and chronic.

SUBFERTILITY: Slight infertility.

SUCKLING: An unweaned foal.

SUMMER SORES: Sites of external inflammation caused by Habronema larvae.

SWABBING: Process of obtaining samples of secretions from the vagina, cervix, uterus, or clitoral fossa using absorptive material.

SYNOVIAL CAPSULE: The cavity containing synovia, or lubricating fluid, found in limb joints or bursae.

SYNOVIAL FLUID: A viscous fluid containing mucin and salts secreted by the synovial membrane to reduce friction in the movement of joints, muscles, or tendons.

SYSTEMIC: Relating to the entire organism as distinguished from any of its individual parts.

SYSTEMIC INFECTION: An infection which invades the entire body; not localized.

TACTILE STIMULATION: Stimulation by touch.

TAIL-END FRACTION: Accessory sex gland secretion which contains very little gel and few sperm cells. It is occasionally collected for semen evaluation.

TEASER: A male horse used to detect heat in mares.

TEAT: The nipple of a mammary gland.

TEAT CANALS: Passageway in the mammary gland which drains one quarter of the gland.

TERATOGEN: An agent that causes abnormal development.

TERATOLOGY: The study of abnormal growth or development.

TESTES: Plural of testis.

TESTIS: The male gonad.

TESTICULAR: Pertaining to the testes.

TESTOSTERONE: The male hormone produced in the testes and responsible for the development and maintenance of secondary sex characteristics.

TETANUS: A disease caused by toxin produced by *Clostridium tetani*. An acute infection in the body caused by the invasion of a disease-producing toxin causing muscular spasms and lockjaw.

THECA EXTERNA: The outer cellular layer of the ovarian follicular wall.

THECA INTERNA: The inner cellular layer of the ovarian follicular wall.

THERAPEUTIC: Relating to the treatment of disease.

THERMOREGULATION: Temperature control.

THRUSH: A foul-smelling infection of the frog and sole of the hoof.

TORSION: The act of twisting; the state of being twisted.

TOXICITY: A quality of being poisonous.

TOXIN: A poisonous substance.

TOXOID: A toxin that has been treated to destroy its toxic property but that still retains its capability to stimulate antibody production.

TRABECULAE: Supportive tissue which traverses the substance of a structure.

TRACE MINERAL: Mineral required by the body in small amounts.

TRACTABILITY: The quality of being easily managed.

TRACTION: Pulling the foal as contractions of delivery occur.

TRANSMITTANCE: Measure of the amount of light passing through a liquid as determined by a spectrophotometer. Designated by "%T."

TRANSITIONAL ANESTRUS: The period which occurs at the beginning and end of the breeding season, marked by erratic estrous cycles.

TUMOR: See "neoplasm."

TUNICA ALBUGINEA: The thin, tough tissue layer which covers the ovary's surface area with the exception of the ovulation fossa.

TUNICA DARTOS: A layer of smooth muscles in the scrotum which aids in heat regulation.

TUNICA VAGINALIS: A membrane enclosing the testis.

TURBID: Cloudy.

TURGID: Swollen; congested.

TWITCH: An instrument used to restrain a horse by placing pressure on the upper lip.

TYMPANY: Resonant tone obtained when an air-distended cavity is struck.

UDDER: The mammary gland.

UMBILICAL CORD: The cord which attaches the placenta to the fetus at the umbilicus, carrying urine and blood between the fetus and the placenta for waste/nutrient transfer.

UMBILICAL RING: Constriction which forms at the junction of the amnionic sac and the embryo.

UNILATERAL CASTRATE: An individual in which only one testicle has been removed.

UNILATERAL CRYPTORCHIDISM: Retention of one testis in the body cavity.

URACHUS: Small vessel in the umbilical cord that connects the fetal bladder to the allantois.

UROGENITAL SINUS: The structure in the early embryo which develops into the terminal portion of the male and female reproductive tracts.

UTERINE ATONY: Absence of uterine tone.

UTERINE BODY: That portion of the uterus which lies between the uterine horns and the cervix.

UTERINE HORN: The portion of the uterus which lies between the fallopian tube and the uterine body. The paired uterine horns are formed by a division of the uterus into two tubular structures.

UTERINE PROLAPSE: Inversion of the uterus, resulting in partial or complete expulsion from the vulva.

UTERUS: The hollow, muscular organ, consisting of cervix, uterine body, and uterine horns.

VAGINA: The genital passageway that extends from the cervix to the vulva.

VAGINITIS: Inflammation of the vagina.

VAS DEFERENS: The excretory duct of the testicle, passing from the testis to the ejaculatory duct.

VASECTOMY: Surgical procedure in which the vas deferens is cut to prevent passage of sperm from the testicle.

VASCULAR: Relating to or containing blood vessels.

VEIN: Blood vessel which carries blood toward the heart.

VENTRAL: Relating to the abdomen.

VENTRICLE: A small cavity; specifically, one of the two lower chambers of the heart.

VESTIBULE: The vaginal area posterior to the hymen.

VESTIGE: A rudimentary or undeveloped part which was well-developed and functional in the embryo.

VIABLE: Capable of living.

VICE: An undesirable habit.

VILLI: Hair-like projections on certain mucous membranes.

VILLOUS: Bearing villi.

VIRUS: A microscopic agent of infectious disease which lacks metabolism and can reproduce only in living tissue or a culture medium.

VITELLINE MEMBRANE: Cellular membrane covering the ovum.

VULVA: The external genitalia of the female, comprised of the labia, the clitoris, the vestibule of the vagina and its glands, and the opening of the urethra and of the vagina.

WANDERER FOAL: A foal suffering from convulsive syndrome caused by a lack of oxygen at birth.

WATER SAC: Common name for the allantochorionic portion of the placenta.

WAVELENGTH: The distance fromone point to the next point in the same phase of a wave.

WAXING: Popular term for the formation of beads of dried colostrum at the ends of the teats.

WEANING: To permanently deprive of milk.

WINDPUFF: Swelling and edema in fetlock area caused by stress or concussion.

WINDSUCKING: A vice in which the horse grasps an object with the incisor teeth, arches the neck, and swallows quantities of air. Popular but incorrect term for pneumovagina.

WINKING: Protrusion of the clitoris between the labia, frequently observed in mares in estrus.

WOBBLER SYNDROME: A disease of young horses exhibiting incoordination and weakness. Associated with lesions of the cervical region and compression of the spinal cord.

WOLFFIAN BODY: One of three excretory organs appearing in the evolution of vertebrates. In young mammalian embryos, it is well-developed and functional for a time before the establishment of a definitive kidney. In older embryos, it undergoes regression as an excretory organ but its duct system is retained in the male as the epididymis and ductus deferens.

WOLFFIAN DUCTS: Duct system in the early embryo which develops into the epididymis, vas deferens, ampulla, and seminal vesicles in the male. In the female, however, the ducts degenerate.

X CHROMOSOME: Sex chromosome found paired in the cells of female animals.

Y CHROMOSOME: Sex chromosome found paired with an X chromosome in the cells of male animals.

YELLOW BODY: See "corpus luteum."

YOLK SAC: Wall formed by the combined endoderm and ectoderm.

ZONA PELLUCIDA: Protective layer which surrounds the ovum.

ZOONOTIC: Pertaining to an infection or infestation which may be transmitted to may or lower vertebrates.

ZYGOTE: Fertilized ovum.

INDEX

G

gametogenesis, 96-98, 642-647
gangrene, 191
gel fraction, 110-111, 344
gel-free volume, 154, 345
genes, 366, 638
genetic potential, 366
genetic variation, 349
genetics, 182-190, 227, 388-389, 637-646
genital horse pox, 248
genital ridge, 379
genital tract
   mare, 208-214
   stallion, 92-105
germ cell layers, 370
germinal cells, 96, 209
gestation, 366-393
gestation length, 392, 601-603
glans clitoris, 214
glans penis, 104
glucose, 509
glycerol, 346
glycogen, 385, 509
goblet cells, 213
goiter, 390
golden slippers, 457
gonadotropin releasing hormone, 96, 221, 287
gonads, embryonic development, 379-381
Graafian follicle, 216, 217, 298
gradual weaning, 583
granulosa cell tumor, 234-235, 236, 253, 264
granulosa cells, 216
grooming, 592-593
guarantees, live foal, 78
gut strangulation, see bowel strangulation
guttural pouch, tympany, 527

H

habitual abortion, 426
Habronema, 199, 620
Haemophilus equigenitalis, 247, 419-420
hair patterns, 31
half year convention, 68
halterbreaking, 555-556
Hancock stain, 160
hand breeding, 309-329
hand feeding, 544-545
handling
   areas, 13
   foals, 554-560
   stallions, 119-126
health records, 36

heart defects, 525
heart, embryonic, 378
heart rate of newborn, 508, 511
heat, see estrus
heat detection, 229, 289-301
   hormone assays, 300
   rectal palpation, 298
   teasing, 291-292
heat lamps, 441, 515, 542
hemacytometer, 166
hematomas, 191, 253, 264
hemoglobin concentration, 117
hemolytic streptococcus, 197
hemorrhage, 483-484
hemospermia, 172, 203
hernias, 187, 387, 529-530, 579
herpesvirus, 202
hippomanes, 456
history, 117-118, 262
hobbles, 311-312
hobby loss provision, 62
hoof care, 138-139, 565-567
hormone therapy, 287
hormones
   assays, 300
   estrous cycle control, 220-221, 287
   estrus detection, 300
   foaling, 443-444
   imbalances, 426
   interstitial cells, 95
   mare infertility, 229-230
   Sertoli cells, 95
   stallion, 95-96
   stallion infertility, 182-183
hormones controlling reproduction, 604
human chorionic gonadotropin, 130, 287
humidity, 11
hydrocephalus, 477, 522
hygiene, 346-347
   breeding, 314
   foaling, 435, 437, 442, 443, 468
   insemination, 331
hymen, 213, 258, 267
hypercarbia, 523
hypoplasia, 188, 235, 255, 427, 522-523
hypothalamus, 95-96, 221
hypothyroidism, 183

I

identification, 27-34
   blood typing, 34
   brands, 31
   chestnuts, 31
   freeze marking, 31
   hair patterns, 31